*The Molecular World*

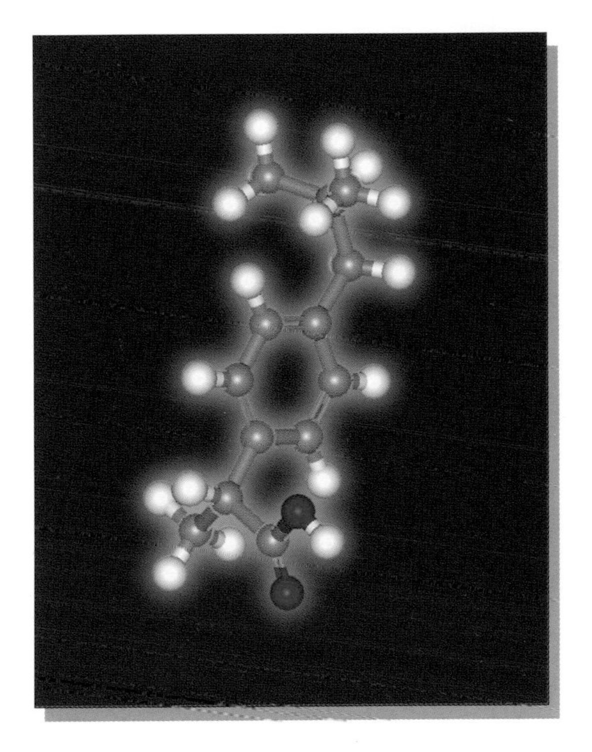

# Mechanism Synthesis

edited by
Peter Taylor

This publication forms part of an Open University course, S205 *The Molecular World*. Most of the texts which make up this course are shown opposite. Details of this and other Open University courses can be obtained from the Call Centre, PO Box 724, The Open University, Milton Keynes MK7 6ZS, United Kingdom: tel. +44 (0)1908 653231, e-mail ces-gen@open.ac.uk.

Alternatively, you may visit the Open University website at http://www.open.ac.uk where you can learn more about the wide range of courses and packs offered at all levels by The Open University.

The Open University, Walton Hall, Milton Keynes, MK7 6AA

First published 2002



Edited, designed and typeset by The Open University.

Published by the Royal Society of Chemistry, Thomas Graham House, Science Park, Milton Road, Cambridge CB4 0WF, UK.

Printed in the United Kingdom by Bath Press Colourbooks, Glasgow.

ISBN 0 85404 695 X

A catalogue record for this book is available from the British Library.

1.1

s205book 7 i1.1

## The Molecular World

This series provides a broad foundation in chemistry, introducing its fundamental ideas, principles and techniques, and also demonstrating the central role of chemistry in science and the importance of a molecular approach in biology and the Earth sciences. Each title is attractively presented and illustrated in full colour.

**The Molecular World** aims to develop an integrated approach, with major themes and concepts in organic, inorganic and physical chemistry, set in the context of chemistry as a whole. The examples given illustrate both the application of chemistry in the natural world and its importance in industry. Case studies, written by acknowledged experts in the field, are used to show how chemistry impinges on topics of social and scientific interest, such as polymers, batteries, catalysis, liquid crystals and forensic science. Interactive multimedia CD-ROMs are included throughout, covering a range of topics such as molecular structures, reaction sequences, spectra and molecular modelling. Electronic questions facilitating revision/consolidation are also used.

The series has been devised as the course material for the Open University Course S205 *The Molecular World*. Details of this and other Open University courses can be obtained from the Course Information and Advice Centre, PO Box 724, The Open University, Milton Keynes MK7 6ZS, UK; Tel +44 (0)1908 653231; e-mail: ces-gen@open.ac.uk. Alternatively, the website at www.open.ac.uk gives more information about the wide range of courses and packs offered at all levels by The Open University.

*Further information about this series is available at www.rsc.org/molecularworld.*

*Orders and enquiries should be sent to*:

Sales and Customer Care Department, Royal Society of Chemistry, Thomas Graham House, Science Park, Milton Road, Cambridge, CB4 0WF, UK

Tel: +44 (0)1223 432360; Fax: +44 (0)1223 426017; e-mail: sales@rsc.org

The titles in *The Molecular World* series are:

*THE THIRD DIMENSION*
edited by Lesley Smart and Michael Gagan

*METALS AND CHEMICAL CHANGE*
edited by David Johnson

*CHEMICAL KINETICS AND MECHANISM*
edited by Michael Mortimer and Peter Taylor

*MOLECULAR MODELLING AND BONDING*
edited by Elaine Moore

*ALKENES AND AROMATICS*
edited by Peter Taylor and Michael Gagan

*SEPARATION, PURIFICATION AND IDENTIFICATION*
edited by Lesley Smart

*ELEMENTS OF THE p BLOCK*
edited by Charles Harding, David Johnson and Rob Janes

*MECHANISM AND SYNTHESIS*
edited by Peter Taylor

## The Molecular World Course Team

**Course Team Chair**
Lesley Smart

**Open University Authors**
Eleanor Crabb (Book 8)
Michael Gagan (Book 3 and Book 7)
Charles Harding (Book 9)
Rob Janes (Book 9)
David Johnson (Book 2, Book 4 and Book 9)
Elaine Moore (Book 6)
Michael Mortimer (Book 5)
Lesley Smart (Book 1, Book 3 and Book 8)
Peter Taylor (Book 5, Book 7 and Book 10)
Judy Thomas (*Study File*)
Ruth Williams (skills, assessment questions)

*Other authors whose previous contributions to the earlier courses S246 and S247 have been invaluable in the preparation of this course:* Tim Allott, Alan Bassindale, Stuart Bennett, Keith Bolton, John Coyle, John Emsley, Jim Iley, Ray Jones, Joan Mason, Peter Morrod, Jane Nelson, Malcolm Rose, Richard Taylor, Kiki Warr.

**Course Manager**
Mike Bullivant

**Course Team Assistant**
Debbie Gingell

**Course Editors**
Ian Nuttall
Bina Sharma
Dick sharp
Peter Twomey

**CD-ROM Production**
Andrew Bertie
Greg Black
Matthew Brown
Philip Butcher
Chris Denham
Spencer Harben
Peter Mitton
David Palmer

**BBC**
Rosalind Bain
Stephen Haggard
Melanie Heath
Darren Wycherley
Tim Martin
Jessica Barrington

**Course Reader**
Cliff Ludman

**Course Assessor**
Professor Eddie Abel, University of Exeter

**Audio and Audiovisual recording**
Kirsten Hintner
Andrew Rix

**Design**
Steve Best
Vicki Eaves
Carl Gibbard
Lee Johnson
Sian Lewis
Jon Owen
John Taylor
Andrew Whitehead
Liz Yeomans

**Library**
Judy Thomas

**Picture Researchers**
Lydia Eaton
Deana Plummer

**Technical Assistance**
Brandon Cook
Pravin Patel

**Consultant Authors**
Ronald Dell (*Case Study:* Batteries and Fuel Cells)
Adrian Dobbs (Book 8 and Book 10)
Chris Falshaw (Book 10)
Andrew Galwey (*Case Study:* Acid Rain)
Guy Grant (*Case Study:* Molecular Modelling)
Alan Heaton (*Case Study:* Industrial Organic Chemistry, *Case Study:* Industrial Inorganic Chemistry)
Bob Hill (*Case Study:* Polymers and Gels)
Roger Hill (Book 10)
Anya Hunt (*Case Study:* Forensic Science)
Corrie Imrie (*Case Study:* Liquid Crystals)
Clive McKee (Book 5)
Bob Murray (*Study File*, Book 11)
Andrew Platt (*Case Study:* Forensic Science)
Ray Wallace (*Study File*, Book 11)
Craig Williams (*Case Study:* Zeolites)

# CONTENTS

## PART 1 CARBONYL COMPOUNDS

*Jim Iley and Roger Hill*

# PART 2 SYNTHETIC APPLICATIONS OF ORGANOMETALLIC COMPOUNDS

*Peter Morrod and Malcolm Rose*

# PART 3 RADICAL REACTIONS IN ORGANIC SYNTHESIS

*Adrian Dobbs*

# PART 4 STRATEGY AND METHODOLOGY IN ORGANIC SYNTHESIS

*Jim Iley, Ray Jones and John Coyle*

# PART 5 SYNTHESIS AND BIOSYNTHESIS: TERPENES AND STEROIDS

*Jim Iley, Chris Falshaw and Richard Taylor*

# CASE STUDY: POLYMER CHEMISTRY

*Bob Hill*

## Part 1

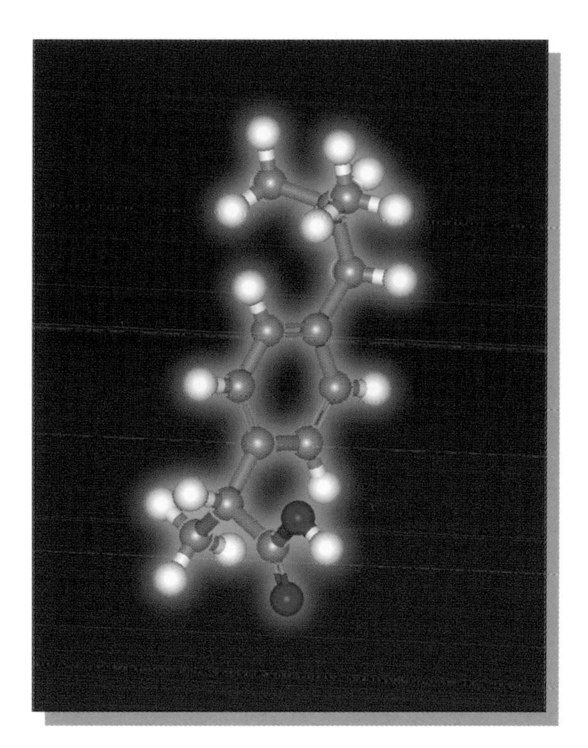

# Carbonyl compounds

edited by Roger Hill and Peter Taylor

*based on* Carbonyl compounds, *by Jim Iley*

# INTRODUCTION: CARBONYL COMPOUNDS IN CONTEXT

# 1

Organic compounds are conveniently subdivided into classes, based on their functional group; each member within a class reacts in the same way with a particular reagent. This Part of the Book explains the chemistry of the carbonyl group (C=O); you can see in Table 1.1 that the carbonyl group turns up in a number of functional groups. You have met many of these before, and, no doubt, have encountered them in everyday life, probably without realizing it (see Box 1.1). Carbonyl compounds are so widespread that some understanding of their behaviour is an essential requirement for every student of organic chemistry.

**Table 1.1** Some common carbonyl functional groups together with spectroscopic data for exemplar compounds

| General formula | Functional group | Example | Name (common name) | δ/p.p.m. | Wavenumber/cm$^{-1}$ |
|---|---|---|---|---|---|
| O R O$^-$ | carboxylate | O $H_3C$ O$^-$ | ethanoate (acetate) | 181.7 | 1 575 |
| O $R^1$ $NR^2R^3$ | amide | O $H_3C$ $NH_2$ | ethanamide (acetamide) | 169.6 | 1 645 |
| O R OH | carboxylic acid | O $H_3C$ OH | ethanoic acid (acetic acid) | 178.1 | 1 710 |
| O $R^1$ $R^2$ | ketone | O $H_3C$ $CH_3$ | propanone (acetone) | 204.1 | 1 710 |
| O R H | aldehyde | O $H_3C$ H | ethanal (acetaldehyde) | 200.4 | 1 725 |
| O $R^1$ $OR^2$ | ester | O $H_3C$ $OCH_3$ | methyl ethanoate (methyl acetate) | 170.7 | 1 738 |
| O R Cl | acid chloride | O $H_3C$ Cl | ethanoyl chloride (acetyl chloride) | 169.9 | 1 815 |
| O O $R^1$ O $R^2$ | acid anhydride | O O $H_3C$ O $CH_3$ | ethanoic anhydride (acetic anhydride) | 166.1 | 1 825, 1 755† |

* This symbol, , indicates that the structure of this compound can be studied as a WebLab ViewerLite™ file on the CD-ROM associated with this Book.

† Acid anhydrides always have two carbonyl absorptions in their IR spectrum.

## Box 1.1 Some commonly encountered carbonyl compounds

### *Ketones in perfumes*

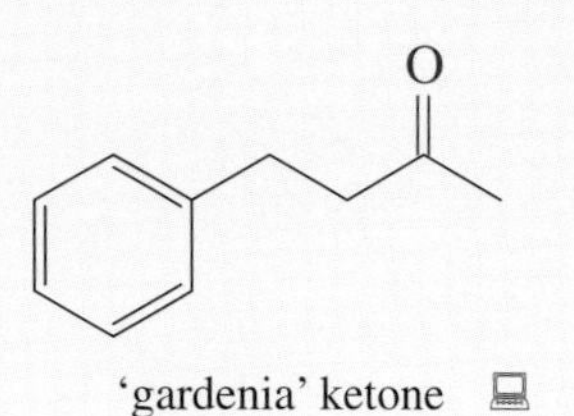


'gardenia' ketone

### *Aldehyde for human vision*

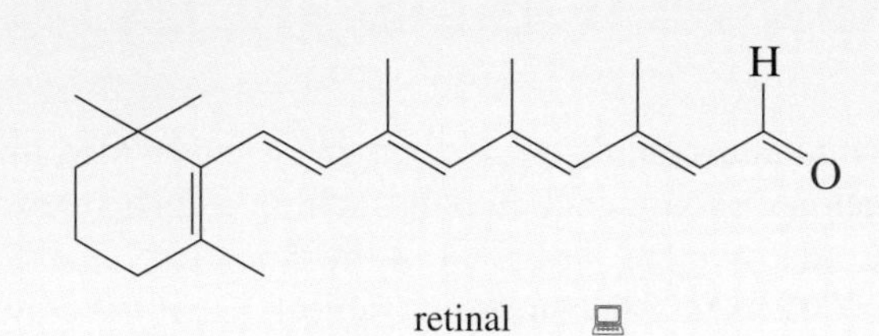


retinal

### *Vegetable oils are esters*

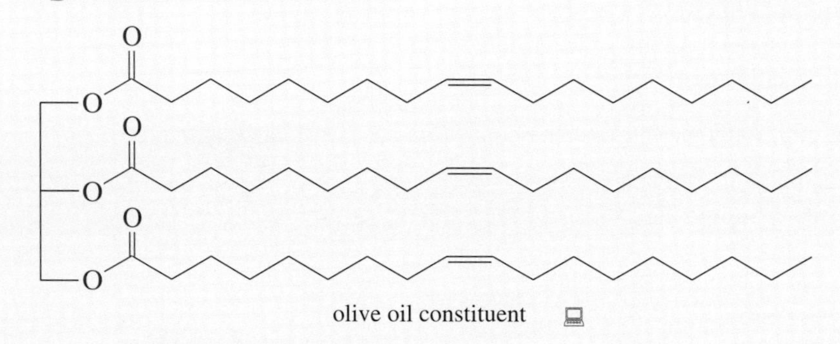


olive oil constituent

### *Amides in pharmaceuticals and proteins*

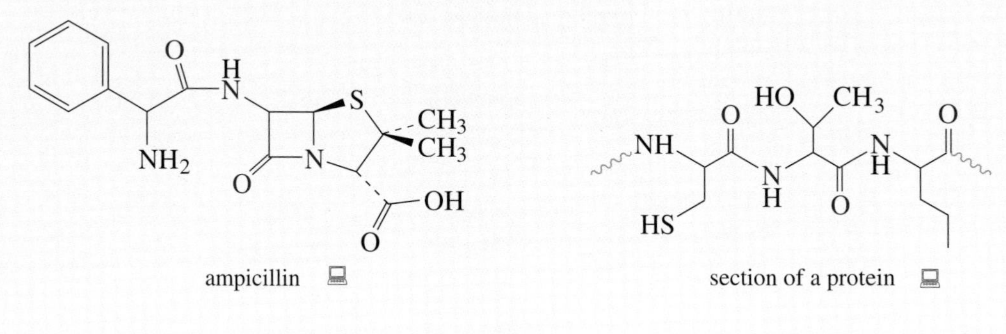


ampicillin

section of a protein

### *Soaps are salts of carboxylic acids*

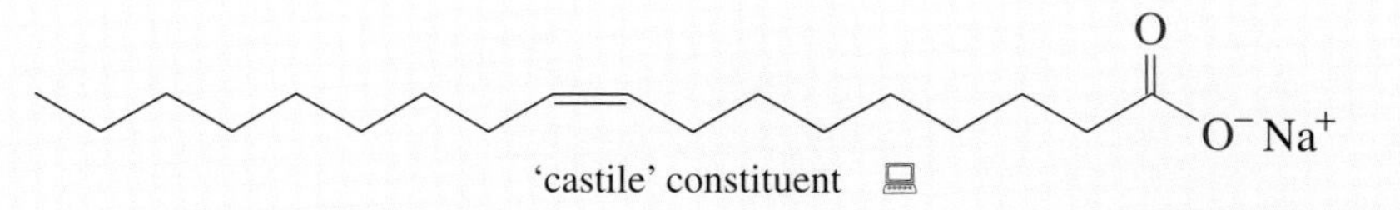


'castile' constituent

### *Aldehydes in most carbohydrates*

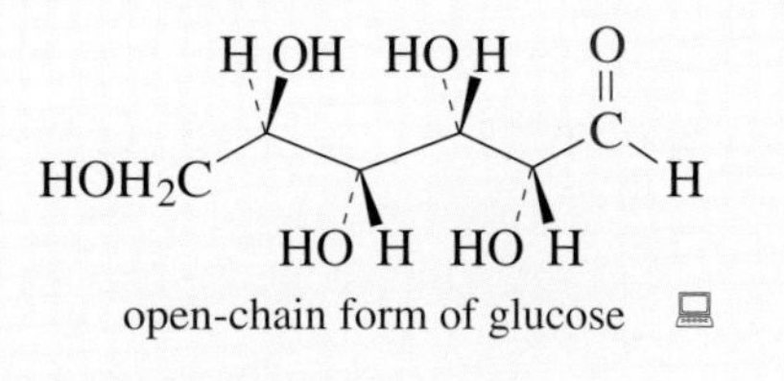


open-chain form of glucose

We'll start our examination of the topic with a close look at the structure of the carbonyl group.

# THE STRUCTURE OF THE CARBONYL GROUP

# 2

X-ray crystallography shows that the carbonyl group is planar. The carbon and oxygen atoms of the carbonyl group, together with the two carbon atoms attached to the carbonyl carbon, all lie in the same plane, with the three bond angles all close to 120° (**2.1**). This implies a particular arrangement of molecular orbitals.

$R^1$ 120° 120° C=O 120° $R^2$

**2.1**

- What hybridization of s and p orbitals is implied at the carbon atom?
- Since the central carbon atom is bound to three other atoms, it must form three $\sigma$ bonds and so it will be $sp^2$-hybridized.

- What about the oxygen atom?
- It forms one $\sigma$ bond to carbon, but has two non-bonding electron pairs, so is also $sp^2$-hybridized.

- So both the oxygen atom and the central carbon atom contribute one electron to the $\sigma$ bond between them, which leaves one over for each. How are these deployed?
- They are each contained in a p orbital perpendicular to the plane and overlapping sideways with each other, forming a $\pi$ bond.

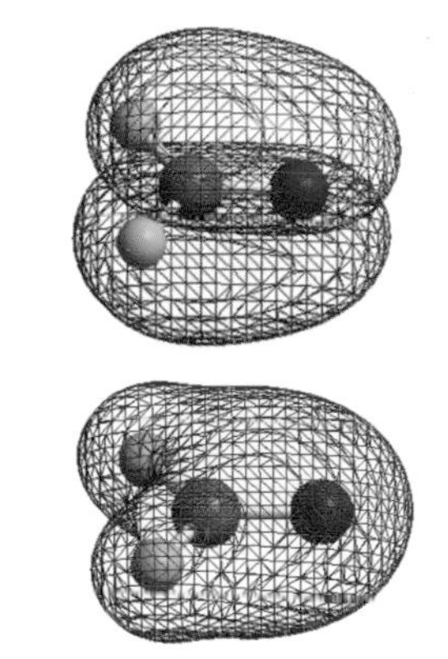


**Figure 2.1**
The orbital overlap in (a) the carbon–oxygen $\pi$ bond of the carbonyl group, and (b) the $\sigma$ frameworkof the carbonyl group.

The bond between carbon and oxygen is thus a double bond, with two electrons in a $\sigma$ bond and two electrons in a $\pi$ bond (Figure 2.1).

The carbonyl bond is polar; that is, the four shared electrons are closer to one atom than to the other, leaving one atom electron-deficient or slightly positive ($\delta+$), and the other with a surplus or slightly negative ($\delta-$).

- Which atom bears the $\delta+$ and why?
- Oxygen is more electronegative than carbon, and claims more than an equal share of the four electrons, leaving the carbon atom electron deficient. This is shown in Figure 2.2.*

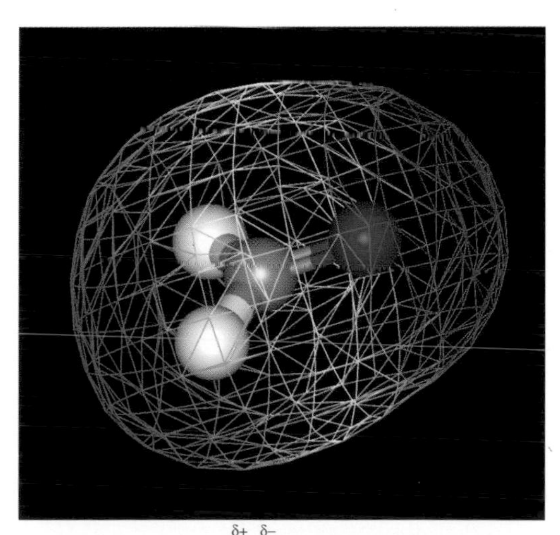

**Figure 2.2**
The charge density distribution in the carbon–oxygen double bond in methanal. The red areas of the surface indicate regions of high electron density and the blue areas, low electron density.

* Although the C=O bond in an organic carbonyl compounds is polarized $\overset{\delta+\ \delta-}{C=O}$, in carbon monoxide the charge distribution is the other way round, $\overset{\delta-\ \delta+}{C=O}$.

This **polarization** of C=O (**2.2**) is extremely useful in understanding the chemical behaviour of carbonyl compounds.

$R^1$ \ $\delta+$ $\delta-$ C=O / $R^2$

**2.2**

the polarization of the carbonyl group

Which of the two atoms would be prone to attack by nucleophiles?

Nucleophiles have an electron-rich centre (they are 'nucleus-liking') and therefore seek to associate with the electron-deficient (positively charged) carbon atom.

What is the simplest electrophile, and how would it seek to associate with a carbonyl compound?

The hydrogen ion (proton) is the simplest electrophile, and would associate with the oxygen atom; that is, the carbonyl oxygen atom can be protonated.

You will see shortly how a large part of the chemistry of carbonyl compounds can be explained by the answers to the last two questions. But the electronic structure of a group also determines its spectroscopic properties, so this is a good place for a brief look in that direction.

Why should you want to know about spectroscopic properties?

The spectra (e.g. infrared and nuclear magnetic resonance) of a class of organic compounds have features common to that class. Conversely, some features of the spectra are diagnostic for that class of compound, so organic chemists use spectroscopy to tell whether or not a compound belongs to that class — that is, whether it contains the relevant group within its molecular structure.

Each of our exemplar compounds in Table 1.1 fall within the range of IR and $^{13}$C NMR values usually quoted for carbonyl groups, namely 1 650–1 850 cm$^{-1}$ and 160–220 p.p.m., respectively. Details for each carbonyl functional group are given in Table 23.1 of the *Data Book*, which is available from the CD-ROM. We quote a range because, although the spectroscopic values differ from one functional group to another, there is a specific range for each functional group depending on the nature of the R group(s). When we use these ranges to determine the structure of an unknown compound, there is some overlap; in such cases we need to bring together all the information we have about a compound before we can be sure of the nature of the functional group.

## 2.1 Summary of Sections 1 and 2

1 The carbonyl group is found in the following classes of compound: aldehydes, ketones, carboxylic acids and their salts, esters, amides, acid halides and acid anhydrides.

2 The carbonyl group is planar, and the three bond angles around the central carbon atom are all close to 120°.

3 The bonding in the carbonyl group involves σ bond formation by overlap of an sp$^2$ hybrid orbital on carbon with an sp$^2$ hybrid orbital on oxygen, and π bond formation by overlap of a p orbital on carbon with a p orbital on oxygen.

4 The carbon–oxygen double bond is polarized such that the carbon atom is slightly positively charged, and the oxygen atom slightly negatively charged.

5 Each carbonyl functional group has a distinctive range of IR and $^{13}$C NMR parameters that can be used to help identify the nature of an unknown carbonyl compound.

### QUESTION 2.1

Which of the following six compounds contain one or more carbonyl groups (see Box 2.1 for information about camphor)? Identify each carbonyl group as ester, amide, etc., and, based on Table 1.1, indicate the wavenumber that you might expect the IR stretching vibration of the group to have.

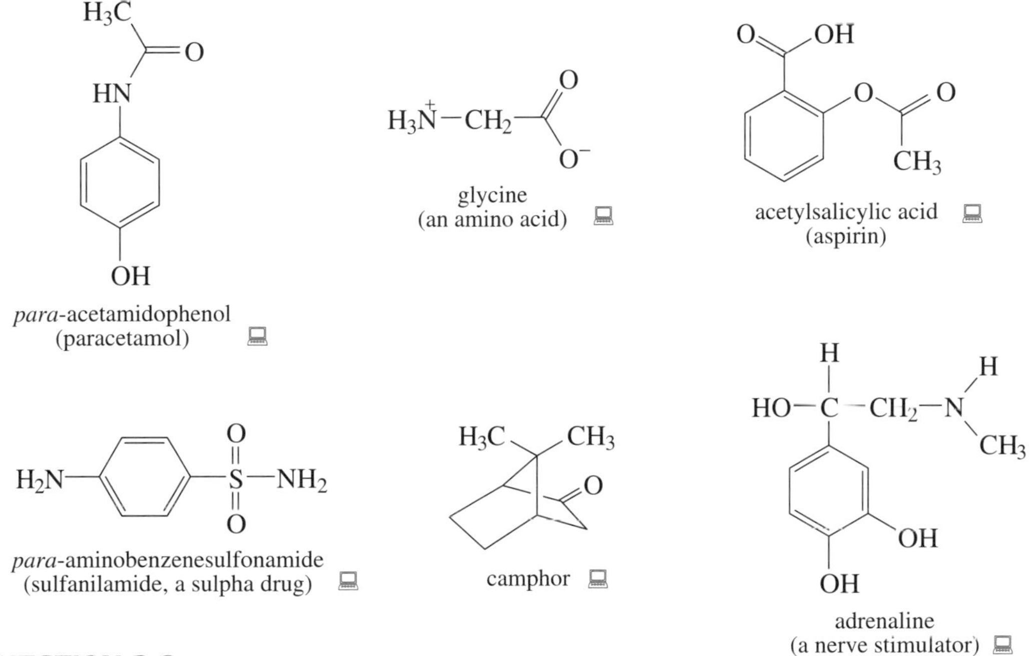


### QUESTION 2.2

Complete Table 2.2 by inserting in the inference column for each of the compounds A–G *either* the likely class of carbonyl compound involved *or* 'carbonyl compound unlikely'.

**Table 2.2** IR and NMR data for compounds A to G

| Compound | Highest-value signal in $^{13}$C NMR spectrum/p.p.m. | Strong absorbance in infrared spectrum/cm$^{-1}$ | Inference |
|---|---|---|---|
| A | 155.8 | 1 530 | |
| B | 202.4 | 1 709 | |
| C | 170.2 | 1 814 | |
| D | 180.1 | 1 705 | |
| E | 124.5 | 1 540 | |
| F | 165.3 | 1 650 | |
| G | 169.9 | 1 735 | |

### STUDY NOTE

A set of interactive self-assessment questions are provided on the *Mechanism and Synthesis* CD-ROM. The questions are scored, and you can come back to the questions as many or as few times as you wish in order to improve your score on some or all of them. This is a good way of reinforcing the knowledge you have gained while studying this Book.

## BOX 2.1 Camphor

Question 2.1 shows that camphor is a ketone. Camphor is obtained from the camphor tree, *Cinnamomum camphora* (Figure 2.3), which grows in Asia and Brazil. It is a very difficult plant to kill, and other plants will not usually grow well around it. The camphor is stored in its trunks and branches. The camphor is obtained by passing steam through chips of the root, stem, or bark so that it co-distils with the steam (so-called *steam distillation*). It is thought to be beneficial in keeping colds and flu away; however, it is poisonous if ingested in large amounts. Camphor is also renowned for its insect-repellent attributes.

**Figure 2.3** A camphor tree.

# NUCLEOPHILIC ATTACK AT THE CARBONYL GROUP

3

## 3.1 An overview: organization through mechanism

You saw in the previous Section that we expect nucleophiles to attack the carbon atom of the carbonyl group. The outcome would be the formation of a new carbon–nucleophile bond:

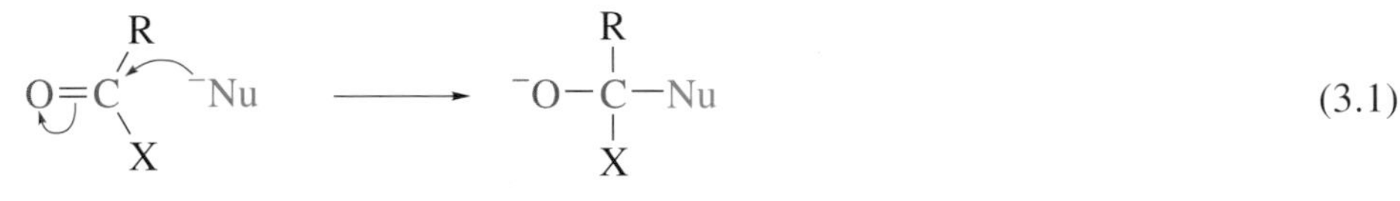


(3.1)

### BOX 3.1 The essentials of curly arrows

Make sure you always use curly arrows with precision, as in Reaction 3.1. Place each one carefully to depict movement of an electron *pair*, with the tail at its origin and the head at its destination. Remember that origin and destination can each have only one of two locations: *between* two atoms (bonding pair) or *on* one atom (an unshared or non-bonding pair). You should know that bonding pairs are depicted as lines in organic structures and that non-bonding pairs on atoms such as nitrogen, oxygen and halogens are usually not shown at all. However, if you are unsure about the non-bonding pairs in reaction mechanisms it is a good idea to draw the non-bonding electrons in as dots, so you can keep track of the electrons.

In Reaction 3.1 we assume the nucleophile is a negatively charged reagent like $HO^-$, and show it forming a new bond by donating one of its electron pairs to the carbon atom of the carbonyl group. In order to maintain an eight-electron noble gas outer shell, however, the carbon atom must simultaneously give up its hold on another electron pair. One of the two electron pairs that make up the double bond now becomes exclusively associated with the oxygen atom, and thus the oxygen becomes negatively charged.

Reaction 3.1 expresses the type of chemistry most frequently encountered with carbonyl compounds. So how does it fit with the orbital picture of the C=O bond you saw in Section 2 (Figure 2.1)? Which of the electron pairs, σ or π, ends up on the oxygen atom?

The product retains a single bond between carbon and oxygen, and as single bonds are always σ bonds, with the electron pair distributed about the internuclear axis, intuition suggests that it's the π electron pair that moves. This is correct, and the reason is that π electrons are more **polarizable** than σ electrons; that is, being further away from the atomic nuclei they are more easily influenced by external agents — more easily 'pushed around', if you like. Important consequences follow from this involvement of the π bond. For a start, we can now begin to envisage nucleophilic attack in three dimensions as in Structure **3.1**, where the broken green line indicates the trajectory of the nucleophile as it approaches the carbonyl compound.

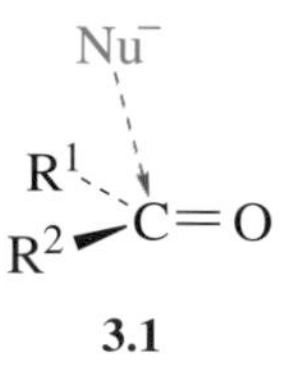


**3.1**

The nucleophile will approach from above (or below) the plane of the carbonyl group and, as its pair of electrons is heading for the slightly electron-deficient carbon atom, and pushing the π electrons towards oxygen, one might expect the nucleophile to try to keep away from the oxygen and its developing negative charge. This is all borne out by experimental and theoretical studies.

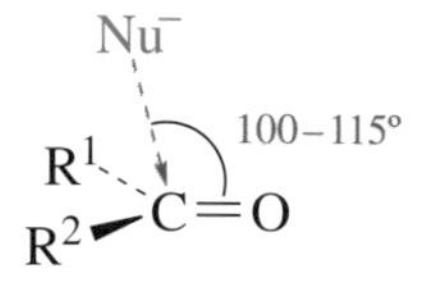


**3.2 a**

angle of approach of a nucleophile to a carbonyl compound

Careful examination of the crystal structures of molecules that contain both a carbonyl group and a nucleophilic group (such as an amino group) shows that the nucleophile approaches the carbonyl group at an angle of about 105° to the plane containing the $R^1$, $R^2$ groups and the C and O atoms. In practice, this angle is seen to vary between 100° and 115° (Structure **3.2a**). This is not very surprising if you consider that in the tetrahedral product this angle must be about 109°.

Not only does the nucleophile approach the carbonyl group at this angle, but it does so from a direction that lies between the $R^1$ and $R^2$ groups (see Structure **3.2b**). Again, this is not too surprising, since this is the direction along which the Nu group lies in the tetrahedral product.

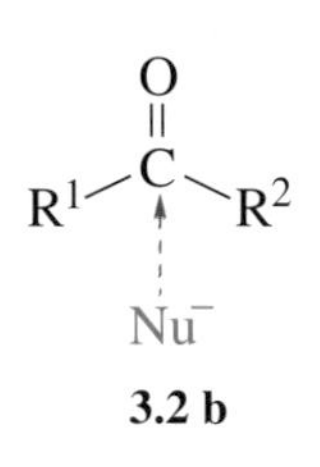


**3.2 b**

As the nucleophile gets closer to the carbonyl carbon atom, the geometry around the carbonyl group resembles more and more that of the product. So, the carbon–oxygen bond gets longer (the C=O bond length in a ketone is 120 pm, compared to a C—O bond length of 141 pm in an alcohol), and the $R^1$ and $R^2$ groups move out of the plane containing the carbonyl C and O atoms, to take up positions nearer to those seen in the tetrahedral arrangement in the product. (The $R^1$ and $R^2$ groups are shown in black in the carbonyl, but in magenta as they move to their positions in the tetrahedral product.)

(3.2)

This anion, however, is not the end product, so you need to consider how this may react further. **3.3** is simply the anion of an alcohol, so one possible process available to it is protonation, to form the corresponding alcohol:

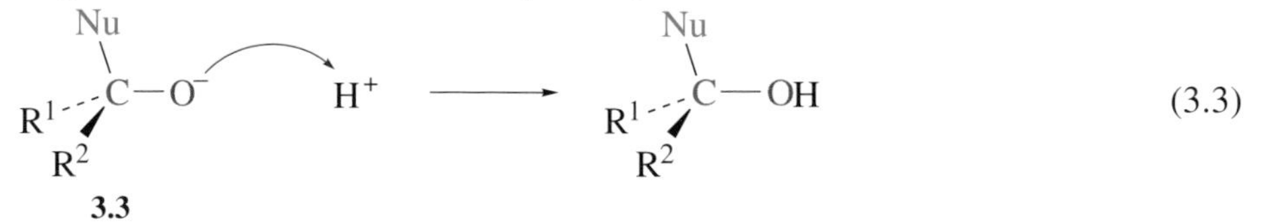


**3.3**

(3.3)

This reaction preserves the tetrahedral centre in the product. However, if instead of an alkyl group, $R^2$, one of the groups attached to the carbonyl is a good leaving group, X, an alternative process (leading to the formation of **3.4**) occurs — decomposition with loss of $X^-$:

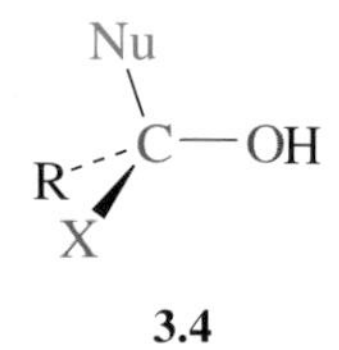


**3.4**

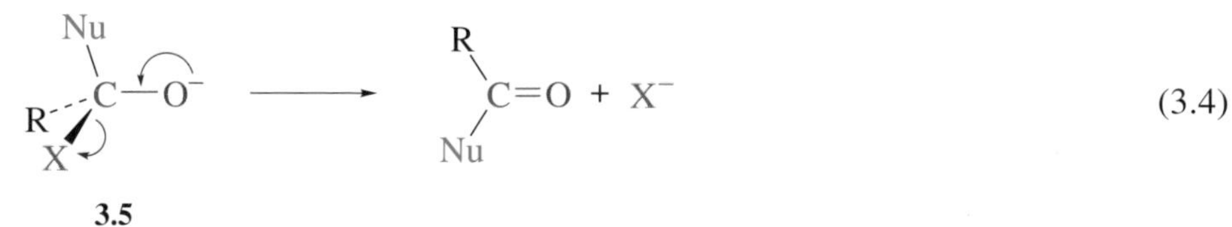


**3.5**

(3.4)

This is analogous to a β-elimination reaction to form an alkene. For a β-elimination reaction, the pair of electrons used to form the carbon–carbon π bond comes from a C—H bond, and the C—X bond breaks heterolytically; that is, the two electrons in the C—X bond both end up on the leaving group X:

$$\text{H—C—C—X} \longrightarrow \text{C=C} + \text{H}^+\text{X}^- \qquad (3.5)$$

In Reaction 3.4, the electrons used to form the carbon–oxygen π bond originate from a non-bonded electron pair on oxygen. Notice that this decomposition of **3.5** regenerates the planar carbonyl group, and that the nucleophile, Nu, has replaced the group X. Looking at the reaction overall, substitution of Nu for X has occurred, but this has taken place by a mechanism involving addition of Nu followed by elimination of X.

Now, a major question arises: 'Do all carbonyl compounds undergo each of these reactions?'. *Certainly, all carbonyl compounds undergo nucleophilic attack to form* **3.5**. However, not all of them undergo decomposition with loss of $X^-$. You can see why with the following argument:

- For each of the following classes of carbonyl groups, identify X in the general formula RCOX: aldehyde, ketone, carboxylic acid, ester, amide, acid chloride and acid anhydride.

- aldehyde, RCHO: X = H

  ketone, $R^1R^2CO$: X = $R^2$

  carboxylic acid, RCOOH: X = OH

  ester, $R^1COOR^2$: X = $OR^2$

  amide, $R^1CONR^2R^3$: X = $NR^2R^3$

  acid chloride, RCOCl: X = Cl

  acid anhydride, $R^1COOCOR^2$: X = $OCOR^2$

- Which of these groups, X, correspond to good leaving groups, and which are poor leaving groups?

- The acids H—H and $R^2$—H are very weak acids, so $H^-$ and $(R^2)^-$ are very poor leaving groups*. Of the rest, $Cl^-$ is a good leaving group, and $R^2CO_2^-$ is a reasonable leaving group. $R^2R^3N^-$, $R^2O^-$ and $HO^-$ are the anions of the weak acids $R^2R^3NH$, ROH and $H_2O$, so they are poor leaving groups. However, protonation will convert **3.6** to **3.7**, **3.8** to **3.9** and **3.10** to **3.11**, generating the good leaving groups $R^2R^3NH$, $R^2OH$ and $H_2O$, respectively.

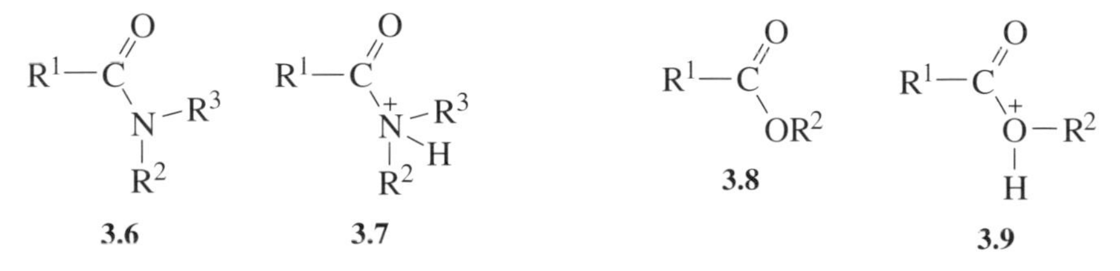


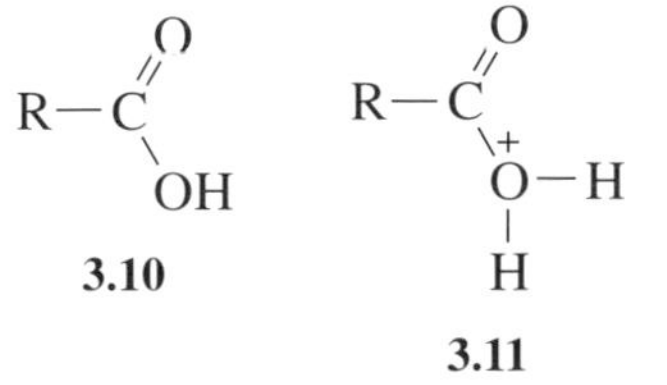


* We have used the ( ) notation for $R^2$ to avoid confusion between a group $R^2$ that bears a single negative charge, $(R^2)^-$, and a group R that bears a double negative charge, $R^{2-}$.

The outcome of this reasoning is as follows. Aldehydes and ketones do *not* possess good leaving groups, whereas, under appropriate conditions, carboxylic acids and their derivatives do.

> Therefore, aldehydes and ketones undergo addition of the nucleophile to the carbonyl group, whereas carboxylic acid derivatives undergo addition followed by loss of the $X^-$ leaving group.

Because of this, we shall discuss the chemistry of these two groups of compounds separately. You should always bear in mind, however, that the reaction of all carbonyl compounds with nucleophiles involves the same first step, namely nucleophilic attack at the carbonyl carbon atom to form a *tetrahedral intermediate*. The different scenarios are summarized in Scheme 3.1.

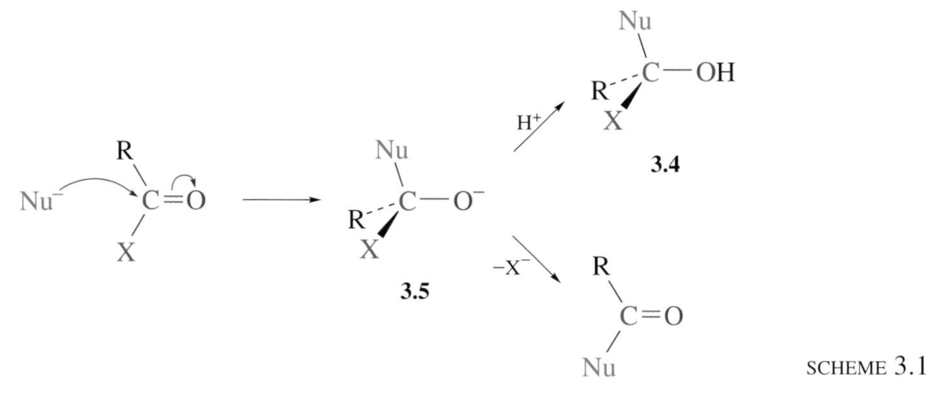


SCHEME 3.1

Take a few minutes to consolidate this important distinction between aldehyde/ketone chemistry and carboxyl chemistry by attempting Question 3.1, which anticipates some reactions we'll discuss in more detail later.

### QUESTION 3.1

Predict the structures of the final products when each of the carbonyl compounds $CH_3CHO$, $CH_3COCH_3$ and $CH_3CO_2CH_3$ is allowed to react with a strong nucleophile $Y^-$, when the initial concentration of $Y^-$ is at least twice that of the carbonyl compound. Assume that the reactions are completed by the addition of $H^+$.

So far we have used $Nu^-$ to represent our general nucleophile, but many common nucleophiles are neutral. Nucleophiles such as $HO^-$ react readily with electrophilic centres like the carbonyl carbon, but so also do $NH_3$, $CH_3OH$ and $H_2O$. This emphasizes that the key feature in a nucleophile is an available electron pair, usually (as in these cases) an unshared pair. So how would you express a reaction with a neutral nucleophile?

You'll notice that all three of our examples have a hydrogen atom attached to the nucleophilic atom. So we may represent these by the general formula :NuH, and its reaction with a carbonyl compound thus:

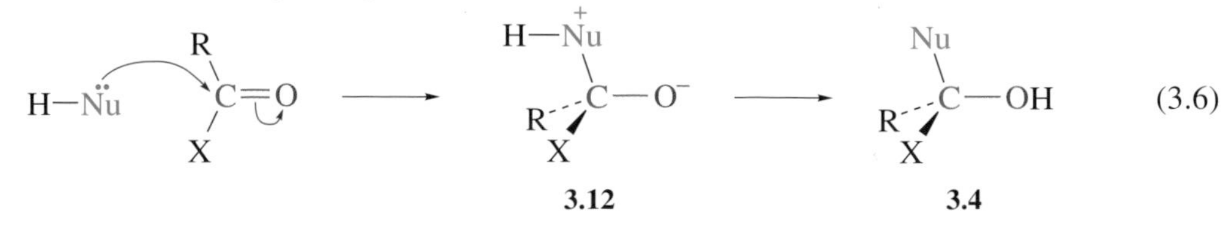


(3.6)

For such nucleophiles, the initial attack generates an intermediate, **3.12**, in which what was the oxygen atom of the carbonyl now carries a negative charge, and the nucleophile atom of NuH carries a positive charge.

However, this intermediate can rearrange to form the product **3.4**, simply by transferring a proton from the positively charged atom of the nucleophile to the oxygen anion. So, attack of :NuH yields the same product, **3.4**, as attack of $Nu^-$ followed by addition of $H^+$. Of course, in carbonyl compounds the intermediate **3.12** can undergo decomposition, with loss of $X^-$, in just the same way as does the intermediate **3.5**.

- Draw an analogous scheme for the formation of the first (addition) product between $CH_3NH_2$ and $C_6H_5COOCH_3$.

- See Scheme 3.2 at the bottom of p. 26.

  (The various types of reaction undergone by carbonyl compounds are summarized in Appendix 1, which also includes some details of reactions you have yet to meet.)

Finally, a word about reactivity. The rate at which the attack of the nucleophile at the carbonyl group takes place depends on the nucleophilicity of the nucleophile and the electrophilicity of the carbonyl carbon atom. Good nucleophiles will react with strong electrophiles rapidly, whereas poor nucleophiles will react only very slowly, if at all, with poor electrophiles. Remember that good nucleophiles tend to be those that are negatively charged: neutral nucleophiles are usually less reactive.

Good electrophiles are those that are particularly electron deficient or positively charged. Therefore, one way of promoting the reaction between a nucleophile and a poorly electrophilic carbonyl compound, or between a poor nucleophile and a carbonyl compound, is to make the carbonyl carbon atom more electrophilic. This is done by employing an acid, which protonates the carbonyl oxygen atom:

$$R(X)C{=}\ddot{O}: \xrightarrow{H^+} R(X)C{=}\overset{+}{\ddot{O}}H \qquad (3.7)$$

The protonated carbonyl group enables the resonance structure **3.13** to make a much greater contribution to the overall structure than does the charge-separated resonance structure **3.14** to the structure of the neutral compound:

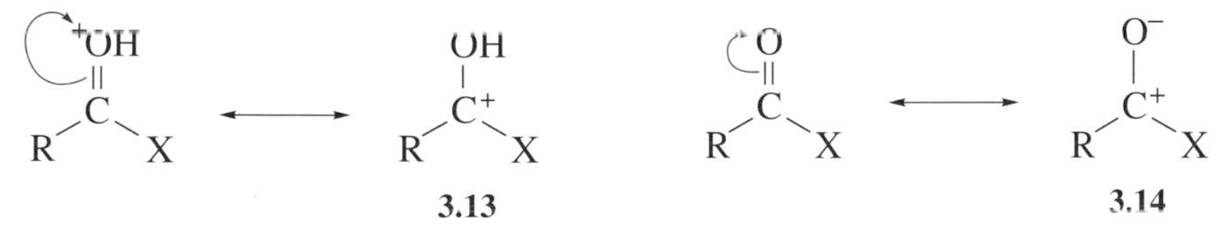

**3.13** **3.14**

Protonation therefore increases the positive charge at the carbonyl carbon atom, making it more electrophilic. In turn, this enables the carbonyl compound to react more rapidly with nucleophiles, even poor ones. You will find shortly that acids are often used to catalyse the reaction between carbonyl compounds and nucleophiles. One of the major reasons for this is the enhancement of electrophilicity described above. This is not the only role of the $H^+$, however; we shall consider its other roles for the specific cases in which they arise.

So much for general principles. Now let's examine some reactions.

## 3.2 Aldehydes and ketones

Our discussion of carbonyl compounds, RCOX, begins with those in which X is a poor leaving group, namely aldehydes and ketones. We shall start with reactions involving good nucleophiles, for example, $R^-$ and $H^-$, and then deal with the poorer nucleophiles, such as ROH and $R^1NHR^2$.

### 3.2.1 Carbon nucleophiles: molecule builders

The formation of carbon–carbon bonds is central to organic chemistry. It is *the* process by which complex organic compounds can be synthesised from simpler ones. Because of this, the reaction of carbon nucleophiles with carbonyl compounds is a particularly important reaction, in that a new carbon–carbon bond is formed.

The three most important types of carbon nucleophile that we consider are: the cyanide ion, $^-CN$; the alkynide anion, $R{-}C{\equiv}C^-$, and organometallic reagents, R—M, where M is a metal (alkynide ions and organometallic reagents will be discussed in much more detail in Part 2 of this Book). All of these are good nucleophiles, and do not require $H^+$ to enhance the reactivity of the carbonyl compound. Indeed, as you will see, the presence of $H^+$ in some of these reactions could stop them happening altogether.

Applying the general mechanism for nucleophilic attack, we can show the reaction of $^-CN$ with propanone as follows:

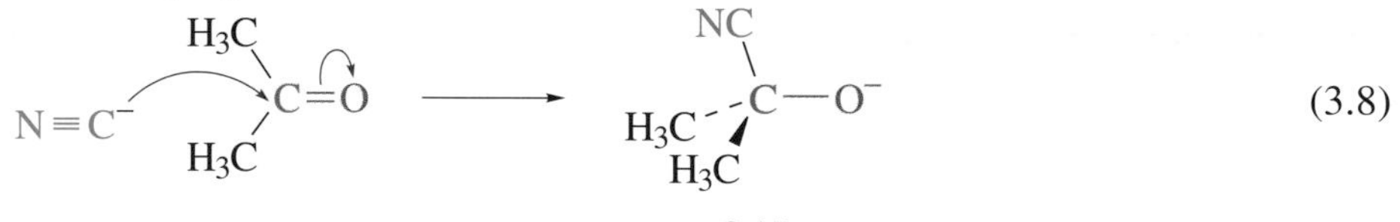


(3.8)

The reaction is, in fact, an equilibrium reaction, because cyanide ion is not only a good nucleophile, but can also act as a leaving group. So **3.15** can decompose by loss of $^-CN$:

$$\underset{\mathbf{3.15}}{(NC)(H_3C)(CH_3)C{-}O^-} \rightleftharpoons NC^- + (H_3C)_2C{=}O \qquad (3.9)$$

The anion **3.15** is derived from an alcohol, and as such will readily react with a proton, $H^+$. If we complete the reaction by adding $H^+$ (which is another equilibrium

*Answer to question on p.25*

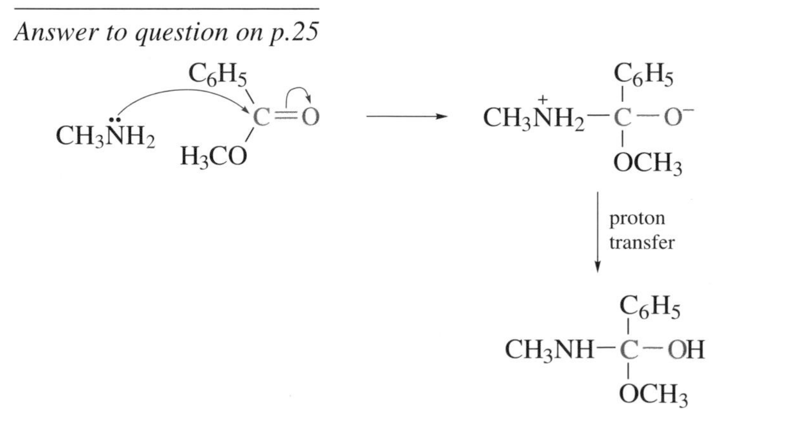


SCHEME 3.2

reaction) we have, in effect, added H—C≡N to the carbonyl group of propanone, to give a type of compound that is called a **cyanohydrin** (a cyanohydrin that occurs in nature is discussed in Box 3.2):

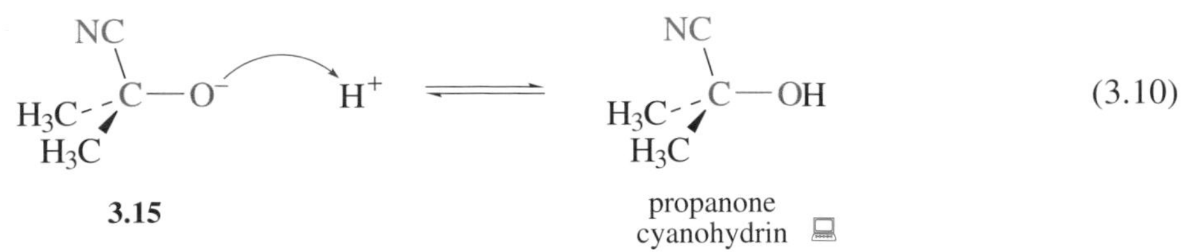


Experimentally, the reaction is usually carried out using an excess of the cyanide salt, for example sodium cyanide, with an amount of acid added that is equal in moles to the amount of carbonyl compound (known as an **equimolar amount**). Since HCN is a weak acid, the addition of acid to the solution of cyanide salt forms HCN:

$$^{-}CN + H^+ \rightleftharpoons HCN \qquad (3.11)$$

However, because the cyanide ion is present in excess, there is still plenty of cyanide ion left to react with the propanone to give **3.15**. This is then protonated by $H^+$ (remember that although HCN is a weak acid it still forms *some* $H^+$). An equimolar amount of the acid is used because, if it is used in excess, it is found to retard the reaction, since it decreases the amount of cyanide ion available for the first step.

Although the reactions are equilibria, the cyanohydrin can usually be isolated as a pure compound because it separates out from the solution as an oil. According to Le Chatelier's principle, this removes the product from the solution and drives the equilibria to the right — that is, favouring product formation.

## BOX 3.2 Cassava plants

Cassava is a shrubby, tropical, plant (Figure 3.1), which is used for making tapioca. However, in parts of Africa cassava is one of the staple foods. There are two types of cassava — a bitter and a sweet species. The bitter type of cassava requires special preparation through grating, pressure and heat to make it safe to eat. This is because it contains a sugar derivative of a cyanohydrin (**3.16**).

**Figure 3.1**
A cassava plant.

Enzymatic hydrolysis of **3.16** leads to a cyanohydrin, which can break down to give hydrogen cyanide and propanone; up to 50 mg of hydrogen cyanide can be generated per 100 g of cassava! At this concentration, it is not safe to eat the cassava, and so **3.16** has to be decomposed by treatment before cooking. Compounds such as **3.16** are found in many plant families such as Rosaceae and Leguminoseae, particularly the kernels of almonds, apricots, cherries, peaches and apples. In fact, acute intoxication and death have been reported in children following the ingestion of apricot stones.

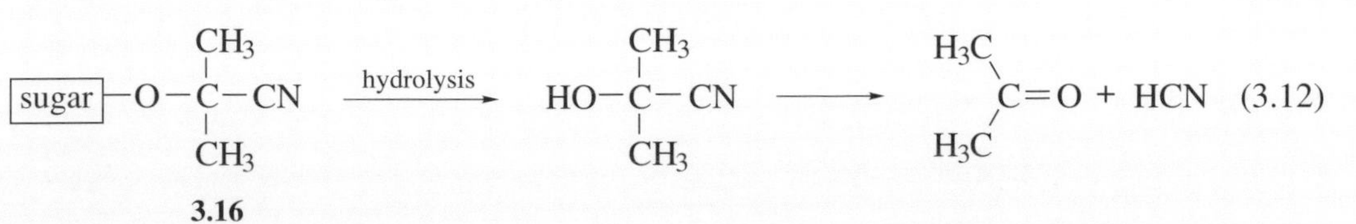

### QUESTION 3.2

Draw a full curly arrow representation of cyanohydrin formation with 2-methylpropanal, **3.17**. (There is no need to use stereostructures.)

$(H_3C)_2CH{-}CHO$

**3.17**

Alkynide ions can be formed by the reaction of an alkyne with a strong base like sodium amide:

$$R{-}C{\equiv}C{-}H + Na^{+}\,{}^{-}NH_2 \longrightarrow R{-}C{\equiv}C^{-}\,Na^{+} + NH_3 \qquad (3.13)$$

Just like other nucleophiles, alkynide ions can react with aldehydes or ketones to give a tetrahedral intermediate followed by protonation, for example:

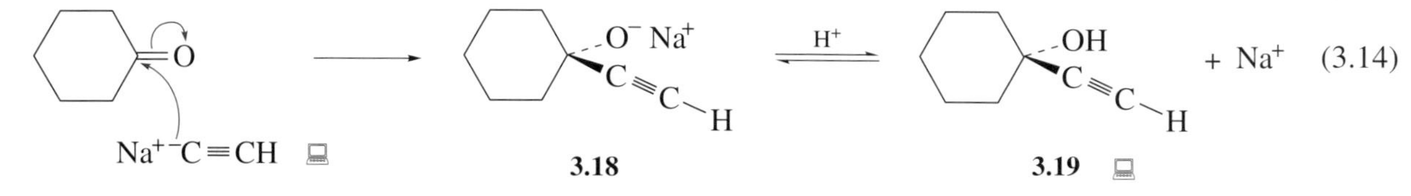


Such reactions are important because nucleophilic attack of ethynide ion on the ketone leads to formation of a carbon–carbon bond. Once the sodium salt of the oxoanion **(3.18)** is completely formed, acid or water is added to the reaction mixture to give the desired product.

- Why is water added after formation of **3.18**? Why not put the reagents all in the reaction mixture to begin with?

- The formation of ethynide ion from ethyne requires a strong base, which indicates that ethynide ion must be a reasonably good base itself. Thus, if acid or water were added at the beginning, the ethynide ion would react with the acid to form ethyne:

$$HC{\equiv}C^{-}\,Na^{+} + H^{+} \longrightarrow HC{\equiv}CH + Na^{+} \qquad (3.15)$$

### QUESTION 3.3

Draw a full curly arrow representation of the synthesis of the alcohol **3.20** from a carbonyl compound and an alkynide reagent.

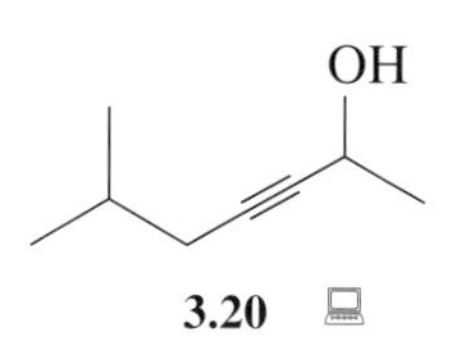


**3.20**

As you will read in Part 2, organometallic reagents, written as M—R, are a source of nucleophilic R⁻, and these react with aldehydes and ketones in the same way as the alkynide ion:

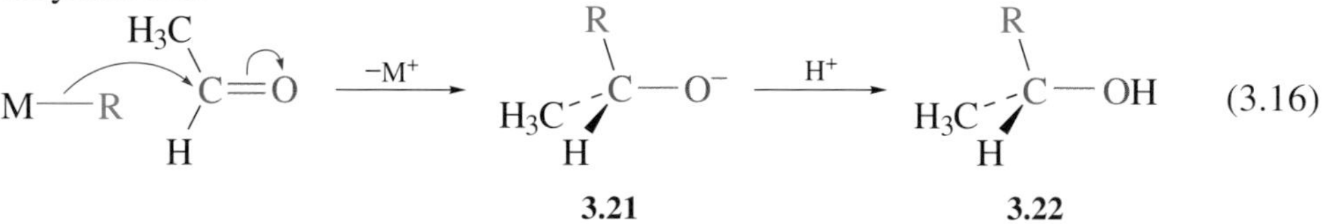


Subsequent treatment of **3.21** with acid liberates the product alcohol **3.22**.

Again, the acid has to be added after the nucleophile has reacted with the aldehyde or ketone, because the organometallic reagent will react with a proton to generate a non-nucleophilic alkane:

$$M{-}R + H^{+} \longrightarrow H{-}R + M^{+} \qquad (3.17)$$

Carbon nucleophiles react with ketones to give tertiary alcohols, with aldehydes to give secondary alcohols, and with methanal (formaldehyde, $H_2CO$) to give primary alcohols. Here are some general examples (the intermediate alkoxy ions are not shown):

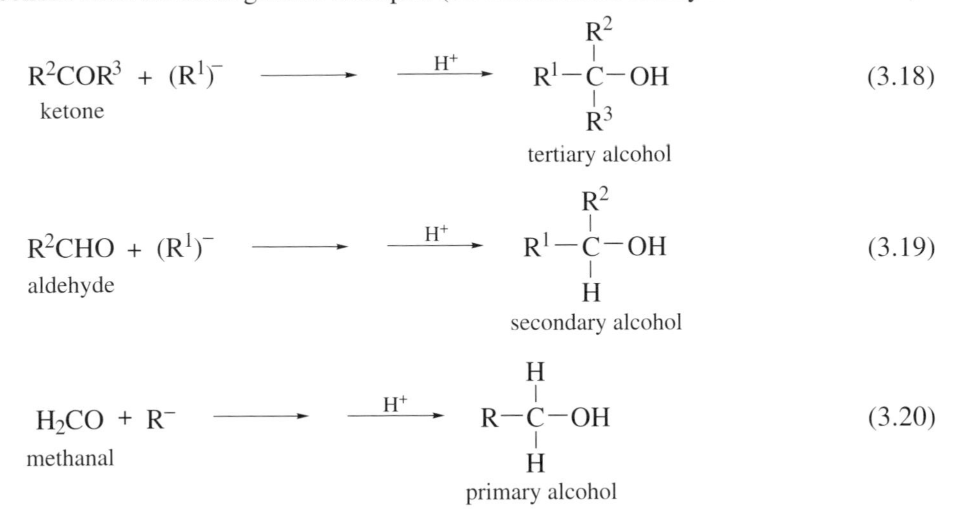


Note, however, that unlike cyanohydrin formation, the reactions of alkynides and organometallic reagents are essentially irreversible. This is because the reverse reaction would involve ejection of $HC{\equiv}C^-$ or $R^-$ as leaving groups. Since these carbon nucleophiles are anions of very weak acids, they are consequently very poor leaving groups, and so the back reaction does not occur.

QUESTION 3.4

See if you can identify the nucleophiles and aldehydes or ketones that could be used to synthesise each of the following alcohols.

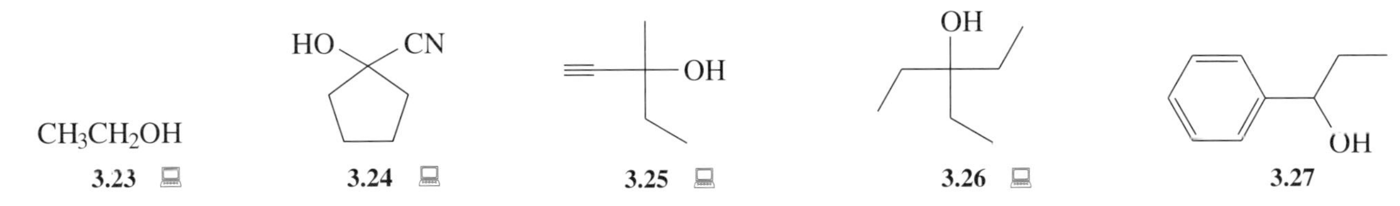


## 3.2.2 Hydrogen nucleophiles: lithium tetrahydridoaluminate (lithium aluminium hydride)

You've seen how we can generate a carbon nucleophile by selecting an appropriate reagent. Similarly, a nucleophile behaving like $H^-$ can be made available in several ways. Here we focus on just one example, lithium aluminium hydride, $LiAlH_4$.

- Before considering this reagent, how would you describe the electronic structure of $H^-$? How do you justify its negative charge?

- $H^-$ is a hydrogen atom to which one electron has been 'added'. The two electrons now fill its 1s electron shell, giving it the electronic configuration of the noble gas, helium. Having only one proton but two electrons leads to the negative ion, $H^-$.

In fact, $LiAlH_4$ does not provide free $H^-$ as an independent nucleophile, but the outcome of its reaction with an aldehyde or a ketone is the same as that expected from nucleophilic attack by $H^-$. Although the actual mechanism is more complex, we can write its reactions as follows:

$$H_3\bar{Al}{-}H + R^1R^2C{=}O \xrightarrow{-AlH_3} HR^1R^2C{-}O^- \xrightarrow{H^+/H_2O} HR^1R^2C{-}OH \qquad (3.21)$$

(Notice that the curly arrow is drawn from the Al—H σ bond, not from the negative charge — another example where using charges with curly arrows would mislead. Think 'electron pairs' not 'negative charges'!)

Because it is a very reactive reagent, being both a good nucleophile and a good base, $LiAlH_4$ must be used in a dry inert solvent such as diethyl ether. Reaction of $LiAlH_4$ with an aldehyde or ketone generates the lithium salt of the corresponding alcohol, from which the alcohol can be formed by the subsequent addition of aqueous acid (second step of Reaction 3.21).

- Do you think the first step is reversible like that with $^-CN$ or essentially irreversible like that with R—M?

- $H^-$ is the anion of $H_2$, considered as an acid ($H^+H^-$). $H_2$ is a *very* weak acid, and therefore $H^-$ is a *very* poor leaving group. So once the C—H bond is formed, it will not break again by the departure of $H^-$.

Like the reaction of carbon nucleophiles with aldehydes and ketones, the product of this reaction is also an alcohol, in this case a secondary alcohol from a ketone, and a primary alcohol from an aldehyde:

$$R^1R^2CO \xrightarrow[\text{(ii) } H^+/H_2O]{\text{(i) } LiAlH_4} R^1R^2CHOH \qquad (3.22)$$

$$R^1CHO \xrightarrow[\text{(ii) } H^+/H_2O]{\text{(i) } LiAlH_4} R^1CH_2OH \qquad (3.23)$$

- You have now met two methods of preparing both primary and secondary alcohols from carbonyl compounds. What are they?

- Primary alcohols may be made from the reaction of either (i) a carbon nucleophile and methanal, or (ii) lithium aluminium hydride and an aldehyde. Secondary alcohols can be made from (i) a carbon nucleophile and an aldehyde, or (ii) lithium aluminium hydride and a ketone.

### QUESTION 3.5

Ethanal ($CH_3CHO$) has a resonance in its $^{13}C$ NMR spectrum at 200 p.p.m., which is a doublet in the off-resonance spectrum. However, after an acid work-up, the reaction between ethanal and lithium aluminium hydride yields a product that shows a $^{13}C$ NMR signal at 57 p.p.m., which is a triplet in the off-resonance spectrum. Write a mechanism for the reaction, and explain the change in the $^{13}C$ NMR spectrum.

### 3.2.3 Oxygen nucleophiles: hydrates and acetals

Oxygen and nitrogen compounds are nucleophilic by virtue of their unshared electron pairs, and the reactions they undergo with the carbonyl group form a very important part of the chemistry of aldehydes and ketones. Nevertheless we should remember that these neutral nucleophiles are less reactive than the anionic carbon and hydrogen nucleophiles we have discussed in Section 3.2.2.

We start this Section with oxygen nucleophiles, the most familiar of which is water:

$$H_2O{:} + (H_3C)_2C{=}O \rightleftharpoons (H_3C)_2C(\overset{+}{O}H_2){-}O^- \qquad (3.24)$$

Notice that in using one of its non-bonded electron pairs to form the bond to the carbonyl carbon atom, the oxygen atom of water becomes positively charged and the carbonyl oxygen atom becomes negatively charged. It is always true that the sum of the charges on one side of an equilibrium equation should balance the sum of the charges on the other; here the sum is zero. It is good bookkeeping practice to check this.

The product of this reaction can readily transfer a proton from the positively charged oxygen atom to the negatively charged oxygen atom. Another water molecule is usually involved in this process, this molecule initially acting as a base to remove a proton:

$$H_2O{:} + (H_3C)_2C(\overset{+}{O}H_2){-}O^- \rightleftharpoons H_3O^+ + (H_3C)_2C(HO){-}O^- \qquad (3.25)$$

then, acting as the acid, $H_3O^+$, it 'protonates' the alcohol anion:

$$(H_3C)_2C(HO){-}O^- + H{-}\overset{+}{O}H_2 \rightleftharpoons (H_3C)_2C(HO){-}OH + H_2O \qquad (3.26)$$

propane-2,2-diol

**Proton transfers** like this are extremely rapid in aqueous solutions. Often, chemists tend to ignore the role of water in such reactions, and for brevity write:

$$R^1R^2C(\overset{+}{O}H_2){-}O^- \rightleftharpoons R^1R^2C(HO){-}OH \qquad (3.27)$$

This is fine, so long as you remember how the transfer of a proton actually takes place.

Propane-2,2-diol is usually called propanone (acetone) hydrate, and hydrates are significant products when aldehydes or ketones are in contact with water. Generally they cannot be isolated, however, for they are thermodynamically less stable than their carbonyl precursors, so equilibrium favours the carbonyl form (Box 3.3).

## BOX 3.3 Relative thermodynamic stability of aldehydes, ketones and their hydrates

Consider the bond energies involved in propanone and its hydrate:

$$(H_3C)_2C{=}O + H_2O \longrightarrow (H_3C)_2C(OH)_2 \qquad (3.28)$$

In the case of propanone and water, for example, the starting materials contain six C—H bonds, two C—C bonds, one C=O bond and two O—H bonds. The product, however, contains six C—H bonds, two C—C bonds, two C—O bonds and two O—H bonds. In effect, then, a C=O bond has been replaced by two C—O bonds:

$$C_2C{=}O + H{-}O{-}H \rightleftharpoons C_2C(O{-}H)_2 \qquad (3.29)$$

The energy of a C=O bond is about 770 kJ mol$^{-1}$, and that of *two* C—O bonds is about 716 kJ mol$^{-1}$ (each C—O bond contributes 358 kJ mol$^{-1}$). For this reason the carbonyl compound is *more* stable than the hydrate, but note that 54 kJ mol$^{-1}$ is a small difference, which may be overcome by other factors, such as hydrogen-bonding effects and inductive effects.

For example, although the equilibrium constant for the formation of propanone hydrate at room temperature is very small ($10^{-3}$), that for the hydrates of propanal, $CH_3CHO$ and formaldehyde, HCHO, are about 1 and $10^3$, respectively, indicating that, in water, formaldehyde is almost entirely hydrated.

### QUESTION 3.6

The active ingredient of a sleeping draught, chloral, is the well-known hydrate of trichloroethanal, $CCl_3CHO$ (the history of its use is the subject of Box 3.4).

Write a mechanism for its formation, abbreviating the proton transfer step as suggested earlier.

Noting the trend in equilibrium constants in Box 3.3, speculate on what aspect of its structure might lead to chloral's unusual stability. (The equilibrium constant for its formation is $10^4$.)

### QUESTION 3.7

Write a mechanism, with abbreviated proton transfer steps, that accounts for the following experimental observation.

When propanone, $CH_3COCH_3$, is mixed with excess water that has the oxygen atom replaced by the heavier isotope $^{18}O$ (i.e. $H_2{}^{18}O$ rather than $H_2{}^{16}O$) and then recovered, it is found by mass spectrometry that the sample has become largely $CH_3C^{18}OCH_3$.

## BOX 3.4 A Mickey Finn

Chloral hydrate, **3.28**, is the oldest of the hypnotic (sleep-inducing) depressants. It was discovered in 1832 by Justus von Liebig while examining the effect of chlorine on ethanol. Its hypnotic powers were discovered in 1869 by Otto Liebriech, a professor of pharmacy in Berlin.

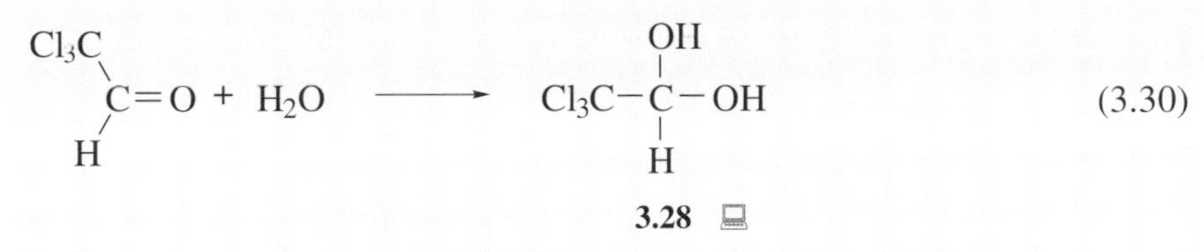

(3.30)

Since it could be administered orally and was cheap, it became widely used to produce sleep (it takes effect in about 30 minutes). It has little effect on the respiratory system and blood pressure. However, the margin between an effective sleeping draft and a lethal solution is very close! It is also addictive, as the Pre-Raphaelite painter Dante Gabriel Rossetti found to his cost (Figure 3.2).

Vita Sackville-West wrote in 1928:

> I am so sleepy with chloral simmering in my spine that I can't write, nor yet stop writing — I feel like a moth, with heavy scarlet eyes and a soft cape of down…

In Evelyn Waugh's largely autobiographical novel *The Ordeal of Gilbert Pinfold*, the 50-year-old novelist Pinfold takes chloral, among other things, to ease the discomfort of rheumatism and insomnia. However, the large amount of drugs he was taking and the sudden withdrawal of them caused a variety of mental problems.

Such side-effects have meant that it has been largely replaced by less addictive hypnotics.

A solution of chloral in alcohol has become known as a 'Mickey Finn'. In 1896, Mickey Finn was the proprietor of the Lone Star Saloon and Palm Garden restaurant in a particularly seedy part of Chicago. He mixed chloral with drinks, which were plied to customers by the local girls. The sleepy customers were then robbed of their clothes and belongings before being thrown out into the back alley. The saloon was closed down in 1903, but Finn escaped prosecution!

**Figure 3.2** *The Bower Meadow* (1872) by Dante Gabriel Rossetti (1828–82). Rossetti used chloral hydrate to combat insomnia, taking 10 grains of it with whisky around the time he was working on this painting; by the time he died, his addiction had increased the intake to 180 grains a day. (The grain is an old pharmacy unit: 1 grain ≡ 64.8 mg.)

The addition of water to the carbonyl group of ketones and aldehydes is catalysed by both hydroxide ion and acid. **Base catalysis** by hydroxide ion is readily understood because $^{-}OH$ is a better nucleophile than $H_2O$, and so it, rather than water, attacks the carbonyl group:

$$HO^{-} + R^1R^2C{=}O \rightleftharpoons R^1R^2C(OH){-}O^{-} \qquad (3.31)$$

The anion formed in this way can then attack a water molecule to generate another hydroxide ion, which is why $^{-}OH$ acts as a catalyst in this reaction:

$$R^1R^2C(OH){-}O^{-} + H{-}OH \rightleftharpoons R^1R^2C(OH){-}OH + {}^{-}OH \qquad (3.32)$$

Of course, this last step is yet another example of proton transfer between electronegative atoms.

**Acid catalysis** occurs because carbonyl compounds can act as bases. In acidic solution the carbonyl compound can be protonated (Section 3.1). Attack at the carbonyl carbon atom by a water molecule will be more rapid for the protonated than for the unprotonated substrate, since the carbon atom is more electrophilic, as shown by the positive charge on the carbonyl carbon in one of the resonance forms (**3.30**) of the protonated molecule.

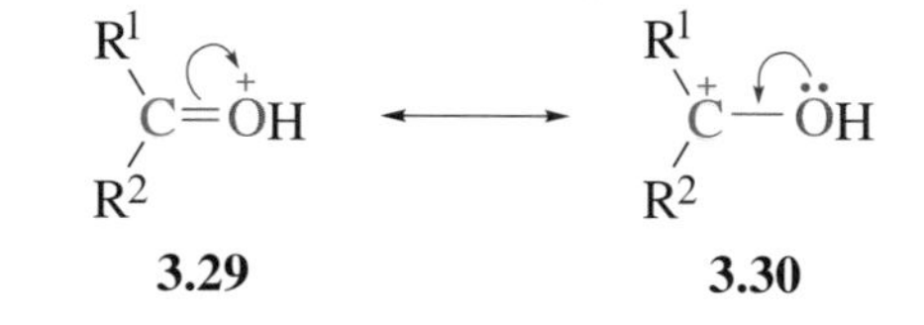

**3.29** **3.30**

Nucleophilic attack by water generates the intermediate species **3.31**, and in the final step $H^+$ is regenerated, showing that the acid, too, acts as a catalyst:

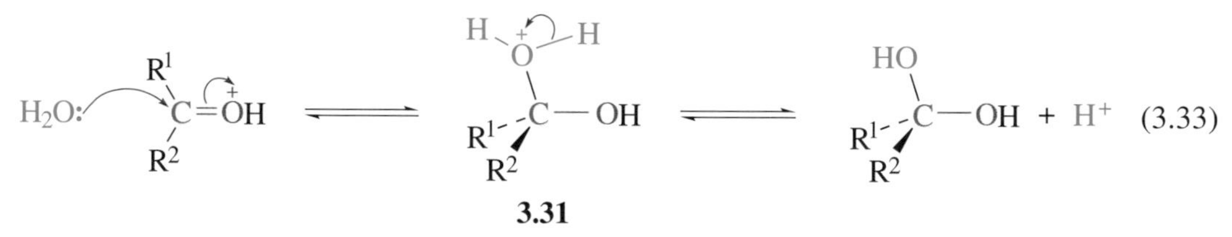

**3.31** (3.33)

An important feature of this sequence of reactions is that the addition of a proton to a carbonyl compound yields a cation that is a more powerful electrophile than the carbonyl group itself. This allows reaction with weak nucleophiles to occur, which would otherwise not occur in the absence of acid.

So much for water. What about alcohols, ROH? Well, they can attack aldehydes and ketones in a very similar way, and, again, the overall reaction is reversible, involving a series of equilibria. Here's an example with benzaldehyde and ethanol in the presence of concentrated HCl as catalyst:

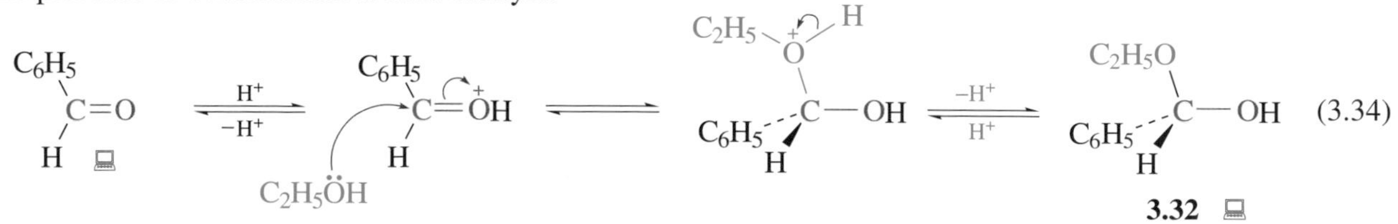

**3.32** (3.34)

The product of Reaction 3.34 is called a **hemiacetal** (Structure **3.32**). The use of the prefix 'hemi', meaning half, will probably not be obvious to you. The reason is that under the conditions employed, this is not the final product, which is called an **acetal**. Reaction 3.34 is reversible, and you can see that the first step in the regeneration of the aldehyde from the hemiacetal is protonation of the oxygen atom that is attached to the ethyl group. What might be the consequence of protonating the other oxygen atom?

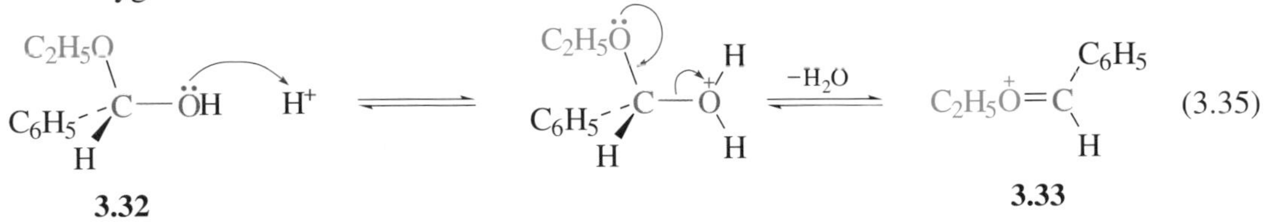


You can see that protonation forms water, a good leaving group, creating at the same time the strongly electrophilic ion, **3.33** which is an easy target for attack by another molecule of ethanol. After loss of a proton the final product, the acetal **3.34** emerges:

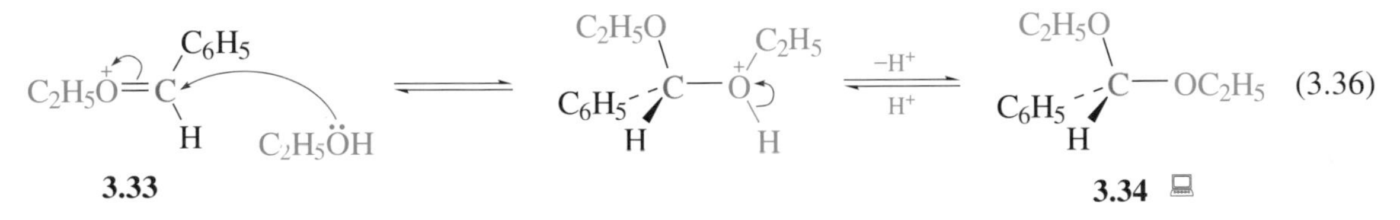


So acetals have the general structure **3.35**, and their usual precursors, hemiacetals, have the general structure **3.36**.

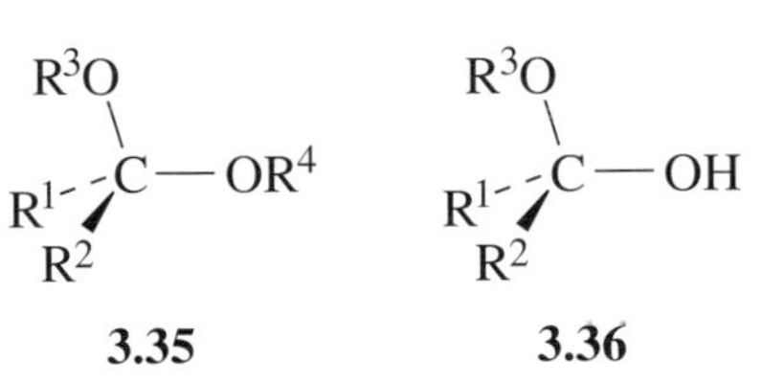


### QUESTION 3.8

Write down the structures of the acetals and hemiacetals from the following pairs of reactants:

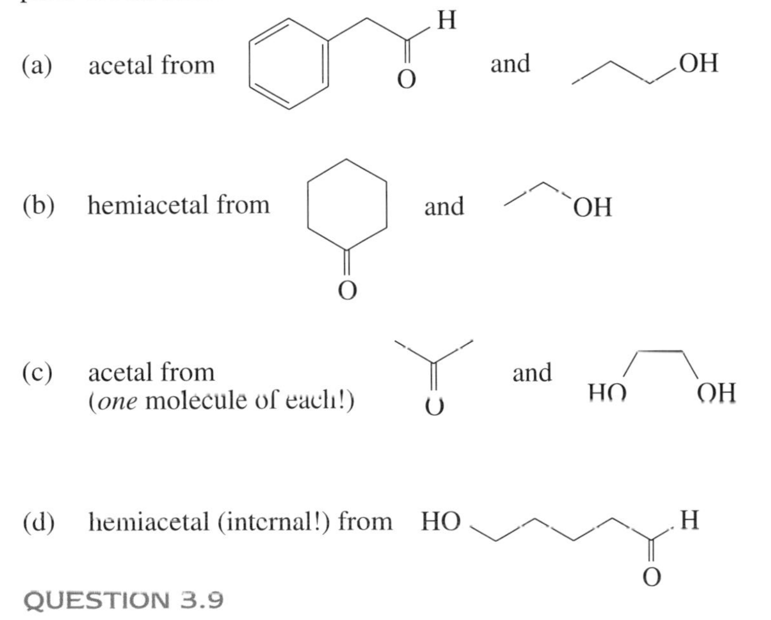


### QUESTION 3.9

By referring to the mechanisms of hemiacetal and acetal formation illustrated in the text, see if you can write curly arrow mechanisms for the formation of (i) the acetal in Question 3.8c and (ii) the hemiacetal in Question 3.8d, assuming hydrogen ions are available in both cases.

*It's important to remember that the whole sequence of acetal formation is reversible.* Therefore, if you want to make an acetal, you would arrange for the carbonyl compound to react in the presence of excess alcohol using an anhydrous acid catalyst such as dry HCl gas. Correspondingly, if you wanted to regenerate a carbonyl compound from an acetal, you would use water as the solvent so that the acetal will be hydrolysed back to the aldehyde or ketone.

Nucleophilic attack by alcohols to form hemiacetals and acetals has important consequences for the chemistry of aldehydes and ketones. Two prominent examples are firstly in synthetic methodology are as a useful strategy for temporarily protecting* a carbonyl group from attack while carrying out a reaction elsewhere in the same molecule (Box 3.5), and secondly in the fundamental chemistry of carbohydrates, a huge class of compounds found in biological systems (Box 3.6).

## BOX 3.5 Protection of carbonyl groups by acetal formation

Acetal formation is very important to the synthetic chemist. Quite often, a reagent that is required to carry out a particular chemical reaction will also attack any carbonyl group present in the molecule. For example, it would be impossible to make compound **3.38** from the reaction of **3.37** with lithium aluminium hydride, because both the ketone carbonyl group and the ester group would react with this reagent:

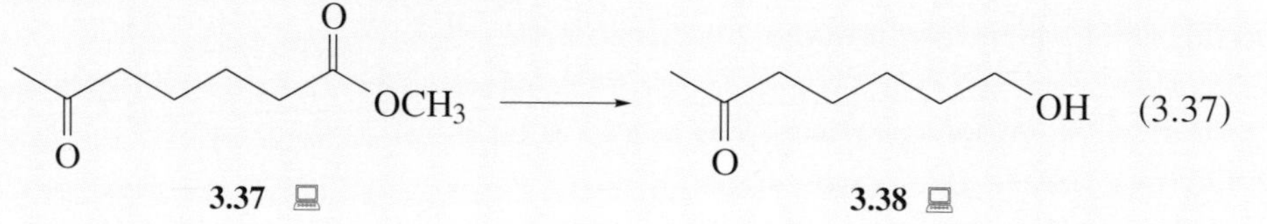


(The reaction of esters with $LiAlH_4$ to give alcohols is described in Section 3.3.4.) To prevent the ketone group from reacting, it must somehow be protected. One of the best ways of doing this is to form the acetal **3.39**, using ethane-1,2-diol (**3.40**).

HO OH

**3.40**

The cyclic acetal is stable as long as aqueous acid conditions are avoided, so the reaction of $LiAlH_4$ with the cyclic acetal **3.39** proceeds to form the alcohol **3.41**, from which the hydroxyketone **3.38** can easily be regenerated by aqueous acid hydrolysis:

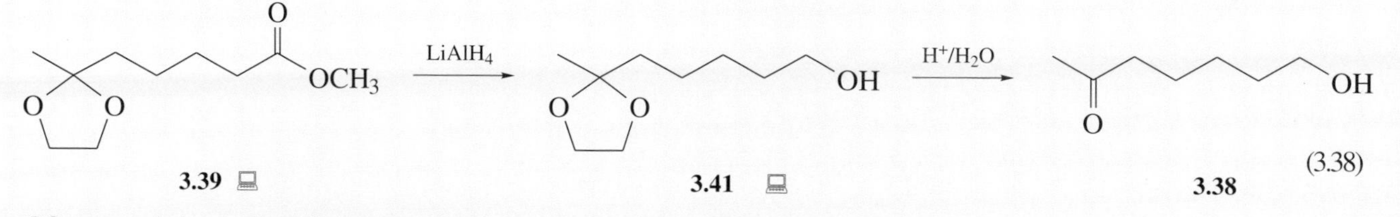


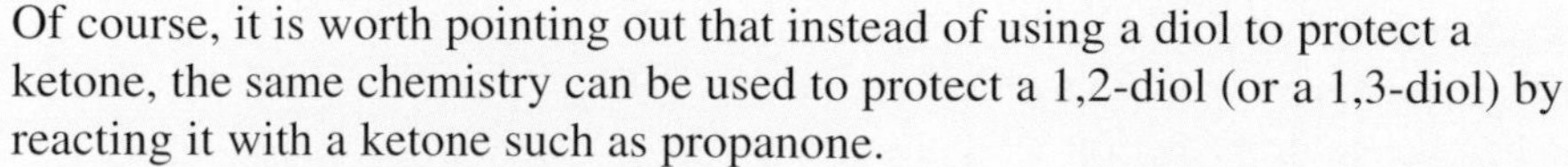

Of course, it is worth pointing out that instead of using a diol to protect a ketone, the same chemistry can be used to protect a 1,2-diol (or a 1,3-diol) by reacting it with a ketone such as propanone.

* The issue of protection in organic synthesis is discussed in detail in Part 4 of this Book.

QUESTION 3.10

Compound **3.42** is used as a perfume. From which carbonyl compound is this compound derived? Write a mechanism for the acid-catalysed formation of compound **3.42** from the corresponding carbonyl compound.

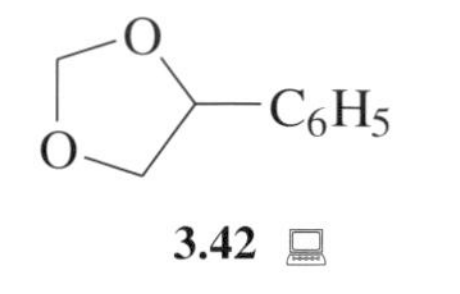


**3.42**

## BOX 3.6 Carbohydrates as aldehydes and hemiacetals

Hemiacetal formation is fundamental to the chemistry of carbohydrates. These compounds, such as glucose, sucrose, starch and cellulose, contain aldehyde or ketone groups, and several hydroxyl groups, all within the same molecule. Glucose, for example, can be drawn as shown in Structure **3.43**. As you can see, there are four chiral centres in the molecule, and so the compound is optically active.

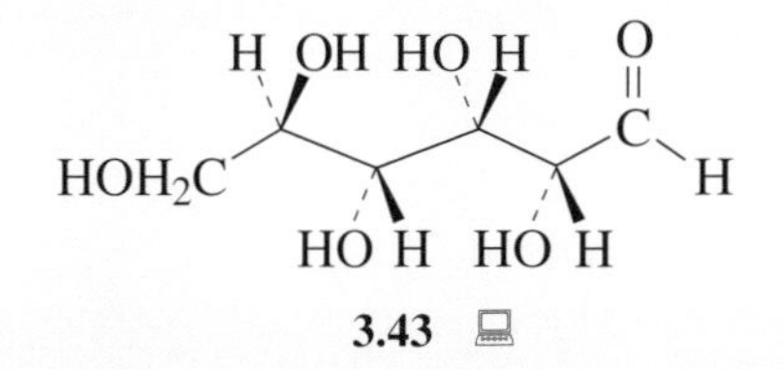


**3.43**

The particular form drawn here is called D-glucose (you needn't worry about this notation). In fact, D-glucose does not exist as the open-chain structure as we have drawn it. The hydroxyl group shown in red below can attack the aldehyde group to form a cyclic hemiacetal (even without acid catalysis):

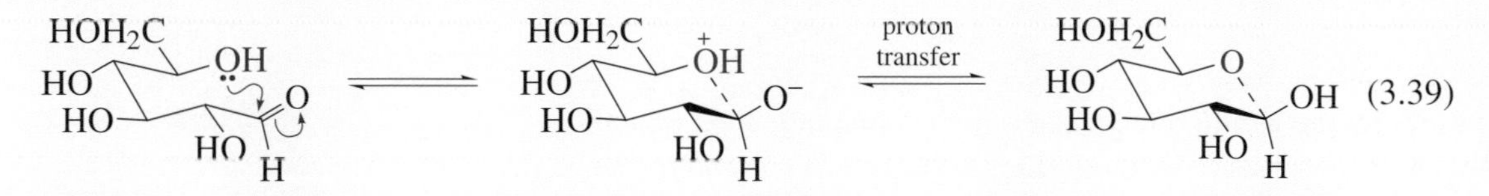


(3.39)

Two isomers can be formed by this cyclization. The two forms are called β-D-glucose (**3.44**) and α-D-glucose (**3.45**), and they may be represented as:

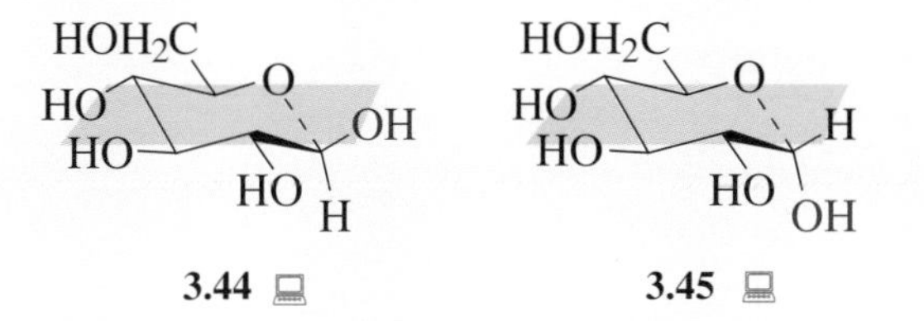


Are these two compounds enantiomers?

No. The mirror image of β-D-glucose cannot be superimposed on α-D-glucose. They are therefore diastereomers.

The only difference between the two compounds is that the hemiacetal OH group (in red) in β-D-glucose (**3.44**) lies *in* the plane of the ring, whereas in α-D-glucose (**3.45**) it lies *below* the ring plane (blue). β-D-glucose is the thermodynamically more stable isomer: if either isomer is dissolved in water, a mixture of the two forms, consisting of 62 per cent β and 38 per cent α, always results. This arises because the two forms can interconvert by re-forming the non-cyclic aldehyde and cyclizing in the opposite manner. The process of interconversion between the α form and the β form of glucose is also catalysed by either acid or base. The cyclization reaction shown in Reaction 3.39 is extremely important because it determines the structure of nearly all the carbohydrates, including sucrose, starch and cellulose.

**QUESTION 3.11**

Explain how base ($^{-}$OH) catalyses the interconversion of β-D-glucose and α-D-glucose.

This completes our study of the nucleophilic reactions of water and alcohols on aldehydes and ketones. Now we turn to those of nitrogen nucleophiles before moving on to the other major class of carbonyl compounds — carboxylic acids and their derivatives.

## 3.2.4 Nitrogen nucleophiles: the formation of imines

Our description of the chemistry of aldehydes and ketones has progressed from their reactions with powerful carbon and hydride nucleophiles to those with weaker oxygen nucleophiles. We now want you to consider nucleophiles intermediate in reactivity, the nitrogen nucleophiles. You will see right away that a distinctive kind of product arises. But don't worry: we are still looking at the outcome of nucleophilic attack on a carbonyl and the initial stages are very similar.

Here's an example. Benzaldehyde reacts with aniline in the presence of a low concentration of acid to give an **imine**, a compound containing a C=N bond:

$$(C_6H_5)(H)C{=}O + C_6H_5NH_2 \xrightarrow{H^+ \text{(catalyst)}} (C_6H_5)(H)C{=}NC_6H_5 + H_2O \qquad (3.40)$$

imine

See if you can predict the analogous product that is obtained when pentan-3-one, $CH_3CH_2COCH_2CH_3$ reacts with ethylamine, $C_2H_5NH_2$, in the presence of a water scavenger like anhydrous magnesium sulfate.

See Reaction 3.46 at the bottom of p. 40.

Notice that, in both reactions, the carbonyl oxygen atom is completely removed from the product, and the C=O bond is replaced by a C=N bond. This looks very different to the hemiacetals we obtained when alcohols reacted with aldehydes and ketones, but in fact the mechanism for the formation of imines is very similar to that of hemiacetal formation — at least in the initial stages.

We'll start in the usual manner for carbonyl compounds, with the nucleophilic nitrogen attacking the electrophilic carbon atom of the carbonyl to give a tetrahedral intermediate:

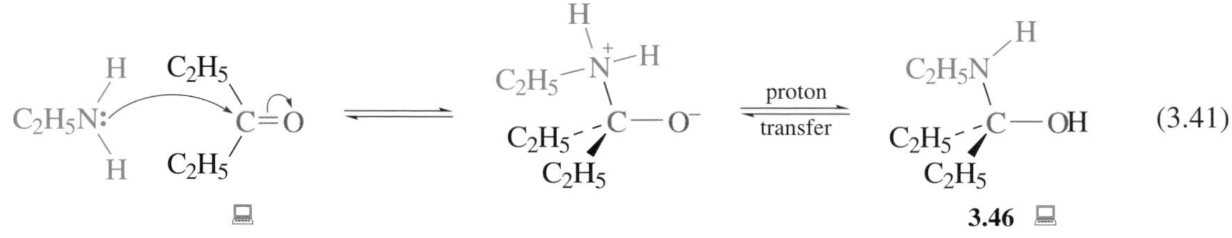


Compounds with structures like **3.46** are called **carbinolamines**; you can see that they are analogous to the hemiacetals obtained when the nucleophile is an alcohol. Indeed, they have the general structure **3.4** of the primary product of attack by any nucleophile. But there is a significant difference between carbinolamines and hemiacetals that accounts for the appearance of an imine as the final product.

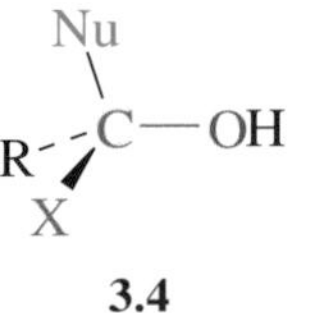


**3.4**

Compare the structures of the carbinolamine **3.46**, the hemiacetal **3.47**, and the imine **3.48**:

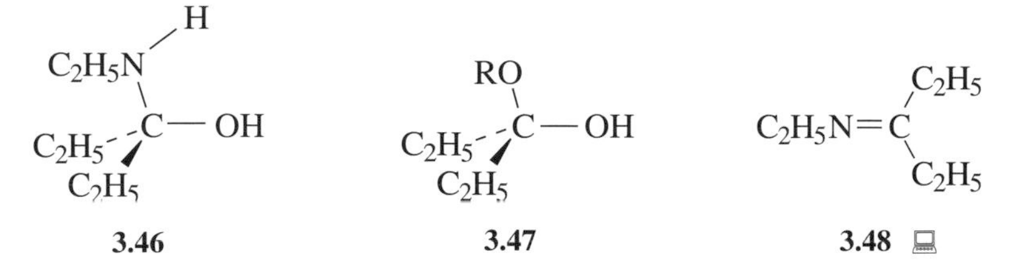


What is the key difference between the first two, and what type of reaction converts **3.46** to **3.48**?

The key difference is that the nitrogen atom in **3.46** bears a hydrogen that is not available in the analogous position on a hemiacetal. This makes the elimination of water possible to give the imine.

A carbinolamine will undergo a β-elimination, which, in fact, occurs in two steps. The first retains the analogy with hemiacetals (Reaction 3.35). The non-bonded electrons move to form a double bond to carbon, this time expelling $HO^-$ rather than $H_2O$. It's the next step that makes the difference. Whereas the reactive oxygen analogue from the hemiacetal **3.32** is attacked at the blue carbon by a nucleophile, the intermediate from a carbinolamine can form the relatively stable imine **3.48** simply by losing a proton from the positively charged nitrogen atom:

$C_2H_5\overset{+}{O}=C(C_6H_5)H$

**3.33**

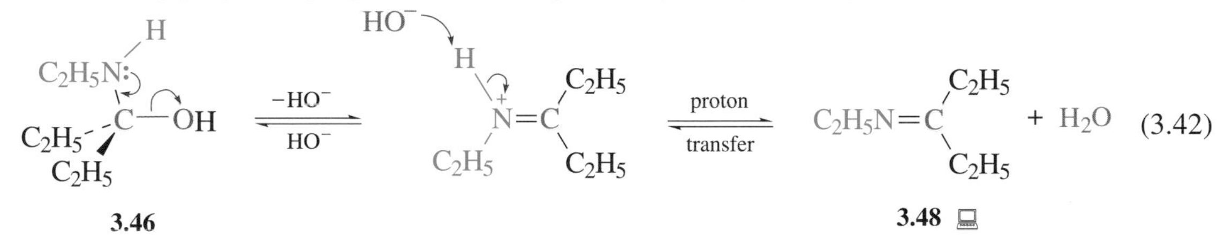


Because these reactions involve addition of the nucleophile followed by elimination of water, they are often referred to as **addition–elimination reactions**. They are equilibria, and the presence of a water scavenger ensures that the water produced in the elimination step is removed from the reaction.

What effect will this have on the equilibria?

The equilibria will shift, to favour formation of the product imine **3.48**.

Often, however, the presence of an acid is required for the imine to be formed. The role of the acid here is not to protonate the carbonyl group, since, in general, amines are sufficiently nucleophilic to attack the carbonyl carbon atom. Rather, the acid is needed to catalyse the dehydration (that is, loss of water) of the carbinolamine. The acid protonates the hydroxyl group to form $-\overset{+}{O}H_2$, which is a much better leaving group than $^-OH$:

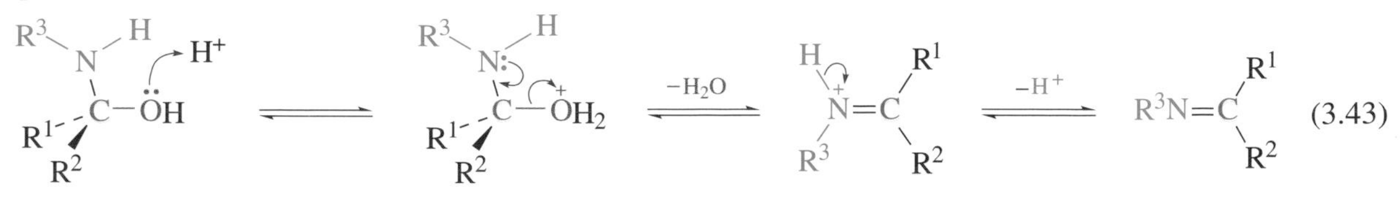

Of course, the nitrogen atom of the amino group in the carbinolamine can also be protonated, but this only helps to catalyse the decomposition of the carbinolamine back to starting materials. Remember that the role of a catalyst is to speed up the attainment of equilibrium, it does *not* affect the position of the equilibrium: it speeds up both the forward and back reaction.

Strong acids can actually inhibit the reaction, because they protonate the amine, so eliminating its nucleophilic character:

$$RNH_2 + H^+ \rightleftharpoons R\overset{+}{N}H_3 \qquad (3.44)$$

Thus, the concentration of the acid has to be quite carefully controlled; the optimum pH is about 4.5.

QUESTION 3.12

The most widely used reaction for detecting an aldehyde or ketone chemically is illustrated below with ethanal:

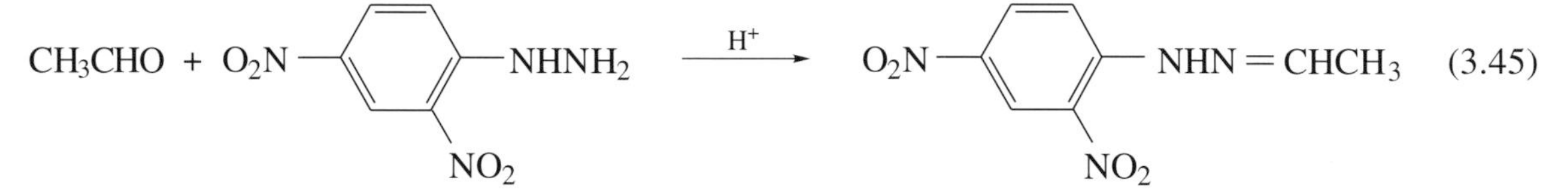


The reaction is carried out in ethanol in the presence of hydrochloric or sulfuric acid, and the product, called a 2,4-dinitrophenylhydrazone (or colloquially, a 2,4-DNP derivative), appears as yellow or orange crystals (see Figure 3.3, p. 42).

Write out a full curly arrow mechanism for the formation of ethanal DNP.

The chemical derivative of an aldehyde described in Question 3.12 is just one of a number of nitrogen nucleophiles with an $NH_2$ group that may be used as a means of detecting or identifying aldehydes by the distinctive melting temperatures of their derivatives. Others are mentioned in Box 3.7. In particular, you should notice that they are all *primary* amines, because *two* hydrogen atoms must be transferred to displace the oxygen atom of the carbonyl group as water.

Before we end our discussion of aldehydes and ketones, we shall point out a useful distinction between them in the next Section.

## 3.2.5 Reactivity of aldehydes and ketones: a comparison

Aldehydes and ketones generally undergo the same type of reactions; they both take part in reactions that involve addition of a nucleophile, and they both undergo imine formation. However, they do so at different rates. In general, aldehydes are much more reactive than ketones. Partly, this is because there is a steric difference between the two types of compound.

What steric reason makes aldehydes more reactive than ketones?

---

*Answer to question on p. 38*

$$(C_2H_5)_2C{=}O + C_2H_5NH_2 \xrightarrow{-H_2O} (C_2H_5)_2C{=}NC_2H_5 \qquad (3.46)$$

For both aldehydes and ketones, one of the groups bonded to the carbonyl carbon atom is an alkyl or aryl group. However, in aldehydes, the other group bonded to the carbonyl carbon atom is a hydrogen atom, whereas ketones have another alkyl or aryl group bonded to the carbonyl carbon atom. A hydrogen atom is much smaller than an alkyl group, so there is less steric hindrance to nucleophilic attack at an aldehyde carbonyl group.

There is also a difference in the reactivity of aldehydes and ketones due to electronic effects, but we shall not pursue this further here. It is sufficient for you to know that aldehydes are more reactive than ketones. For simple compounds containing only an aldehyde or a ketone functional group, this difference in reactivity is unimportant. However, when a compound contains both types of functional group, the greater reactivity of aldehydes must be borne in mind.

### 3.2.6 Summary of Sections 3.1 and 3.2

1. Nucleophiles attack the carbon atom of the carbonyl group, and the carbon–oxygen $\pi$ bond breaks during the reaction.
2. Nucleophilic attack on RCOX occurs at the carbonyl carbon atom, and results in the formation of a tetrahedral alcohol product. If X or XH is a good leaving group, this product can decompose by loss of X, so that the overall reaction is replacement of X by the nucleophile. For aldehydes and ketones, X is a very poor leaving group; for carboxylic acid derivatives, X/XH is a good leaving group.
3. The reactivity of the carbonyl compound towards nucleophiles can be increased by protonation.
4. Carbon nucleophiles react with aldehydes and ketones by carbon–carbon bond formation, to yield alcohols. This is an important reaction for synthesis.
5. Lithium aluminium hydride also yields alcohols from its reaction with aldehydes and ketones.
6. With ketones and aldehydes, water forms a hydrate, which can sometimes be isolated; its presence can be demonstrated using $^{18}O$-labelled water. The reaction is catalysed by acid or base.
7. Acids catalyse the addition of other oxygen nucleophiles, for example alcohols, to aldehydes and ketones. The initial product is a hemiacetal, and the final product an acetal.
8. Many carbohydrates exist as internal hemiacetals, and this plays a fundamental part of the chemistry of this important class of biological compounds.
9. Diols can be used to form a cyclic acetal, which can be used to protect the carbonyl group from attack by other reagents in the absence of acid.
10. Nitrogen nucleophiles undergo addition–elimination reactions with ketones and aldehydes. This results in the loss of the carbonyl oxygen atom, and the formation of a C=N group.
11. Aldehydes are more reactive than ketones.

A diagram summarizing the reactions of aldehydes and ketones is given in Appendix 1.

## BOX 3.7 Chemical characterization of aldehydes and ketones

Aldehydes and ketones with low relative molecular mass are often volatile liquids, difficult to characterize by boiling temperature, but they can easily be converted into crystalline derivatives by reaction with nitrogen nucleophiles. Favourites that have become established through long practice are the **semicarbazones** and **2,4-dinitrophenylhydrazones (2,4-DNPs)** (Table 3.1 and Figure 3.3). It is usually easy to obtain these compounds in a pure state by recrystallization, and their sharp melting temperatures allow the original aldehyde or ketone to be identified by comparison with those of known compounds.

In addition, 2,4-DNP derivatives offer a convenient way of separating the components of a mixture of aldehydes and ketones. Rather than attempt to do this directly by fractional distillation, a chemist might take advantage first of the easy separation of 2,4-DNP derivatives by column chromatography (they are yellow–orange, so this could be done visually). As their formation from aldehydes and ketones is reversible, hydrolysis of the separated derivatives will then regenerate the original carbonyl compounds.

The oxime and hydrazone derivatives in Table 3.1 are less useful for characterization, but are valued intermediates in organic synthesis.

**Figure 3.3** The 2,4-dinitrophenylhydrazone formed from benzaldehyde.

**Table 3.1** Nitrogen nucleophilic reagents that undergo addition–elimination reactions with $R^1R^2C{=}O$

| Nucleophile | Structure | Product structure | Product name |
|---|---|---|---|
| amine | $R^3NH_2$ | $R^1R^2C{=}NR^3$ | an imine |
| hydrazine | $H_2N-NH_2$ | $R^1R^2C{=}NNH_2$ | a **hydrazone** |
| 2,4-dinitrophenylhydrazine | $NH_2$–HN at C1 of benzene ring; $NO_2$ at C2; $NO_2$ at C4 | $N{=}CR^1R^2$–HN at C1 of benzene ring; $NO_2$ at C2; $NO_2$ at C4 | a 2,4-dinitrophenylhydrazone (2,4-DNP) |
| hydroxylamine | $H_2NOH$ | $R^1R^2C{=}NOH$ | an **oxime** |
| semicarbazide | $H_2N-N(H)-C(=O)-NH_2$ | $R^1R^2C{=}N-N(H)-C(=O)-NH_2$ | a semicarbazone |

QUESTION 3.13

How would you make the alcohol **3.49**, from a carbonyl compound?

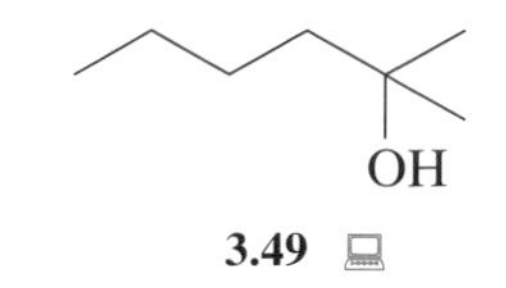


**3.49**

QUESTION 3.14

How would you carry out the following transformation, which is part of a synthesis of a contraceptive hormone (don't worry if a mixture of stereoisomers would be obtained)?

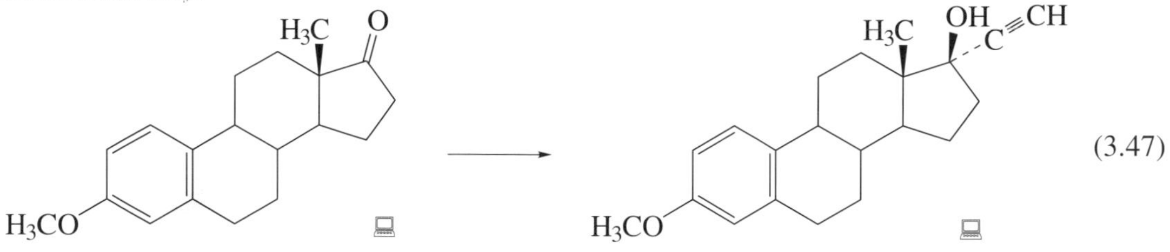


(3.47)

QUESTION 3.15

Write down the structure of the product you would expect from the reaction between 1-phenylethanone (acetophenone), $C_6H_5COCH_3$, and semicarbazide in the presence of acid. Write a mechanism to explain its formation.

# 3.3 CARBOXYLIC ACID DERIVATIVES

## 3.3.1 Introduction: organization through mechanism revisited

In our overview of nucleophilic attack at the carbonyl group, you saw that a nucleophile (for example, $Nu^-$) would attack the carbonyl carbon atom of the compound RCOX, to form the tetrahedral intermediate **3.4**. Moreover, we pointed out that carboxylic acid derivatives will have a different chemistry from aldehydes and ketones, because X is a good leaving group for carboxylic acid derivatives. So the product of addition can decompose with loss of $X^-$ (or XH):

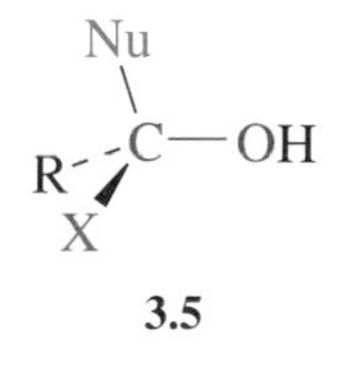


**3.5**

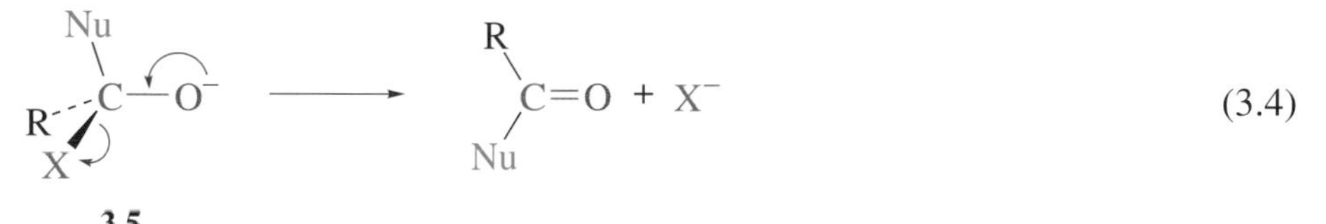


(3.4)

Now we shall see if our expectation is borne out.

## 3.3.2 Oxygen nucleophiles: hydrolysis and esterification

With water or an alcohol, $R^1OH$, as nucleophile, acid chlorides (X = Cl) generally react rapidly (sometimes violently), acid anhydrides ($X = OCOR^2$) react less quickly, but esters ($X = OR^2$), amides ($X = NR^2$) and carboxylic acids (X = OH) generally do not react at all, unless either an acid or base catalyst is present. We'll consider first the more-reactive carboxylic acid derivatives, the acid chlorides and anhydrides.

Applying the earlier principles, you should be expecting an acid chloride like ethanoyl chloride ($CH_3COCl$) to react with an alcohol like ethanol ($C_2H_5OH$) in two steps; a nucleophilic addition followed by an elimination to give a product in which the C=O group has been restored. Here's the first step:

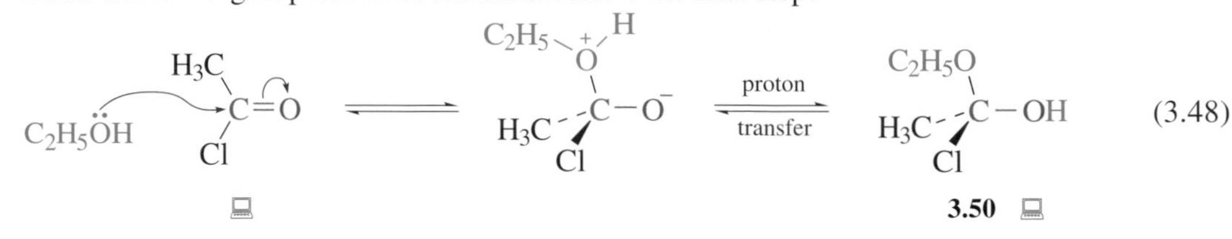


(3.48)

The product of this reaction, **3.50**, can then eliminate either HCl or $C_2H_5OH$; of course, elimination of $C_2H_5OH$ will simply regenerate starting materials.

- Which do you think will be eliminated preferentially, HCl or $C_2H_5OH$?

- HCl will be eliminated preferentially, because it is a much stronger acid than $C_2H_5OH$; $Cl^-$ is therefore a better leaving group than $C_2H_5O^-$.

So we complete the reaction between ethanol and ethanoyl chloride by showing the elimination of hydrogen chloride:

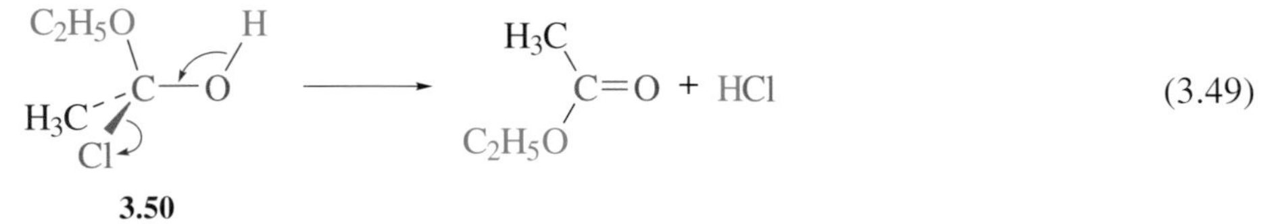


(3.49)

Although $Cl^-$ is a good nucleophile, the final step of this reaction is essentially irreversible, because $Cl^-$ is a much better leaving group than $C_2H_5O^-$. The product of this substitution reaction, in which Cl has been substituted by $C_2H_5O$, is an ester.

### QUESTION 3.16

(a) Using the above example as a model, write a two-step curly arrow mechanism for the reaction between water and ethanoic anhydride, $CH_3COOCOCH_3$.

(b) Predict the products from the reaction between the unsymmetrical anhydride, $CH_3COOCOC_2H_5$, and ethanol (no mechanism required).

So acid chlorides react with water to produce carboxylic acids, and with alcohols to produce esters; anhydrides react with water to produce carboxylic acid(s), and with alcohols to produce ester(s) *and* carboxylic acid(s):

$$RCOCl + H_2O \longrightarrow RCOOH + HCl \quad (3.50)$$

$$R^1COCl + R^2OH \longrightarrow R^1COOR^2 + HCl \quad (3.51)$$

$$R^1COOCOR^2 + H_2O \longrightarrow R^1COOH + R^2COOH \quad (3.52)$$

$$R^1COOCOR^2 + R^3OH \longrightarrow R^1COOR^3 + R^2COOH + R^2COOR^3 + R^1COOH \quad (3.53)$$

Remember (Question 3.16b) that an unsymmetrical anhydride can react with an alcohol to give *four* products, because the alcohol can attack at either carbonyl group of the anhydride. Of course, for the more common anhydrides, in which $R^1$ is the same as $R^2$, the products comprise just one carboxylic acid and one ester.

The compounds formed in the first step in these reactions between oxygen nucleophiles and acid chlorides or anhydrides all have a similar tetrahedral structure; either **3.5** if the nucleophile is charged, $Nu^-$ or **3.4** (after proton transfer) if the nucleophile is neutral, **:NuH**.

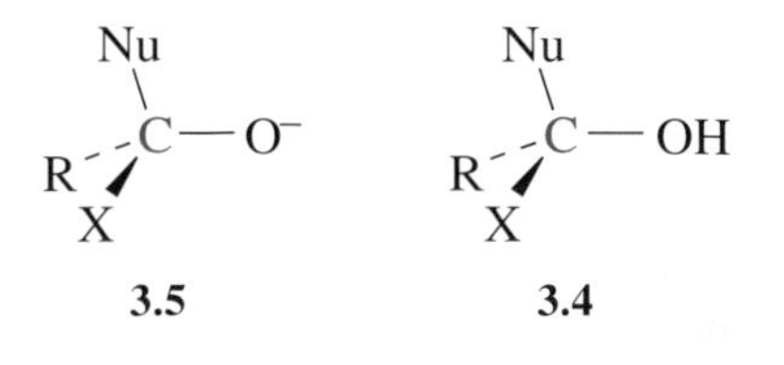

It is usually not possible to isolate these compounds because the elimination of $X^-$ or HX occurs very easily. So they are more properly considered as intermediates in the overall process, and are usually referred to as **tetrahedral intermediates**.

Just as acid catalyses the reaction of $H_2O$ and ROH with ketones and aldehydes, so it will also catalyse the reactions of $H_2O$ and ROH with the less-reactive carboxylic acid derivatives — esters and amides, and carboxylic acids themselves. For the reaction of carboxylic acid derivatives with water, one of the products is always a carboxylic acid, whereas with an alcohol, one of the products is always an ester. We can summarize some of these reactions by the following equations:

$$R^1{-}C({=}O){-}OR^2 + H_2O \xrightleftharpoons{H^+\ \text{catalyst}} R^1{-}C({=}O){-}OH + R^2OH \qquad (3.54)$$

$$R^1{-}C({=}O){-}OR^2 + R^3OH \xrightleftharpoons{H^+\ \text{catalyst}} R^1{-}C({=}O){-}OR^3 + R^2OH \qquad (3.55)$$

$$R^1{-}C({=}O){-}NR^2R^3 + H_2O \xrightleftharpoons{H^+\ \text{catalyst}} R^1{-}C({=}O){-}OH + R^2R^3NH \qquad (3.56)$$

$$R^1{-}C({=}O){-}NR^2R^3 + R^4OH \xrightleftharpoons{H^+\ \text{catalyst}} R^1{-}C({=}O){-}OR^4 + R^2R^3NH \qquad (3.57)$$

The hydrolysis of esters to acids under these conditions generally occurs more readily than does the hydrolysis of amides.

### QUESTION 3.17

Draw the tetrahedral intermediates for each of the following reactions:

(a) $$C_6H_5COCl + CH_3OH \longrightarrow C_6H_5COOCH_3 + HCl \qquad (3.58)$$

(b) $$CH_3CONHCH_3 + H_2O \xrightleftharpoons{H^+} CH_3COOH + CH_3NH_2 \qquad (3.59)$$

(c) $$\text{(cyclic anhydride, five-membered ring with two C=O and ring O)} + CH_3OH \xrightleftharpoons{H^+} HOOC{-}CH_2CH_2{-}COOCH_3 \qquad (3.60)$$

(d) $$HO{-}CH_2CH_2CH_2CH_2{-}C({=}O){-}OH \xrightleftharpoons{H^+} \text{(six-membered lactone, O=C in ring with ring O)} \qquad (3.61)$$

Experimental observations have shown that the reactivity of carboxylic acid derivatives with water or alcohols (with or without $H^+$ catalysis) *decreases* in the following order:

| acid chloride | > | acid anhydride | > | ester | > | amide |
|---|---|---|---|---|---|---|
| (X = Cl) | | ($X = R^2COO$) | | ($X = R^2O$) | | ($X = R^1R^2N$) |

It turns out that this is the general order of reactivity with all nucleophiles, not just those based on oxygen.

How is this explained? Well, clearly it has to do with the carboxylic acid derivative, not the nucleophile. And one thing that becomes apparent when you compare them is that the ability of $X^-$ as a leaving group from the tetrahedral intermediates **3.4** and **3. 5** decreases in the same order. Remember that the stronger the acid XH, the better is the leaving group $X^-$. This leads to the observed order of reactivity where the leaving group abilities are

$$Cl^- > R^2CO_2^- > R^2O^- > R^2R^3N^-$$

So you know that carboxylic acid derivatives react with nucleophiles:

- in a two-step reaction via a tetrahedral intermediate;
- to give, as a final product, a carboxylic acid or one of its derivatives;
- in the following order of reactivity: chloride > anhydride > ester > amide.

You also know how to present the mechanisms of the reactions of acid chlorides and anhydrides with ROH using curly arrows, and that, in principle, similar reactions should occur with esters and amides. But we have said that the latter, being less reactive, require the assistance of acid catalysis. How does this work? We'll look first at the reaction of water with an ester — that is, **ester hydrolysis**.

The first step is protonation of the carbonyl oxygen atom. Let's take as an example the hydrolysis of ethyl ethanoate (ethyl acetate). Protonation of the ester makes the carbonyl group much more electrophilic:

$$H_3C(C_2H_5O)C{=}\ddot{O} + H^+ \rightleftharpoons H_3C(C_2H_5O)C{=}\overset{+}{\ddot{O}}H \longleftrightarrow H_3C(C_2H_5O)\overset{+}{C}{-}\ddot{O}H \quad (3.62)$$

This enables water to attack the positively charged carbon atom, forming the charged intermediate **3.51**, which contains three electronegative oxygen atoms, so a proton can be transferred between them very easily. Since the solvent for the reaction is water, it is almost certain that the first proton transfer occurs to a solvent water molecule, to give an uncharged tetrahedral intermediate **3.52**, with concomitant formation of $H_3O^+$. Now, any of the three oxygen atoms in **3.52** is able to acquire a proton from $H_3O^+$. However, protonation of either of the two OH groups regenerates **3.51**, from which the starting ester will be formed by loss of water. Alternatively, a proton can be transferred to the $OC_2H_5$ oxygen atom, to form **3.53**. This intermediate can lose a molecule of $C_2H_5OH$, to generate the carboxylic acid, $CH_3COOH$, and $H^+$ (as $H_3O^+$); in other words, $H^+$ acts as a catalyst:

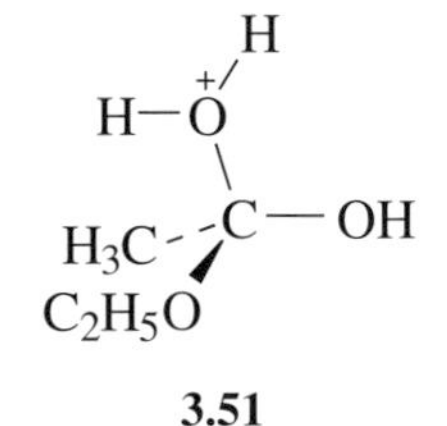




SCHEME 3.3

Often, as a shorthand way of writing the mechanism, proton transfer is shown to occur directly between **3.51** and **3.53**:

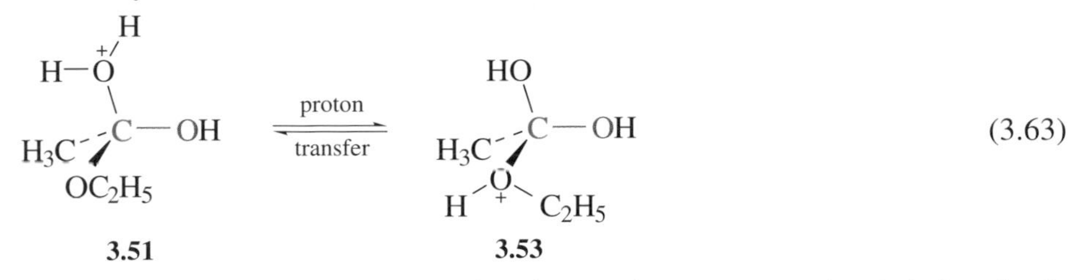

You can use this shorthand, but remember that a solvent water molecule is involved in the proton transfer.

These reactions are all reversible. So, in the presence of a strong acid, ethanol will react with ethanoic acid to give ethyl ethanoate, and we are now considering not ester hydrolysis but ester formation, or **esterification**. Protonation of the ethanoic acid is followed by attack of ethanol:

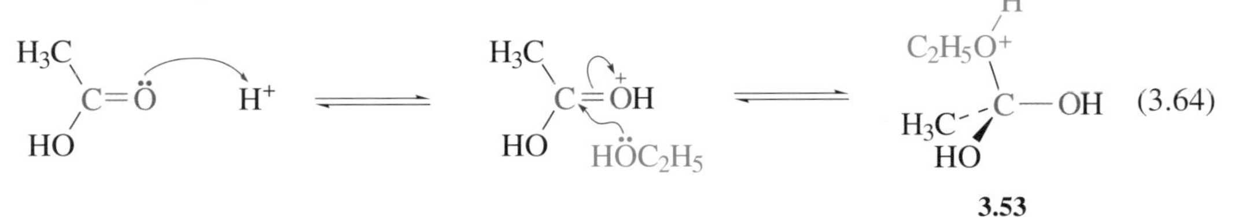

Elimination of either $H_2O$ or $C_2H_5OH$ can occur from **3.53**; elimination of ethanol would yield the starting materials, but elimination of water gives the ester:

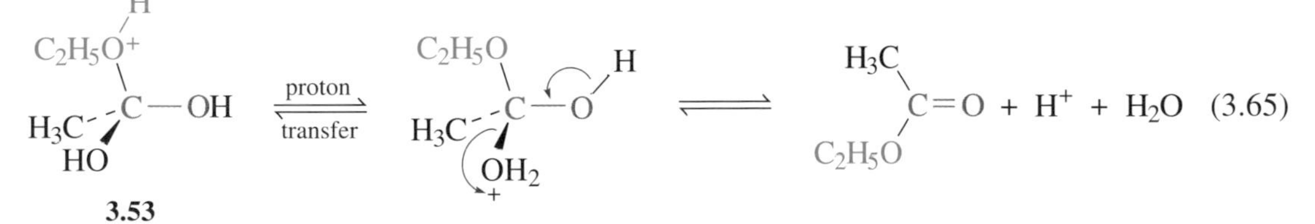

Again, proton transfer to the OH group produces the better leaving group, $H_2O$. Considering both forward and reverse reactions, the overall reaction involving ethanol and ethanoic acid may be written as:

$$\mathrm{CH_3C(=O)OC_2H_5} + \mathrm{H_2O} \underset{}{\overset{\mathrm{H^+}}{\rightleftharpoons}} \mathrm{CH_3C(=O)OH} + \mathrm{C_2H_5OH} \quad (3.66)$$

Notice that all these acid-catalysed reactions — of the carboxylic acid and carboxylic acid derivatives — follow the same type of pathway namely:

- protonation of the carbonyl group;
- attack by the nucleophile;
- a proton transfers to the leaving group;
- the leaving group departs.

You should make sure you remember these four steps!

### QUESTION 3.18

Review the last few paragraphs and, without further reference, see if you can write a curly arrow mechanism for the acid-catalysed esterification of benzoic acid, $C_6H_5CO_2H$, with methanol, $CH_3OH$.

It is important to emphasize the significance of the acid catalyst in these processes. The addition of a trace of a strong acid to a mixture of water and carboxylic acid ester only accelerates the *rate* at which equilibrium between the reactants and the products is established; it does not alter the amount of each compound in the equilibrium mixture.

The equilibrium constant for Reaction 3.66 is given by

$$K = \frac{[CH_3}{[CH_3}$$

Because there is little difference between the Gibbs functions of the reactants and the products (they have similar types and numbers of bonds), the numerical value of many such equilibrium constants for a variety of acids, alcohols and esters is around unity at 25 °C. In order for the reaction to go to completion in either direction, so that an acid or an ester can be prepared by this means, a way must be found of shifting the position of the equilibrium.

- Give two simple ways in which the equilibrium can be shifted so as to favour the formation of the acid.

- We can either remove the ethanol, or add a considerable excess of water. The latter method is usually employed.

Conversely, if we want to form an ester from the parent carboxylic acid, the necessary alcohol should be present in excess, usually as the solvent.

That is the position using acid catalysis and neutral nucleophiles. But sodium hydroxide and potassium hydroxide also react with esters, amides, acid anhydrides and acid chlorides. The following are typical examples:

$$C_6H_5COOCH_3 + HO^- \longrightarrow C_6H_5CO_2^- + CH_3OH \qquad (3.67)$$

$$C_6H_5CONH_2 + HO^- \longrightarrow C_6H_5CO_2^- + NH_3 \qquad (3.68)$$

$$CH_3COOCOCH_3 + 2HO^- \longrightarrow 2CH_3CO_2^- + H_2O \qquad (3.69)$$

These reactions are essentially irreversible and the mechanism shows why. Let's look at the reaction of methyl benzoate with aqueous sodium hydroxide. As always, the first step is the attack of the nucleophile $HO^-$ on the carbonyl carbon atom, forming a tetrahedral intermediate. This can either eject $HO^-$, to give the starting materials, or eject the $CH_3O^-$ anion, to give the acid:

$$HO^- + (C_6H_5)(CH_3O)C{=}O \rightleftharpoons (HO)(C_6H_5)(CH_3O)C{-}O^- \rightleftharpoons (C_6H_5)(HO)C{=}O + CH_3O^- \qquad (3.70)$$

The reaction is completed by the basic methoxide ion receiving a proton from the benzoic acid:

$$C_6H_5COOH + CH_3O^- \rightleftharpoons C_6H_5CO_2^- + CH_3OH \qquad (3.71)$$

This last step is essentially irreversible, because methanol is a much weaker acid than benzoic acid. This makes base hydrolysis of an ester an irreversible process, and, for this reason, ester hydrolysis is usually carried out under basic conditions.

- Is $HO^-$ acting as a catalyst in the base hydrolysis of esters?

- No, because it is not regenerated during the reaction. The reaction is *faster* than with water because $HO^-$ is a better nucleophile than $H_2O$.

So, in summary, esters are best *formed* by allowing the carboxylic acid to react in the alcohol as solvent and in the presence of a strong acid like HCl. Esters are best *hydrolysed*, to a carboxylic acid and an alcohol, by treatment with, typically, aqueous sodium hydroxide. This last reaction forms the basis of a large tonnage manufacturing process (Box 3.8).

## 3.3.3 Nitrogen nucleophiles: formation of amides

Carboxylic acid derivatives react with nitrogen nucleophiles in much the same way as with oxygen nucleophiles. In general, nitrogen nucleophiles are more reactive than oxygen nucleophiles so acid catalysis is not usually needed.

In common with all reactions of carbonyl compounds, the reactions involve initial attack at the carbonyl carbon atom; for example

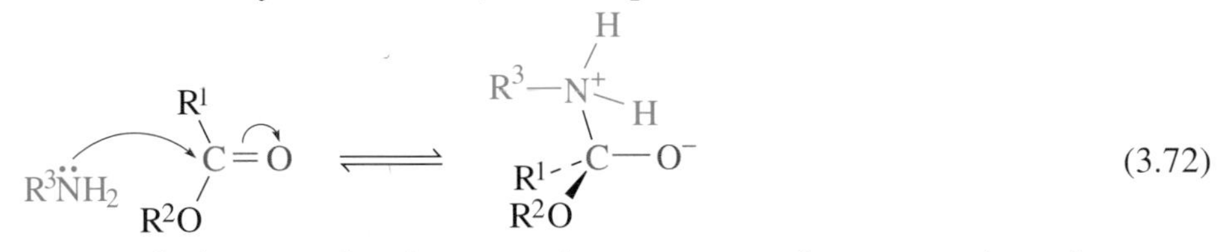

(3.72)

Proton transfer between the nitrogen and oxygen atoms forms an uncharged tetrahedral intermediate, **3.54**, from which the product, an amide, is formed (in this case by loss of an alcohol):

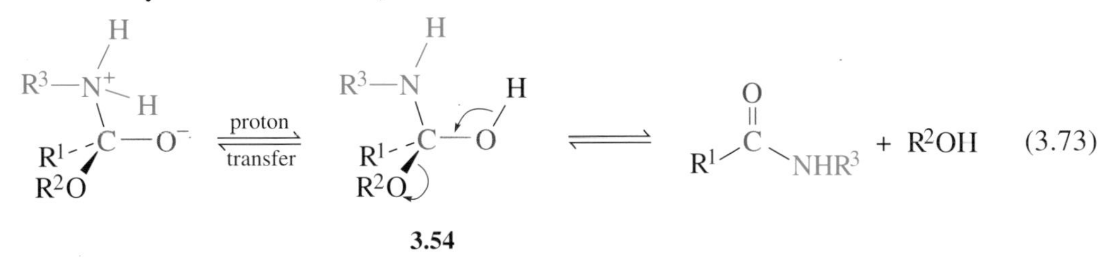


(3.73)

- We have just shown an amine reacting with an ester. If the reaction were with an acid chloride or an acid anhydride or another amide, would you expect these carboxyl derivatives to be more or less reactive towards the same amine?

- They should follow the same order of reactivity as shown with other nucleophiles (e.g. water), namely:

  acid chloride > anhydride > ester > amide

### QUESTION 3.19

(a) Predict the products of the reaction between aniline, $C_6H_5NH_2$, and benzoyl chloride, $C_6H_5COCl$, and suggest why the maximum yield of amide requires *two* moles of amine for every mole of acid chloride.

(b) Write a curly arrow mechanism for the reaction.

Have you noticed we have omitted one carboxyl derivative in this discussion about nitrogen nucleophiles? We've left out the parent, the carboxylic acid itself! And the reason is that, at room temperature, carboxylic acids do not react with amines to give amides.

## BOX 3.8 Fats and soaps

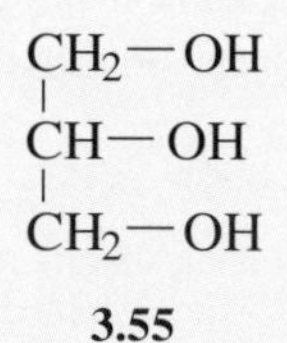

**3.55**

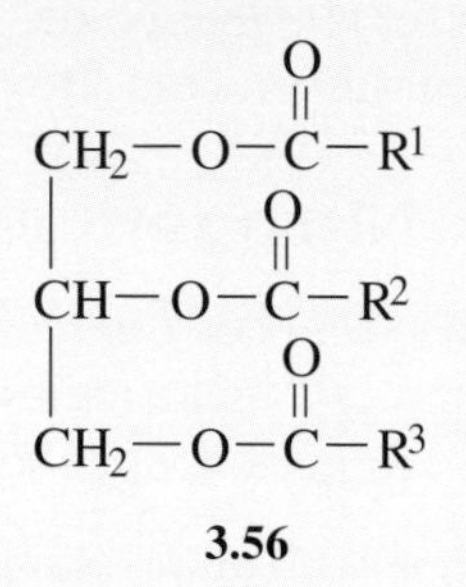

**3.56**

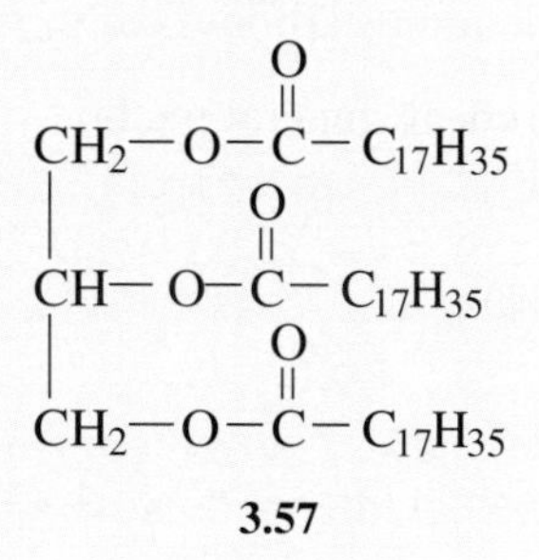

**3.57**

The base hydrolysis of esters has an important industrial application — the manufacture of soaps. The desire to keep clean is quite old, and soap-making is one of the oldest known chemical syntheses: at least 2 000 years ago people discovered that if they boiled goat fat with potash from the ashes of wood fires, they would get soaps. Today we use other animal and vegetable oils and fats, which are all esters of long-chain carboxylic acids (fatty acids). In particular, they are esters of the alcohol propane-1,2,3-triol (glycerol, **3.55**), which contains three OH groups, and so can form three ester linkages. The fats and oils derived from this alcohol have the general formula **3.56**. Each alkyl group usually contains twelve or more carbon atoms. For example, in glyceryl tristearate (**3.57**), a fat found in beef dripping, all the alkyl groups are $C_{17}H_{35}$. These esters also form a large part of the energy-providing foods in the human diet. They are insoluble in water, and can be stored in the body, where they often act as insulating layers under the skin.

Soaps are made by hydrolysing these fatty acid esters using sodium hydroxide or potassium hydroxide. The products of the reaction are glycerol and the sodium or potassium salts of the long-chain carboxylic acids; the latter are the **soaps** (Figure 3.4):

$$\begin{array}{l} CH_2{-}O{-}C({=}O){-}R^1 \\ | \\ CH{-}O{-}C({=}O){-}R^2 \\ | \\ CH_2{-}O{-}C({=}O){-}R^3 \end{array} + 3NaOH \longrightarrow \begin{array}{l} CH_2OH \\ | \\ CHOH \\ | \\ CH_2OH \end{array} + \underbrace{\left[\begin{array}{l} R^1{-}C({=}O){-}O^- Na^+ \\ R^2{-}C({=}O){-}O^- Na^+ \\ R^3{-}C({=}O){-}O^- Na^+ \end{array}\right]}_{\text{soap}} \qquad (3.74)$$

Potassium salts form the soft soaps, since they are more soluble. Soaps made from olive oil (**3.55**, in which $R^1 = R^2 = R^3 = Z$-oleyl, **3.58**) are the Castile soaps (Box 1.1). The process of soap making is given the name **saponification**, (from the Latin *sapo* meaning 'soap' and *facio* meaning 'to make'); this term is now used more generally for the hydrolysis of esters by dilute alkali.

$$CH_3(CH_2)_6CH_2{-}CH{=}CH{-}CH_2(CH_2)_6{-} \quad (Z)$$

**3.58**

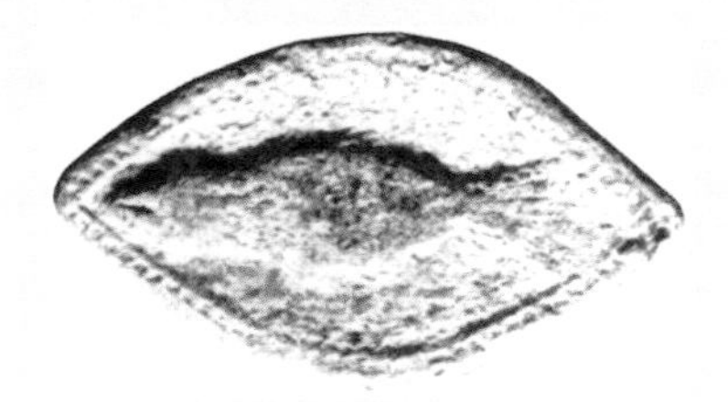

**Figure 3.4**
A bar of soap made in Tudor times. Soap was an expensive commodity and thus people rarely washed!

### QUESTION 3.20

Draw a curly arrow mechanism for the saponification of the last ester link in the formation of Castile soap; that is, assume that the hydrolysis of two links has been completed.

- If you added an amine, which is a base, to ethanoic acid, what product would you get (recall Question 3.19)?
- The carboxylic acid, as its name suggests, acts as an acid, and will protonate the base, in this case an amine, to form an ammonium ion. So, an alkyl ammonium ethanoate would be produced:

$$CH_3COOH + RNH_2 \rightleftharpoons CH_3CO_2^{-}R\overset{+}{N}H_3 \quad (3.75)$$

- The formation of the quaternary ammonium salt is accompanied by the evolution of heat. By applying Le Chatelier's principle, how might we reverse this reaction?
- As heat is evolved in the forward reaction, we should be able to make the reverse reaction favourable by supplying heat.

Temperatures in excess of 200 °C are usually required, and then a different reaction between the acid and amine occurs. This gives products from the attack of the nucleophilic nitrogen atom at the carbonyl carbon atom, a process that affords a good, cheap general method for the synthesis of amides; for example

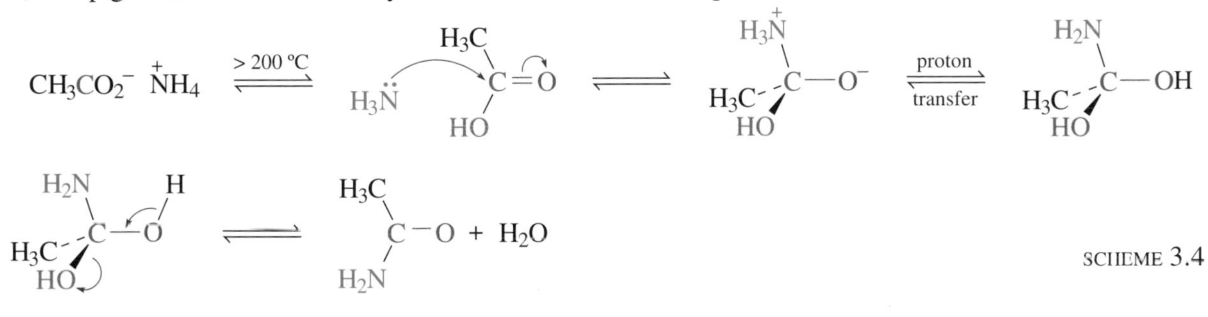


SCHEME 3.4

Amides are ubiquitous in nature, most notably as the dominant functional group in proteins, but they have also been widely incorporated in synthetic materials (see Box 3.9).

We can now summarize reactions between nitrogen nucleophiles and carboxylic acid derivatives as follows:

$$R^1COCl + 2R^2NH_2 \longrightarrow R^1CONHR^2 + R^2\overset{+}{N}H_3\,Cl^- \quad (3.76)$$

$$R^1COOCOR^1 + 2R^2NH_2 \longrightarrow R^1CONHR^2 + R^1CO_2^{-}\overset{+}{N}H_3R^2 \quad (3.77)$$

$$R^1COOR^2 + R^3NH_2 \longrightarrow R^1CONHR^3 + R^2OH \quad (3.78)$$

$$R^1CONR^2R^3 + R^4NH_2 \longrightarrow R^1CONHR^4 + R^2R^3NH \quad (3.79)$$

$$R^1COOH + R^2NH_2 \longrightarrow R^1CONHR^2 + H_2O \quad (3.80)$$

## BOX 3.9 Nylon-6,6 — a polyamide

As discussed in the Case Study at the end of this book, nylon is used for the manufacture of a wide range of products (Figure 3.5). It is a synthetic polymer formed by joining monomers together via amide bonds. For example, the reaction between an acid chloride and an amine has most interesting consequences when the acid chloride molecule contains two COCl groups (for example, hexane-1,6-dioyl dichloride), and the amine contains two $NH_2$ groups (for example, hexane-1,6-diamine):

$$\underset{\text{hexane-1,6-dioyl dichloride}}{Cl{-}C({=}O){-}CH_2(CH_2)_2CH_2{-}C({=}O){-}Cl} + \underset{\text{hexane-1,6-diamine}}{H_2N(CH_2)_6NH_2} \longrightarrow Cl{-}C({=}O){-}CH_2(CH_2)_2CH_2{-}C({=}O){-}NH(CH_2)_6NH_2 + HCl \qquad (3.81)$$

The initial product of this reaction retains one amino group and one acid chloride group. It follows that both groups can undergo a further reaction with more starting materials. The product of this reaction would be

$$H_2N(CH_2)_6HN{-}C({=}O){-}CH_2(CH_2)_2CH_2{-}C({=}O){-}NH(CH_2)_6HN{-}C({=}O){-}CH_2(CH_2)_2CH_2{-}C({=}O){-}Cl$$

**3.59**

which can react again and again, and so build up a giant polymer molecule. The name of this polymer is nylon-6,6, one of a family of extremely important synthetic materials used in the textile industry and elsewhere.

On an industrial scale, however, there are problems using a corrosive reactant like hexane-1,6-dioyl dichloride and producing a byproduct like HCl.

- Suggest a strategy for the production of nylon-6,6 that might avoid these problems.

- Use the dicarboxylic acid $HOOC(CH_2)_4COOH$ and heat it with hexane-1,6-diamine.

In fact nylon-6,6 is manufactured by heating the *salt* formed between the diamine and the diacid at over 200 °C. By using the salt, the 1 : 1 ratio of reagents needed for this reaction is assured. The byproduct is then water.

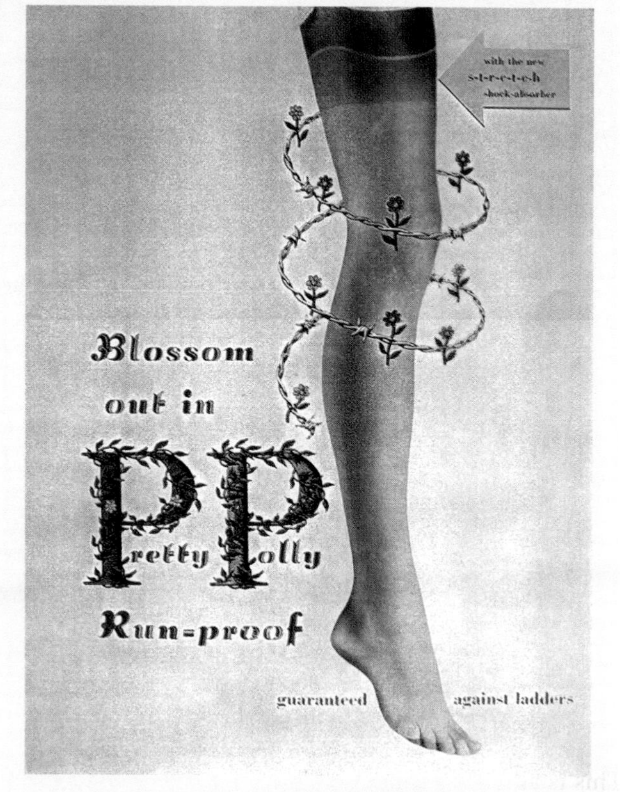


**Figure 3.5** Stockings were one of the common uses of nylon.

### 3.3.4 Hydrogen nucleophiles: a route to primary alcohols or primary amines

All carboxylic acid derivatives react with lithium aluminium hydride, $LiAlH_4$. Here, for example, is the first step in the reaction with the ester, ethyl ethanoate:

$$H_3\bar{Al}{-}H + (H_3C)(C_2H_5O)C{=}O \longrightarrow (H)(H_3C)(C_2H_5O)C{-}O^- + AlH_3 \qquad (3.82)$$

- Which group is the best leaving group in the tetrahedral intermediate?
- The ethoxide anion.

So the second step of the reaction would be written:

$$(H)(H_3C)(C_2H_5O)C{-}O^- \longrightarrow (H_3C)(H)C{=}O + C_2H_5O^- \qquad (3.83)$$

- Well, do you think these are the final products? What else might happen?
- The product is an aldehyde, which we know can react further with $LiAlH_4$ (Section 3.2.2).

In fact, aldehydes react faster than esters, so the aldehyde is never isolated from the reduction of carboxylic acid derivatives with $LiAlH_4$.

Therefore, we must now write a third step:

$$H_3\bar{Al}{-}H + (H_3C)(H)C{=}O \longrightarrow (H)(H_3C)(H)C{-}O^- \xrightarrow{H^+/H_2O} CH_3CH_2OH \qquad (3.84)$$

So reduction of an ester with $LiAlH_4$, followed by acid hydrolysis, produces a primary alcohol. Likewise, so does reduction of an acid chloride, anhydride or carboxylic acid.

**QUESTION 3.21**

Compound **3.60** was needed to establish the configuration at one of the chiral centres in a natural product from the sweet potato. Write the structure of a dicarboxylic acid that may be reduced to **3.60** with lithium aluminium hydride.

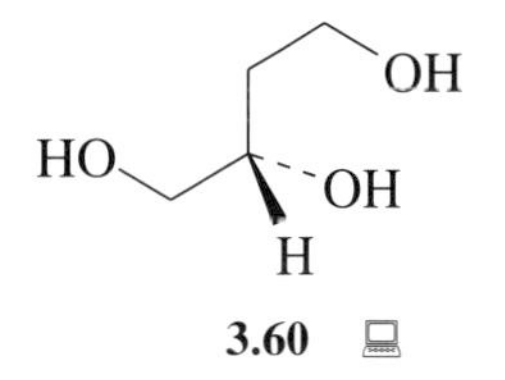

**3.60**

Although chlorides, anhydrides, esters and acids all give primary alcohols with $LiAlH_4$, amides chart a different course. The carbonyl group does become a methylene group ($CH_2$), but the product is an amine, not an alcohol:

$$R^1CONHR^2 \xrightarrow{LiAlH_4} R^1CH_2NHR^2 \qquad (3.85)$$

This is almost certainly due to an intermediate imine being formed in the following way:

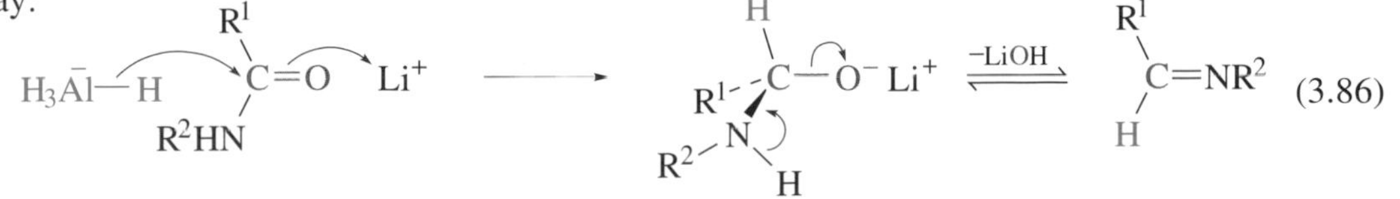

The imine can then be reduced to the corresponding amine, in the same way as an aldehyde is reduced to an alcohol:

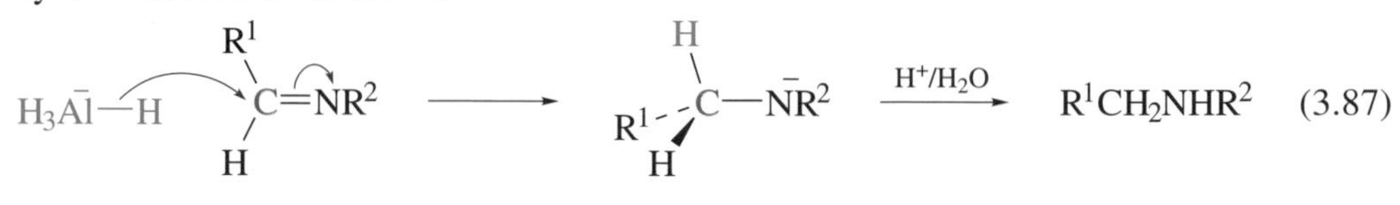


Carbon nucleophiles such as organometallic reagents react with carboxylic acid derivatives in a similar way to lithium aluminium hydride.

You will study these reactions at some length in Part 2 of this Book, but you may like to test your skill in presenting nucleophilic reactions at a carbonyl group by trying to anticipate (using curly arrows) how the reaction between a reagent $CH_3—M$ and the ester $CH_3COOCH_3$ might proceed!

See Scheme 3.6 at the bottom of p. 56.

## 3.3.5 Summary of Section 3.3

1. Carboxylic acid derivatives (RCOX) undergo substitution reactions because they contain the group X, which can act as a leaving group.
2. The reactivity of carboxylic acid derivatives follows the order:

   chloride > anhydride > ester > amide
3. The reaction of carboxylic acid derivatives with water and alcohols is catalysed by acids; with water, a carboxylic acid is produced, and with an alcohol an ester is produced.
4. Carboxylic acid derivatives can also be hydrolysed by base. However, in this reaction the base does not act as a catalyst, because it is used up. The reaction proceeds faster with $HO^-$, because it is a better nucleophile than $H_2O$.
5. Soaps are made by the hydrolysis of fats and oils, using sodium hydroxide or potassium hydroxide.
6. The reaction between carboxylic acid derivatives and nitrogen nucleophiles produces amides. The reaction is the basis of the manufacture of nylon.
7. Lithium aluminium hydride reacts with carboxylic acid derivatives to give primary alcohols. However, amides generally yield amines.

A comprehensive summary of the reactions of carboxylic acid derivatives is presented schematically in Appendix 2 (p. 58).

### QUESTION 3.22

The formation of the hydrogen carbonate ion (**3.61**) from carbon dioxide, O=C=O, is an important biological reaction. Write a mechanism for the formation of hydrogen carbonate ion by the attack of $HO^-$ on $CO_2$.

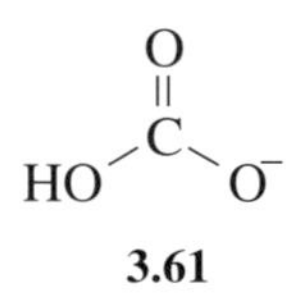


**3.61**

### QUESTION 3.23

Methyl benzoate, in which the carbonyl oxygen atom was labelled with oxygen-18 (**3.62**), was hydrolysed in aqueous sodium hydroxide. The reaction was allowed to proceed to 50% completion. The remaining ester was then isolated and found to have lost some of its $^{18}O$ label. Use equations to account for this observation.

$^{18}O$ = C($C_6H_5$)($OCH_3$)

**3.62**

### QUESTION 3.24

Show the mechanism of the reaction of methyl benzoate with hydroxylamine, $NH_2OH$, and hence predict the structure of the product.

### QUESTION 3.25

A careless chemist 'burnt' holes in a nylon lab-coat by spilling aqueous sulfuric acid on it. By considering a small unit of nylon-6,6, for example **3.63**, write down a reasonable equation to explain the chemistry involved.

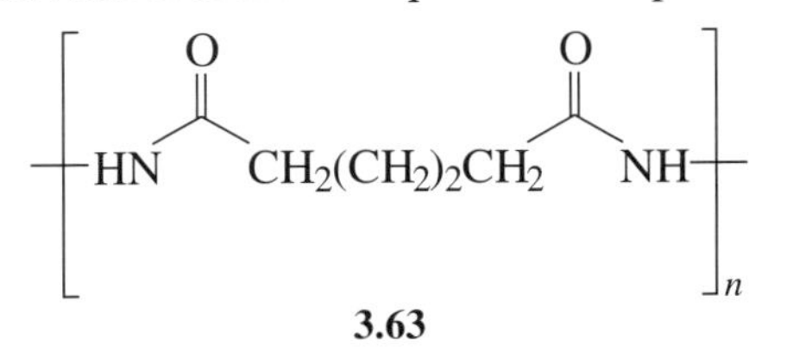


### QUESTION 3.26

Suggest reagents and reaction conditions for each of steps 1–3 in Scheme 3.6. Why couldn't the starting material be treated with the reagent for step 2 to give the final product directly?

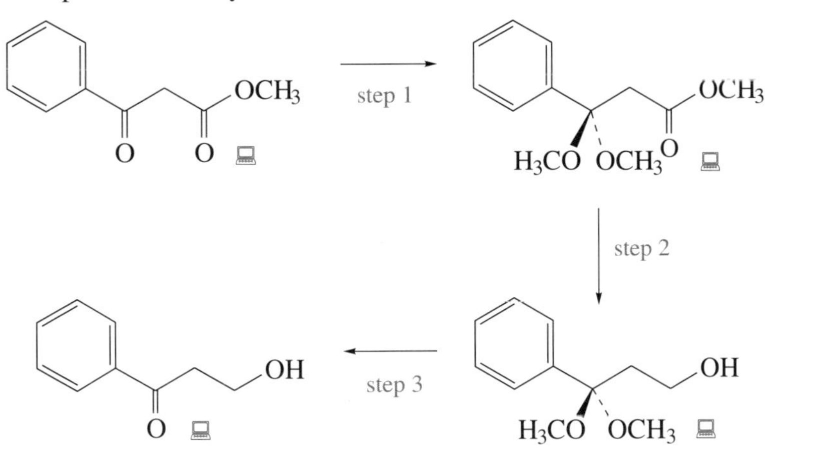

SCHEME 3.5

# 4 CONCLUSION

You have met a lot of chemistry in Part 1 of this Book. But we hope that the common elements of mechanism here have made it manageable. Being familiar with this linking theme and being able to correctly illustrate the bond changes in reactions using curly arrows are perhaps the most important outcomes of your study. Remembering the reactions of particular compounds then becomes unnecessary. The challenging part is being aware of relative reactivities among both carbonyl compounds and among nucleophiles, but even here you've seen how some general principles reduce the need to memorize.

We hope the references we have been able to make to biological and industrial topics, though necessarily brief, have convinced you of the widespread importance of carbonyl compounds. Moreover, the carbonyl group is often called 'the centrepiece of organic chemistry'. Not only is it the most commonly encountered and synthetically versatile functional group, but it also serves as a model for all functional groups involving a π bond between dissimilar atoms.

In Part 2 you will meet carbonyl compounds again and again, and will see how creatively and usefully the principles you've just been studying can be exploited.

---

### COMPUTER ACTIVITY 4.1 Reprise of carbonyl chemistry

In this activity, on the CD-ROM associated with this Book, you will be able to revise your understanding of the carbonyl chemistry you have learned in Part 1. In particular, it revises the mechanisms of addition, addition–elimination and substitution reactions of carbonyl compounds, which is brought together in Appendix 3.

This Activity should take about 45 minutes.

---

*Answer to question on p. 54*

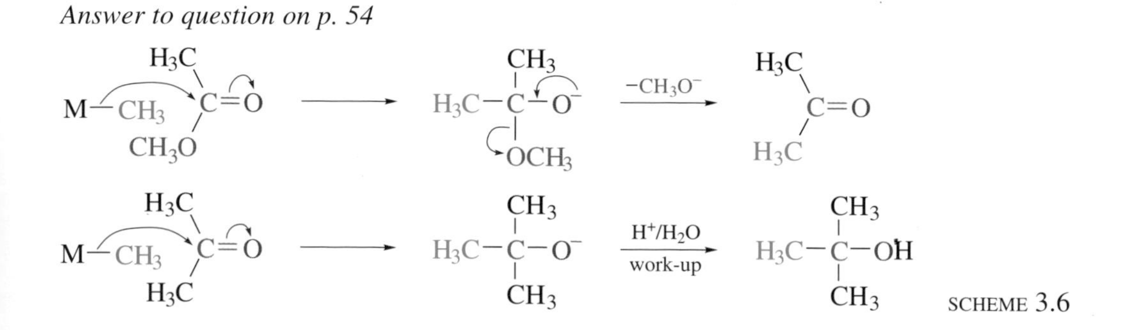


SCHEME 3.6

# APPENDIX 1 SUMMARY OF THE REACTIONS OF ALDEHYDES AND KETONES

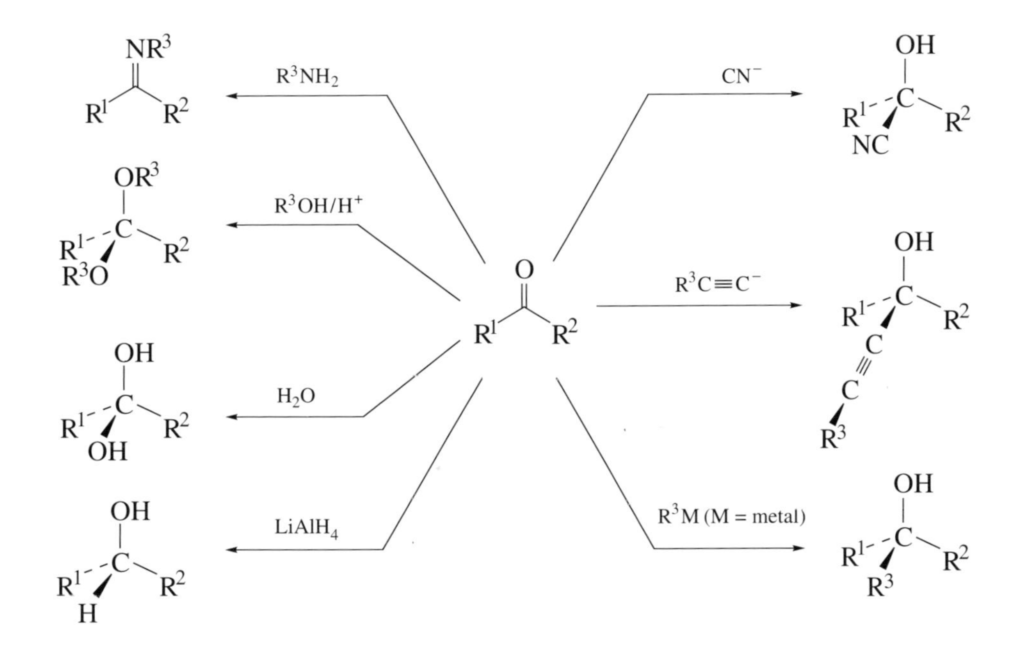

# APPENDIX 2
# SUMMARY OF THE REACTIONS OF CARBOXYLIC ACID DERIVATIVES

The reactions of carboxylic acid derivatives with carbon nucleophiles is discussed in Part 2, and is not discussed here.

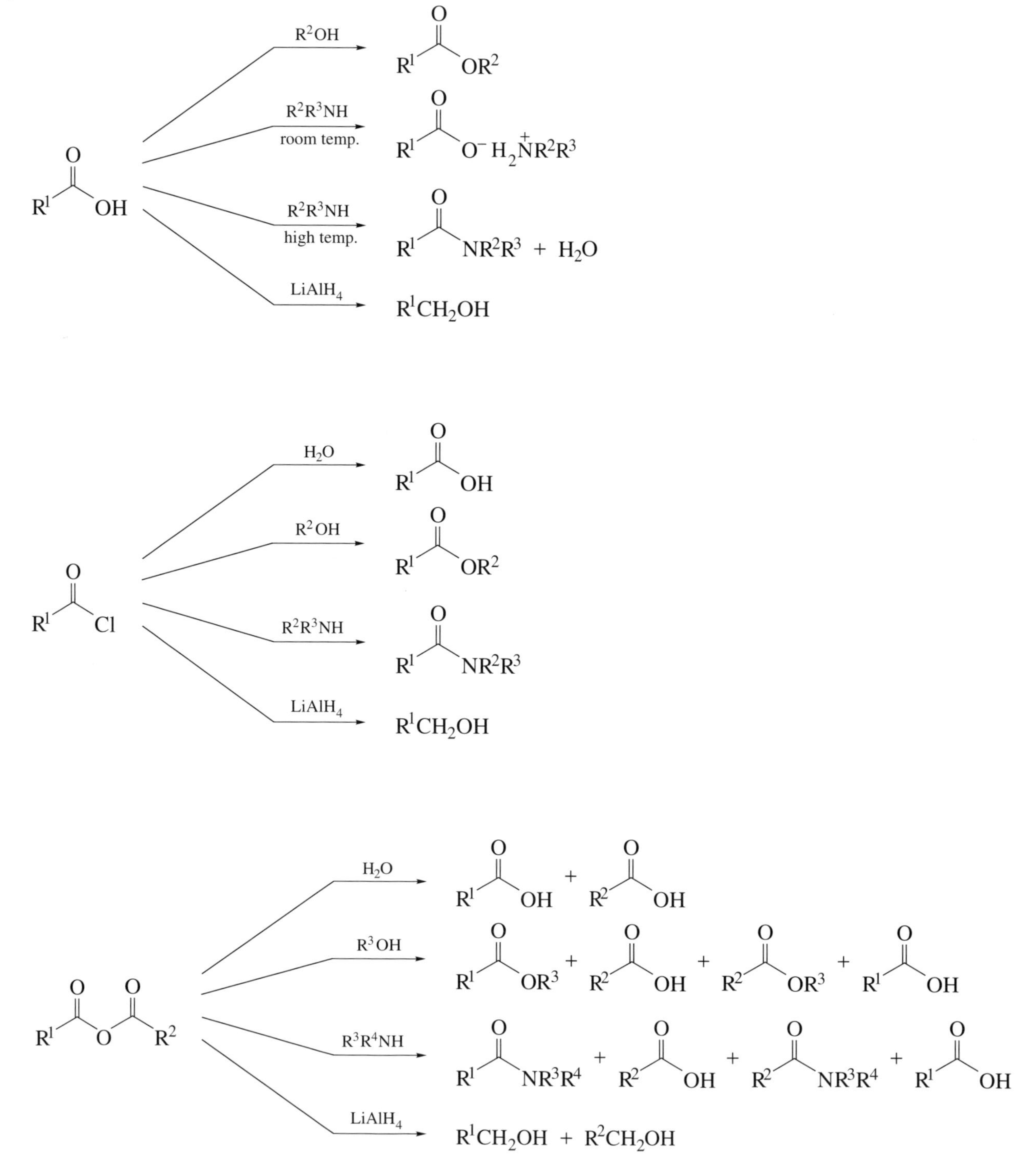

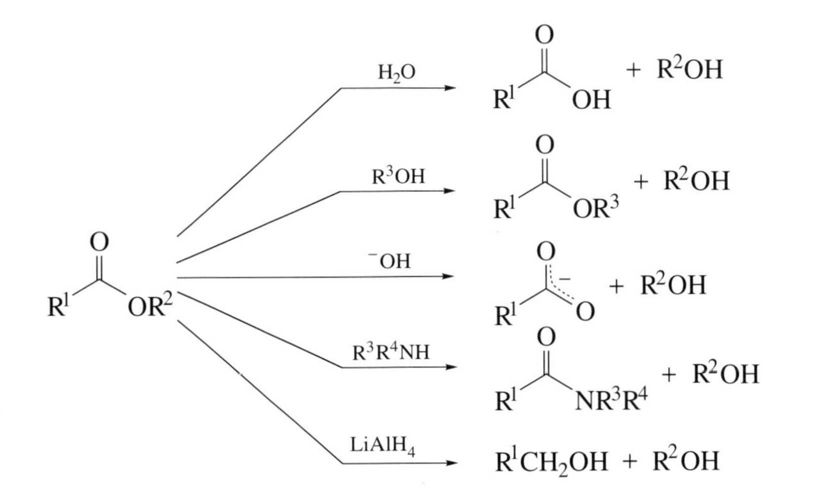
$R^1COOR^2$
$H_2O$
$R^1COOH + R^2OH$
$R^3OH$
$R^1COOR^3 + R^2OH$
$^-OH$
$R^1COO^- + R^2OH$
$R^3R^4NH$
$R^1CONR^3R^4 + R^2OH$
$LiAlH_4$
$R^1CH_2OH + R^2OH$

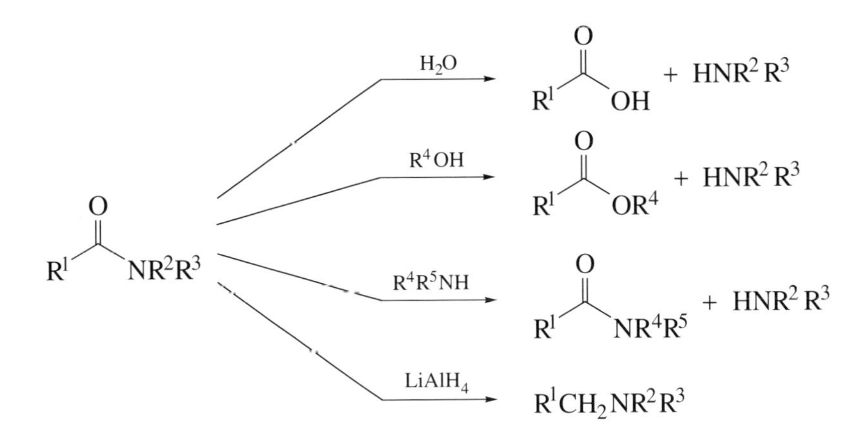
$R^1CONR^2R^3$
$H_2O$
$R^1COOH + HNR^2R^3$
$R^4OH$
$R^1COOR^4 + HNR^2R^3$
$R^4R^5NH$
$R^1CONR^4R^5 + HNR^2R^3$
$LiAlH_4$
$R^1CH_2NR^2R^3$

# APPENDIX 3 SUMMARY OF CARBONYL COMPOUND REACTION TYPES

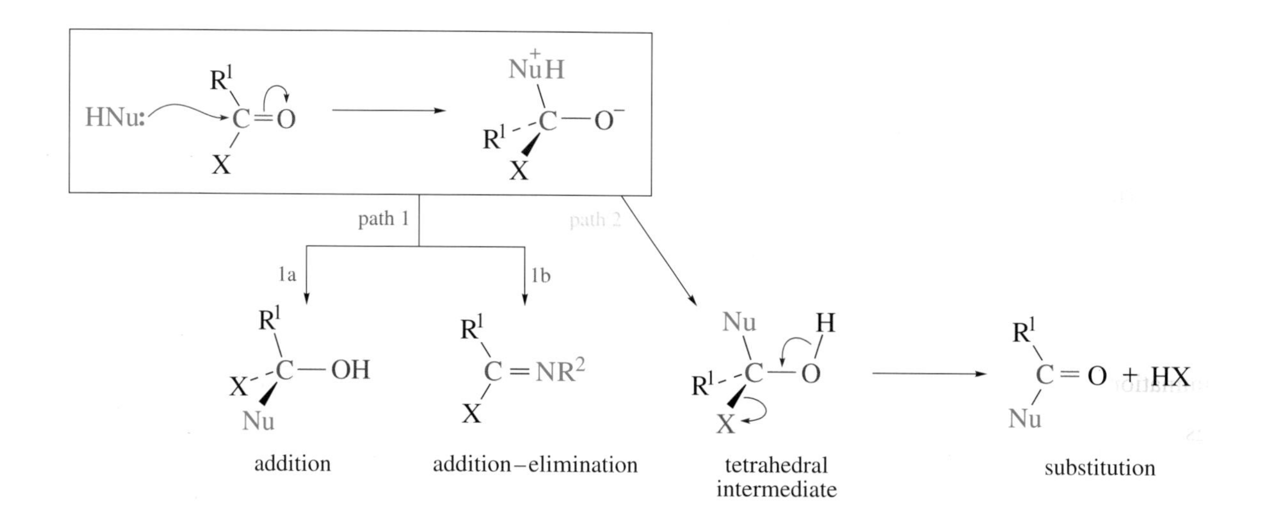


## Notes

Path 1 occurs for aldehydes (X = H) and ketones ($X = R^2$).

Nucleophiles that follow path 1a are:

Nu = $H_2O$, $R^2OH$, $R^2R^3NH$ and HCN

$Nu^-$ = $(R^2)^-$ (for example, $RC{\equiv}C^-$), $^-C{\equiv}N$ and $H^-$ (for example, $LiAlH_4$)

Nucleophiles that follow path 1b are:

NuH = $R^2NH_2$, $HONH_2$, $NH_2NH_2$, $NH_2CONHNH_2$ and $R^2NHNH_2$

Path 2 occurs only for carboxylic acids (X = OH) and their derivatives: esters ($X = OR^2$), anhydrides ($X = OCOR^3$), amides ($X = NR^2R^3$) and acid chlorides (X = Cl).

Note the similarity in structure between the tetrahedral intermediate and the addition compound from path 1a.

Nucleophiles that follow path 2 are:

NuH = $H_2O$, $R^2OH$, $R^2NH_2$ and $R^2R^3NH$

$Nu^-$ = $HO^-$, $(R^2)^-$ and $H^-$ (for example, $LiAlH_4$)

The reactivity of RCOX decreases in the order

$X = Cl > OCOR^2 > OR^2 > NR^2R^3$

# LEARNING OUTCOMES

When you have completed *Mechanism and Synthesis: Part 1 — Carbonyl compounds*, you should be able to do the following things:

1 Recognize valid definitions of, and use in a correct context, the terms, concepts and principles listed in the following Table. (All Questions)

List of scientific terms, concepts and principles used in *Carbonyl compounds*

| Term | Page number | Term | Page number |
|---|---|---|---|
| acid catalysis | 34 | hemiacetal | 34 |
| acetal | 35 | hydrazone | 42 |
| addition–elimination reaction | 39 | imine | 38 |
| base catalysis | 34 | oxime | 42 |
| carbinolamines | 38 | polarization | 18 |
| cyanohydrin | 27 | polarizable electrons | 21 |
| 2,4-dinitrophenylhydrazone | 42 | proton transfer | 31 |
| equimolar amount | 27 | semicarbazone | 42 |
| ester hydrolysis | 46 | soap saponification | 50 |
| esterification | 47 | tetrahedral intermediate | 45 |

2 Identify compounds that contain, or are derived from, the carbonyl group, describe the bonding in such compounds, and assign IR and $^{13}$C NMR data to the group. (Questions 2.1, 2.2, 3.5 and 3.10)

3 Demonstrate improved skills in the use of curly arrows to depict reaction mechanisms by applying them correctly to the nucleophilic reactions of carbonyl compounds. (Questions 3.1, 3.2, 3.3, 3.5, 3.6, 3.7, 3.9, 3.10, 3.11, 3.12, 3.14, 3.15, 3.16, 3.18, 3.19, 3.20, 3.22, 3.23, 3.24 and 3.25)

4 Illustrate both common and distinctive features of the reactions between nucleophiles and aldehydes and ketones, and between nucleophiles and carboxylic acids and their derivatives. (Questions 3.1, 3.7 and 3.17)

5 Predict the expected product of the reaction between a specified carbonyl compound and a given nucleophilic reagent under stated conditions. (Questions 3.1, 3.5, 3.8, 3.16, 3.17, 3.20, 3.24 and 3.26)

6 Suggest the structure of a carbonyl compound that may be used to prepare a given product directly by reaction with a nucleophile, and identify both the nucleophile and suitable reaction conditions. (Questions 3.3, 3.4, 3.10, 3.13 and 3.21)

7 Given the structure of an appropriate acetal-protected ketone, carbohydrate, characterization derivative of an aldehyde or ketone, polyamide, fat or soap, show how it illustrates a typical feature of carbonyl chemistry. (Questions 3.9, 3.11, 3.12, 3.15, 3.19 and 3.25)

8 Suggest appropriate experimental conditions for achieving a specified nucleophilic reaction of a carbonyl compound. (Questions 3.14 and 3.26)

# QUESTIONS: ANSWERS AND COMMENTS

## QUESTION 2.1 (*Learning Outcome 2*)

Paracetamol contains one carbonyl group, which is attached to a nitrogen and is therefore an amide. From Table 1.1 you should predict that the IR spectrum would show an absorption at about 1 645 $cm^{-1}$. The actual value is 1 656 $cm^{-1}$.

Glycine contains one carbonyl group, as part of a carboxylate anion, which according to Table 1.1 should absorb at about 1 575 $cm^{-1}$ in the IR spectrum. The actual value is 1 600 $cm^{-1}$.

Aspirin contains two carbonyl groups. One is part of a carboxylic acid group (**Q.1**), the other part of an ester group (**Q.2**). The IR absorption would be predicted to occur at about 1 710 $cm^{-1}$ and 1 740 $cm^{-1}$, respectively. The actual values are 1 685 $cm^{-1}$ and 1 750 $cm^{-1}$.

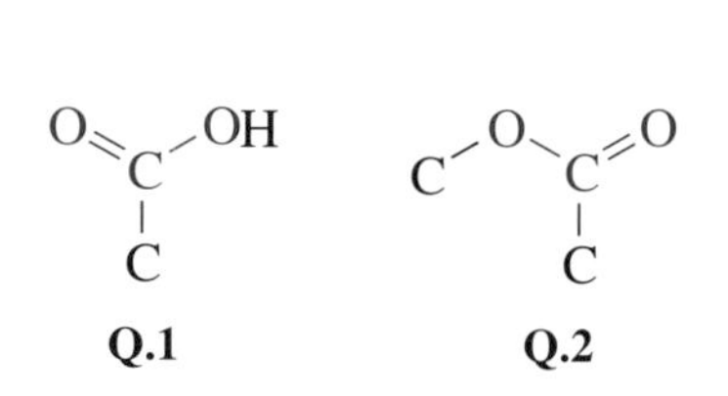


**Q.1** **Q.2**

Sulfanilamide contains no carbonyl groups. However, it does contain a sulfonyl group, in which oxygen atoms are double bonded to a *sulfur* atom (rather than a carbon atom).

Camphor contains one carbonyl group, as part of a ketone group (**Q.3**), which should absorb at about 1 710 $cm^{-1}$. The actual value is 1 735 $cm^{-1}$.

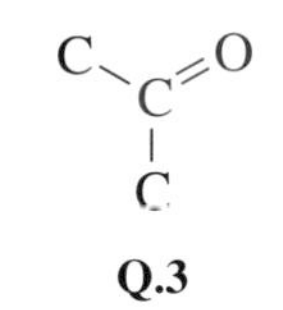


**Q.3**

Adrenaline contains no doubly bonded C=O groups. It contains one C—OH secondary alcohol group, one C—NHR amine group and two phenol OH groups.

## QUESTION 2.2 (*Learning Outcome 2*)

Looking for values around those given in Table 1.1 we get: A, carbonyl compound unlikely; B, ketone; C, acid chloride; D, carboxylic acid; E, carbonyl compound unlikely; F, amide; G, ester. Don't worry if you were 'inaccurate' in evaluating the data. The degree of variation within a class is included in the source you will normally use for spectroscopic identification. This exercise merely demonstrates the principle of diagnostic analysis. It also emphasizes the value of combining more than one kind of evidence in structure determination.

## QUESTION 3.1 (*Learning Outcomes 3, 4 and 5*)

The first two compounds, the aldehyde $CH_3CHO$ and the ketone $CH_3COCH_3$, would give the following addition products (after protonation) with one mole equivalent of $Y^-$:

$H_3C-C(H)(Y)-OH$ **Q.4**

$H_3C-C(CH_3)(Y)-OH$ **Q.5**

The second mole of $Y^-$ would remain unreacted and become HY in the work-up. The addition product from the ester $CH_3CO_2CH_3$, however, would decompose to give another carbonyl compound which could form a second addition product with a second mole equivalent of $Y^-$.

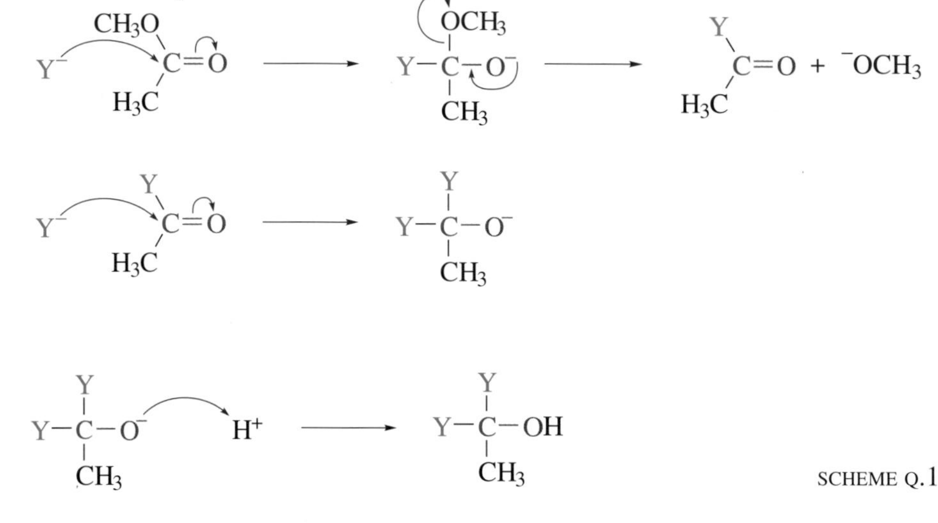

SCHEME Q.1

## QUESTION 3.2 (Learning Outcome 3)

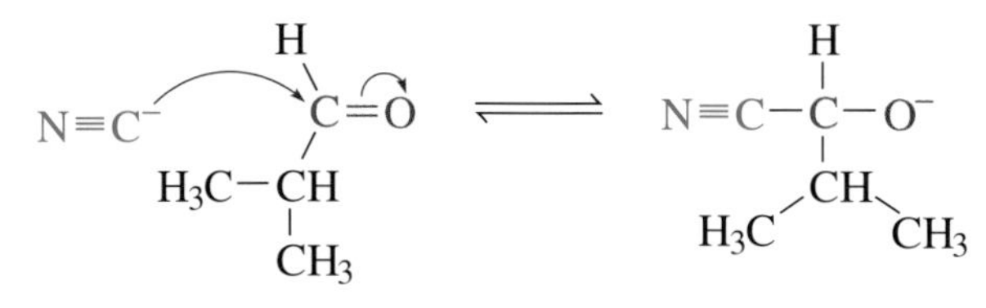

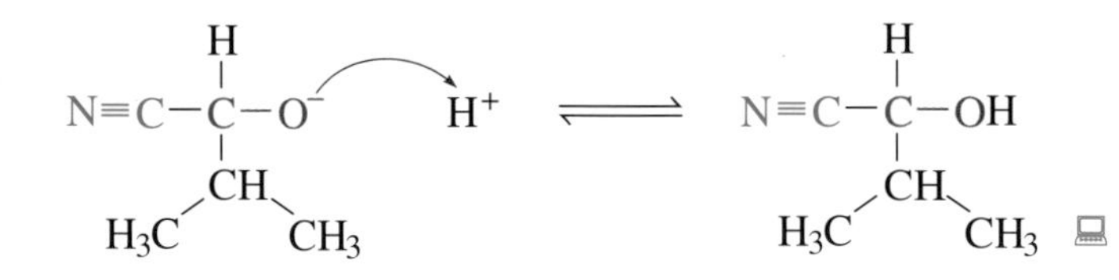

SCHEME Q.2

## QUESTION 3.3 (Learning Outcomes 3 and 6)

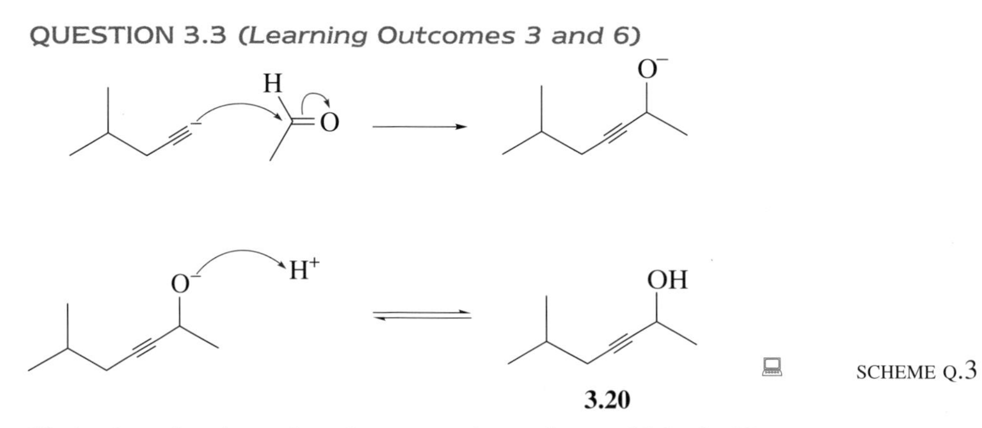

SCHEME Q.3

Notice here that the carbonyl compound must be an aldehyde (H atoms are not shown in the skeletal representation of the structure).

## QUESTION 3.4 (Learning Outcome 6)

**3.23** $CH_3{-}M$ and $H_2CO$.

**3.24** $CN^-$ and cyclopentanone (=O)

**3.25** $HC{\equiv}C^-$ and $CH_3COCH_2CH_3$.

**3.26** $CH_3CH_2{-}M$ and $CH_3CH_2COCH_2CH_3$.

**3.27** $C_6H_5{-}CHO$ and $CH_3CH_2{-}M$ or $C_6H_5{-}M$ and $CH_3CH_2CHO$

## QUESTION 3.5 (Learning Outcomes 2, 3 and 5)

Ethanal is reduced by $LiAlH_4$ to give ethanol on acid work-up:

$$H_3\bar{A}l{-}H + H_3C(H)C{=}O \longrightarrow H_3C{-}CH_2{-}O^- \xrightarrow{H^+/H_2O} H_3C{-}CH_2{-}OH \quad (Q.1)$$

The signal at 200 p.p.m. in the $^{13}C$ NMR spectrum of ethanal is produced by the carbon nucleus of the carbonyl group. It is a doublet because there is one hydrogen atom attached to this carbon atom. In ethanol, this carbon nucleus is more shielded, and so the chemical shift is smaller (57 p.p.m.). There are now two attached hydrogen atoms, so the signal is a triplet in the off-resonance spectrum.

## QUESTION 3.6 (Learning Outcome 3)

$$H_2\ddot{O} + H(Cl_3C)C{=}O \rightleftharpoons H_2\overset{+}{O}{-}CH(CCl_3){-}O^- \xrightarrow[\text{transfer}]{\text{proton}} HO{-}CH(CCl_3){-}OH \quad (Q.2)$$

Box 3.3 shows an equilibrium trend in favour of the hydrate:

$HCHO > CH_3CHO > CH_3COCH_3$

and we can place $Cl_3CCHO$ before methanal. This suggest that replacing alkyl substituents on the carbonyl carbon atom by hydrogen atoms seem to favour the hydrate; you may recall that hydrogen is less electron donating than an alkyl group.

$Cl_3CCHO > HCHO > CH_3CHO > CH_3COCH_3$

increasing electron donation to the carbonyl →

A carbon atom bearing three (electron withdrawing) chlorine atoms would be a very poor electron donor. These results suggest that it is the (inductive) supply or withdrawal of electrons to the carbonyl that determines the position of the equilibria in this series, and thus the position of $Cl_3CHO$ before methanal in the series.

## QUESTION 3.7 (*Learning Outcomes 3 and 4*)

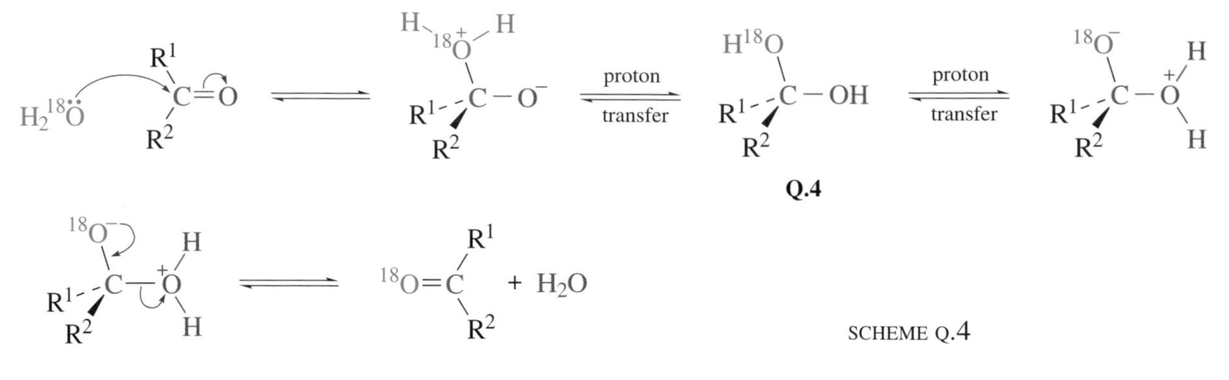


SCHEME Q.4

At the hydrate stage, **Q.4**, there are two pathways open for re-formation of the ketone: either transfer of a proton from the $^{16}O$ atom to the $^{18}O$ atom, with subsequent loss of $H_2{}^{18}O$ and formation of the original $^{16}O$ carbonyl compound; or transfer of a proton from the $^{18}O$ atom to the $^{16}O$ atom, with subsequent loss of $H_2{}^{16}O$ and formation of an $^{18}O$-labelled carbonyl group. As all the steps are equilibria, a situation will eventually be reached where the ratio of $^{16}O$ to $^{18}O$ in the ketone depends on the molar ratio of propanone to water. If the water is in a great excess, most of the propanone will contain $^{18}O$ rather than $^{16}O$.

## QUESTION 3.8 (*Learning Outcome 5*)

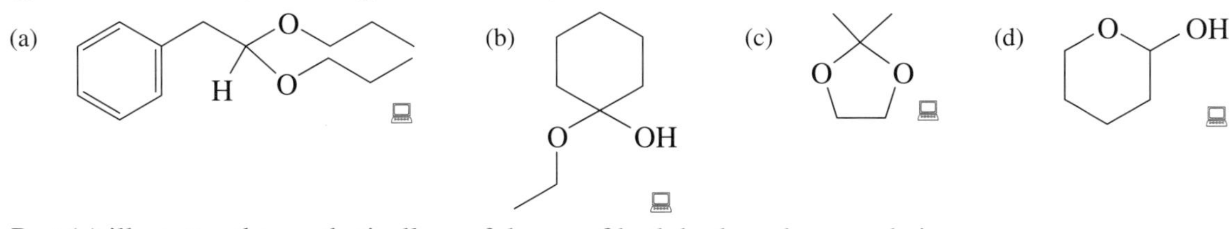


Part (c) illustrates the synthetically useful case of both hydroxyl groups being provided by a single molecule of the alcohol (see Box 3.5) and part (d), which is an internal hemiacetal formation, a process that is fundamental to the chemistry of a major class of biological compounds, the carbohydrates (see Box 3.6).

## QUESTION 3.9 (*Learning Outcomes 3 and 7*)

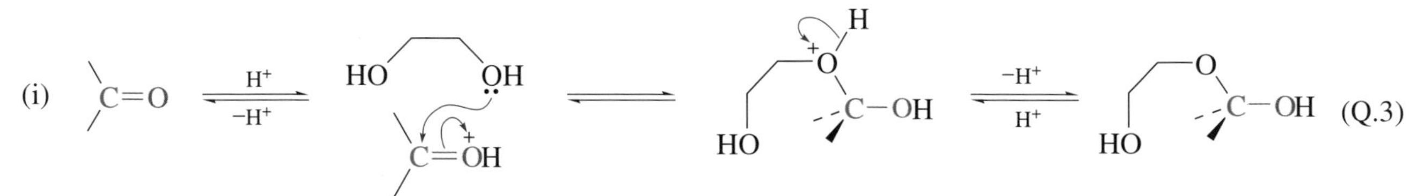


In the presence of acid, the hemiacetal OH group can be protonated, which enables a molecule of water to be lost, to form the intermediate **Q.5**. The second hydroxyl group of ethane-1,2-diol can then attack to form a cyclic acetal.

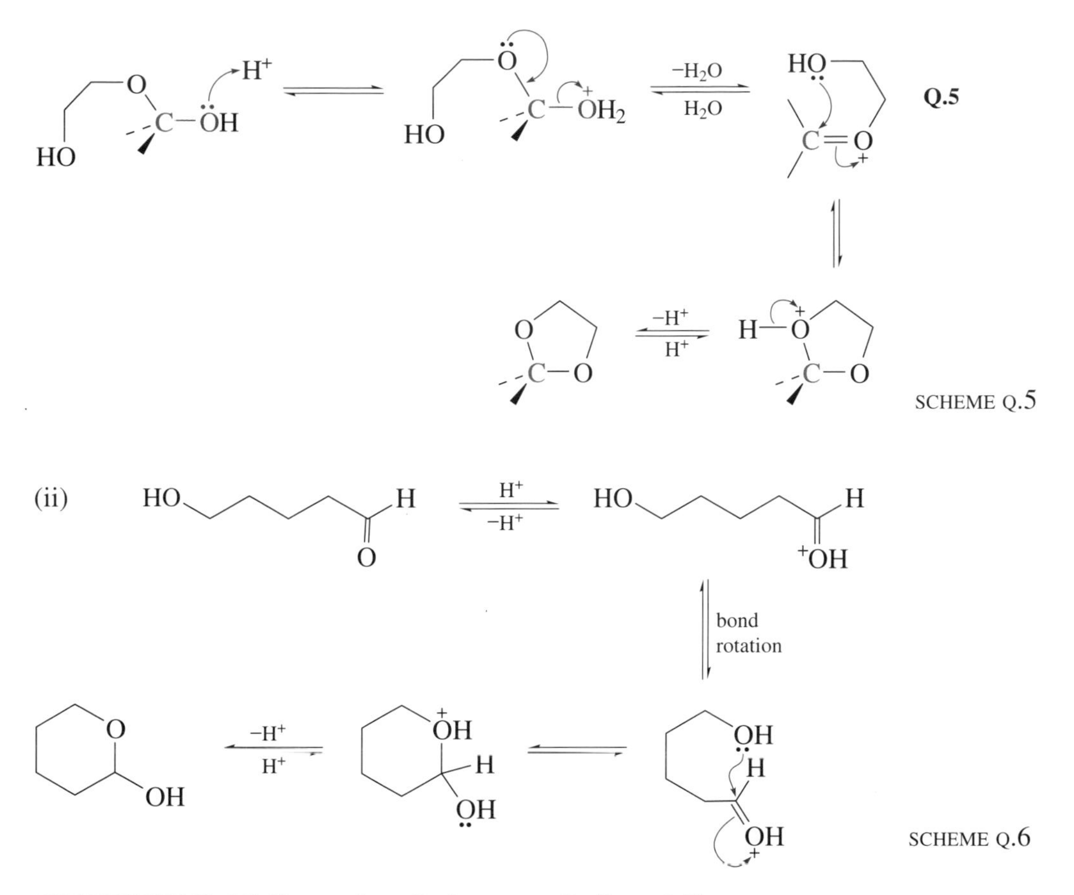


SCHEME Q.5

SCHEME Q.6

### QUESTION 3.10 (*Learning Outcomes 2, 3 and 6*)

Compound **3.42** is an acetal of methanal. It is formed by the acid-catalysed reaction between the diol 1-phenylethane-1,2-diol and methanal:

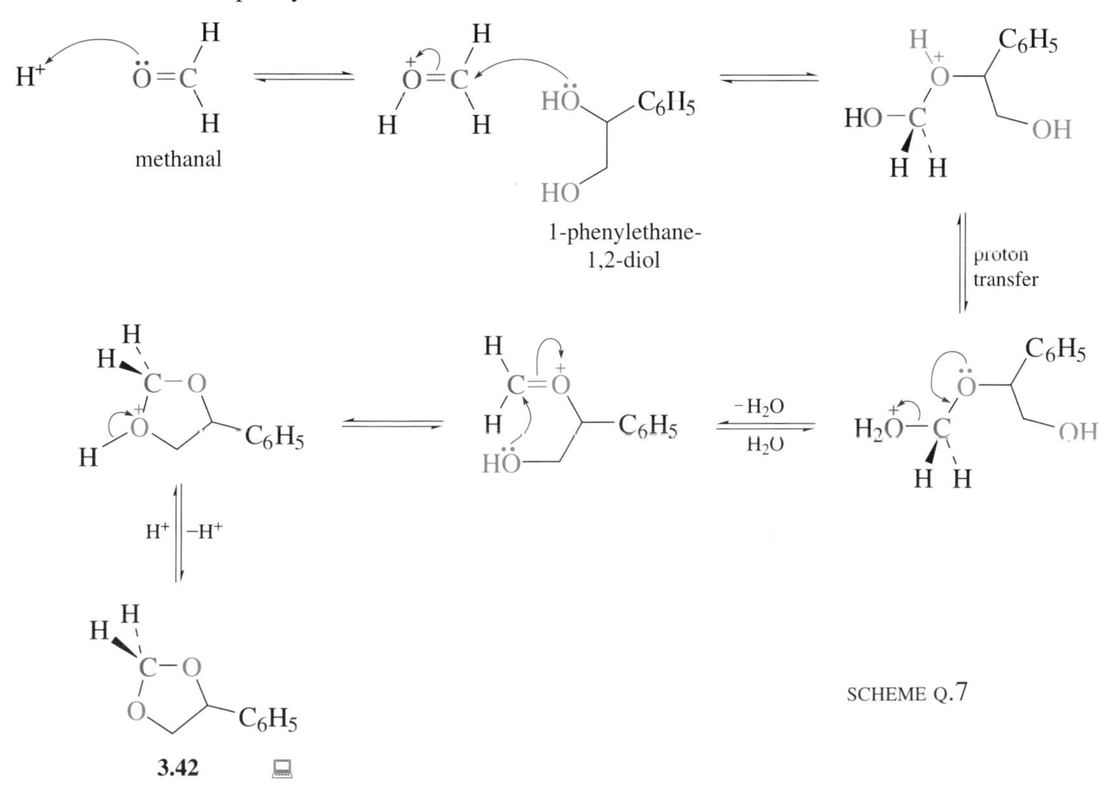


SCHEME Q.7

QUESTION 3.16 (*Learning Outcomes 3 and 5*)

(a) Water attacks one of the anhydride carbonyl groups. This is followed by a proton transfer:

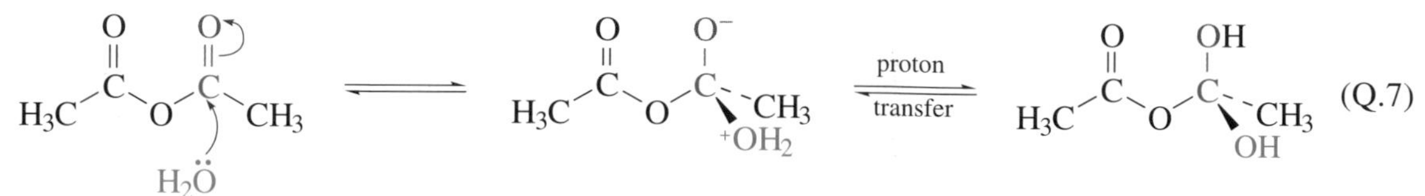


Elimination of either $H_2O$ or $CH_3COOH$ is possible; $CH_3COO^-$ is a better leaving group than $HO^-$, so $CH_3COOH$ is formed:

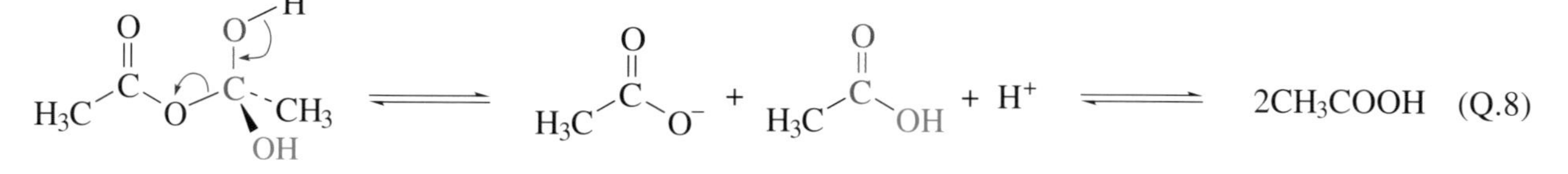


(b) Did you spot the complication here? There are four products if you make the reasonable assumption that there is little or no discrimination between the two carbonyl carbon atoms, allowing attack at C-1 as in **Q.7** or at C-2 as in **Q.8**:

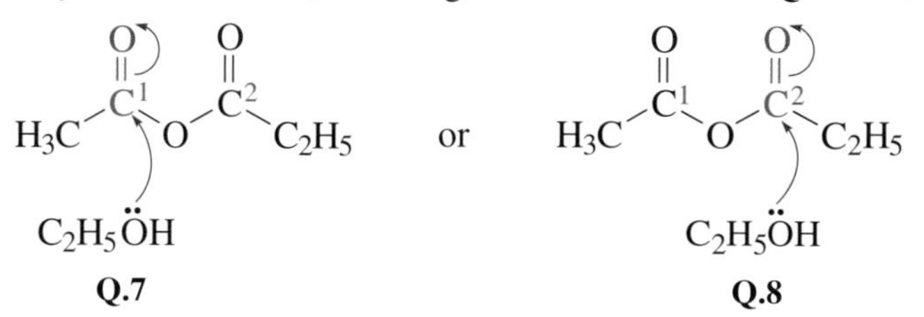


The products are therefore $CH_3COOC_2H_5$ and $C_2H_5COOH$, or $C_2H_5COOC_2H_5$ and $CH_3COOH$.

QUESTION 3.17 (*Learning Outcomes 4 and 5*)

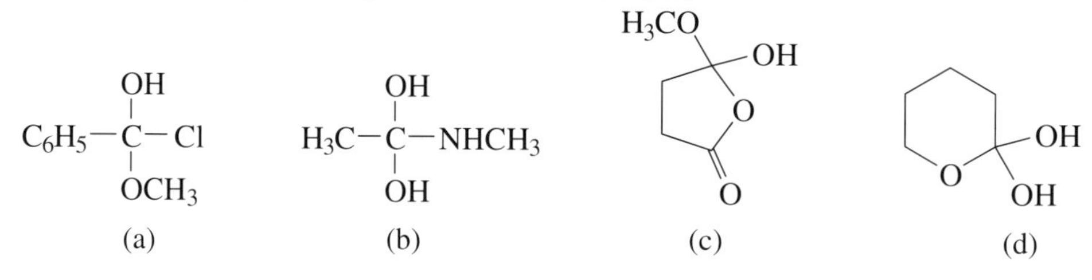


QUESTION 3.18 (*Learning Outcome 3*)

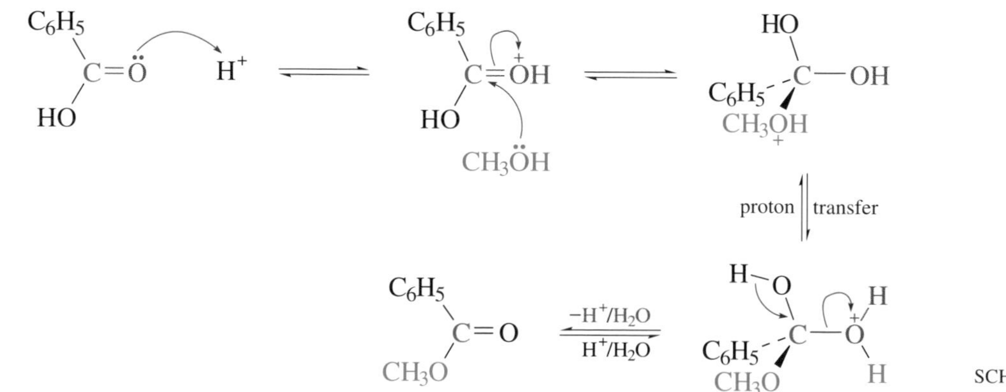


SCHEME Q.11

### QUESTION 3.19 (*Learning Outcomes 3 and 7*)

At first, this question may prompt you to suggest that the products should be $C_6H_5CONHC_6H_5$ and HCl. This is fine as far as it goes, but the amine is not only a nucleophile but also a base. So it will react with the HCl to give a salt:

$$C_6H_5NH_2 + HCl \rightleftharpoons C_6H_5\overset{+}{N}H_3\,Cl^- \qquad (Q.9)$$

The quaternary ammonium salt is no longer a nucleophile, so if this equilibrium is biased towards the salt — for example, by precipitation — we would lose one mole of reactant for every molecule of HCl produced. Hence we should allow for this by using excess amine:

$$C_6H_5COCl + 2C_6H_5NH_2 \longrightarrow C_6H_5CONHC_6H_5 + C_6H_5\overset{+}{N}H_3\,Cl^- \qquad (Q.10)$$

(b) $C_6H_5\ddot{N}H_2$ + $C_6H_5$–C(=O)–Cl ⇌ $C_6H_5\overset{+}{N}H_2$–C($C_6H_5$)(Cl)–$O^-$ ⇌ $C_6H_5\overset{+}{N}H(H)$–C(=O)$C_6H_5$ + $Cl^-$

$-H^+$ ↓

$C_6H_5NH$–C(=O)$C_6H_5$ + HCl

SCHEME Q.12

### QUESTION 3.20 (*Learning Outcomes 3 and 5*)

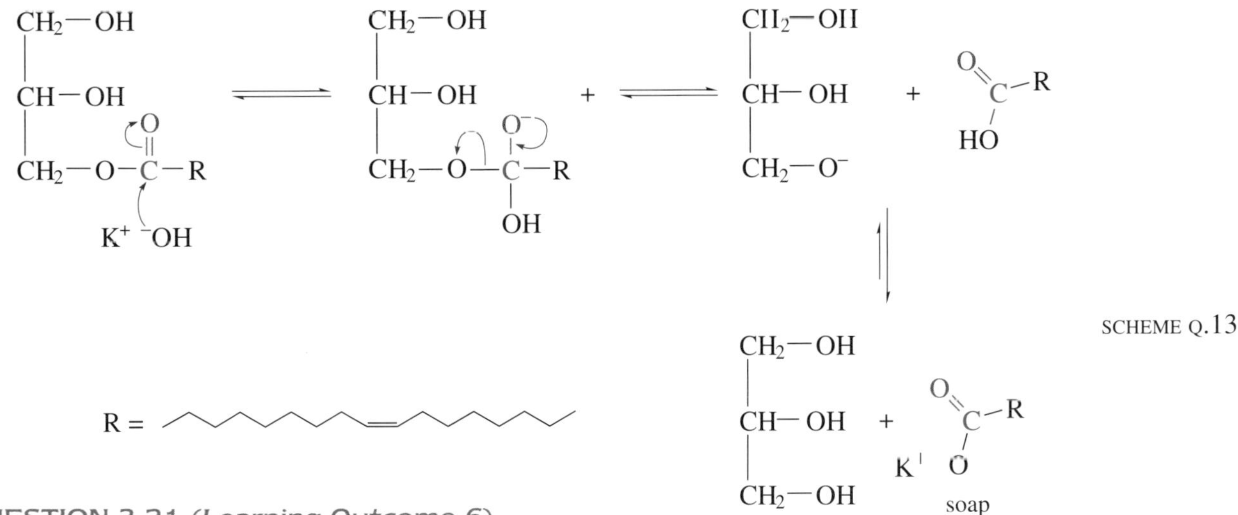


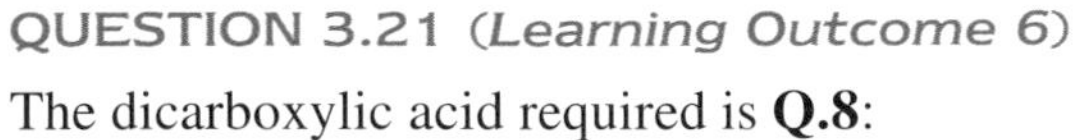


SCHEME Q.13

### QUESTION 3.21 (*Learning Outcome 6*)

The dicarboxylic acid required is **Q.8**:

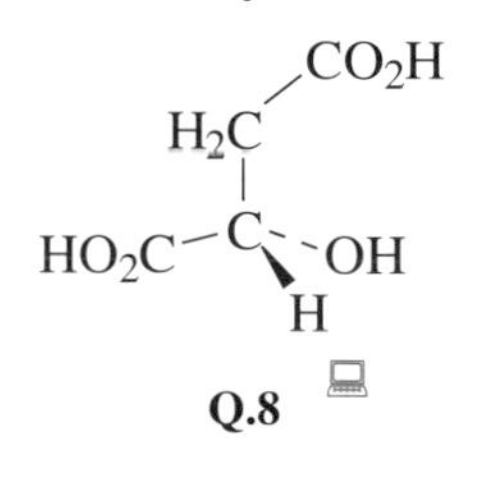


**Q.8**

Notice that $LiAlH_4$ reduction of a carbonyl function necessarily generates a primary alcohol. So a hydrocarbon chain is extended by one methylene group, and the secondary hydroxyl group cannot be involved.

### QUESTION 3.22 (*Learning Outcome 3*)

Hydroxide anion attacks the carbon atom of carbon dioxide, to produce the hydrogen carbonate ion directly:

$$O{=}C{=}O + HO^- \rightleftharpoons HO{-}C({=}O){-}O^- \qquad (Q.11)$$

### QUESTION 3.23 (*Learning Outcome 3*)

Formation of the tetrahedral intermediate proceeds as usual:

$$HO^- + C_6H_5(CH_3O)C{=}^{18}O \rightleftharpoons C_6H_5(CH_3O)C(OH){-}^{18}O^- \qquad (Q.12)$$

This can decompose to products by ejecting $CH_3O^-$, or revert to starting materials by ejecting $H^{16}O^-$. Alternatively, proton transfer can occur between the two unsubstituted oxygen atoms:

$$C_6H_5(CH_3O)C(OH){-}^{18}O^- \xrightleftharpoons[\text{transfer}]{\text{proton}} C_6H_5(CH_3O)C(O^-){-}^{18}OH \qquad (Q.13)$$

This new tetrahedral intermediate can also decompose to products by ejecting $CH_3O^-$, or to starting materials by ejecting $H^{18}O^-$:

$$C_6H_5(CH_3O)C(O^-){-}^{18}OH \rightleftharpoons C_6H_5(CH_3O)C{=}O + H^{18}O^- \qquad (Q.14)$$

The ester molecules formed in the above step therefore do not contain an $^{18}O$ atom.

### QUESTION 3.24 (*Learning Outcomes 3 and 5*)

In the same way that amines are more nucleophilic than alcohols, the nitrogen atom in hydroxylamine is more nucleophilic than the oxygen atom. Attack of the nitrogen atom on the carbonyl carbon atom of methyl benzoate produces a tetrahedral intermediate, **Q.9**:

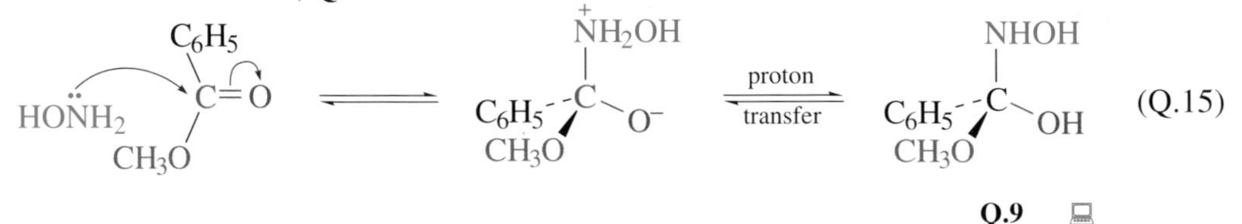


Elimination of methanol yields the hydroxyamide product:

$$C_6H_5(CH_3O)C(NHOH){-}O{-}H \xrightleftharpoons[\text{transfer}]{\text{proton}} C_6H_5(HOHN)C{=}O + CH_3OH \qquad (Q.16)$$

**Q.9**

## QUESTION 3.25 (*Learning Outcomes 3 and 7*)

Scheme Q.14 summarizes the sequence of reactions. First, the acid protonates an amide carbonyl oxygen atom; attack by water follows. Proton transfer to the adjacent amino nitrogen atom generates a better leaving group, such that decomposition of the tetrahedral intermediate causes a break in the polymer strand. Subsequent hydrolyses in a similar manner cause further breaks in the polymer, which are manifested as holes.

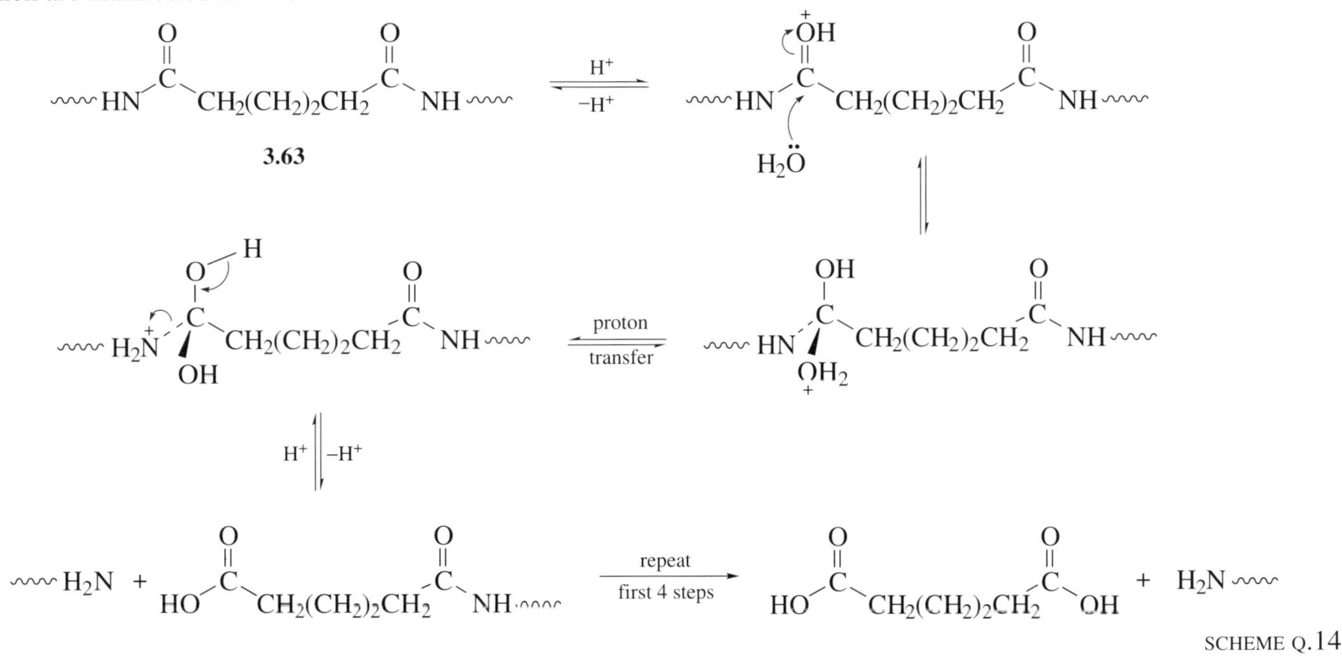


SCHEME Q.14

## QUESTION 3.26 (*Learning Outcomes 5 and 8*)

Stage 1 is the formation of an acetal, which requires dry HCl gas as a catalyst, and methanol as a reagent and solvent.

Stage 2 is the reduction of an ester by $LiAlH_4$.

Stage 3 is the regeneration of a ketone from the acetal, which requires aqueous acid.

Lithium aluminium hydride could not be used to reduce the starting material directly to the product, because it reduces a ketone group to an alcohol so it would give **Q.10**.

$C_6H_5CH(OH)CH_2CH_2OH$

**Q.10**

# ACKNOWLEDGEMENTS

Grateful acknowledgement is made to the following sources for permission to reproduce material in this book:

*Figure 2.3*: Bob Gibbons/Oxford Scientific Films; *Figure 3.1*: Malcolm Coe/Oxford Scientific Films; *Figure 3.2*: © The Bridgeman Art Library/Manchester City Art Galleries; *Figure 3.4*: Unilever Historical Archives; *Figure 3.5*: Courtesy of Pretty Polly.

*Part 2*

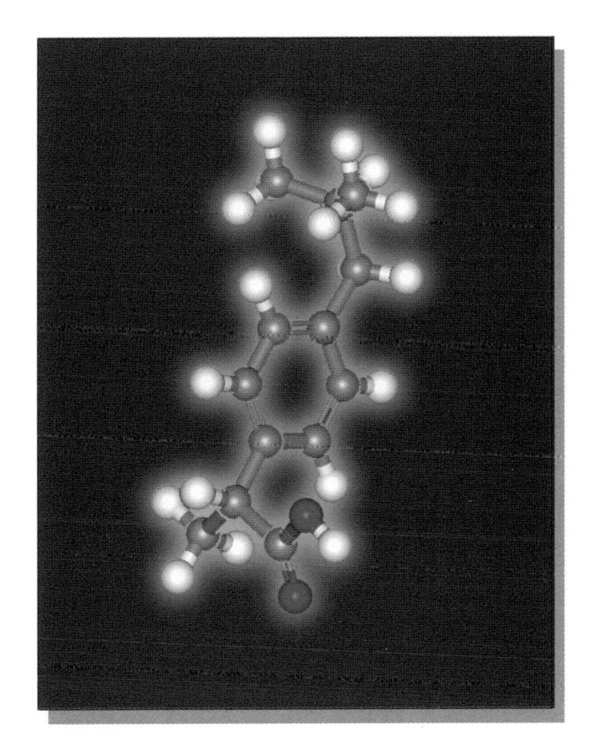

# Synthetic applications of organometallic compounds

edited by Peter Taylor

*based on* Synthetic applications of organometallic compounds,

*by Peter Morrod and Malcolm Rose*

# 1 INTRODUCTION

**Organometallic compounds** are most simply defined as compounds whose molecules possess direct carbon-to-metal bonds. Although this definition is not completely satisfactory, it is adequate for our present study. For the purpose of the discussion here, the semi-metal boron is included in the same category as 'proper' metals. The study of organometallic chemistry has progressed extremely rapidly in recent years, forming a major bridge between the organic and inorganic branches of chemistry.

- How will the electrons be distributed in a metal–carbon bond? What will be the partial charges on each atom?

- In general, metals are less electronegative than carbon. Thus, the electrons in the carbon–metal bond will be concentrated on the carbon atom, leading to polarization in the sense $\overset{\delta-}{C}-\overset{\delta+}{M}$ — the reverse of what we have seen for carbonyl chemistry in Part 1.

The size of the charge separation depends on the electronegativity. Thus, the alkali metals in Group I, such as sodium, form organometallic compounds, in which the bond is substantially ionic with most of the electron density around the carbon atom. However, less electronegative metals, such as lead in Group IV, form predominantly covalent bonds with carbon, in which the charge separation is less. Some of these covalent organometallic compounds are very inert, whereas the more ionic organometallics are extremely reactive towards water, oxygen and almost all organic solvents (other than ethers and hydrocarbons). In the next few Sections we shall focus predominantly on these more reactive organometallic compounds, because compounds with high reactivity generally enter into a wide variety of reactions, and are of value in synthetic work — particularly for **carbon–carbon bond-forming reactions**. Generally, to form a new C—C bond (shown in green below), we need a carbon nucleophile and an electrophilic carbon centre:

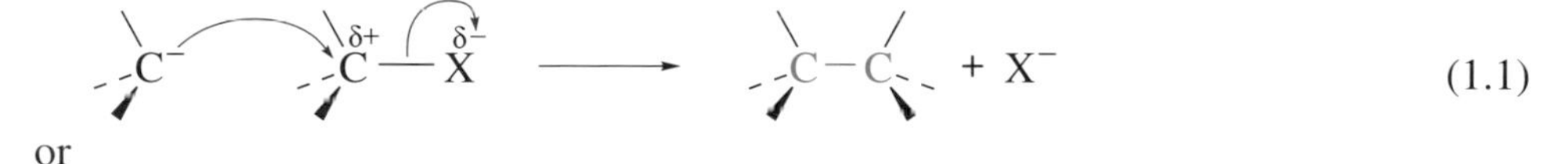

(1.1)

or

$$\gt C^- \; + \; \gt\overset{\delta+}{C}=\overset{\delta-}{O} \longrightarrow \gt C-C\lt O^- \qquad (1.2)$$

nucleophile electrophile

You are probably familiar with carbon electrophiles with leaving groups, X = Cl, Br, I, etc., and electrophilic carbonyl compounds. However, you have come across few carbon nucleophiles apart from $NC^-$ and $RC\equiv C^-$, and, of course, the benzene ring and carbon–carbon multiple bonds. In fact, organometallic reagents provide a wide range of carbon nucleophiles. In these reagents, R—M (where M is a metal), we generally find that the organic part, R, has nucleophilic character; in other words, it behaves as $R^-$. Given that R can be any one of a number of aliphatic or aromatic groups, organometallic compounds are extremely versatile for constructing carbon skeletons in synthesis.

It is important to note that we cannot do justice to the immense breadth of organometallic chemistry. A comprehensive account would have to cover, among many other topics, the organometallic compounds that are used as catalysts for many processes in industry. For example, the reaction of carbon monoxide, hydrogen and propene to give $C_4$ aldehydes is a major industrial process that involves an organorhodium species in the catalysis:

$$CH_3CH{=}CH_2 + CO + H_2 \xrightarrow{\text{'rhodium catalyst'}} \underset{\displaystyle CHO}{CH_3\underset{|}{C}HCH_3} + \underset{\text{major product}}{CH_3CH_2CH_2CHO} \qquad (1.3)$$

Organorhodium compounds are also used in the synthesis of ethanoic acid from methanol and carbon monoxide. The polymerization of ethene to give polythene involves an organotitanium compound in the catalysis. So, the treatment here is necessarily very selective, and concentrates only on those organometallic reagents (not catalysts) that are of greatest current synthetic utility in *organic* chemistry.

# ORGANOMAGNESIUM HALIDES

## 2.1 Formation

When a solution of a haloalkane or halobenzene in dry* diethyl ether (ethoxyethane, $C_2H_5OC_2H_5$) is allowed to stand over magnesium turnings, a vigorous reaction takes place. After going cloudy, the solution begins to boil, and the magnesium gradually dissolves. The resulting solution is known as a **Grignard reagent** (or just 'a **Grignard**'), after Victor Grignard who in 1912 received the Nobel Prize for Chemistry for discovering these reagents (Box 2.1, p. 81). They are among the most useful and versatile reagents available to the organic chemist, because they are very reactive and comparatively easy to prepare directly from metallic magnesium:

$$RX + Mg \xrightarrow{\text{dry ether}} RMgX \quad (R = \text{alkyl or aryl group; } X = Cl, Br \text{ or } I) \qquad (2.1)$$

The Grignard reagent has the general formula RMgX, but its exact molecular structure in ether is actually much more complicated than this (Figures 2.1 and 2.2), involving molecules of ether as well. Indeed, for some Grignard reagents, there may be no RMgX molecules present at all. However, we can use the formula RMgX with some confidence, since it successfully accounts for the products of the reactions that this class of organometallic compound undergoes.

**Figure 2.1** The X-ray crystal structure of phenylmagnesium bromide. The magnesium atom is shown in green, and is shown attached to a phenyl ring (grey), a bromine (brown), and two diethyl ether molecules via the oxygen atoms (red).

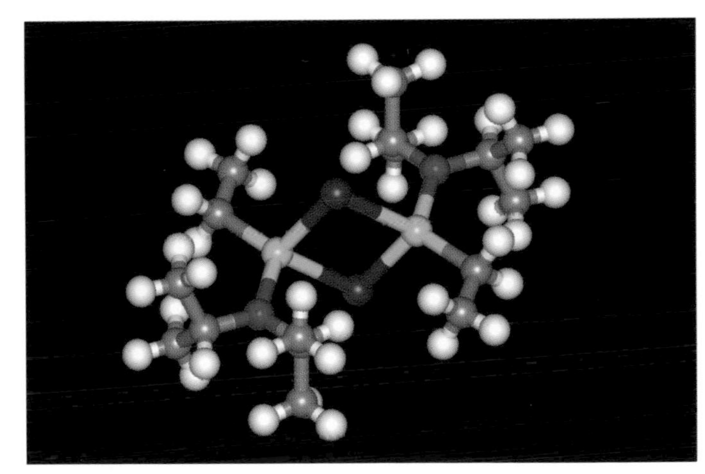

**Figure 2.2** The X-ray crystal structure of ethylmagnesium bromide; notice the dimeric form compared with the monomeric one in Figure 2.1. Again the magnesium atoms are shown in green and the bromine atoms in brown. Each magnesium atom is attached to one diethyl ether molecule, and the two bromines form a bridge between the two magnesiums.

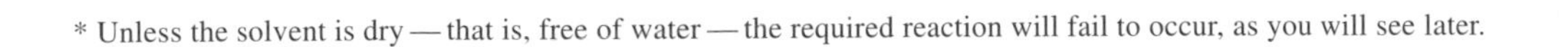

* Unless the solvent is dry — that is, free of water — the required reaction will fail to occur, as you will see later.

We shall not discuss the mechanism of the reaction between the haloalkane and the magnesium metal, because it is not well understood; suffice it to say that the magnesium inserts into the R—X bond, forming a covalent bond with the carbon atom of the R group that was previously bonded to the halogen atom. Hence, there is no structural change in the alkyl or aryl group that was present in the original haloalkane prior to reaction with the magnesium. For example, 1-bromopropane yields propylmagnesium bromide, and 2-bromopropane yields 1-methylethyl-magnesium bromide:

$$\underset{\text{1-bromopropane}}{CH_3CH_2CH_2Br} + Mg \xrightarrow{\text{dry ether}} \underset{\text{propylmagnesium bromide}}{CH_3CH_2CH_2MgBr} \qquad (2.2)$$

$$\underset{\text{2-bromopropane}}{(H_3C)_2CHBr} + Mg \xrightarrow{\text{dry ether}} \underset{\text{1-methylethylmagnesium bromide}}{(H_3C)_2CHMgBr} \qquad (2.3)$$

Although aromatic chloro, bromo and iodo compounds do not often undergo the same types of reaction as do aliphatic haloalkanes, the formation of Grignard reagents is an exception. Thus, bromobenzene readily yields the Grignard reagent phenylmagnesium bromide:

$$\underset{\text{bromobenzene}}{C_6H_5Br} + Mg \xrightarrow{\text{dry ether}} \underset{\text{phenylmagnesium bromide}}{C_6H_5MgBr} \qquad (2.4)$$

Similarly, vinyl halides react to form vinyl Grignard reagents:

$$\underset{\substack{\text{bromoethene}\\ \text{(vinyl bromide)}}}{CH_2{=}CHBr} + Mg \xrightarrow{\text{dry ether}} \underset{\substack{\text{ethenyl magnesium bromide}\\ \text{(vinyl magnesium bromide)}}}{CH_2{=}CHMgBr} \qquad (2.5)$$

## 2.2 Reaction with aldehydes and ketones

If we did not know the types of compound that would react with an organo-magnesium halide, we could make predictions by considering the distribution of electrons in the carbon–magnesium bond. Let's assume that the bond is fully ionized ($R^{-}\,{}^{+}MgBr$), even though this is not actually the case.

- Would the ion $R^-$ be classed as an electrophile or a nucleophile?

- $R^-$ is a nucleophile, because it is electron-rich: it has a pair of non-bonded electrons ready to form a bond to an electrophile.

This kind of carbon nucleophile is called a **carbanion** (which is shorthand for 'carbon anion'), and provides something of a contrast to the carbocations ($R^+$) that you have met so often before. We shall now look at examples of reactions in which such a nucleophilic carbanion is involved in carbon–carbon bond formation, starting with the reaction of Grignard reagents with aldehydes and ketones to yield alcohols.

## BOX 2.1 François Auguste Victor Grignard

François Auguste Victor Grignard was born in Cherbourg on 6 May 1871. In 1889, after attending local schools, he went into training at Cluny, Burgundy to become a maths teacher. However, when the college closed down, he was transferred to the University of Lyons, where he enrolled in the Science Faculty and obtained a licence in mathematics (1894). He slowly became more and more interested in chemistry, and eventually was appointed as the chief demonstrator ('chef des traveaux') at Lyons in 1898. At this time, Phillippe Barbier had been working on organomagnesium compounds as a development of work carried out by A. M. Saytsev using organozinc compounds. Barbier had published his initial findings and suggested that Grignard improve his previous work. Grignard then spent the next 30 years working on these new reagents. Barbier's original procedure involved adding the alkyl halide to a mixture of the carbonyl compound and magnesium in ether; it was Grignard who developed the initial formation of the organometallic reagent followed by addition of the carbonyl compound. In 1901, he presented his doctoral thesis, and in 1909 moved to become Professor of Chemistry at Nancy. In 1912, Grignard was awarded the Nobel Prize for Chemistry, for his organomagnesium research, much to Barbier's chagrin, who wrote:

> from the scientific viewpoint I must consider myself the originator of the very basis of the reaction.

In 1919, Grignard returned to Lyons as Professor of Chemistry. Grignard married Augustine Marie Boulant, an old friend from Cherbourg, and had one son. During the First World War, Grignard had to guard a railway bridge in Normandy, before being moved into munitions. Grignard died on 13 December 1935.

**Figure 2.3**
François Auguste Victor Grignard (1871–1935).

- What functional group do both aldehydes and ketones contain?
- The carbonyl group.

- In which direction is the C=O bond polarized?
- It is polarized $\overset{\delta+}{C}=\overset{\delta-}{O}$, because oxygen is more electronegative than carbon.

We can now understand why a Grignard reagent (with $C^{\delta-}$) reacts with an aldehyde or ketone (with $C^{\delta+}$) to form a new C—C bond. But before we represent the reaction with curly arrows, we have to remember that the nucleophile $R^-$ does not exist as such in a solution of RMgX. In recognition of the fact that the electrons forming the new C—C bond come from the R—Mg bond and not really from $R^-$, we usually draw the attack on an electrophile (E) as

R—MgX   E

rather than

$R^-$   E

Now, let's look at the reaction of phenylmagnesium bromide, $C_6H_5MgBr$, with ethanal (acetaldehyde):

$$C_6H_5{-}MgBr \quad + \quad (H_3C)(H)C{=}O \longrightarrow C_6H_5{-}C(CH_3)(H){-}O^- \overset{+}{M}gBr \qquad (2.6)$$

This is exactly the kind of nucleophilic attack that occurs in the first step of all nucleophilic additions to aldehydes to form a tetrahedral product; in this case it is a magnesium salt of a weakly acidic alcohol. It is easily converted into the alcohol itself by addition of water:

$$C_6H_5{-}C(CH_3)(H){-}O^- \overset{+}{M}gBr \xrightarrow{H_2O} C_6H_5{-}C(CH_3)(H){-}OH + Mg(OH)Br \qquad (2.7)$$

Because the Mg(OH)Br that is formed is a gelatinous material, the alcohol is difficult to isolate. A dilute acid (for example, HCl or $H_2SO_4$) is commonly used instead of water, so that a water-soluble magnesium salt is formed and any excess Grignard reagent is also destroyed:

$$Mg(OH)Br + H^+ \longrightarrow Mg^{2+} + Br^- + H_2O \qquad (2.8)$$

- What class of alcohol is obtained from the reaction of aldehydes with Grignard reagents, such as in Reaction 2.6?

- There is one hydrogen atom on the carbon atom that carries the OH group, and by definition this is a secondary alcohol. ($RCH_2OH$ is a primary alcohol, $R_2CHOH$ a secondary alcohol, and $R_3COH$ a tertiary alcohol.)

In general, aldehydes yield secondary alcohols, with one important exception. Methanal (formaldehyde) yields primary alcohols ($RCH_2OH$):

$$R{-}MgBr \quad + \quad (H)(H)C{=}O \longrightarrow R{-}C(H)(H){-}O^- \overset{+}{M}gBr \xrightarrow{H^+/H_2O} RCH_2OH + Mg^{2+} + Br^- \qquad (2.9)$$

It's as well to be aware of exceptions to general rules, but in this case the reaction is also of some synthetic importance, since it provides a way of lengthening the carbon chain of an alcohol by one carbon atom. The procedure used is to convert the original alcohol into a haloalkane, then to form a Grignard reagent from this haloalkane, and then to treat the Grignard with methanal:

$$ROH \xrightarrow{SOCl_2} RCl \xrightarrow[ether]{Mg} RMgCl \xrightarrow[(ii)\ H^+/H_2O]{(i)\ H_2CO} RCH_2OH \qquad (2.10)$$

The first step of this procedure, using thionyl chloride ($SOCl_2$), may be new to you, but you do not need to know the details. You will meet it again in Part 4 of this Book.

Ketones also react with Grignard reagents, and there is another general result to be discovered here. Let's look at the reaction of butylmagnesium bromide with propanone (acetone) as an example:

$$C_4H_9{-}MgBr \quad + \quad (H_3C)(H_3C)C{=}O \longrightarrow C_4H_9{-}C(CH_3)(CH_3){-}O^- \overset{+}{M}gBr \xrightarrow{H^+/H_2O} C_4H_9{-}C(CH_3)(CH_3){-}OH + Mg^{2+} + Br^- \qquad (2.11)$$

Here, the nucleophilic addition of the Grignard reagent yields a tertiary alcohol. Ketones, then, yield tertiary alcohols, and there are no exceptions this time.

> In summary, then, carbonyl compounds react with Grignard reagents to form alcohols as follows:
>
> methanal gives primary alcohols
>
> aldehydes give secondary alcohols
>
> ketones give tertiary alcohols

When trying to prepare a particular alcohol, how do we decide which Grignard reagent, and which carbonyl compound, to use? We have to consider the structure of the alcohol required. Of the alkyl or aryl groups attached to the carbon atom bearing the OH group, one must come from the Grignard reagent, and the other two (including any hydrogen atoms) must come from the carbonyl compound. Most alcohols can be obtained from more than one combination of reagents. Consider, for example, the synthesis of 2-phenylhexan-2-ol. As shown in Scheme 2.1, we could make this from a butyl Grignard reagent and an aromatic ketone, 1-phenylethanone (acetophenone), by route (a), or from a methyl Grignard and a more complex aromatic ketone by route (b), or from a phenyl Grignard reagent and a six-carbon aliphatic ketone (hexan-2-one), as in route (c).

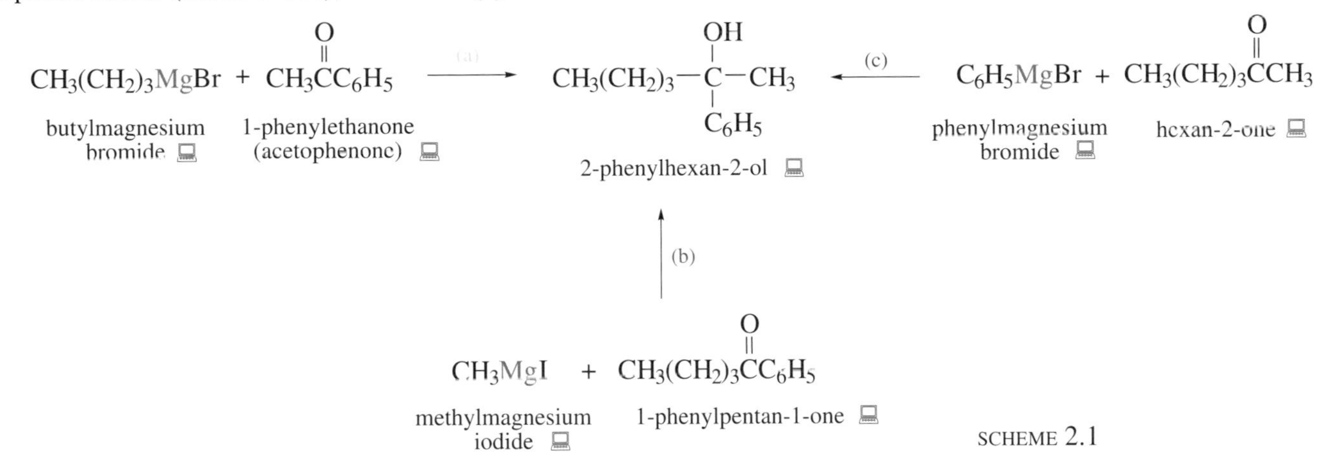


SCHEME 2.1

Notice that in route (b) the Grignard reagent chosen is methylmagnesium *iodide*. This is because the starting material, iodomethane, is the only halomethane that is a liquid at room temperature. However, in general, the bromide Grignard reagents are easier to handle.

The choice between the various potential routes often comes down to which compounds are more readily available in the laboratory as starting materials. In this synthesis, acetophenone and bromobutane are readily available as are hexan-2-one and bromobenzene, but not 1-phenylpentan-1-one.

How would you make 1-phenylcyclohexanol (**2.1**) from a ketone using a Grignard reaction?

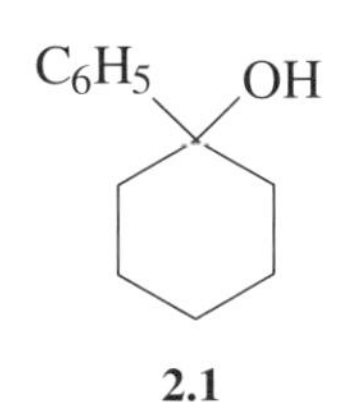


**2.1**

- The more obvious strategy is to react phenylmagnesium bromide with cyclohexanone.

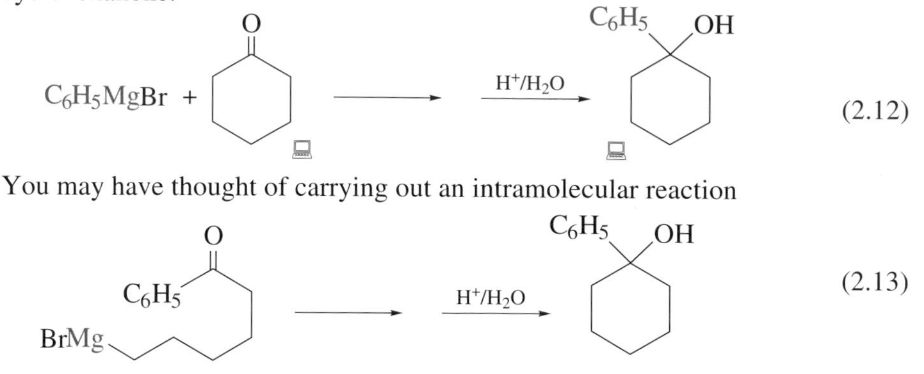

You may have thought of carrying out an intramolecular reaction

Although this route looks OK on paper, it is complicated by competing *inter*molecular processes; that is, the Grignard reagent of one molecule can react with the carbonyl group of a different molecule.

As you will see in Part 4 of this Book, the preferred route to 1-phenylcyclohexanol is to react phenylmagnesium bromide with cyclohexanone (Reaction 2.12), since this involves building the molecule from two relatively simple fragments, whereas the intramolecular route requires the preparation of quite a complicated starting material.

## 2.3 Reaction with carboxylic acid derivatives

Carboxylic acid derivatives also have a C=O group in their structures (Part 1), but these carbonyl groups behave rather differently in their reactions with nucleophiles. Attack of the nucleophile leads to a tetrahedral intermediate, which can then react further. In these reactions, Grignard reagents behave just like any other nucleophile.

Carboxylic acid derivatives such as esters, acid chlorides and acid anhydrides (but not amides) all react with Grignard reagents, $R^1MgX$, to give overall substitution; that is, in compounds such as $R^2COX$, the group X (X = $OR^3$, Cl or $OCOR^3$) is replaced by the nucleophilic alkyl or aryl group of the organometallic compound:

$$R^1MgBr + R^2(X)C{=}O \longrightarrow R^2(R^1)C{=}O + MgXBr \qquad (2.14)$$

- What would be the first step in the reaction mechanism?
- The first step is nucleophilic attack by the $R^1$ group of the Grignard reagent at the carbon atom of the carbonyl group, to give a tetrahedral intermediate:

$$R^1{-}MgBr \;\; R^2(X)C{=}O \longrightarrow R^1{-}C(R^2)(X){-}O^- \; \overset{+}{M}gBr \qquad (2.15)$$

The reaction continues, with the loss of the leaving group, X, to yield a ketone:

$$R^1{-}C(R^2)(X){-}O^- \overset{+}{M}gBr \longrightarrow R^2(R^1)C{=}O + MgXBr \qquad (2.16)$$

Overall, this is a nucleophilic substitution reaction, which is typical of many reactions of nucleophiles with carboxylic acid derivatives.

- Why does this reaction not stop at this stage?
- The product at this stage is a ketone, and ketones readily react with Grignard reagents, forming the salt of an alcohol:

$$R^1{-}MgBr \quad R^2(R^1)C{=}O \longrightarrow R^1{-}C(R^2)(R^1){-}O^- \overset{+}{M}gBr \qquad (2.17)$$

The net result is the formation of a tertiary alcohol from an acid derivative:

$$R^2(X)C{=}O + 2R^1MgBr \longrightarrow \xrightarrow{H^+/H_2O} R^1{-}C(R^2)(R^1){-}OH \qquad (2.18)$$

For example:

$$C_6H_5CH_2COOCH_3 + 2C_6H_5MgBr \longrightarrow \xrightarrow{H^+/H_2O} C_6H_5CH_2{-}C(C_6H_5)_2{-}OH \qquad (2.19)$$

Note that two of the groups attached to the alcohol carbon atom originate from the Grignard reagent, $R^1MgBr$, and hence must be the same, so that not all tertiary alcohols can be made in this way. Also note that *two* moles of Grignard reagent are needed to take the reaction to completion. In principle it may be possible to control the reaction conditions (by using just one mole of the Grignard reagent) so that the reaction stops at the ketone stage; however, this is only rarely successful, because ketones react so readily with Grignard reagents. Thus, the reaction between a carboxylic acid derivative and a Grignard reagent is best regarded as a method of preparing certain tertiary alcohols; there are better ways of making ketones from carboxylic acid derivatives.

## 2.4 Reaction with other electrophiles

### 2.4.1 Reaction of a Grignard reagent with nitriles — the preparation of ketones

We have just seen that ketones are formed as intermediates in the reaction of a Grignard reagent with a carboxylic acid derivative, but we can't *prepare* ketones this way because they react further with more Grignard reagent to give an alcohol. One strategy for isolating the ketone is to use a less reactive organometallic compound, which reacts with carboxylic acid derivatives but not ketones. We shall meet one such organometallic reagent — an organocopper reagent — in Section 4. However, ketones can be prepared from Grignard reagents using another route that is not at all obvious at first sight. It is the reaction of a Grignard reagent with a nitrile, $R^2{-}C{\equiv}N$; for example,

$$CH_3MgI + C_6H_5CN \longrightarrow (H_3C)(C_6H_5)C{=}O \qquad (2.20)$$

Even allowing for the fact that no attempt has been made to balance this equation (or indeed to show all the reactants), it looks rather odd. The nitrogen atom has disappeared from the product, and a ketone has been produced, despite the fact that Grignard reagents react readily with ketones! There is, of course, a perfectly reasonable explanation.

- How is the $-C{\equiv}N$ group of nitriles polarized?
- The functional group $-C{\equiv}N$ is polarized in a similar manner to the carbonyl group ($\overset{\delta+}{C}{\equiv}\overset{\delta-}{N}$).

Because of this polarization (shown in Figure 2.4), nitriles can undergo addition reactions with Grignard reagents. The first step in the reaction yields a magnesium salt of an imine; for example,

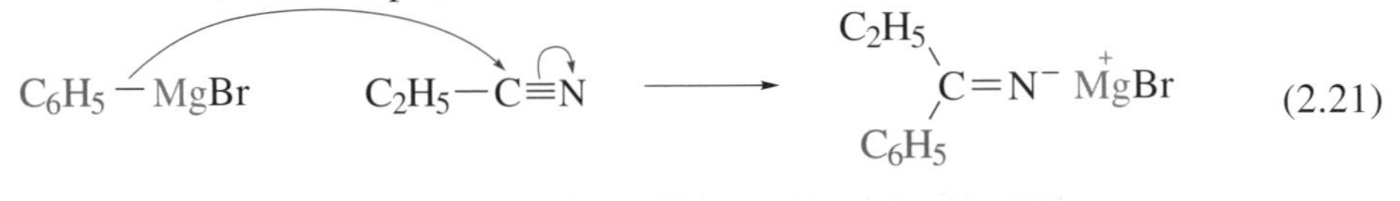


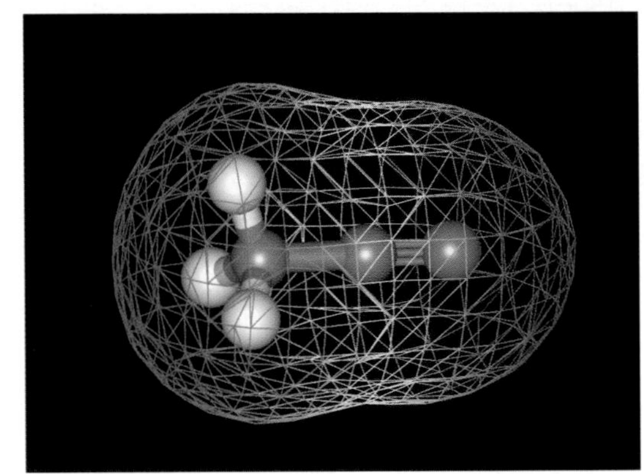

**Figure 2.4**
The charge density distribution in ethanenitrile, $CH_3CN$. The red areas of the surface indicate regions of high electron density and the blue areas, low electron density.

However, in this intermediate no leaving group is available, and because it is negatively charged, it does not react further with the nucleophilic Grignard reagent. On working up the reaction mixture with aqueous acid, any excess of the Grignard reagent is destroyed, and the salt forms an imine. The imine is unstable under the aqueous acidic conditions of the work-up, and rapidly hydrolyses to form a ketone:

$$(C_2H_5)(C_6H_5)C{=}N^-\ \overset{+}{Mg}Br \xrightarrow{H^+/H_2O} \underset{\text{an imine}}{(C_2H_5)(C_6H_5)C{=}NH} + Mg^{2+} + Br^-$$

$$(C_2H_5)(C_6H_5)C{=}NH \xrightarrow{H^+/H_2O} (C_2H_5)(C_6H_5)C{=}O + NH_3$$

SCHEME 2.2

- So, the disappearance of the nitrogen atom of the C≡N group is due to hydrolysis during work-up. But why doesn't the product ketone react further with the Grignard reagent?

- The ketone is formed *during work-up.* That's *after* any excess of Grignard reagent has been destroyed by aqueous acid. Hence, at this stage, there is no Grignard reagent with which the ketone can react!

## 2.4.2 Reaction of a Grignard reagent with carbon dioxide — the preparation of carboxylic acids

Carbon dioxide contains two carbonyl groups and an electrophilic carbon (Figure 2.5), and so reacts with Grignard reagents just like aldehydes and ketones. This provides a means of preparing carboxylic acids:

$$\mathrm{R{-}MgBr} \quad \mathrm{O{=}C{=}O} \longrightarrow \mathrm{BrMg^{+}\,{}^{-}O{-}C(R){=}O} \xrightarrow{H^+/H_2O} \mathrm{HO{-}C(R){=}O} + Mg^{2+} + Br^- \qquad (2.22)$$

The Grignard reagent attacks the carbon–oxygen double bond, just as in the reaction with aldehydes and ketones, to give a bromomagnesium salt of the carboxylic acid. As with the nitrile intermediate, the product salt does not react with another molecule of RMgX, which is not a strong enough nucleophile to attack a functional group that is negatively charged (like the carboxylate product). Instead, the free acid is liberated by treatment with dilute aqueous acid.

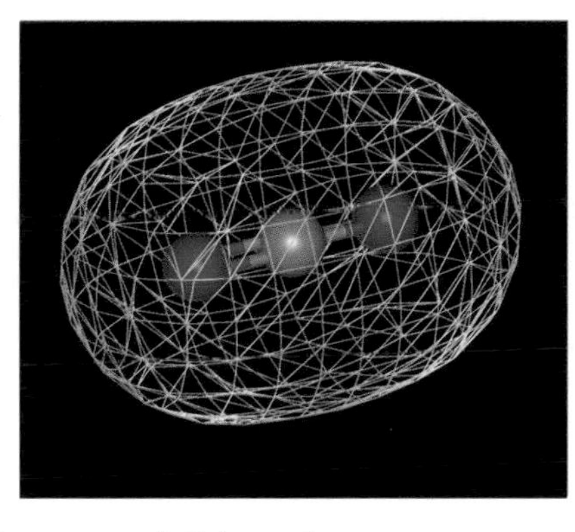

**Figure 2.5**
The charge density distribution in carbon dioxide. The red areas of the surface indicate regions of high electron density and the blue areas, low electron density.

The reaction is carried out by pouring the freshly prepared Grignard reagent on to crushed solid $CO_2$ (dry ice) or by bubbling gaseous $CO_2$ into the ether solution of the Grignard reagent. In the former method the dry ice serves not only as a reagent, but also as a cooling agent, which is required because the reaction is exothermic. The following syntheses illustrate the application of this reaction:

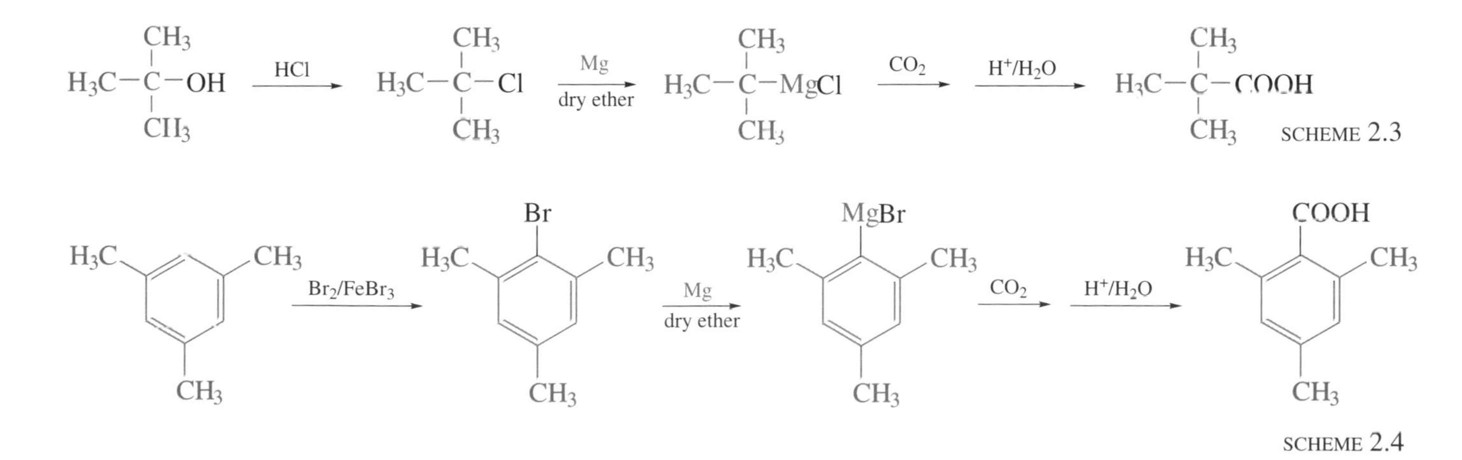


SCHEME 2.3

SCHEME 2.4

### 2.4.3 Reaction of a Grignard reagent with oxiranes — the preparation of primary alcohols

So far, we have seen that Grignard reagents act as nucleophiles, and react with carbon electrophiles such as carbonyls. The one reaction we have not discussed is the reaction with C—X electrophiles such as haloalkanes; this is because such reactions do not generally occur:

$$\text{R—MgBr} \quad \text{C—Br} \not\longrightarrow \text{no reaction} \qquad (2.23)$$

- In fact, we have already seen that Grignard reagents don't react with haloalkanes. What is this evidence (think about how Grignard reagents are formed)?
- Grignard reagents are formed from haloalkanes, so if they did react with each other we would not get a very good yield of organometallic reagent!

Thus, it is not possible to form a carbon–carbon bond using Grignard reagents and haloalkanes (although, as we shall see later this can be achieved using a different organometallic reagent). However, Grignard reagents do react with oxiranes. If oxirane (ethylene oxide) itself is used, a primary alcohol is formed having two more carbons than the starting Grignard reagent:

$$\text{R—MgBr} \quad \text{H}_2\text{C}\overset{\text{O}}{—}\text{CH}_2 \longrightarrow \text{RCH}_2\text{CH}_2\text{O}^- + \text{Mg}^+ \text{Br}^- \xrightarrow{\text{H}^+/\text{H}_2\text{O}} \text{RCH}_2\text{CH}_2\text{OH} + \text{Mg}^{2+} + \text{Br}^- \qquad (2.24)$$

This provides a method for extending a carbon chain by two carbons: starting with the haloalkane, we can form the Grignard reagent and react it with oxirane to give the alcohol, which can be converted back to the corresponding halo derivative:

$$\text{R—Cl} \xrightarrow{\text{Mg/ether}} \text{R—MgCl} \xrightarrow[\text{(ii) H}^+/\text{H}_2\text{O}]{\text{(i) H}_2\text{C}\overset{\text{O}}{—}\text{CH}_2} \text{RCH}_2\text{CH}_2\text{OH} \xrightarrow{\text{SOCl}_2} \text{RCH}_2\text{CH}_2\text{Cl} \qquad (2.25)$$

### 2.4.4 Reaction of a Grignard reagent with acids

As we have seen, Grignard reagents are useful in synthesis because of their high reactivity towards electrophilic carbon atoms in organic molecules, leading to the formation of new carbon–carbon bonds. However, they also react with other types of compound, and their reaction with water to form a hydrocarbon, RH, is typical of the behaviour of Grignard reagents towards acids:

$$\text{R—MgX} \quad \text{H—OH} \longrightarrow \text{RH} + \text{Mg(OH)X} \qquad (2.26)$$

- As what type of compound is the Grignard reagent behaving in this reaction?
- The Grignard reagent is acting as a base, abstracting a proton from the weakly acidic water molecule.

Grignard reagents are very strong bases, and they will react with a wide range of weakly acidic compounds that contain a hydrogen atom attached to an oxygen or a nitrogen atom. So not only will Grignard reagents react with water (which is why the reagents and apparatus must be scrupulously dry when a Grignard reagent is prepared), but also with any Brønsted acid*, such as ammonia, amines and alcohols:

$$RMgX + NH_3 \longrightarrow RH + Mg(NH_2)X \qquad (2.27)$$

$$RMgX + CH_3OH \longrightarrow RH + Mg(OCH_3)X \qquad (2.28)$$

If we were interested in making a hydrocarbon from a Grignard reagent, one acid is as good as another, and water would be the most readily available acid. However, the preparation of hydrocarbons by this method is of very limited usefulness.
The most important implication of this type of reaction is (i) the need to keep the apparatus and reagents dry for the preparation of many organometallic compounds, and (ii) the limitations placed on the compounds that can be used to make Grignard reagents and to react with them. For example, an organic halogen compound that contains another functional group capable of reacting with a Grignard reagent cannot be used successfully to prepare a Grignard reagent. It would be possible to carry out the following reaction:

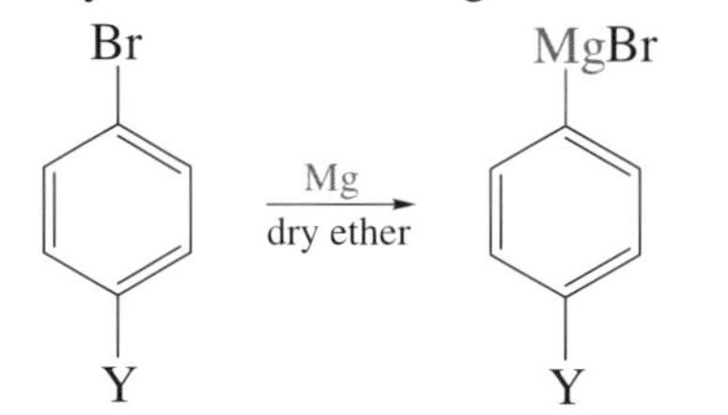


(2.29)

if Y = alkyl, aryl or –OR (ether), but not if

$$Y = -\overset{O}{\overset{\|}{C}}-R,\ -\overset{O}{\overset{\|}{C}}-OH,\ -\overset{O}{\overset{\|}{C}}-OR,\ -C{\equiv}N,\ -OH,\ -NH_2,\ -SO_3H \text{ or } -NO_2$$

By the same token, the aldehyde (or other compound) with which a Grignard reagent is to react may not contain other groups that are reactive towards a Grignard reagent. It is always found that the reaction of the Grignard reagent as a base with an —O—H or an $-\underset{|}{N}-H$ group is faster than addition (as a nucleophile) to a carbonyl group. So attempts to make RMgX react with the carbonyl group in either of the compounds **2.2** and **2.3** would not be successful. These may seem like severe limitations, and they are. Nevertheless, the number of acceptable combinations is so great that the Grignard reagent is one of the organic chemist's most valuable synthetic tools. As we shall see later in this Book, the considerations described here must be taken into account in any kind of organic synthesis; we must not restrict our attention to the group we happen to be interested in, but must look for possible interference by other functional groups.

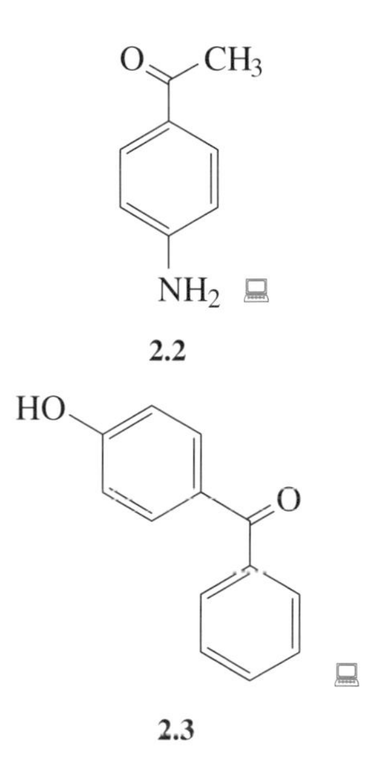


**2.2**

**2.3**

As a final example of a group of acids that react with Grignard reagents, let's look at carboxylic acids, RCOOH. These contain both a carbonyl group and an acidic proton; a Grignard reagent always preferentially reacts with the latter.

- What are the products of such a reaction?

- The products are a hydrocarbon, and a magnesium salt of the carboxylic acid:

$$R^1{-}MgBr \quad + \quad R^2C(=O){-}O{-}H \longrightarrow R^1H + R^2CO_2^- \overset{+}{M}gBr \qquad (2.30)$$

* One from which a proton can be removed.

As we noted earlier in the reaction of RMgX with carbon dioxide, the resulting carboxylate anion is not as reactive towards nucleophiles as other carbonyl compounds are. This is because the electrophilicity of the carboxylate carbon atom is reduced by the negative charge, as shown in Figure 2.6. Also, the salt often precipitates out from ether solution. So it seems that this reaction will not be of any use for converting carboxylic acids into new and useful products. But wait for Section 3 on organolithium compounds!

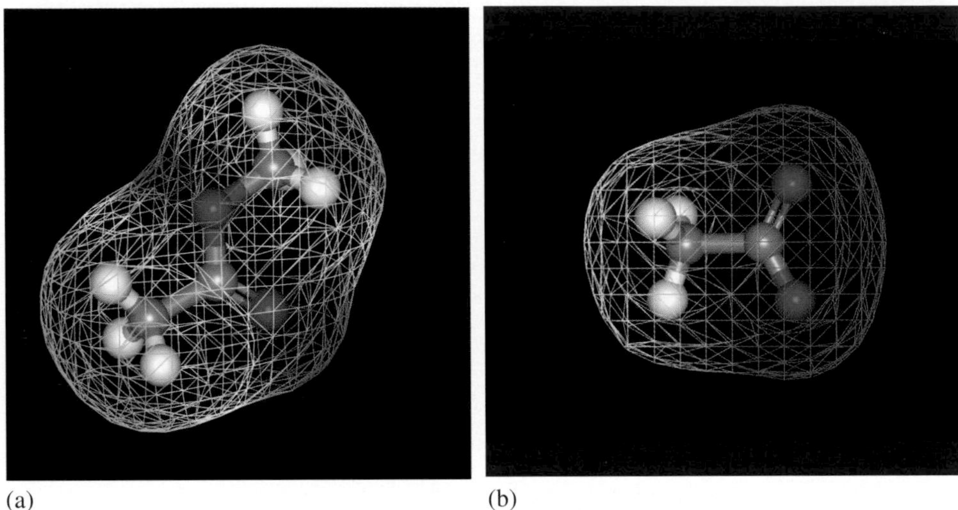


**Figure 2.6**
The charge density distribution in (a) methyl ethanoate, $CH_3COOCH_3$ and (b) sodium ethanoate, $CH_3COO^- Na^+$. The red areas of the surface indicate regions of high electron density and the blue areas, low electron density. Notice that the positive charge density on the carbonyl carbon is much less in the carboxylate salt than in the ester, as indicated by the lack of blue coloration.

## 2.5 Summary of Section 2

1 A Grignard reagent is prepared by the reaction of an alkyl or aryl halide with magnesium in dry ether:

$$RX + Mg \xrightarrow{\text{dry ether}} RMgX \text{ (R = alkyl or aryl group)} \quad (2.31)$$

2 Grignard reagents react with aldehydes or ketones to yield alcohols:

$$R^1MgX + R^2R^3C{=}O \longrightarrow \xrightarrow{H^+/H_2O} R^1{-}C(R^2)(R^3){-}OH \quad (R^2, R^3 = \text{H, alkyl or aryl}) \quad (2.32)$$

Methanal (formaldehyde) gives a primary alcohol, other aldehydes give secondary alcohols, and ketones give tertiary alcohols.

3 Two molar equivalents of Grignard reagent react with carboxylic acid derivatives (except amides) to give tertiary alcohols:

$$2R^1MgBr + R^2(X)C{=}O \longrightarrow \xrightarrow{H^+/H_2O} R^1{-}C(R^2)(R^1){-}OH \quad (X = \text{Cl, OR}^3\text{, OCOR}^3) \quad (2.33)$$

4 Ketones can be prepared by the reaction of Grignard reagents with nitriles:

$$R^1MgX + R^2{-}C{\equiv}N \longrightarrow \xrightarrow{H^+/H_2O} R^1R_2C{=}O \quad (2.34)$$

5 Carboxylic acids can be prepared by the reaction of Grignard reagents with carbon dioxide:

$$RMgX + CO_2 \longrightarrow RCOO^- \xrightarrow{H^+/H_2O} RCOOH \qquad (2.35)$$

6 Grignard reagents do not react with haloalkanes, but do react with oxiranes to form a primary alcohol, and extend the carbon chain by two carbons:

$$R{-}MgX \xrightarrow[\text{(ii) } H^+/H_2O]{\text{(i) } H_2C{-}CH_2 \text{ (bridged by O)}} RCH_2CH_2OH \qquad (2.36)$$

7 In all the above reactions, the Grignard reagent is acting as a carbon nucleophile, attacking an electrophilic carbon centre, to form a new carbon–carbon bond.

8 Grignard reagents also act as strong bases, even reacting with compounds containing weakly acidic —OH or —N(—)—H groups; for example

$$RMgBr + H_2O \longrightarrow RH + Mg(OH)Br \qquad (2.37)$$

This leads to the need for dry conditions when the reagents are prepared, and it also limits the range of compounds that can be used to make, or to react with, a Grignard reagent. Carboxylic acids are a particular group of such compounds that give hydrocarbons with Grignard reagents.

---

### EXERCISE 2.1 Summary of organometallic reactions 1

Look back at the reactions of Grignard reagents that were introduced in Section 2 and summarize them using general equations which highlight the other reactant, conditions and products. Compare your summary with that given in Part A of the Appendix.

---

### QUESTION 2.1

For each of the compounds (i)–(iv), indicate which carbonyl compound, and which Grignard reagent, could be used for its preparation.

(i) $(CH_3)_2CHCH_2COOH$

(ii) $H_3C{-}C(CH_3)(H){-}OH$

(iii) $CH_3CH_2{-}C(OH)(CH_3){-}CH_2CH_2CH_3$ (three possible combinations)

(iv) $C_6H_5CH_2OH$

that we have studied so far. Hence, we shall only consider one special case of organosodium chemistry (Section 3.3), the alkynides (acetylides), which have synthetic utility, and can be made and handled safely.

## 3.2 Formation and reaction of organolithium compounds

The most convenient method for preparing organolithium compounds is the reaction of lithium with a haloalkane or halobenzene (RX, where X = Cl, Br, I but not F), just like the reaction of RX with magnesium to form Grignard reagents. Ethers, particularly diethyl ether, and hydrocarbons are used as solvents for the synthesis of organolithium compounds; as in the preparation of Grignard reagents, the ether must be dry. An example of the formation of organolithium compounds is

$$CH_3I + 2Li \xrightarrow{\text{dry ether}} CH_3Li + LiI \qquad (3.3)$$

Organolithium reagents are very reactive: they even react with the gases in the air, so the reactions need to be carried out under an unreactive atmosphere — usually argon.

Although they are less bulky than Grignard reagents, the structure of organolithium compounds is not simple and can involve the formation of dimers and trimers, etc. (Figure 3.1).

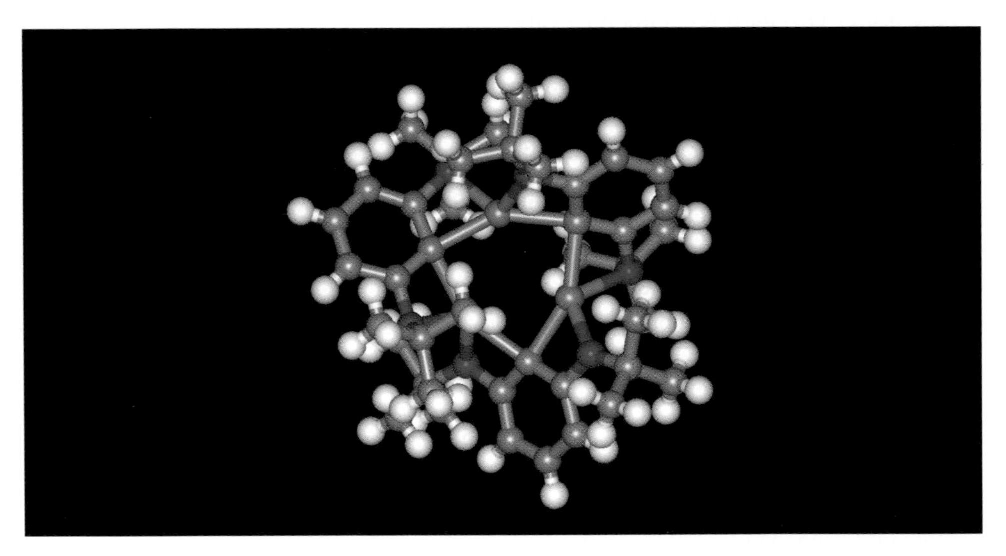

**Figure 3.1**
The crystal structure of an aryllithium (the lithium is purple):

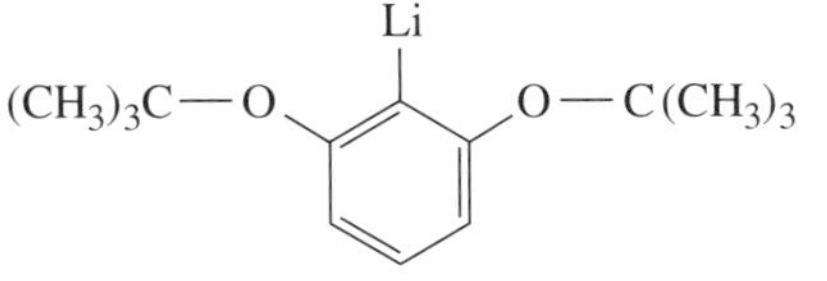


Notice the trimeric form taken by this compound as a result of the bulky alkoxy groups attached to the benzene ring.

After the reagent RLi has been prepared, it is not isolated, but used without further treatment, except perhaps to decant the solution from any excess metal or sediment. Luckily, organolithium reagents, like Grignard reagents, do not react further with the haloalkanes from which they are prepared*.

In terms of their reactions with carbonyl compounds, alkyllithium and aryllithium reagents behave in much the same way as organomagnesium compounds, but usually with increased reactivity. Partly, this is because lithium is less electronegative than magnesium, so the C—Li bond is even more polar (greater ionic

* Organolithium compounds do undergo lithium/halogen exchange with other haloalkanes, and this provides an important method of preparing some organolithiums; for example

$$Bu—Li + R—Br \longrightarrow R—Li + Bu—Br$$

character) than the C—Mg bond. In other words, an organolithium compound is more accurately represented as $R^- Li^+$ than is an organomagnesium compound by $R^- \overset{+}{Mg}Br$. As the R group has greater carbanion character in RLi, it is more nucleophilic. Indeed, all the reactions we discussed for Grignard reagents in Section 2 can be carried out with the appropriate organolithium compound. For example:

$$R^1{-}Li + R^2(X)C{=}O \longrightarrow R^1{-}C(R^2)(X){-}O^- \; Li^+ \xrightarrow{-LiX} R^2(R^1)C{=}O \qquad (3.4)$$

$$R^1{-}Li + R^2(R^1)C{=}O \longrightarrow R^1{-}C(R^2)(R^1){-}O^- \; Li^+ \xrightarrow{H_2O} R^1{-}C(R^2)(R^1){-}OH \qquad (3.5)$$

However, there is one important instance, in which the increased reactivity of the organolithium reagents makes a difference to the observed product, and that is in their reaction with carboxylic acids.

- What are the products of the reaction between $R^1MgX$ and $R^2COOH$?
- The products are a hydrocarbon ($R^1H$) and a carboxylate salt ($R^2CO_2^- \overset{+}{Mg}X$).

With an organolithium compound, a very similar first step occurs:

$$R^1{-}Li + R^2(H{-}O)C{=}O \longrightarrow R^2(Li^+\,{}^-O)C{=}O + R^1H \qquad (3.6)$$

The lithium salt remains in solution, but, in the presence of a second mole of the organolithium compound, further attack takes place at the carbonyl carbon atom of the carboxylate anion. This is in spite of the fact that the anion is negatively charged, and would not normally be expected to react with a nucleophile. The product is a dianion—that is, an anion carrying two negative charges:

$$R^1{-}Li + R^2(Li^+\,{}^-O)C{=}O \longrightarrow R^1{-}C(R^2)(O^-\,Li^+){-}O^- \; Li^+ \qquad (3.7)$$

- The carboxylic acid has reacted with two moles of the organolithium reagent, but the mechanism is different in each step. What is the difference?
- In the first step the 'carbanion' (organolithium compound) is acting as a base, removing the proton from the OH group of the acid. In the second step the 'carbanion' is acting as a nucleophile, attacking the carbon atom of a carbonyl group.

After reaction with the organometallic reagent is complete, acid hydrolysis of the dilithium salt gives a diol, which is the hydrate of a ketone, and decomposes to give the ketone:

$$R^1{-}C(R^2)(O^-\,Li^+){-}O^- \; Li^+ \xrightarrow{H^+/H_2O} R^1{-}C(R^2)(OH){-}OH \rightleftharpoons R^2(R^1)C{=}O + H_2O \qquad (3.8)$$

- Why doesn't further reaction occur between the ketone and more of the organolithium compound?

- The ketone is formed in a separate hydrolysis step after reaction with the organolithium compound is complete. If there were any excess $R^1Li$, it would be destroyed as soon as the water is added, in a reaction analogous to that between Grignard reagents and water.

We are now in a position to understand the reaction of carbon dioxide with an excess of an organolithium compound. In this reaction a ketone is formed rather than the expected carboxylic acid:

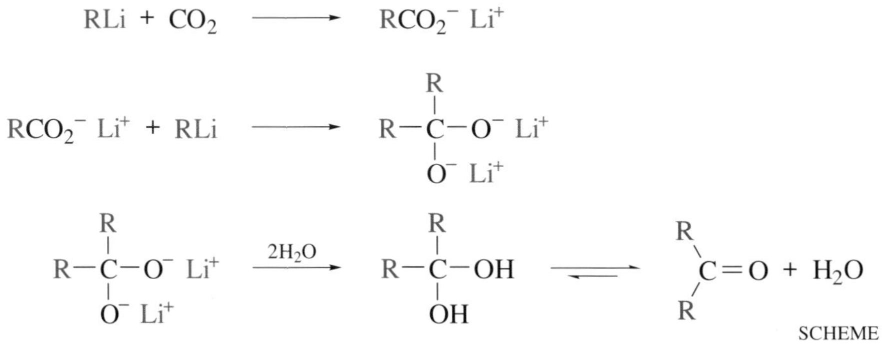

$$RLi + CO_2 \longrightarrow RCO_2^- Li^+$$

$$RCO_2^- Li^+ + RLi \longrightarrow R_2C(O^- Li^+)_2$$

$$R_2C(O^- Li^+)_2 \xrightarrow{2H_2O} R_2C(OH)_2 \rightleftharpoons R_2C{=}O + H_2O$$

SCHEME 3.1

- Can *any* ketone be made from carbon dioxide by this route?
- No. The method is restricted to symmetrical ketones, since both the R groups come from the same organometallic compound.

Grignard reagents and organolithium comounds also find application in the synthesis of other organometallic comounds (Box 3.1).

## BOX 3.1 Organometallic reagents in the synthesis of other organometallic compounds

As well as being useful reagents for organic synthesis, Grignard reagents and organolithiums are important in inorganic synthesis as well. They are used in what are known as *metathetical reactions*, in which a metal halide is reacted with the Grignard reagent or organolithium to give a new organometallic compound and a magnesium or lithium halide:

$$R{-}MgBr + M{-}X \longrightarrow R{-}M + MgXBr \quad (3.9)$$

$$R{-}Li + M{-}X \longrightarrow R{-}M + LiX \quad (3.10)$$

Such metathetical reactions occur with a range of organometallic compounds and the halogen ends up combined with the less electronegative metal and the organic group with the more electronegative metal. Since Grignard reagents and organolithiums are easy to prepare and not very electronegative, these reagents are widely used to make organometallic compounds in the laboratory; for example

$$CdCl_2 + 2C_2H_5MgBr \longrightarrow (C_2H_5)_2Cd + MgCl_2 + MgBr_2 \quad (3.11)$$

$$HgCl_2 + 2CH_3MgBr \longrightarrow (CH_3)_2Hg + MgCl_2 + MgBr_2 \quad (3.12)$$

$$ZnBr_2 + 2C_6H_5MgBr \longrightarrow (C_6H_5)_2Zn + 2MgBr_2 \quad (3.13)$$

$$SiCl_4 + 4C_2H_5MgCl \longrightarrow (C_2H_5)_4Si + 4MgCl_2 \quad (3.14)$$

Organolithium reagents are more reactive and so, as in organic chemistry, they are used where complete substitution of the halogen by the alkyl group is required but cannot be achieved using a Grignard reagent:

$$TiCl_3 + 2RMgCl \longrightarrow R_2TiCl + 2MgCl_2 \quad (3.15)$$

$$TiCl_3 + 3RLi \longrightarrow R_3Ti + 3LiCl \quad (3.16)$$

# 3.3 Formation and reaction of organosodium compounds

By analogy with the formation of Grignard reagents and organolithium compounds, we might expect **organosodium compounds** to be made by reacting a haloalkane with sodium; however, this reaction does not happen. Unlike organolithiums and Grignard reagents, organosodiums are so reactive that they react with the haloalkane that is used to form them to give the symmetrical alkane R—R, this is known as the *Wurtz reaction*, after its discoverer, the French chemist Charles Adolph Wurtz (Box 3.2); for example

$$2CH_3Br \xrightarrow{Na} H_3C-CH_3 \qquad (3.17)$$

$$2(CH_3)_2CHBr \xrightarrow{Na} (CH_3)_2CH-CH(CH_3)_2 \qquad (3.18)$$

The preferred method of making organosodium compounds is by using a strong base, $B^-$, to remove a proton from a C—H bond.

$$RCH_3 + Na^+B^- \longrightarrow RCH_2^- + Na^+ + BH \qquad (3.19)$$

However, this requires the reactant to be weakly acidic, which is not the case for most hydrocarbons! But, as Table 3.1 (p. 98) shows, there is one important exception, alkynes.

A hydrogen atom attached to a triply bonded carbon atom, as in ethyne, or indeed any *terminal alkyne* (RC≡CH), shows very slight acidic properties.

What is the hybridization of the carbon atoms in ethyne, H—C≡C—H?

They are each bonded to two other atoms and so are sp hybridized.

In general, a hydrogen atom attached to an sp-hybridized carbon, as in an alkyne, is more acidic ($K_a = 10^{-25}$) than a hydrogen attached to an $sp^2$-hybridized carbon atom, as in an alkene ($K_a = 10^{-44}$), which in turn is more acidic than a hydrogen attached to an $sp^3$-hybridized carbon as in an alkane ($K_a = 10^{-50}$). Based on these values, an alkyne is $10^{25}$ times more acidic than a hydrocarbon, but $10^9$ less acidic than water!

## BOX 3.2 Charles Adolph Wurtz

Charles Adolph Wurtz was born in Wolfsheim in Alsace in 1817. He started studying medicine but changed to chemistry to study with Justus von Liebig in Giessen. He later moved to Paris, becoming assistant to J. B. Dumas and then succeeding him as Professor of Chemistry at the Sorbonne (1853). He discovered simple amines, and developed a reaction for making hydrocarbons by treating haloalkanes with sodium, which became known as the *Wurtz reaction* (for example, Reactions 3.17 and 3.18).

**Figure 3.2** Charles Adolph Wurtz 1817–1884.

**Table 3.1** $K_a$ values for some organic acids ($R{-}H \rightleftharpoons R^- + H^+$); remember, the smaller the value of $K_a$, the weaker the acid

| Compound, R—H | Anion, $R^-$ | $K_a$/mol dm$^{-3}$ |
|---|---|---|
| $CH_3COOH$ | $CH_3COO^-$ | $10^{-5}$ |
| $H_2O$ | $HO^-$ | $10^{-16}$ |
| $RC{\equiv}CH$ | $RC{\equiv}C^-$ | $10^{-25}$ |
| $NH_3$ | $NH_2^-$ | $10^{-38}$ |
| $CH_2{=}CH_2$ | $CH_2{=}CH^-$ | $10^{-44}$ |
| $CH_3{-}CH_3$ | $CH_3{-}CH_2^-$ | $10^{-50}$ |

Alkynes are sufficiently acidic for them to react with a strong base such as sodamide, $NaNH_2$ (which is the sodium salt of the weak acid ammonia, $H{-}NH_2$), to yield ammonia and the corresponding sodium **alkynide**:

$$RC{\equiv}CH + Na^{+}\,{}^{-}NH_2 \longrightarrow RC{\equiv}C^{-}Na^{+} + NH_3 \qquad (3.20)$$

Note that we still need a strong base to remove the proton; sodium hydroxide cannot be used since it is not sufficiently basic. As evidence of this, Table 3.1 shows that the acidities of water, ethyne and ammonia are in the order:

$$H_2O > RC{\equiv}CH > NH_3$$

Since $RC{\equiv}CH$ is more acidic than $NH_3$, $RC{\equiv}CH$ will protonate $NH_2^-$, leaving the alkynide ion, $RC{\equiv}C^-$, as shown in Reaction 3.20. Water is more acidic than $RC{\equiv}CH$, and so $H_2O$ will protonate $RC{\equiv}C^-$:

$$H_2O + RC{\equiv}C^- \longrightarrow RC{\equiv}CH + HO^- \qquad (3.21)$$

Since the equilibrium of this reaction lies well to the right we can't use the back reaction to form $RC{\equiv}C^-$.

We now turn to the reactions of these organosodium compounds. Unlike Grignard reagents and organolithiums, alkynylsodiums react with both haloalkanes and carbonyl compounds (remember in the equations below that $R^1$ is alkynyl not alkyl).

*Reaction with haloalkane*

$$Na^+(R^1)^- \; R^2CH_2{-}Br \longrightarrow R^1{-}CH_2{-}R^2 + Na^+Br^- \qquad (3.22)$$

*Reaction with carbonyl compounds*

$$Na^+(R^1)^- \; R^2R^3C{=}O \longrightarrow R^1{-}C(R^2)(R^3){-}O^-\,Na^+ \xrightarrow{H^+/H_2O} R^1{-}C(R^2)(R^3){-}OH \qquad (3.23)$$

Sodium alkynides are used in the synthesis of longer-chain alkynes; for example,

$$Na^+ + HC{\equiv}C^- \; CH_3CH_2{-}Br \longrightarrow HC{\equiv}CC_2H_5 + Na^+Br^- \qquad (3.24)$$

but-1-yne

$$Na^+ + C_2H_5C{\equiv}C^- \; CH_3{-}Br \longrightarrow C_2H_5C{\equiv}CCH_3 + Na^+Br^- \qquad (3.25)$$

pent-2-yne

This reaction involves substitution of alkynide ion for the bromide in the halo-alkane. It results from attack by the nucleophilic alkynide ion on an electrophilic carbon (a typical $S_N2$ reaction), and provides us with yet another example of an organometallic compound being used for carbon–carbon bond formation.

Sodium ethynide is a salt of the weak acid ethyne, and hence the ethynide ion is a strong base — stronger than hydroxide ion. Since hydroxide ion causes elimination from haloalkanes*, it is not surprising that the more basic ethynide ion can also cause elimination:

$$HC{\equiv}C^- + H{-}\overset{|}{\underset{|}{C}}{-}\overset{Br}{\underset{|}{C}}{-} \longrightarrow \rangle C{=}C\langle + HC{\equiv}CH + Br^- \quad (3.26)$$

In practice, only primary halides give good yields of the substitution product, namely the longer-chain alkyne. With secondary and tertiary haloalkanes, elimination predominates to such an extent that the method is essentially useless for extending the chain by a substitution reaction.

Now let's consider the second of our reaction types, Reaction 3.23. Ethynides also react with aldehydes and ketones to form alcohols. The reaction is analogous to those we have seen for Grignard reagents, except that now, instead of an alkyl or aryl nucleophile attacking the carbonyl functional group, we have an alkynide nucleophile:

$$R^3C{\equiv}C^- + \overset{R^1}{\underset{R^2}{C}}{=}O \longrightarrow R^3C{\equiv}C{-}\overset{R^1}{\underset{R^2}{C}}{-}O^- \xrightarrow{H^+/H_2O} R^3C{\equiv}C{-}\overset{R^1}{\underset{R^2}{C}}{-}OH \quad (3.27)$$

This reaction forms a new carbon–carbon bond and yields a bifunctional compound that can undergo further transformations, so it is useful in synthesis.

## 3.4 Summary of Section 3

1 An organolithium compound is prepared by the reaction of a haloalkane (or halobenzene) with lithium in dry ether, or a hydrocarbon solvent, in an inert atmosphere:

$$RX + 2Li \longrightarrow R^-Li^+ + Li^+X^- \quad (3.28)$$

2 Organolithium compounds are more reactive than Grignard reagents, and can be used as alternative reagents for carrying out the reactions discussed in Section 2. They are the reagents of choice for preparing sterically hindered tertiary alcohols from ketones, and for preparing ketones from carboxylic acids.

3 In contrast to Grignard reagents, organolithium compounds react with $CO_2$ to form ketones (only symmetrical ketones can be made by this method):

$$CO_2 + 2RLi \longrightarrow \longrightarrow \xrightarrow{2H_2O} \rightleftharpoons R{-}\overset{O}{\overset{\|}{C}}{-}R \quad (3.29)$$

* Elimination reactions are the subject of Part 3 of *Chemical Kinetics and Mechanism*[1].

4 Alkylsodium compounds are too reactive to be of much synthetic value, but sodium alkynides are used to achieve carbon–carbon bond formation with haloalkanes and aldehydes or ketones:

$$R^1C{\equiv}C^- \, Na^+ + R^2X \longrightarrow R^1C{\equiv}CR^2 + Na^+ X^- \qquad (3.30)$$

$$R^1C{\equiv}C^- \, Na^+ + \underset{R^3}{\overset{R^2}{\diagdown}}C{=}O \longrightarrow \xrightarrow{H_2O} R^1C{\equiv}C{-}\underset{R^3}{\overset{R^2}{\underset{|}{\overset{|}{C}}}}{-}OH + Na^+ \, {}^-OH \qquad (3.31)$$

5 A sodium alkynide is prepared by the action of the strong base $NaNH_2$, on an alkyne:

$$RC{\equiv}CH + Na^+ \, {}^-NH_2 \rightleftharpoons RC{\equiv}C^- Na^+ + NH_3 \qquad (3.32)$$

---

**EXERCISE 3.1 Summary of organometallic reactions 2**

Look back at the reactions of the organometallic compounds that were introduced in Section 3, and summarize them using general equations that highlight the other reactant, conditions and products. Compare your summary with that given in Part 2 Appendix B.

---

**QUESTION 3.1**

What is the product of the reaction between benzoic acid, $C_6H_5COOH$ (💻), and excess methyllithium, $CH_3Li$? Write out the mechanism for this reaction.

**QUESTION 3.2**

Explain why the alcohol **3.1** cannot be prepared from cyclohexanone, **3.2**, using a Grignard reagent, and suggest an organometallic reagent that could be used to make compound **3.1** from **3.2**.

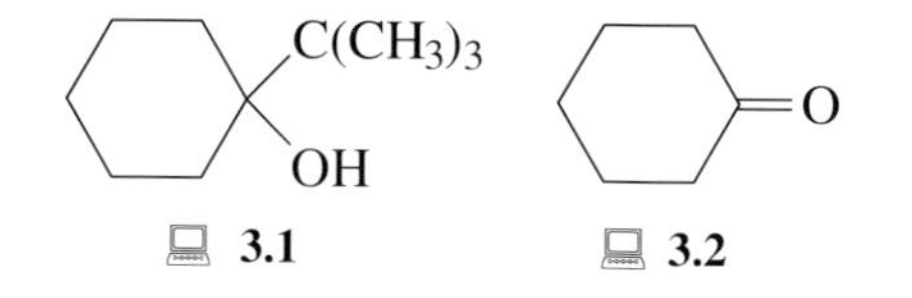


**QUESTION 3.3**

Using curly arrows, write out the mechanisms of the most likely reactions between: (i) $HC{\equiv}C^- Na^+$ and $(CH_3CH_2)_3CBr$; and (ii) $HC{\equiv}C^- Na^+$ and $CH_3CH_2CHO$.

# ORGANOCOPPER COMPOUNDS

4

## 4.1 Formation

In the previous Section, we outlined the formation of alkyl- and aryllithium reagents from haloalkanes and lithium metal in dry ether. Alkyl and aryllithium reagents are important as nucleophiles and bases in synthetic organic chemistry, as we have seen, but they are also valuable for preparing organocopper compounds called **lithium dialkylcopper reagents**. (Although we shall refer to them as 'di*alkyl*copper' compounds, you should assume the same pattern of reactivity applies to lithium di*aryl*copper reagents.)

- Write an equation for the formation of methyllithium from iodomethane.

- $$CH_3I + 2Li \xrightarrow{\text{dry ether}} CH_3Li + Li^+ I^- \qquad (4.1)$$

If the methyllithium formed is added to copper(I) iodide, lithium dimethylcopper is formed:

$$2CH_3Li + CuI \xrightarrow{\text{dry ether}} \underset{\text{lithium dimethylcopper}}{(CH_3)_2Cu^- \ Li^+} + Li^+ I^- \qquad (4.2)$$

You do not have to worry about the exact bonding in lithium dialkylcopper reagents (they will be represented simply in the form $R_2CuLi$ from hereon), because, like Grignard reagents, the precise molecular structure is complex. Another similarity with Grignard reagents is that lithium dialkylcopper compounds are too reactive to isolate, so they are used directly as the ether solution in which they are formed.

## 4.2 Reactions

- Is the R group nucleophilic or electrophilic in Grignard reagents (RMgBr) and organolithium reagents (RLi)?

- In all cases, it can be thought of as a nucleophile, $R^-$.

The $R^1$ groups in $(R^1)_2CuLi$ are also nucleophilic, and the most important reactions of lithium dialkylcopper reagents are with electrophiles of the type $R^2X$ and $R^2COX$ (X = Cl, Br or I), where the alkyl group is substituted for the halide X.

- Assuming that $(CH_3CH_2)_2CuLi$ is a source of nucleophilic $C_2H_5^-$, predict the product of its reaction with $C_6H_5CH_2Br$.

- In this case, the leaving group is $Br^-$, and the product is $C_6H_5CH_2CH_2CH_3$.

The full reaction sequence would be as follows:

$$2C_2H_5Br \xrightarrow{Li} 2C_2H_5Li \xrightarrow[-LiI]{CuI} (C_2H_5)_2CuLi \xrightarrow{C_6H_5CH_2Br} C_6H_5CH_2CH_2CH_3 \quad (4.3)$$

None of the intermediate products would be isolated.

Although the mechanism for such a reaction is not fully understood, you can think of it as a transfer of $C_2H_5$ with its pair of electrons from copper to the alkyl group of the haloalkane:

$$(C_2H_5)_2Cu^- Li^+ + C_6H_5CH_2{-}Br \longrightarrow C_2H_5CH_2C_6H_5 + C_2H_5Cu + Li^+Br^- \quad (4.4)$$

This type of reaction is important because it generates a new, larger carbon skeleton from two smaller ones. Overall, $R^1X$ and $R^2X$ have been joined together in a **coupling reaction** to give $R^1R^2$. In the example above, $R^1X$ is $C_2H_5Br$ and $R^2X$ is $C_6H_5CH_2Br$, so the product $R^1R^2$ is $C_2H_5CH_2C_6H_5$. Organocopper reagents are vital for this type of coupling, because Grignard and organolithium reagents do not generally react with haloalkanes. Sodium alkynides do undergo reaction with a haloalkane, but lithium dialkylcopper reagents provide another way of extending the carbon chain, in which the R group is not limited to alkynes.

Returning to the general scheme for preparing unsymmetrical alkanes, $R^1{-}R^2$, from organocopper reagents, the sequence of steps is:

$$2R^1Li + CuX \longrightarrow (R^1)_2CuLi + Li^+X^- \quad (4.5)$$

$$(R^1)_2CuLi + R^2X \longrightarrow R^1{-}R^2 + R^1Cu + Li^+X^- \quad (4.6)$$

This procedure is very versatile. The $R^2$ group can contain many functional groups that remain inert during the coupling reaction. For example, alkenes, ketones, carboxylate salts, tertiary amides and esters are not affected by the reaction conditions, so their presence in the $R^2$ group does not inhibit coupling reactions. For example, menthone (**4.2**), a terpene (a class of compound discussed in Part 5) found in peppermint (Figure 4.1), can be made from 2-bromo-5-methylcyclohexanone (**4.1**):

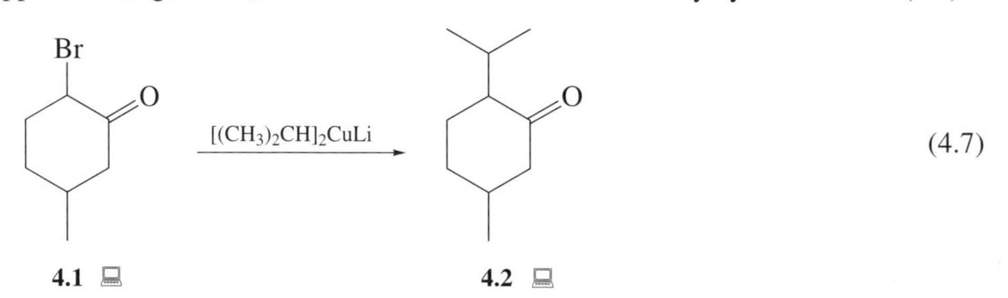

**Figure 4.1**
A peppermint plant.

At this point (if not before!), questions about the reactivity of organocopper and other organometallic reagents may have occurred to you. For example, why does $(R^1)_2CuLi$ react with $R^2X$, whereas $R^1Li$ does not? And, why does $R^1Li$ react with ketones, but $(R^1)_2CuLi$ does not?

Such differences in reactivity are difficult to explain. Certainly we do not expect you to tackle the explanations here. One key difference is that lithium and magnesium are main group elements, whereas copper is a transition element. For now, you should just accept that each type of organometallic reagent has its own characteristic reactivity profile, without being concerned about the reasons. To help you with these different reactivities (a knowledge of which is crucial for the application of organometallic compounds in synthesis), we shall review them in Section 5. For now, we shall complete our consideration of organocopper reagents.

- Muscalure is a sex hormone of the housefly, and even small amounts of it strongly attract houseflies. The efficiency of a fly-trap (Figure 4.2) is enhanced considerably by lacing it with muscalure. A key step in the commercial synthesis of muscalure relies on an organocopper coupling reaction:

$$[CH_3(CH_2)_4]_2CuLi + \underset{\mathbf{4.3}}{\overset{CH_3(CH_2)_6CH_2 \quad CH_2(CH_2)_7Br}{\underset{H \qquad H}{C=C}}} \longrightarrow \text{muscalure} \qquad (4.8)$$

What is the structure of the product, muscalure?

- See Structure **4.4** at the bottom of p. 104.

Note that neither the ketone group in menthone nor the double bond in the *cis*-bromoalkene **4.3** take part in Reactions 4.7 and 4.8, respectively.

**Figure 4.2**
A muscalure fly trap. Muscalure lures the fly into the trap, where it becomes attached to the sticky surface.

- Lithium dialkylcopper compounds also undergo substitution reactions with acid chlorides. Predict the product of the reaction between $C_6H_5COCl$ and $(C_3H_7)_2CuLi$.

- The organocopper reagent can be considered as a source of $C_3H_7^-$, which acts as a nucleophile towards $C_6H_5COCl$, in which $Cl^-$ is the leaving group. The final product is a ketone, because, once formed, $R_2CuLi$ is not reactive enough to attack a ketone:

$$C_6H_5COCl + (C_3H_7)_2CuLi \longrightarrow C_6H_5COC_3H_7 + C_3H_7Cu + Li^+ Cl^- \qquad (4.9)$$

## 4.3 Summary of Section 4

1 A lithium dialkyl (or diaryl, or divinyl) copper compound is prepared from a haloalkane, vinyl halide or halobenzene, via an organolithium compound:

$$RX + 2Li \longrightarrow RLi + Li^+X^- \qquad (4.10)$$

$$2RLi + CuX \longrightarrow R_2CuLi + Li^+X^- \qquad (4.11)$$

2 The alkyl or aryl group of the organocopper reagent acts as a carbon nucleophile, attacking an electrophilic carbon centre to form a new carbon–carbon bond.

3 Unsymmetrical alkanes can be prepared by the reaction of $(R^1)_2CuLi$ with haloalkanes:

$$(R^1)_2CuLi + R^2X \longrightarrow R^1{-}R^2 + R^1Cu + Li^+X^- \qquad (4.12)$$

The group $R^2$ may contain many functional groups that do not interfere with the reaction, for example

$$\mathrm{>C{=}C<} \qquad \mathrm{{-}C({=}O){-}} \qquad -COO^- \qquad -COOR^3 \qquad \mathrm{-C({=}O){-}N<}$$

4 Lithium dialkylcopper reagents can be used to make ketones from acid chlorides:

$$R^2COCl \xrightarrow{(R^1)_2CuLi} R^2COR^1 \qquad (4.13)$$

---

**EXERCISE 4.1 Summary of organometallic reactions 3**

Look back at the reactions of the lithium dialkylcopper reagents that were introduced in Section 4, and summarize them using general equations that highlight the other reactant, conditions and products. Compare your summary with that given in Part C of the Appendix.

---

**QUESTION 4.1**

Suggest how heptan-2-one, $CH_3CH_2CH_2CH_2CH_2COCH_3$ (🖳), can be prepared from $BrCH_2CH_2CH_2COO^-Li^+$, using two different organocopper reagents. (*Hint* Three reactions are needed, and the middle one does not involve an organometallic reagent, but converts a carboxylic acid into an acid chloride.)

---

*Answer to question on p.103*

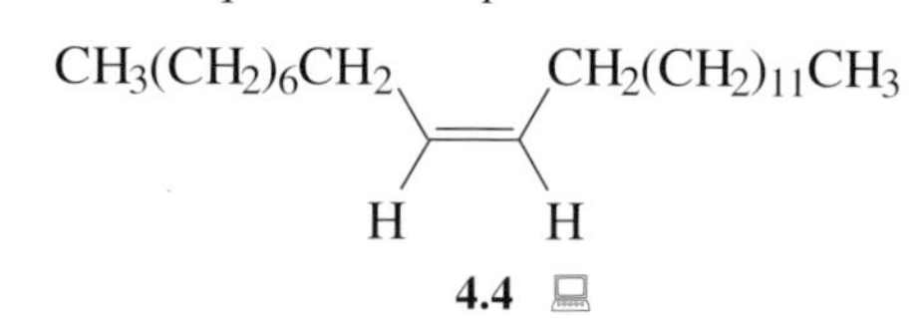

# REVIEW OF THE REACTIVITIES OF ORGANOMETALLIC REAGENTS

5

This short Section summarizes the reactivities of all of the organometallic compounds that you have met so far. It has already been noted that the various organometallic reagents react with a bewildering array of different electrophiles. We cannot easily explain the reactivity profiles of different organometallic compounds, but you must know which reagent reacts with which electrophile if you are to exploit these important reactions in synthesis. To aid you in this task, and to reduce some of the memory work needed, the reactivities of each organometallic reagent are summarized in Table 5.1 and Table 32.1 of the *Data Book* (available from the CD-ROM). Note, though, that the Table gives the reactivities of organometallic reagents, not the products of reactions. It is up to you to learn the outcome of each reaction. Before going on to the next Section, make sure you understand (and agree with!) the important information contained in this Table.

**Table 5.1** The reactivities of organometallic compounds with some classes of compound

| Class of compound | Organometallic reagent | | | |
|---|---|---|---|---|
| | RMgX[a] | RLi | RNa | $R_2CuLi$ |
| $R^1CHO$, $R^1R^2CO$ | ✔ | ✔ | ✔ | × |
| $CO_2$ | ✔ | ✔ | × | × |
| $R^1COO^-$ | × | ✔ | × | × |
| $R^1CN$ | ✔ | ✔ | × | × |
| $R^1COOR^2$ | ✔ | ✔ | × | × |
| $R^1COOCOR^2$ | ✔ | ✔ | × | × |
| $R^1COX$ | ✔ | ✔ | × | ✔ |
| $R^1CONR^2R^3$ | × | × | × | × |
| $R^1X$ | × | × | ✔ | ✔ |

[a] X = Cl, Br or I

✔ = reaction of synthetic utility.

× = no reaction, slow reaction, a reaction of little synthetic utility, or a reaction beyond the scope of this Book.

Let's now go on to the final class of organometallic reagent considered: organoboron compounds. They are the odd-one-out in this Book, because their primary role involves a functional group interconversion, rather than a C—C bond-forming reaction.

# 6 ORGANOBORON COMPOUNDS

In this final Section of Part 2 we turn our attention away from carbon–carbon bond formation, to outline the important role of organoboron chemistry in synthesis. Boron is in Group III of the Periodic Table, and its three valence electrons form three predominantly covalent bonds in compounds of the type $BY_3$; typical examples are boron trifluoride, $BF_3$ (🖫), boric acid, $B(OH)_3$ (🖫), and trimethylborane, $(CH_3)_3B$ (🖫). This last example is a trialkylboron compound, $R_3B$; and a feature of boron chemistry is that the simple alkyl and aryl derivatives of boron will adopt a structure involving just six bonding electrons around the boron atom (**6.1**).

**6.1**

This bonding arrangement makes boranes, $R_3B$, and especially the unsubstituted borane, $BH_3$ very reactive. $BH_3$ exists as the dimer $B_2H_6$, but its chemistry here can be satisfactorily explained by assuming that the species involved is '$BH_3$' *.

- The boron atom in $BH_3$ has fewer than the desired octet of valence electrons. Will $BH_3$ act as a nucleophile or an electrophile?
- The $BH_3$ molecule will tend to react with nucleophiles because they can supply a pair of electrons to the boron atom, thus completing its octet of electrons. $BH_3$ is therefore an electrophile.

In the **hydroboration reaction**, $BH_3$ acts as an electrophile, adding across the double bond of an alkene:

$$RCH{=}CH_2 + BH_3 \xrightarrow{\text{dry ether}} RCH_2CH_2BH_2 \qquad (6.1)$$

The reaction continues with two more molecules of alkene until the appropriate trialkylboron compound is formed:

$$RCH_2CH_2BH_2 + RCH{=}CH_2 \xrightarrow{\text{dry ether}} (RCH_2CH_2)_2BH \qquad (6.2)$$

$$(RCH_2CH_2)_2BH + RCH{=}CH_2 \xrightarrow{\text{dry ether}} (RCH_2CH_2)_3B \qquad (6.3)$$

- What do we mean by 'adds across the double bond'?
- Each of these three addition steps results in a hydrogen atom becoming attached to one of the carbon atoms of the carbon–carbon double bond, and the boron atom becoming attached to the other.

For the moment, let's just look at the first stage in the addition of $BH_3$ across a double bond. Let's view the mechanism of this electrophilic addition reaction as a standard two-step process†. As we have already implied in the reactions above,

* Boron hydrides are discussed in *Elements of the p Block*[2].

† The mechanisms of addition reactions are discussed in *Alkenes and Aromatics*[3].

the electrophilic boron atom adds preferentially to the less-substituted carbon atom of an unsymmetrical alkene:

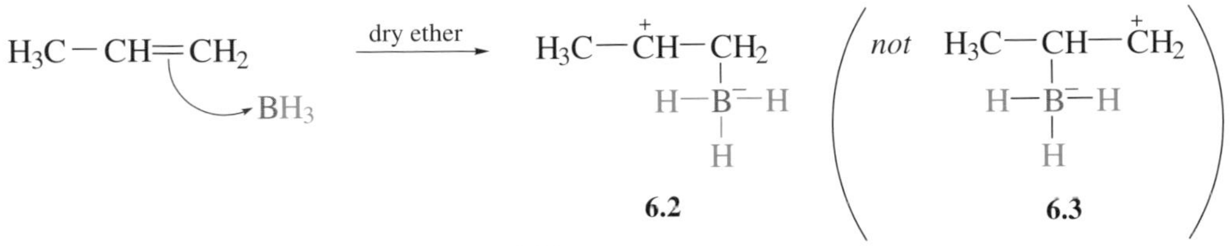

$$\mathrm{H_3C{-}CH{=}CH_2} + \mathrm{BH_3} \xrightarrow{\text{dry ether}} \mathrm{H_3C{-}\overset{+}{C}H{-}CH_2{-}\overset{-}{B}H_3}\ (\textbf{6.2}) \quad \left( not\ \mathrm{H_3C{-}CH(\overset{-}{B}H_3){-}\overset{+}{C}H_2}\ (\textbf{6.3}) \right) \qquad (6.4)$$

- Why is Structure **6.2** favoured over **6.3**?

- Remember the order of stability of carbocations: tertiary > secondary > primary. Structure **6.2** is favoured because it is a secondary carbocation, whereas Structure **6.3** is a primary carbocation.

It is therefore not surprising that the electrophilic boron atom should attack a carbon–carbon double bond in such a way that the positive charge can develop on the carbon atom that is best able to accommodate it. Thus, where possible, attack of the electrophilic boron on a carbon–carbon double bond will lead to:

> a tertiary carbocation in preference to a secondary or primary carbocation
> *or*
> a secondary carbocation in preference to a primary carbocation

This should be familiar to you as the 'mechanistic Markovnikov rule' of electrophilic addition to double bonds.

The first stage of the reaction is completed by transfer of hydride ion ($H^-$) from boron to carbon:

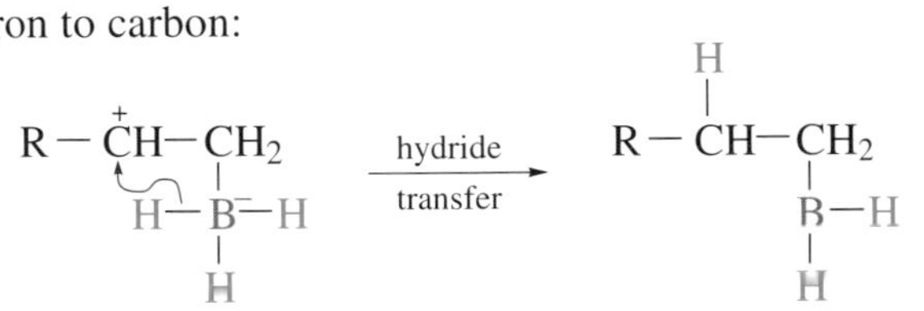

$$\mathrm{R{-}\overset{+}{C}H{-}CH_2{-}\overset{-}{B}H_3}\ (\textbf{6.2}) \xrightarrow{\text{hydride transfer}} \mathrm{R{-}CH(H){-}CH_2{-}BH_2} \quad (\text{that is, } \mathrm{RCH_2CH_2BH_2}) \qquad (6.5)$$

Further investigation has shown that a different mechanism is perhaps more likely, but the one outlined here successfully predicts the orientation of addition in terms with which you should be familiar. There's almost certainly more to the selective orientation of addition of $BH_3$ to a double bond than we have discussed here, but application of the mechanistic Markovnikov rule correctly predicts the products observed; for example

$$3\mathrm{CH_3CH{=}CH_2} + \mathrm{BH_3} \longrightarrow \mathrm{(CH_3CH_2CH_2)_3B} \quad (\text{not } [\mathrm{(CH_3)_2CH}]_3\mathrm{B}) \qquad (6.6)$$

The alkylboron compounds that are prepared in this way can be converted into other classes of organic compounds. In a second reaction, which is usually carried out in the reaction vessel in which the alkylboron compound was prepared, the boron atom is replaced by other functional groups as shown in Box 6.1 (p.109). For the moment we shall concentrate on the reaction with hydrogen peroxide ($H_2O_2$) in aqueous base (NaOH):

$$3\mathrm{RCH{=}CH_2} + \mathrm{BH_3} \longrightarrow \mathrm{(RCH_2CH_2)_3B}$$
$$\mathrm{(RCH_2CH_2)_3B} + 3\mathrm{H_2O_2} \longrightarrow 3\mathrm{RCH_2CH_2OH} + \mathrm{B(OH)_3}$$
$$\textit{Overall} \quad \mathrm{RCH{=}CH_2} \longrightarrow \mathrm{RCH_2CH_2OH}$$

SCHEME 6.1

Overall, the general reaction that takes place is the addition of H—OH across the double bond of an alkene. For completion, we will show a possible mechanism for this final step:

$$H_2O_2 + {}^{-}OH \rightleftharpoons HOO^{-} + H_2O$$

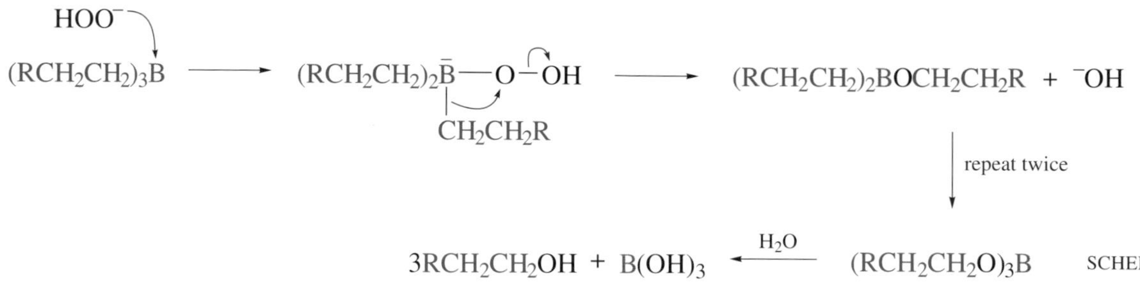

$$(RCH_2CH_2)_3B \xrightarrow{HOO^-} (RCH_2CH_2)_2\bar{B}(CH_2CH_2R){-}O{-}OH \longrightarrow (RCH_2CH_2)_2BOCH_2CH_2R + {}^{-}OH$$

$$\xrightarrow{\text{repeat twice}} (RCH_2CH_2O)_3B \xrightarrow{H_2O} 3RCH_2CH_2OH + B(OH)_3$$

SCHEME 6.2

What type of product is formed from the hydration of a terminal alkene?

The product is a primary alcohol:

$$3RCH{=}CH_2 + BH_3 \xrightarrow{\text{dry ether}} (RCH_2CH_2)_3B \xrightarrow[NaOH]{H_2O_2} 3RCH_2CH_2OH \quad (6.7)$$

What product would be formed if the hydration of the alkene had been carried out using aqueous acid rather than $BH_3$?

The major product would be the secondary alcohol:

$$R{-}CH{=}CH_2 \xrightarrow{H_2O/H^+} R{-}CH(OH){-}CH_3 \quad (6.8)$$

Whereas acid-catalysed hydration gives the Markovnikov product, formation of the alcohol by way of hydroboration apparently gives the anti-Markovnikov product. Until hydroboration was discovered, it was difficult to prepare primary alcohols from such alkenes. Now it is straightforward, and hydroboration is an invaluable part of the organic chemist's synthetic repertoire (see Box 6.1).

Remember that the reason we get this unusual orientation via hydroboration is because, essentially in the final step, the hydroxyl group simply replaces the boron atom in the trialkylboron compound. So the position of the OH group in the final product is the carbon atom to which the boron electrophile becomes attached — that is, the least-substituted carbon. In the acid-catalysed reaction, the electrophile still becomes attached to the least-substituted carbon atom, again on the grounds of carbocation stability, but here the electrophile is $H^+$. The nucleophilic water will then attack the carbocation, leaving the OH group on the site that accommodated the positive charge — that is, the most-substituted carbon.

In summary, the alcohols obtained from alkenes by hydration with aqueous acid and by the hydroboration route are different. The former is the Markovnikov product, and the latter, overall, is anti-Markovnikov. However, both results are readily explained by the 'mechanistic Markovnikov rule', which takes account of (i) the identity of the attacking electrophile in the first step, and (ii) the relative stability of the resulting carbocation.

## BOX 6.1 Versatile boron chemistry

The reaction of $BH_3$ with an alkene gives an organoborane reagent that can be reacted with a reagent X—Y to give a range of different compounds, where essentially the boron atom is replaced by the group Y. The whole sequence is:

$$3RCH{=}CH_2 + BH_3 \longrightarrow (RCH_2CH_2)_3B$$

$$(RCH_2CH_2)_3B + 3XY \longrightarrow 3RCH_2CH_2Y + BX_3$$

$$\textit{Overall} \quad RCH{=}CH_2 \longrightarrow RCH_2CH_2Y \qquad \text{SCHEME 6.3}$$

With hydration, XY is HO—OH. Reaction of the organoborane with ethanoic acid (Y = H, X = $CH_3COO$) gives:

$$RCH{=}CH_2 \longrightarrow RCH_2CH_2H \qquad (6.9)$$

which is effectively the addition of H—H across the double bond—that is, hydrogenation.

Reaction of the trialkylorganoborane with carbon monoxide, followed by treatment with hydrogen peroxide, gives a tertiary alcohol, where all three R groups from the organoborane end up attached to a carbon atom!

$$R_3B \xrightarrow[\text{(ii) } H_2O_2/NaOH]{\text{(i) CO}} R_3COH \qquad (6.10)$$

In fact this reaction has been developed further using the reagent 9-borabicyclo[3.3.1]nonane, 9-BBN (**6.4**), so that only one R group is transferred to the carbon monoxide to give primary alcohols and aldehydes.

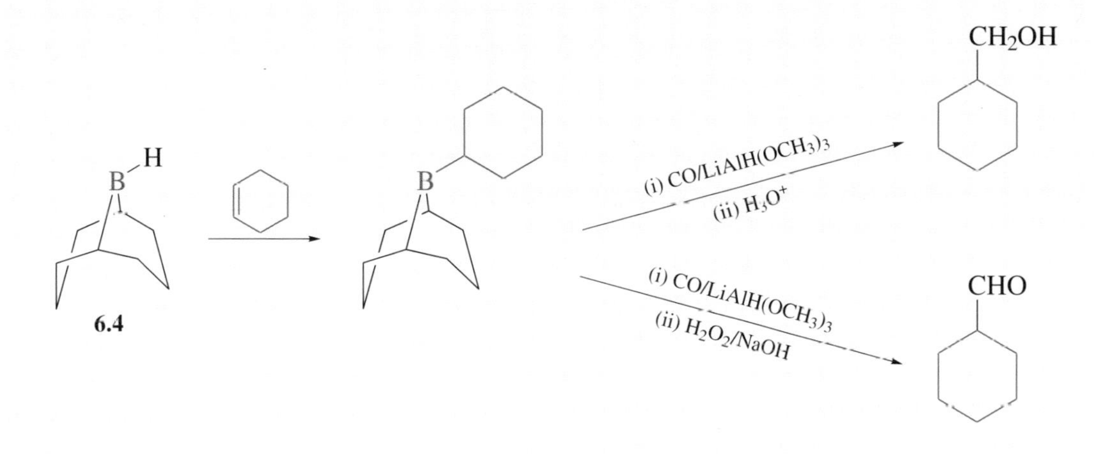


SCHEME 6.4

## 6.1 Summary of Section 6

1 Alkylboron compounds are important intermediate compounds in hydroboration, which provides a synthetic route for the anti-Markovnikov addition of HY to an alkene:

$$RCH{=}CH_2 \longrightarrow RCH_2CH_2Y \ (Y = OH) \qquad (6.11)$$

2 The preparation of primary alcohols is the most important hydroboration reaction, and the orientation of this addition results from initial attack on the alkene by electrophilic boron:

$$3RCH{=}CH_2 + BH_3 \xrightarrow{\text{dry ether}} (RCH_2CH_2)_3B \xrightarrow[\text{NaOH}]{H_2O_2} 3RCH_2CH_2OH \qquad (6.12)$$

### QUESTION 6.1

What are the major products that you would expect from the following reactions?

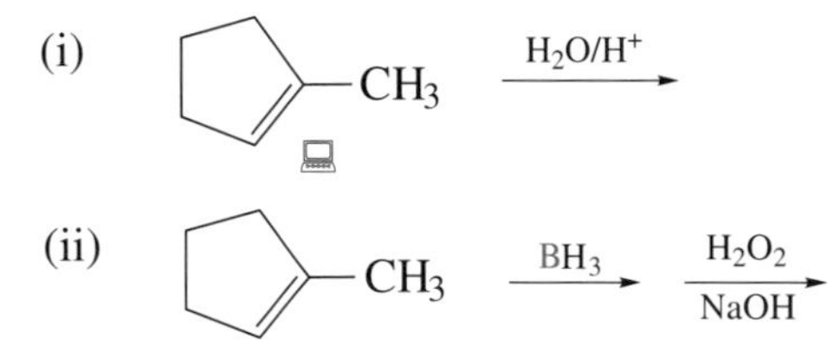


# EPILOGUE TO PART 2

As we stated at the outset, the treatment of organometallic chemistry given here is necessarily very selective, and concentrates only on those organometallic reagents that are of greatest current synthetic utility in *organic* chemistry. There are millions of different organometallic compounds, utilizing nearly all the metals in the Periodic Table, so we have only been able to scratch the surface. Even synthetic organic chemistry uses organometallic compounds based on a wide range of different metals. For example, in the next Part of this Book you will be introduced to some organotin compounds. Since organometallic compounds are important in synthesis and catalysis, this area will continue to be the focus of much research for many years to come.

# APPENDIX SUMMARY OF ORGANOMETALLIC REACTIONS USEFUL IN ORGANIC SYNTHESIS

Note that throughout this Appendix, X = Cl, Br or I.

## A Organomagnesium (Grignard) and organolithium reagents*

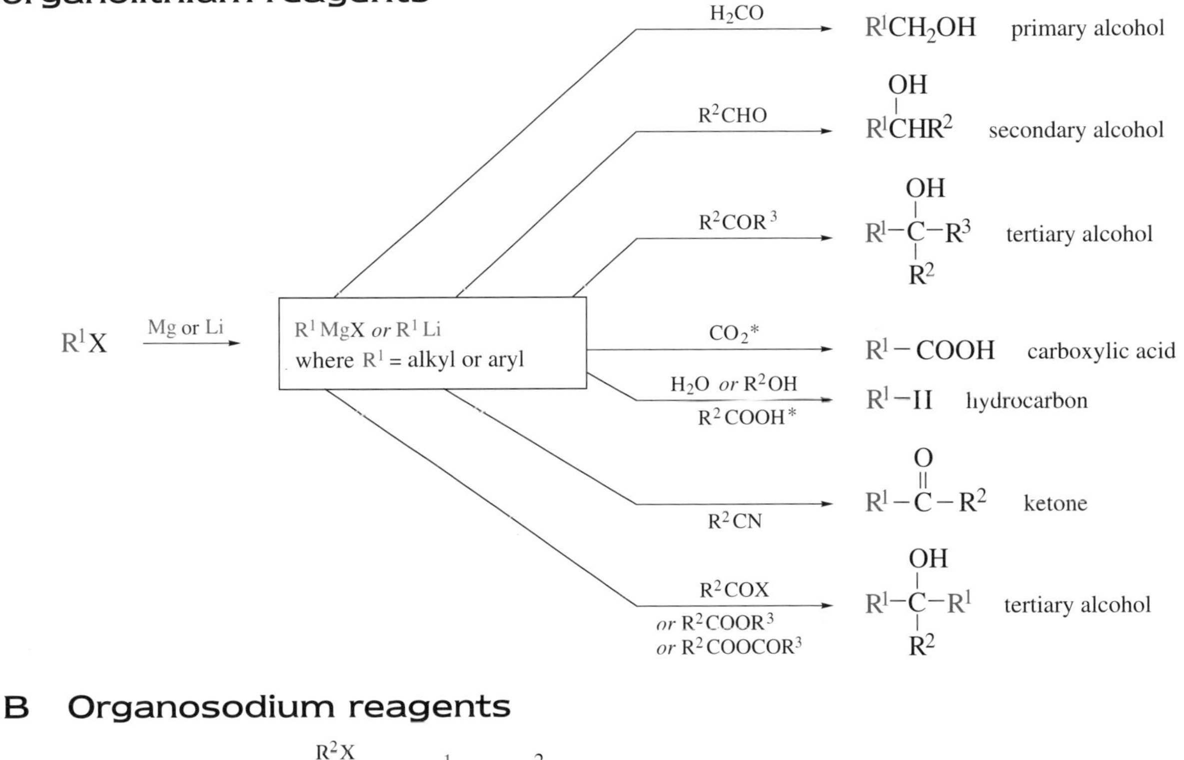


## B Organosodium reagents

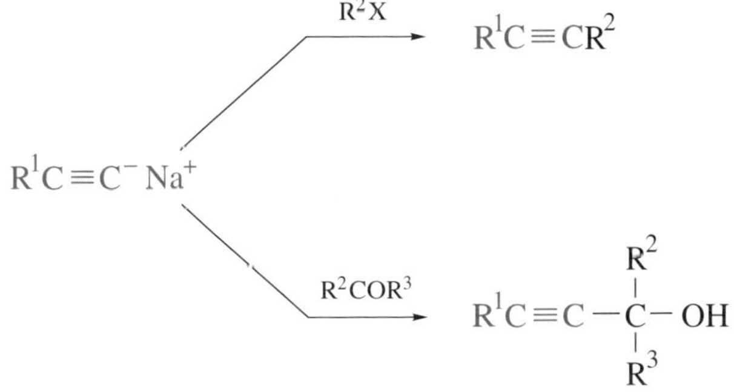


* Owing to the greater activity of organolithium reagents compared with Grignard reagents, further reaction occurs:

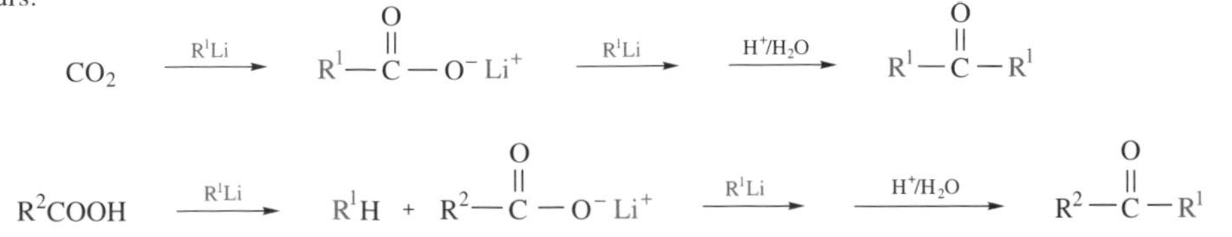

## C Organocopper reagents

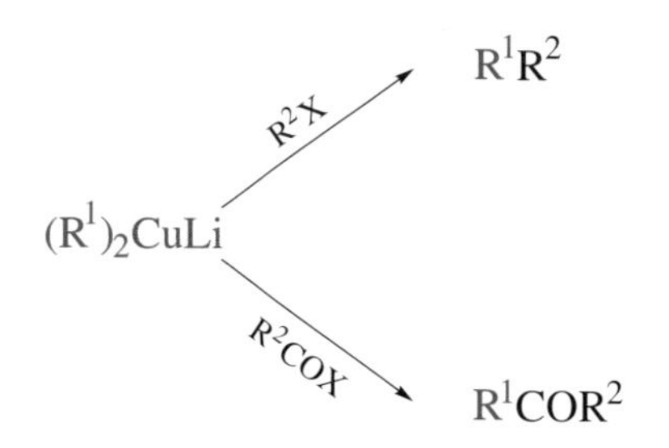


Functional groups that are unaffected by these organocopper reagents (and hence may be present in the R groups):

## D Hydroboration

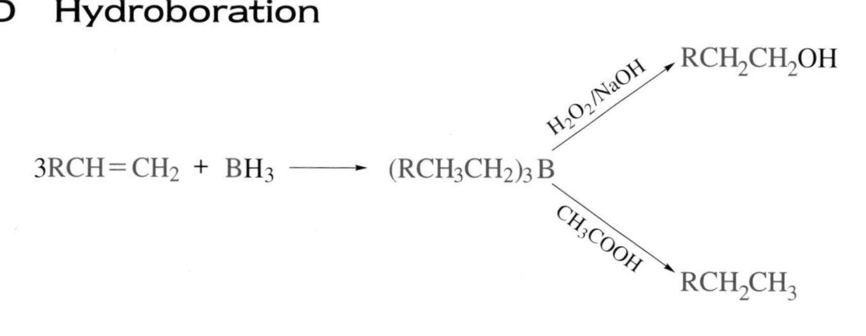

# LEARNING OUTCOMES

When you have completed *Mechanism and Synthesis: Part 2 — Synthetic applications of organometallic chemistry*, you should be able to:

1 Define in your own words, recognize valid definitions of, and use in a correct context, the terms, concepts and principles listed in the following Table. (All questions)

List of scientific terms, concepts and principles used in *Synthetic applications of organometallic chemistry*

| Term | Page | Term | Page |
|---|---|---|---|
| alkynide | 98 | hydroboration | 106 |
| carbanion | 80 | lithium dialkylcopper reagent | 101 |
| carbon–carbon bond-forming reaction | 77 | organolithium compound | 93 |
| coupling reaction | 102 | organosodium compound | 97 |
| Grignard reagent | 79 | organometallic compound | 77 |

2 Suggest suitable reactions for the preparation of organometallic compounds of Mg, Li, Na, Cu or B. (Questions 2.3 and 2.4)

3 Predict the product of the reaction of a Grignard reagent or an organolithium compound with an aldehyde, a ketone, carbon dioxide, a carboxylic acid derivative, a nitrile or a Brønsted acid. (Questions 2.2, 2.3 and 3.1)

4 Suggest ways of making an alcohol, a ketone or a carboxylic acid using a Grignard reagent, an organolithium compound or an organocopper compound. (Questions 2.1, 2.3, 3.2 and 4.1)

5 Use curly arrows to illustrate the mechanism of the reaction of a Grignard reagent or an organolithium compound with an aldehyde, a ketone, carbon dioxide, a carboxylic acid derivative, a nitrile or a Brønsted acid. (Questions 2.2 and 3.1)

6 Predict the product of, and use curly arrows to depict the mechanism of, the reaction between a sodium alkynide with a haloalkane, aldehyde or ketone. (Question 3.3)

7 Predict the product of a coupling reaction between a lithium dialkylcopper compound and a haloalkane. (Question 4.1)

8 Predict the compounds that can be formed from alkenes using organoboron intermediates, and give an explanation of the orientation of the addition products that are obtained in this way. (Question 6.1)

# QUESTIONS: ANSWERS AND COMMENTS

## QUESTION 2.1 (*Learning Outcome 4*)

(i) A carboxylic acid can be made from $CO_2$ and a Grignard reagent with one fewer carbon atom than the required acid:

$$(CH_3)_2CHCH_2MgBr + CO_2 \longrightarrow \xrightarrow{H^+/H_2O} (CH_3)_2CHCH_2COOH \quad (Q.1)$$

(ii) Grignard reagents react with aldehydes to give secondary alcohols. Thus, the reaction of methylmagnesium iodide with ethanal (acetaldehyde) is one possibility:

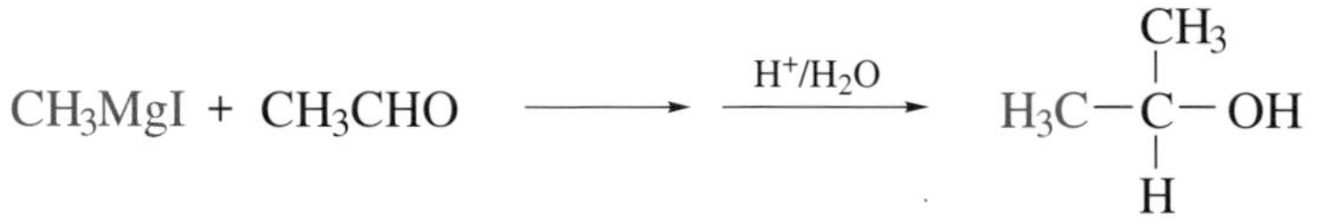

$$CH_3MgI + CH_3CHO \longrightarrow \xrightarrow{H^+/H_2O} H_3C{-}CH(CH_3){-}OH \quad (Q.2)$$

(It is also possible to use two moles of methylmagnesium iodide and an ester of methanoic acid (formic acid).)

(iii) Each of the three possible ways (Scheme Q.1) of making this tertiary alcohol involves the reaction of a Grignard reagent with a ketone. To determine the structures of any pair of these reagents, the procedure is to decide which one of the three alkyl substituents should come from the Grignard; the remaining two alkyl substituents must then come from the ketone.

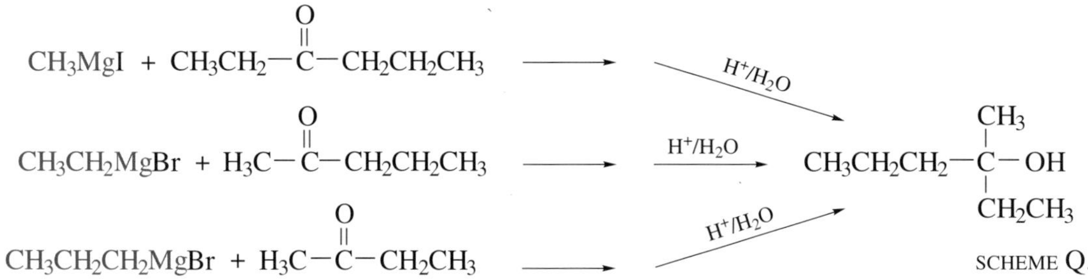

$$CH_3MgI + CH_3CH_2{-}C({=}O){-}CH_2CH_2CH_3 \longrightarrow \xrightarrow{H^+/H_2O}$$

$$CH_3CH_2MgBr + H_3C{-}C({=}O){-}CH_2CH_2CH_3 \longrightarrow \xrightarrow{H^+/H_2O} CH_3CH_2CH_2{-}C(CH_3)(CH_2CH_3){-}OH$$

$$CH_3CH_2CH_2MgBr + H_3C{-}C({=}O){-}CH_2CH_3 \longrightarrow \xrightarrow{H^+/H_2O}$$

SCHEME Q.1

The last of these reactions uses the most readily available materials.

(iv) The product here, benzyl alcohol, is a primary alcohol, and thus can be made by treating phenylmagnesium bromide with methanal (formaldehyde):

$$C_6H_5MgBr + H_2CO \longrightarrow \xrightarrow{H^+/H_2O} C_6H_5CH_2OH \quad (Q.3)$$

## QUESTION 2.2 (*Learning Outcomes 3 and 5*)

(a) Ethyl benzoate is a carboxylic acid derivative, RCOX, in which the X group corresponds to $OC_2H_5$. The mechanism is as follows:

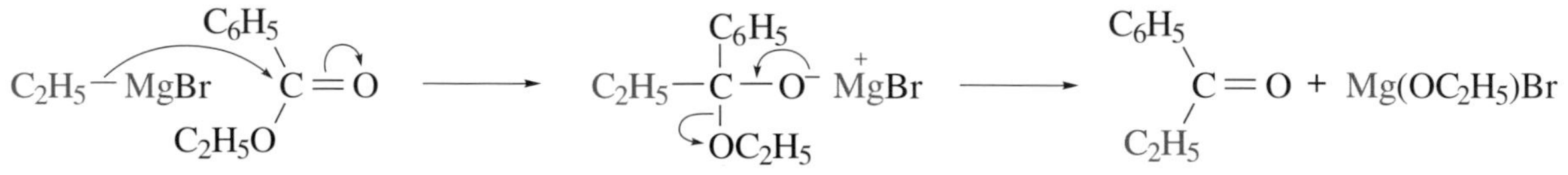

$$C_2H_5{-}MgBr + (C_6H_5)(C_2H_5O)C{=}O \longrightarrow C_2H_5{-}C(C_6H_5)(OC_2H_5){-}O^- \overset{+}{Mg}Br \longrightarrow (C_6H_5)(C_2H_5)C{=}O + Mg(OC_2H_5)Br \quad (Q.4)$$

ethyl benzoate

The ketone thus formed reacts with a second molecule of the Grignard reagent:

$$C_2H_5{-}MgBr + (C_6H_5)(C_2H_5)C{=}O \longrightarrow C_2H_5{-}C(C_6H_5)(C_2H_5){-}O^- \overset{+}{Mg}Br \quad (Q.5)$$

The resulting salt yields a tertiary alcohol on acid work-up:

$$\mathrm{C_2H_5{-}\underset{\displaystyle C_2H_5}{\overset{\displaystyle C_6H_5}{C}}{-}O^-\,\overset{+}{M}gBr \xrightarrow{H^+/H_2O} C_2H_5{-}\underset{\displaystyle C_2H_5}{\overset{\displaystyle C_6H_5}{C}}{-}OH + Mg^{2+} + Br^-} \qquad \text{(Q.6)}$$

3-phenylpentan-3-ol

(b) The corresponding product from excess Grignard reagent and ethyl methanoate is a secondary alcohol:

$$\mathrm{2C_2H_5\overset{+}{M}gBr + HCOOC_2H_5 \longrightarrow \xrightarrow{H^+/H_2O} C_2H_5{-}\underset{\displaystyle H}{\overset{\displaystyle C_2H_5}{C}}{-}OH} \qquad \text{(Q.7)}$$

pentan-3-ol

## QUESTION 2.3 (*Learning Outcomes 2 and 4*)

(i) First prepare a Grignard reagent from the 1-bromobutane:

$$\mathrm{C_4H_9Br + Mg \xrightarrow{dry\ ether} C_4H_9MgBr} \qquad \text{(Q.8)}$$

butylmagnesium bromide

Then treat the Grignard with methanal to yield the primary alcohol:

$$\mathrm{C_4H_9MgBr + H_2CO \longrightarrow \xrightarrow{H^+/H_2O} C_4H_9CH_2OH} \qquad \text{(Q.9)}$$

(ii) The product is a secondary alcohol, and can be obtained by treating the Grignard reagent from (i) with ethanal:

$$\mathrm{C_4H_9MgBr + CH_3CHO \longrightarrow \xrightarrow{H^+/H_2O} C_4H_9{-}\underset{\displaystyle H}{\overset{\displaystyle OH}{C}}{-}CH_3} \qquad \text{(Q.10)}$$

(iii) The product is a ketone—which can be made by treating the Grignard reagent butylmagnesium bromide with a nitrile—ethanenitrile (acetonitrile) in this case:

$$\mathrm{C_4H_9MgBr + CH_3CN \longrightarrow \xrightarrow{H^+/H_2O} (C_4H_9)(H_3C)C{=}O} \qquad \text{(Q.11)}$$

(iv) The carboxylic acid product can be formed by treating the Grignard reagent with carbon dioxide:

$$\mathrm{C_4H_9MgBr + CO_2 \longrightarrow \xrightarrow{H^+/H_2O} C_4H_9COOH} \qquad \text{(Q.12)}$$

(v) The product here is a tertiary alcohol, so it can be prepared from a suitable ketone:

$$\mathrm{C_4H_9MgBr + (C_4H_9)(H_3C)C{=}O \longrightarrow \xrightarrow{H^+/H_2O} C_4H_9{-}\underset{\displaystyle CH_3}{\overset{\displaystyle OH}{C}}{-}C_4H_9} \qquad \text{(Q.13)}$$

If you noticed that two of the alkyl substituents are identical, then you could have suggested an alternative route, namely the reaction of an ethanoate (acetate) ester with two moles of the Grignard reagent:

$$\mathrm{2C_4H_9MgBr + (H_3C)(C_2H_5O)C{=}O \longrightarrow \xrightarrow{H^+/H_2O} C_4H_9{-}\underset{\displaystyle CH_3}{\overset{\displaystyle C_4H_9}{C}}{-}OH} \qquad \text{(Q.14)}$$

### QUESTION 2.4 *(Learning Outcomes 2 and 3)*

(a) The aldehyde must be dry because the presence of any moisture will destroy the Grignard reagent via a reaction in which the alkyl or aryl portion of the Grignard is converted into a hydrocarbon:

$$RMgX + H_2O \longrightarrow RH + Mg(OH)X \qquad (Q.15)$$

(b) The compound 2-bromoethanol, $BrCH_2CH_2OH$, contains two functional groups, and this is the key to the problem. The Grignard reagent is formed as follows:

$$HOCH_2CH_2Br + Mg \xrightarrow{\text{dry ether}} HOCH_2CH_2MgBr \qquad (Q.16)$$

Now the 2-bromoethanol can act as an acid and destroy the Grignard:

$$HOCH_2CH_2MgBr + HOCH_2CH_2Br \longrightarrow CH_3CH_2OH + Mg(OCH_2CH_2Br)Br \qquad (Q.17)$$

The second reaction is faster than the first, so as soon as a molecule of the Grignard reagent is formed, it reacts with the hydroxyl group in another molecule of 2-bromo-ethanol, to yield the undesired product $CH_3CH_2OH$.

### QUESTION 3.1 *(Learning Outcomes 3 and 5)*

From the information in Section 3, you would expect that reaction of this carboxylic acid with the organolithium reagent would yield a ketone, and a mechanism can be written as follows. The first step is analogous to the reaction of an acid with a Grignard reagent:

$$CH_3{-}Li + C_6H_5C(=O){-}O{-}H \longrightarrow CH_4 + C_6H_5C(=O){-}O^-\ Li^+ \qquad (Q.18)$$

Then a second molecule of methyllithium can react with the carboxylic acid salt:

$$CH_3{-}Li + C_6H_5C(=O){-}O^-\ Li^+ \longrightarrow H_3C{-}C(C_6H_5)(O^-\ Li^+){-}O^-\ Li^+ \qquad (Q.19)$$

Note the formation of the new carbon–carbon bond in this step. Acid hydrolysis of the resulting dilithium salt results in the formation of the ketone phenylethanone:

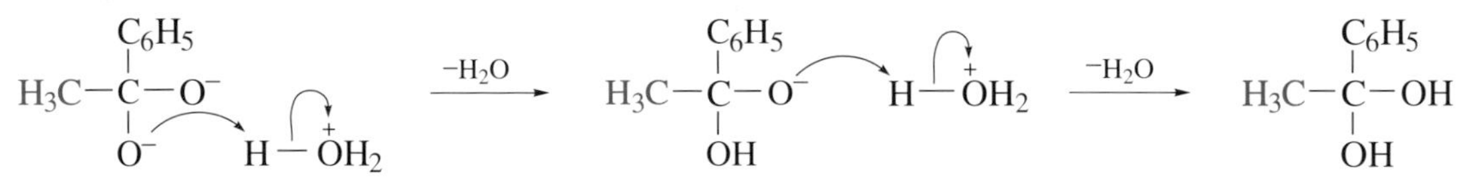

$$H_3C{-}C(C_6H_5)(OH){-}OH \underset{-H^-}{\overset{H^+}{\rightleftharpoons}} H_3C{-}C(C_6H_5)(\overset{+}{O}H_2){-}OH \xrightarrow[-H^+]{-H_2O} C_6H_5(H_3C)C{=}O$$

SCHEME Q.2

### QUESTION 3.2 (*Learning Outcome 4*)

The appropriate Grignard reagent, $(CH_3)_3CMgBr$, is bulky, and its approach to the carbonyl group of cyclohexanone (**3.2**) would be impeded by the C—H bonds in the ring; that is, reaction of cyclohexanone with 1,1-dimethylethylmagnesium bromide is sterically hindered. The corresponding organolithium compound is the reagent of choice:

$$\text{cyclohexanone } (\mathbf{3.2}) + (CH_3)_3CLi \longrightarrow \xrightarrow{H_2O} \text{1-}C(CH_3)_3\text{-1-OH-cyclohexane } (\mathbf{3.1}) \qquad (Q.20)$$

### QUESTION 3.3 (*Learning Outcome 6*)

(i) The haloalkane in this reaction is a tertiary compound, $(C_2H_5)_3CBr$. This means that it is highly susceptible to elimination in the presence of a base. The ethynide ion is therefore most likely to act as a base rather than as a nucleophile with this substrate:

$$HC{\equiv}C^- + H_3C{-}CH(H){-}C(Br)(C_2H_5){-}C_2H_5 \longrightarrow CH_3CH{=}C(C_2H_5)_2\ (\mathbf{Q.1}) + HC{\equiv}CH + Br^- \qquad (Q.21)$$

The alkene **Q.1** is likely to be the predominant product.

(ii) Here, the alkynide ion will act as a nucleophile to give **Q.2**:

$$HC{\equiv}C^- + H(C_2H_5)C{=}O \longrightarrow HC{\equiv}C{-}CH(C_2H_5){-}O^- \xrightarrow{H^+/H_2O} HC{\equiv}C{-}CH(C_2H_5){-}OH\ (\mathbf{Q.2}) \qquad (Q.22)$$

### QUESTION 4.1 (*Learning Outcomes 4 and 7*)

Probably the most likely synthesis of heptan-2-one from $BrCH_2CH_2CH_2COO^-Li^+$ using lithium dialkylcopper reagents is as follows:

$$BrCH_2CH_2CH_2COO^-\,Li^+ \xrightarrow[\text{(ii) } H^+/H_2O]{\text{(i) } (CH_3CH_2)_2CuLi} CH_3CH_2CH_2CH_2CH_2COOH$$

$$CH_3CH_2CH_2CH_2CH_2COOH \xrightarrow{SOCl_2} CH_3CH_2CH_2CH_2CH_2C({=}O)Cl \xrightarrow{(CH_3)_2CuLi} CH_3CH_2CH_2CH_2CH_2C({=}O)CH_3$$

SCHEME Q.3

In the first step, the carboxylate group is inert to the organocopper reagent, so it does not interfere with the coupling reaction that extends the chain by two carbon atoms. The acid group is then converted to the more-reactive acid chloride, which undergoes substitution with $(CH_3)_2CuLi$. The ketone product is inert to further attack by any remaining organocopper reagent.

## QUESTION 6.1 *(Learning Outcome 8)*

(i) In aqueous acid, Markovnikov addition of water occurs:

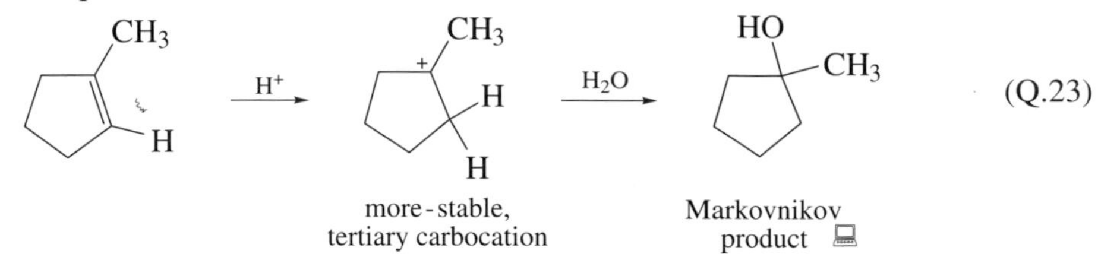


(Q.23)

(ii) This reaction exhibits overall anti-Markovnikov addition of water:

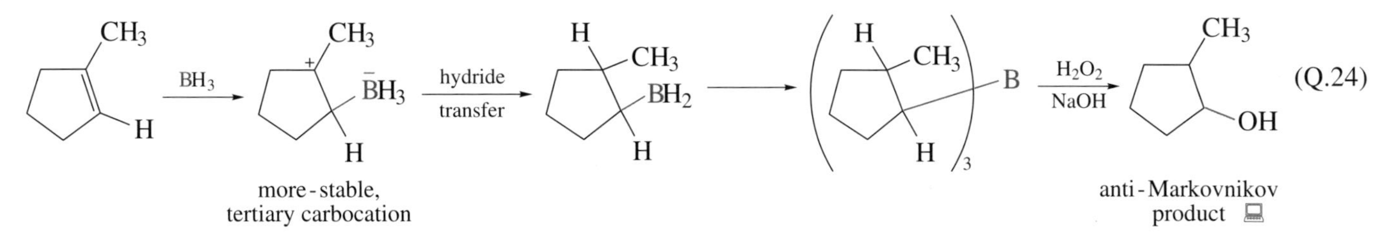

# FURTHER READING

1 M. Mortimer and P. G. Taylor (eds), *Chemical Kinetics and Mechanism*, The Open University and the Royal Society of Chemistry (2002).

2 C. J. Harding, D. A. Johnson and R. Janes (eds), *Elements of the p Block*, The Open University and the Royal Society of Chemistry (2002).

3 P. G. Taylor and J.M.F. Gagan (eds), *Alkenes and Aromatics*, The Open University and the Royal Society of Chemistry (2002).

# ACKNOWLEDGEMENTS

Grateful acknowledgement is made to the following sources for permission to reproduce material in this book:

*Figure 2.3*: © The Nobel Foundation; *Figure 3.2*: reproduced by courtesy of the Library and Information Centre, Royal Society of Chemistry; *Figure 4.1*: Oxford Scientific Films; *Figure 4.2*: Astrid and Hans Frieder-Michler/Science Photo Library.

Every effort has been made to trace all the copyright owners, but if any has been inadvertently overlooked, the publishers will be pleased to make the necessary arrangements at the first opportunity.

*Part 3*

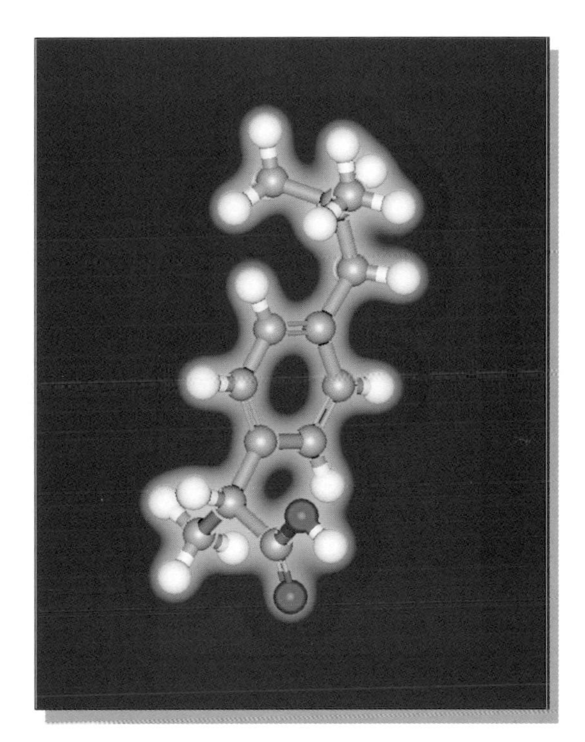

# Radical reactions in organic synthesis

Adrian Dobbs

1

# INTRODUCTION

All the organic reactions that we have studied so far in this Book have been *ionic reactions* — in other words, reactions in which covalent bonds are **cleaved** unevenly or **heterolytically** to produce *ions*. When a covalent bond is broken in such reactions, the pair of electrons moves towards one of the atoms, to give a negative ion (*anion*), leaving an electron-deficient (positively charged) atom at the other centre (*cation*). These ions may be involved in reactions as reactants, intermediates or products.

However, there is another large group of organic reactions which does not involve ions or any form of heterolytic bond cleavage. Instead of forming ions, they involve a process called **bond homolysis** (or **homolytic bond cleavage**). In this process, the two electrons in the covalent bond are split evenly between the two atoms that had previously formed the covalent bond: one electron goes to each. Instead of forming ions, the resulting highly reactive entities are called **radicals**, which are characterized by the fact that they possess *one unpaired electron*; that is, they usually have one electron fewer than the number required for a noble gas electronic configuration. This gives them a very different set of properties and reactions. To highlight these radical species we attach a dot to the symbol; for example, Cl•. In Part 3 of this Book, we are going to consider the chemistry of radicals.

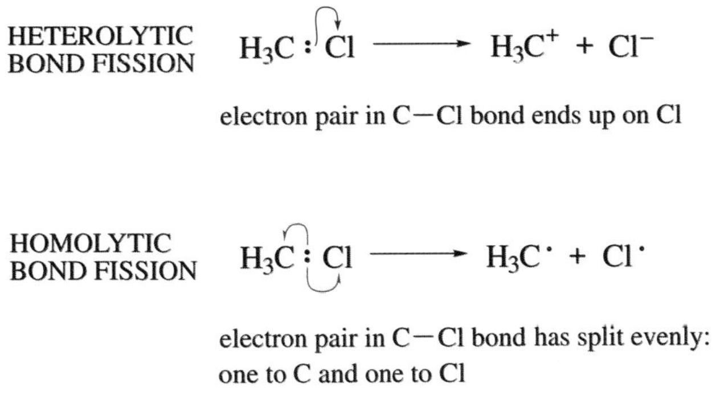

HETEROLYTIC BOND FISSION $H_3C:Cl \longrightarrow H_3C^+ + Cl^-$

electron pair in C—Cl bond ends up on Cl

HOMOLYTIC BOND FISSION $H_3C:Cl \longrightarrow H_3C^{\bullet} + Cl^{\bullet}$

electron pair in C—Cl bond has split evenly: one to C and one to Cl

SCHEME 1.1

One notable feature about Scheme 1.1 is the use of purple single-headed curly arrows. This departure from our previous use of red double-headed curly arrows is discussed in Box 1.1.

## BOX 1.1 Curly arrows in organic reaction mechanisms

In Parts 1 and 2 you have seen many examples of the use of a red 'curly arrow' to show the movement of a pair of electrons during an organic reaction. The electron pair (and thus the arrow) may be from an atom or bond, but always starts from somewhere that is electron rich and goes towards somewhere that is electron deficient. The type of arrow that we draw is very important. For ionic reactions, we made use of a *full* red arrow — that is, an arrow with two tips. This actually tells us something very important about the use of curly arrows, for the number of tips on the end of the arrow is an indication of the number of electrons that is deemed to be moving. Therefore, when we write a full red arrow, we show the movement of a *pair* of electrons.

In radical chemistry, however, we are by definition studying the movement of single electrons (a radical being a species with an unpaired electron). Therefore, we cannot use a full arrow for radical reaction mechanisms. Instead, we use a single-headed purple arrow to specify movement of a single electron. This type of curly arrow is often referred to as a **fishhook** curly arrow.

A 'full' arrow represents the movement of two electrons (that is, an electron pair), and should always be used in ionic reaction mechanisms; for example, $S_N2$

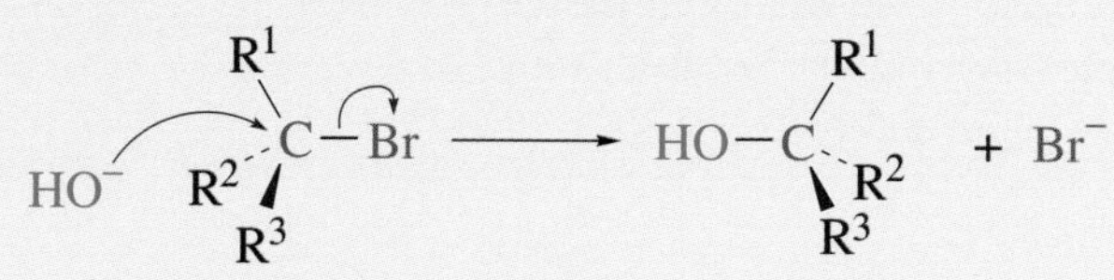


A 'single-tipped', or 'fishhook', arrow represents the movement of one electron (that is, a single electron), and should always be used in radical reaction mechanisms; for example, homolytic bond fission

$$H_3C : Cl \longrightarrow H_3C^{\bullet} + Cl^{\bullet}$$

## 1.1 Bond cleavage, bond dissociation energies and the resultant species

Energy must be supplied in order to break covalent bonds, so if we supply energy to a molecule, what happens? Obviously, one result of supplying energy above a certain threshold will be the cleavage of one, some, or all of the bonds in the molecule. But if bonds are being broken, what factors decide if they will break heterolytically to form ions or homolytically to form radicals?

Although it may be slightly surprising, if you supply enough energy, most bonds will dissociate *homolytically*. We can gain an idea of the likelihood of homolytic bond cleavage from the bond dissociation energy. Consider the atoms of molecular hydrogen or chlorine. When either of these atoms combine to form hydrogen or chlorine molecules, energy is given out: bond formation is an exothermic process. When hydrogen atoms combine, 436 kJ of heat is evolved for every mole of hydrogen molecules that is formed. In a similar way, when chlorine atoms combine, a smaller amount of heat is evolved, 243 kJ mol$^{-1}$:

$$H^{\bullet} + H^{\bullet} \longrightarrow H{-}H \quad \Delta H^{\ominus}_{m} = -436\ \text{kJ mol}^{-1} \tag{1.1}$$

$$Cl^{\bullet} + Cl^{\bullet} \longrightarrow Cl{-}Cl \quad \Delta H^{\ominus}_{m} = -243\ \text{kJ mol}^{-1} \tag{1.2}$$

Now when it comes to breaking these bonds, the energy required is exactly the same as the energy given out when they were formed. Thus, cleavage of the chlorine–chlorine bond is an endothermic process and requires 243 kJ of energy for the homolytic cleavage of one mole of chlorine molecules:

$$H{-}H \longrightarrow H^{\bullet} + H^{\bullet} \quad \Delta H^{\ominus}_{m} = +436\ \text{kJ mol}^{-1} \tag{1.3}$$

$$Cl{-}Cl \longrightarrow Cl^{\bullet} + Cl^{\bullet} \quad \Delta H^{\ominus}_{m} = +243\ \text{kJ mol}^{-1} \tag{1.4}$$

The energy required to break a covalent bond is termed the **homolytic bond dissociation energy**, *D*, and has been determined experimentally for many different bonds. Some of the most important bond dissociation energies of relevance to organic chemistry are given in Table 1.1.

**Table 1.1** Homolytic bond dissociation energies for various elements and covalent bonds: $A{-}B \longrightarrow A^{\bullet} + B^{\bullet}$

| Bond broken | $D^a$/kJ mol$^{-1}$ | Bond broken | $D^a$/kJ mol$^{-1}$ |
|---|---|---|---|
| H–H | 436 | H–OH | 498 |
| O=O | 498 | HO–OH | 144 |
| F–F | 158 | $CH_3$–OH | 383 |
| Cl–Cl | 243 | $CH_3CH_2CH_2$–H | 410 |
| Br–Br | 193 | $H_3C$–CH(–H)–$CH_3$ | 395 |
| I–I | 151 | $(CH_3)_3C$–H | 381 |
| H–F | 568 | $CH_3$–F | 452 |
| H–Cl | 432 | $CH_3$–Cl | 349 |
| H–Br | 366 | $CH_3$–Br | 293 |
| Sn–H | 264 | $CH_3$–I | 234 |
| $CH_3CH_2$–$Sn(CH_3CH_2)_3$ | 239 | $CH_3$–$CH_3$ | 368 |
| $(CH_3CH_2)_3Sn$–Cl | 414 | $CH_3CH_2$–$CH_3$ | 356 |
| Sn–Br | 339 | $PhCH_2O$–$OCH_2Ph$ | 139 |
| Sn–I | 234 | NC–C(CH$_3$)$_2$–N=N–C(CH$_3$)$_2$–CN | 131$^b$ |

$^a$ *D* = homolytic bond dissociation energy.

$^b$ This is not a true homolytic bond dissociation energy since it takes into account the formation of N≡N (see text).

Homolytic bond dissociation energies have found many uses, such as in the estimation of energy changes during reactions. They can also give us important information on the relative stabilities of organic compounds and, in particular, of radicals.

- From Table 1.1, put the following bonds in order of strength: primary C—H, secondary C—H, tertiary C—H

- Table 1.1 shows that:

$$\underset{\text{primary}}{CH_3CH_2CH_2{-}H} \longrightarrow CH_3CH_2CH_2^{\bullet} + H^{\bullet} \qquad D = 410\,\text{kJ mol}^{-1} \quad (1.5)$$

$$\underset{\text{secondary}}{(CH_3)_2CH{-}H} \longrightarrow (CH_3)_2\dot{C}H + H^{\bullet} \qquad D = 395\,\text{kJ mol}^{-1} \quad (1.6)$$

$$\underset{\text{tertiary}}{(CH_3)_3C{-}H} \longrightarrow (CH_3)_3C^{\bullet} + H^{\bullet} \qquad D = 381\,\text{kJ mol}^{-1} \quad (1.7)$$

In other words, a tertiary C—H bond is slightly weaker than a secondary C—H bond, which is in turn slightly weaker than a primary C—H bond. Thus, carbon–hydrogen bond strengths vary in the order

primary > secondary > tertiary

As a general rule, more energy is required to form a primary radical than a secondary or tertiary one. Consider the following two examples:

$$CH_3CH_2CH_2^{\bullet} + H^{\bullet} \xleftarrow[D\,=\,410\text{ kJ mol}^{-1}]{\text{primary}} H_3C{-}\underset{\displaystyle H}{\underset{|}{CH}}{-}CH_2{-}H \xrightarrow[D\,=\,395\text{ kJ mol}^{-1}]{\text{secondary}} (CH_3)_2\dot{C}H + H^{\bullet} \quad (1.8)$$

These two reactions are similar in that they both involve propane as the starting material and both involve hydrogen abstraction to give an alkyl radical and a hydrogen atom. However, in the propane molecule there are two different sites where it is possible to form a radical — at a primary centre and at a secondary one. These two reactions differ in the amount of energy required to form the radicals. As we have just seen, more energy will be required to produce the primary radical than the secondary one. The more energy that is required to form a radical, the less stable the resultant radical will be. Therefore, the primary radical that is formed in Reaction 1.8 (left) is less stable than the secondary one (right) since greater energy is required to form it. Similarly, since it is even easier to form a tertiary radical, this is by far the most stable radical of the three types. This gives us an overall sequence for radical formation and stability:

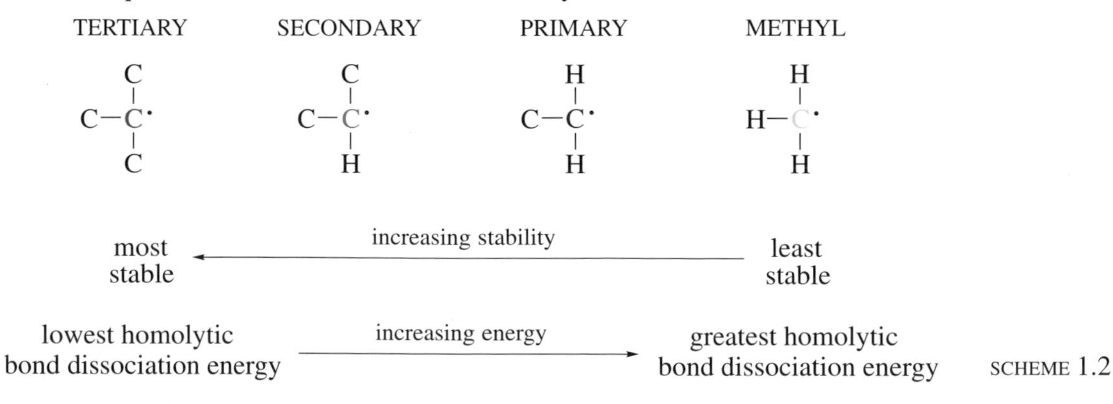


SCHEME 1.2

- Are there any obvious similarities or differences between radicals and carbocations?
- The order of stability for carbocations is identical to that of radicals.

- What reason for this can you suggest?
- The carbon bearing the odd electron in the radical is electron deficient: it is missing one electron, and so has similarities to the central carbon in a carbocation. Therefore, any group or atom that is electron releasing (such as an alkyl group) will help to stabilize a radical. The more electron-releasing substituents there are on a radical, the more stable it will be. Hence, tertiary radicals are the most stable, just like tertiary carbocations.

There is, however, an added complication when considering the stability of radicals, and that is that they may also be stabilized by an adjacent electron-withdrawing group.

- Give some examples of electron-withdrawing groups.
- The following are electron-withdrawing groups: —CN, —COR, —COOR, $-NO_2$.

These groups are able to stabilize the radical by delocalization of the unpaired electron through a $\pi$ system.

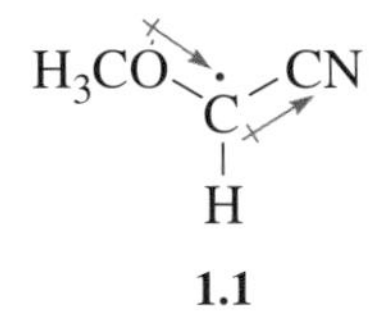


**1.1**

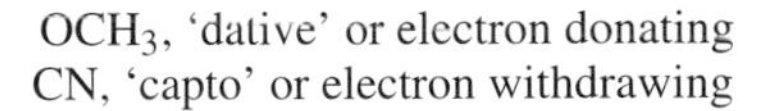

$OCH_3$, 'dative' or electron donating
CN, 'capto' or electron withdrawing

As we have seen, radicals are stabilized by having either electron-donating or electron-withdrawing groups attached to them. In fact, the ideal situation for a radical is to have one of each! Electron-withdrawing groups are sometimes termed captive groups and electron donating groups are termed dative groups. When a radical possesses one group of each type, it is often referred to as a **captodative radical** (for example, **1.1**) and is particularly stable.

### QUESTION 1.1

Suggest some examples of combinations of substituents that may generate captodative radicals.

A particularly common captodative radical is the 2-cyanopropyl radical, **1.2**. This is formed by the homolytic bond cleavage of a molecule called **α,α′-azo-*bis*-*iso*-butyronitrile or AIBN**. This is an azo compound; in other words, it contains a nitrogen–nitrogen double bond. The carbon–nitrogen bonds in this molecule are very weak, since there is a large, and favourable, thermodynamic driving force for these bonds to cleave homolytically. The consequence is liberation of nitrogen gas and formation of the highly stabilized 2-cyanopropyl radical:

$$\text{AIBN} \longrightarrow 2\ (H_3C)_2\dot{C}CN\ (\mathbf{1.2}) + N{\equiv}N \qquad (1.9)$$

Since this reaction occurs so readily and, as Table 1.1 shows, requires considerably less energy input to form the radical (compared with the formation of a normal alkyl radical from a C—H bond, for example), compounds such as AIBN are often used as **initiators** to 'start off' radical reactions, as we shall see later.

The other principal class of compounds to be employed to start off radical reactions are peroxides, and here the most common one is dibenzoyl peroxide, **1.3**:

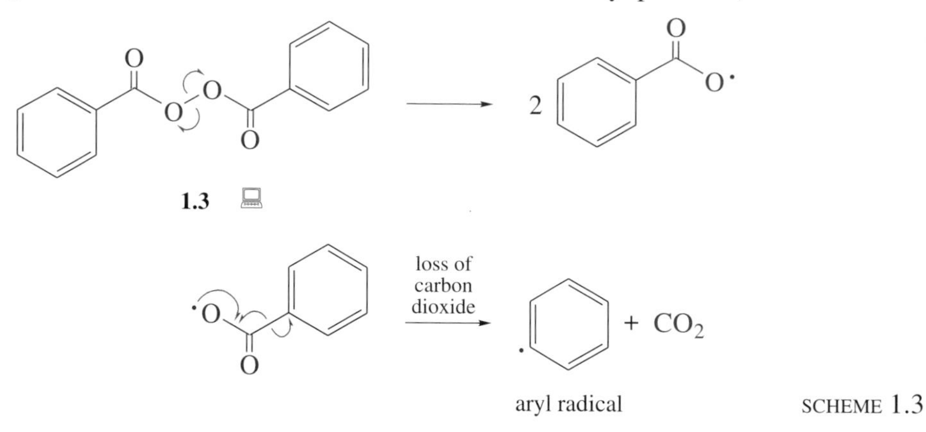


SCHEME 1.3

With compounds such as benzoyl peroxide, homolytic bond cleavage of the weak O—O bond occurs first, and this is followed by the thermodynamically favoured loss of the stable molecule carbon dioxide to generate an aryl radical, which then starts the radical reaction.

The importance of such compounds to start off radical reactions will become clear in due course. However, the important point to remember about such compounds at this stage is that they all possess particularly weak bonds which are susceptible to homolytic bond cleavage.

## 1.2 Radicals and charges

The final question that we must address in this introductory Section is why, if a radical has an unpaired electron, is it not charged? The most convenient way to answer this question is by performing an electron count on the carbon atom (Figure 1.1). The simplest example is chloromethane, in which the central carbon atom possesses two 1s electrons and a half share of eight valence electrons, which are shared with three hydrogen atoms and one chlorine atom. This gives a total of six electrons (two (from the C 1s orbital) + four (from a half share of eight)), and, since carbon has six protons, the atom is neutral.

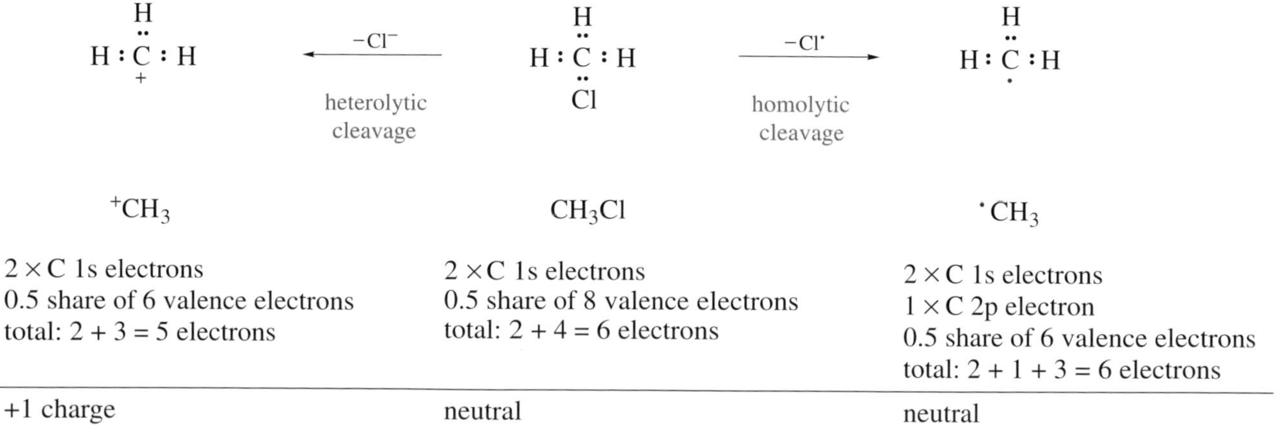


**Figure 1.1** Proof that a radical is not charged.

When heterolytic bond cleavage occurs to form a carbocation, the carbon still has its two 1s electrons, but instead of having a half share of eight valence electrons, now it only has a half share of six valence electrons (that is, three electrons). This gives the carbon a total of five electrons, and because there are still six protons, an overall charge of +1.

Now, consider the homolytic bond cleavage. Carbon still has its two 1s electrons and a half share of six valence electrons. However, it also now has one lone electron all to itself (the unpaired electron). Therefore, the radical has two (from 1s) + three (from half share of six electrons) + one (unpaired electron), giving a total of six electrons, and therefore the atom is again neutral.

A similar electron count may be performed for all carbon radicals, and the net result is always the same: all carbon radicals are electrically neutral. As they are not charged, this will obviously have a profound influence on their reactions and behaviour, as you will learn as you study Part 3. Much of the early work on organic radicals was done by Moses Gomberg (Box 1.2, p. 130).

## 1.3 Summary of Section 1

1 Radicals are species that possess an unpaired electron.

2 Full (double-headed) curly arrows are used to show the movement of pairs of electrons in reaction mechanisms. Single-headed (or fishhook) arrows are used to show the movement of single electrons.

3 Bond dissociation energies give an indication of the strength of various covalent bonds.

4 A primary carbon–hydrogen bond is stronger than a secondary, which in turn is stronger than a tertiary bond.

5 A tertiary carbon radical is more stable than a secondary one, which in turn is more stable than a primary or methyl radical.

6 Both electron-withdrawing and electron-donating groups stabilize radicals; a combination of both confers extra stabilization to a radical, known as a captodative radical.

7 Initiators are compounds with particularly weak bonds that readily dissociate to form stable radicals, and so start off a radical reaction or process.

8 Carbon radicals are electrically neutral but very reactive species.

### QUESTION 1.2

Which of the following are stabilized or captodative radicals?

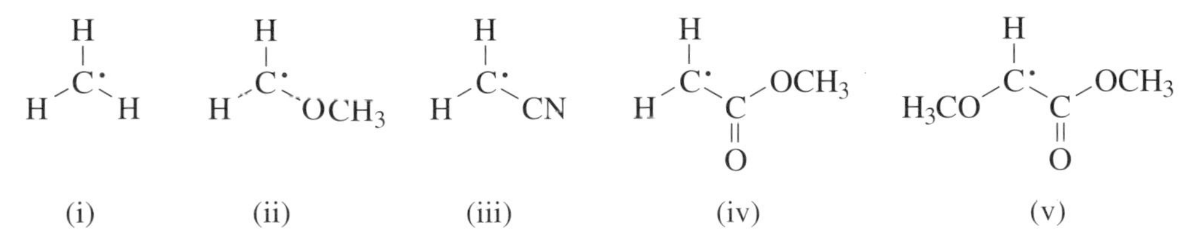


### QUESTION 1.3

Use the representative values in Table 1.1 to suggest an order of bond strengths for the following molecules:

H—H, Br—Br, F—F, H—Br, O=O, H—Cl

## BOX 1.2 Moses Gomberg

Moses Gomberg (Figure 1.2) was born on 8 February 1866 in the small town of Elizabetgrad, Russia. His father, George was accused of anti-Tsarist activities, and in 1884 was forced to flee with his family to Chicago in the USA. After a period of great hardship Moses entered the University of Michigan at Ann Arbor and graduated with a B.Sc in 1890; two years later he received an M.Sc. He obtained a PhD in 1894 under the supervision of Professor A. B. Prescott. Gomberg then travelled, spending 1896–7 in A. von Baeyer's laboratory in Germany, and then a further term in Heidelberg working with Victor Meyer, where he was the first chemist to prepare tetraphenylmethane successfully.

**Figure 1.2**
Moses Gomberg, 1866–1947.

On his return to Michigan, Gomberg continued in this field of work and turned his attention to the synthesis of hexaphenylethane. He attempted to do this by the reaction of triphenylmethyl chloride with zinc. To his surprise, however, he obtained a bright yellow compound, which was highly reactive, reacting with oxygen to give triphenylmethyl peroxide and with halogens to give the triphenylmethyl halide (Scheme 1.4). From these results, Gomberg postulated the formation of a triphenylmethyl radical (rather than a particularly reactive hexaphenylethane molecule, Figure 1.3).

**515. M. Gomberg: Triphenylmethyl, ein Fall von dreiwerthigem Kohlenstoff.**

[Vorläufige Mittheilung.]

(Eingegangen am 1. October; mitgetheilt in der Sitzung vom 8. October von Hrn. R. Stelzner.)

**Figure 1.3** The title heading to Gomberg's first paper on radical formation. ['Triphenylmethyl, a case of trivalent carbon' (preliminary announcement). Received on 1 October; announced at the meeting on 8 October by Mr R. Stelzner.]

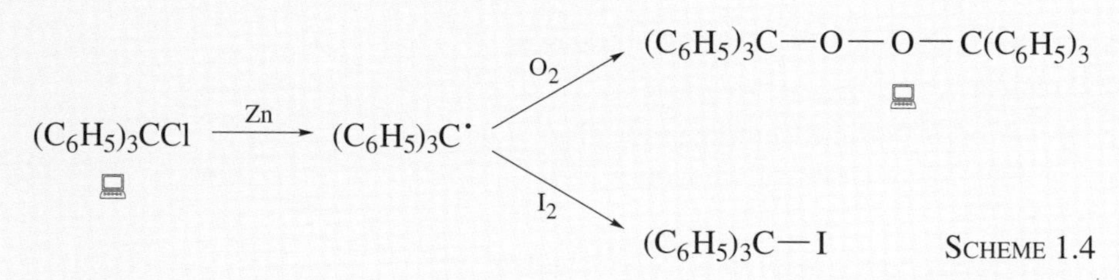


SCHEME 1.4

Gomberg concluded his original paper saying 'I wish to reserve this field of study for myself'. How little did he suspect what a difference to modern chemistry his discovery would make!

Gomberg retired from chemistry in 1936 and died on 12 February 1947 at Ann Arbor.

The year 2000 marked the centenary of the discovery and characterization of the first radical, which was commemorated in a memorial stone (Figure 1.4).

**Figure 1.4** The memorial stone to Gomberg's discovery of radicals at the University of Michigan.

### QUESTION 1.4

Use the representative values in Table 1.1 to suggest an order of bond strengths for the highlighted bonds in the following compounds:

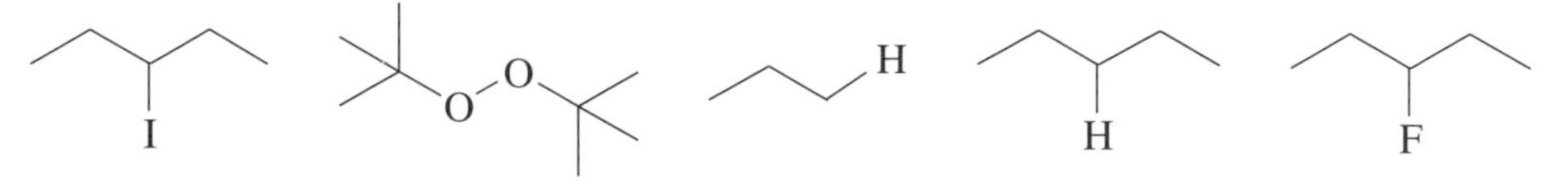

# REACTIONS OF RADICALS

2

So far, you have learnt something about the nature and generation of radicals. From this, it would be easy to imagine that it might be very difficult to control the chemical reactions of such reactive and relatively unstable species. Fortunately, from the chemist's point of view, this is not so. There are three general types of reaction available for a radical and we shall consider each in turn.

1 RADICAL – RADICAL COUPLING REACTION

radical + radical ⟶ non-radical (coupled product)

2 RADICAL CHAIN REACTIONS

radical X + non-radical reactant A ⟶ radical Y + non-radical product P

radical Y + non-radical reactant B ⟶ radical X + non-radical product Q

3 RADICAL FRAGMENTATION

radical ⟶ new radical + fragment molecule

SCHEME 2.1

# RADICAL–RADICAL COUPLING REACTIONS (RADICAL COMBINATIONS) 3

If there are a large number of radicals moving about very rapidly either in solution or in the gas phase, it is inevitable that two of them will eventually encounter each other and their electrons pair up. There is an energy release associated with this pairing of electrons. However, there are a number of reasons why this type of radical reaction is relatively uncommon, and these should become clear as we look at some examples of **radical–radical couplings** or dimerizations.

## 3.1 The pinacol coupling

The most famous example of a radical–radical coupling reaction is known as the **pinacol coupling reaction**. One of the methods for the generation of radicals is that of single electron transfer, most commonly from a metal to a neutral molecule. Treatment of a carbonyl compound with a metal in solution generates a **radical anion** that is commonly known as a **ketyl radical**, **3.1**:

$$R^1C(=O)R^2 \xrightarrow[\text{e.g. Na, Mg, Zn}]{\text{metal (M)}} R^1\dot{C}(O^-)R^2 + M^+ \qquad (3.1)$$

**3.1**

This involves a carbon-centred radical that is neutral and an oxygen anion.

Metals may often be considered as a source of electrons, because they readily release an electron. This electron then attacks the carbonyl group to form the ketyl radical. Notice the precise use of curly arrows in this mechanism (Box 1.1) — the purple fishhook for movement of single electrons, and the full red arrow for movement of an electron pair, such as in the carbon–oxygen double bond:

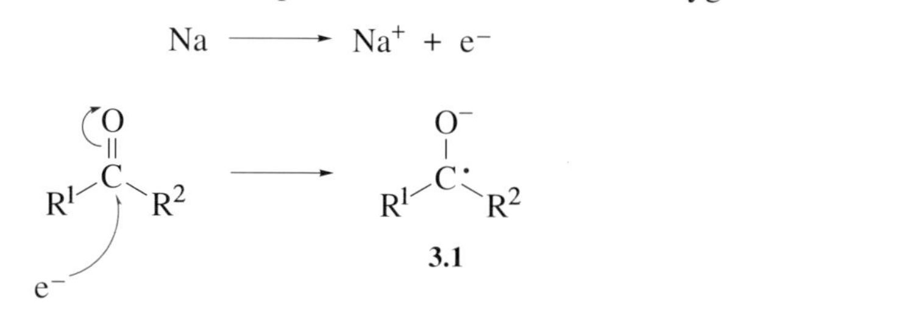


SCHEME 3.1

What happens next depends on the exact reaction conditions.

With a monovalent metal (for example sodium) in a **protic solvent** (that is, a solvent that can donate a hydrogen atom, such as ethanol), a metal alkoxide anion is first generated. On addition of an acid, this gives an alcohol; that is, we get simple reduction of the carbonyl compound.

If a divalent metal is used (for example, magnesium or zinc) in an **aprotic solvent** (that is, one that cannot donate a hydrogen atom, such as benzene or diethyl ether), there are no readily available hydrogens for the reduction to occur, so the concentration of the ketyl radicals increases. In addition, in order to satisfy its own valence requirements, the divalent metal will coordinate to two of these radicals, placing them in close proximity to one another, forming a *diradical*. So once the two ketyls have been bound to the metal the radicals combine (dimerize). Protonation during work-up then gives a 1,2 diol; for example

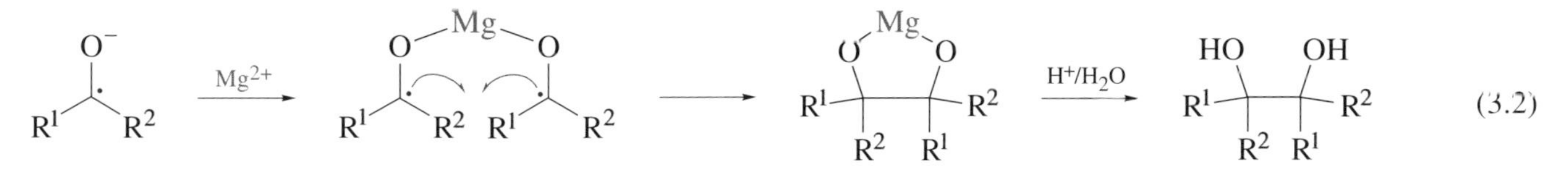


(3.2)

The key feature of this reaction (the pinacol reaction) is the use of a metal that is capable of forming two or more strong metal–oxygen bonds at the same time. The pinacol reaction is general for most aldehydes and ketones, but is particularly rapid and high yielding for aromatic ketones, because the intermediate ketyl radical is especially stable.

- Predict the product of the following reaction:

$CH_3COO$ O $\xrightarrow[C_6H_6]{Mg}$ (3.3)

- This is a straightforward application of the pinacol reaction: treatment of an arylalkyl ketone with a divalent metal in an aprotic solvent. A single electron transfer occurs in the first step to generate the ketyl radical, followed by coordination to the metal centre. Finally, radical dimerization occurs to give the product shown in Scheme 3.2.

  The product is an important intermediate in the laboratory synthesis of oestrogen hormones (Box 3.1).

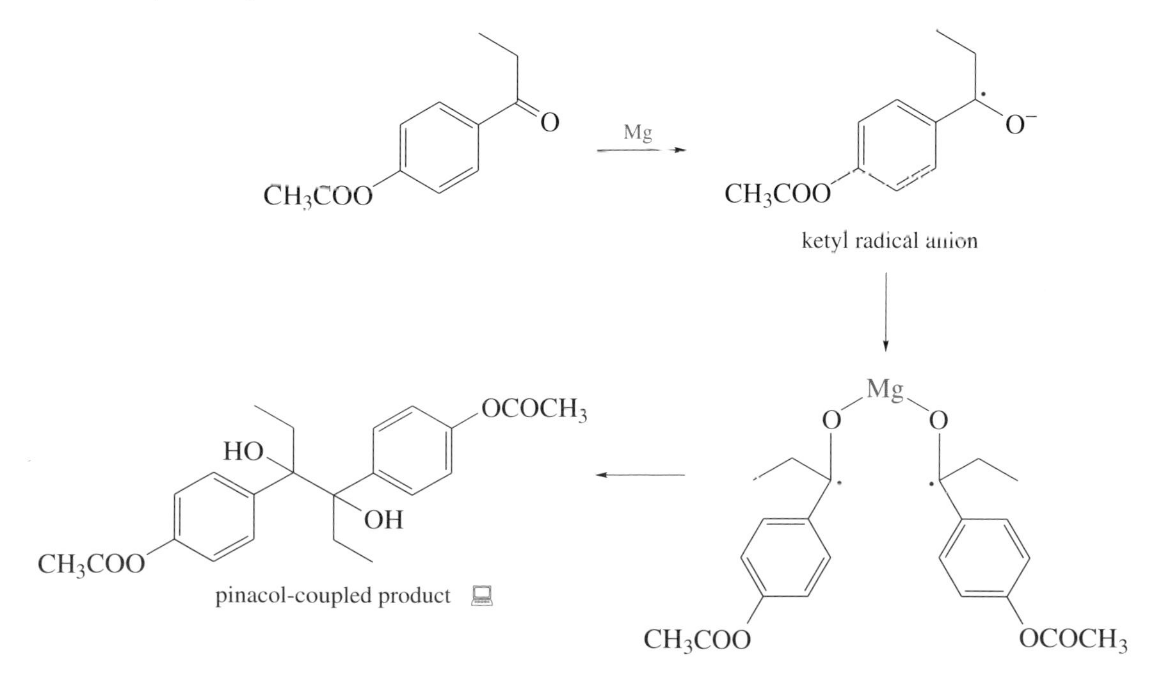


SCHEME 3.2

## BOX 3.1 Oestrogens

The oestrogens are female sex hormones responsible for the development and maintenance of sex organs and sexual characteristics in the female. They belong to a class of natural product compound known as **steroids**, which you will learn more about in Part 5 of this Book. There are three naturally occurring oestrogens, the two most common of which are oestradiol (**3.3**) and oestratriol (**3.4**):

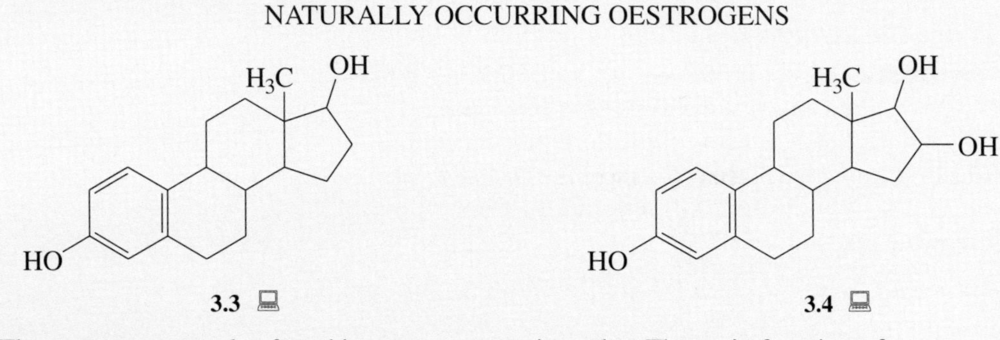


The oestrogens are also found in trace amounts in males. The main function of the oestrogens is (in conjunction with other hormones) the regulation of the menstrual cycle.

Synthetic oestrogens have found many uses, but most noticeably in correcting menstruation problems, in the 'morning after' pill and as a key component in hormone replacement therapy (HRT; see Figure 3.1). Ethynyloestradiol is the main synthetic oestrogen used in HRT, where it is believed to help in the prevention of osteoporosis (bone weakening), and the prevention of certain cancers; it also reduces the risk of cardiovascular disease (the single biggest cause of death in Western women):

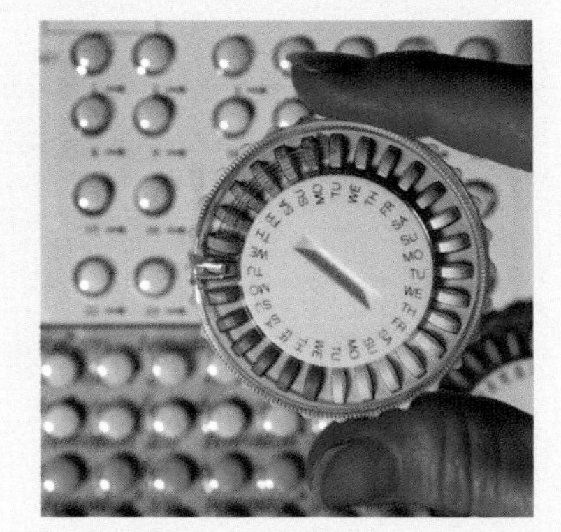

**Figure 3.1** A selection of HRT pills.

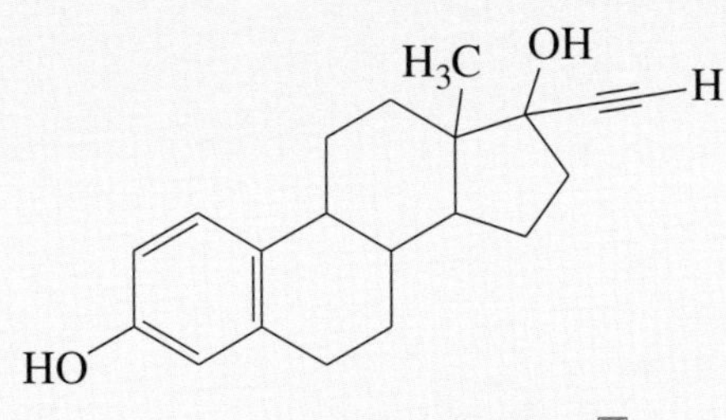


ethynyloestradiol

A number of synthetic oestrogens can be made using the pinacol reaction to form a carbon–carbon bond:

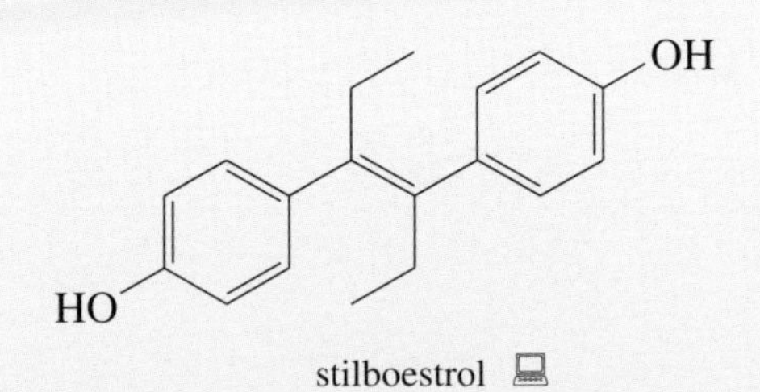


stilboestrol

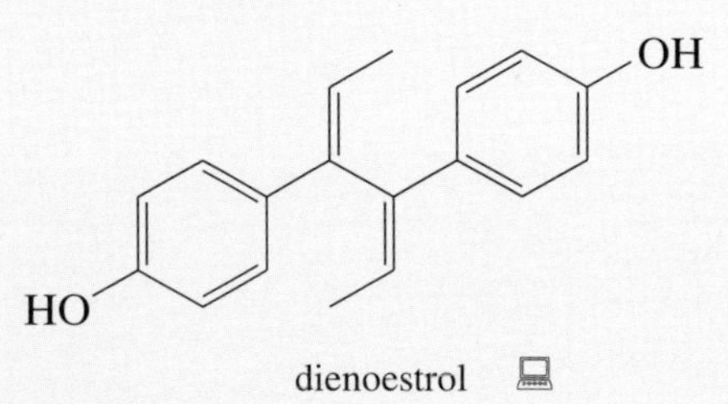


dienoestrol

QUESTION 3.1

Predict the outcome of the following reaction:

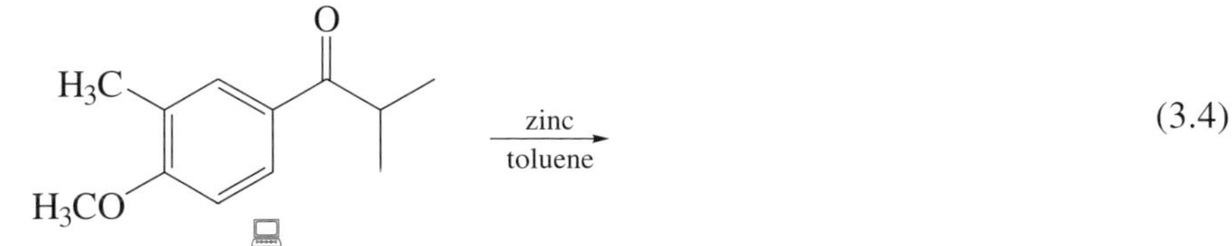


(3.4)

Just as in Question 3.1, the pinacol coupling has found uses in a number of natural product syntheses, the most famous of which is its use in the total synthesis of the anti-cancer agent taxol (Box 3.2). One of the interesting steps in this synthesis is the pinacol coupling that occurs between the two aldehydes at positions 9 and 10 rather than two ketones, but it still proceeds in good yield:

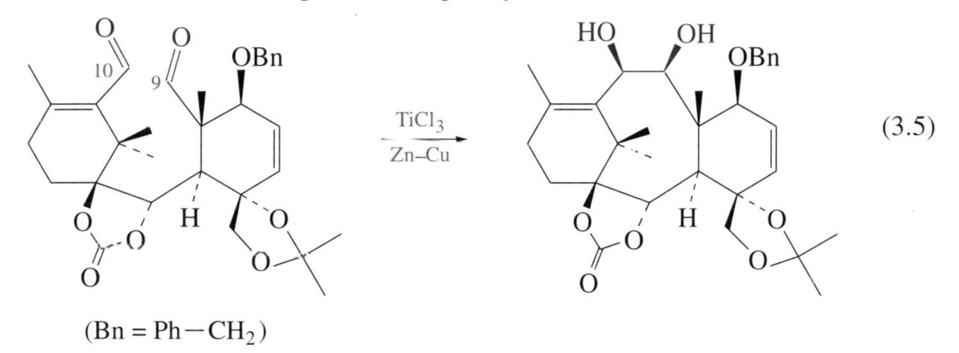


(3.5)

- What other interesting feature is there in this reaction that we have not come across before in this Book?
- As well as the reaction occurring between two aldehydes, both aldehydes (and reacting centres) are in the *same* molecule; that is, it is an **intramolecular reaction**.

There are two important variations of the pinacol coupling which we briefly need to consider.

## 3.1.1 The McMurry reaction

- Predict the product of the following reaction:

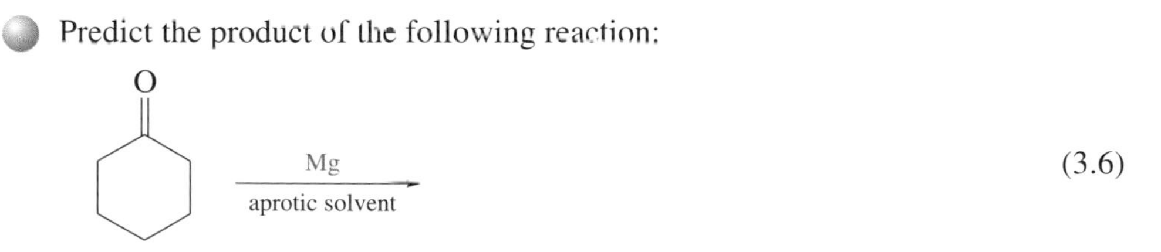


(3.6)

- The answer is the straightforward pinacol-coupled 1,2-diol:

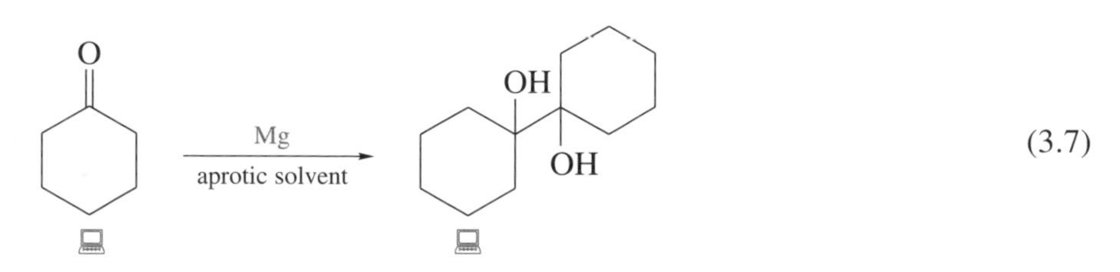


(3.7)

## BOX 3.2 Taxol

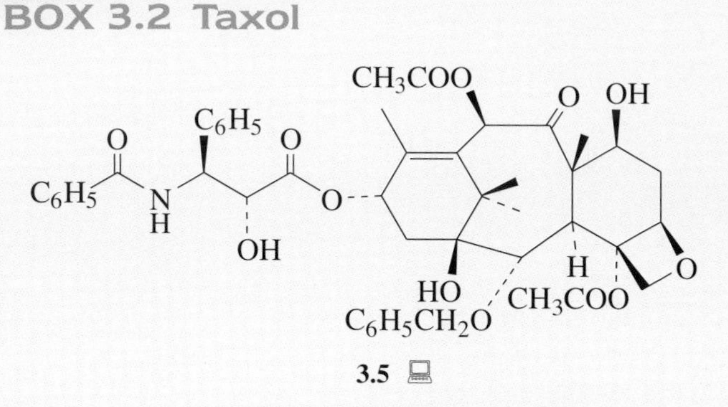

**Figure 3.2** The Pacific Yew tree.

Rarely has a molecule received such media attention as taxol (**3.5**) has in recent years. The history of this complex molecule dates back to 1962, when A. Barclay collected bark from the Pacific Yew tree (*Taxus brevifolia*, Figure 3.2) as part of a programme aimed at the discovery of new anticancer agents. One of the many compounds that were isolated in minute quantities was taxol. Such is the rarity of this compound, it has been estimated that one 100-year old yew tree would only yield 300 mg of taxol in its lifetime.

By 1991, taxol had been hailed as a 'celebrity molecule' and a 'miracle drug' against cancer due to its by-then well-established and unique mode of biological action against a range of cancers. However, the supply problem still remained. A long argument raged between doctors, environmentalists and politicians concerning the ethical and practical dilemmas concerning obtaining, isolating and administering taxol to patients. All agreed, however, that an alternative source of taxol, in place of the Pacific Yew tree was required. Various studies and approaches were undertaken, including harvesting special crops of the renewable European Yew tree (which was also found to contain taxol), cellular culture production and complete chemical synthesis. The synthesis of complex natural products like taxol, a molecule with many rings and chiral centres, remains a fundamental challenge for organic chemists. It took ten chemists, working together under the direction of Professor Kyriacos Nicolaou at the Scripps Research Institute in California, three and a half years to complete the first total synthesis of taxol! Taxol was given approval by the American Food and Drug Administration (FDA) for the treatment of ovarian cancer in 1992, and is currently being further investigated for the treatment of many other forms of cancer including breast, lung and melanoma.

However, if we replace magnesium with titanium, a different product is observed, since further reaction takes place between the metal and the diol:

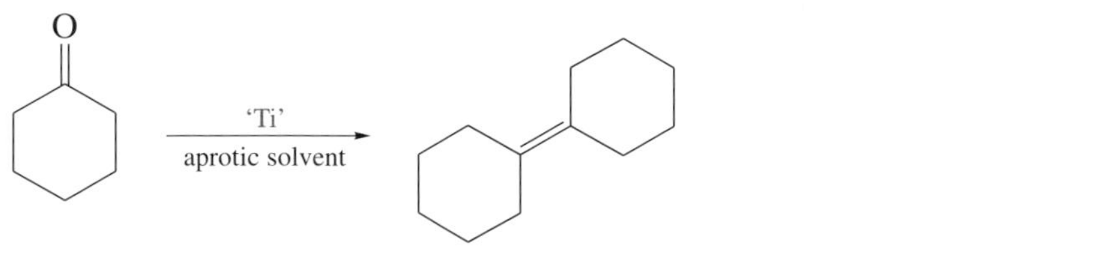


(3.8)

The product of this reaction is an alkene. The first feature of this reaction is that if the reaction is kept cold and reaction times are kept short, the 1,2-diol product is observed, suggesting that this is the first product formed. Longer reaction times or warmer reaction temperatures then give the alkene as the major product. The active metal, 'Ti', is formed by the reduction of titanium trichloride (in which titanium is in oxidation number +3) by lithium aluminium hydride ($LiAlH_4$). The titanium(0) species that is formed is the source of electrons for ketyl radical formation. Coordination to the titanium and normal pinacol coupling is believed to occur, but rather than the reaction stopping here, the titanium then deoxygenates the alcohol via a mechanism which is not fully understood. This is known as the **McMurry reaction**:

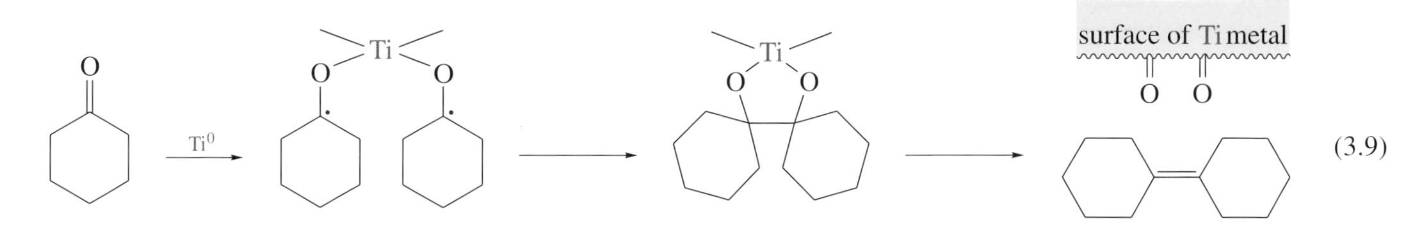


## 3.1.2 The acyloin reaction

The second variation on the theme is known as the **acyloin reaction**, which is identical to pinacol coupling except that it involves the reaction of two esters instead of two aldehydes or ketones. Metals again act as the electron donors. The reaction may proceed between two separate ester molecules, but it has also been found to be particularly useful as a ring-forming (*cyclization*) reaction. The intramolecular reaction of two ester groups, under ionic reaction conditions, to give medium to large (7–12-membered) rings is almost impossible, but radical reactions provide a possible route:

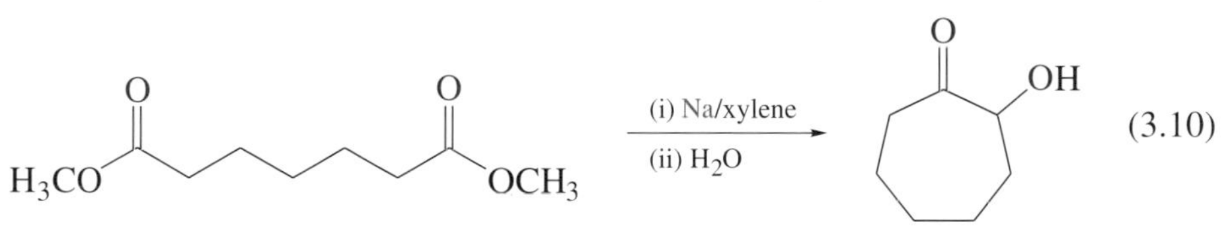


The mechanism of this reaction is not known with any certainty, although it is believed to occur at the metal surface (Figure 3.3), accounting for the radical–radical coupling nature of the reaction. However, the first step is ketyl radical formation in an exactly analogous way to the pinacol reaction:

**Figure 3.3**
The acyloin reaction takes place at the surface of the finely divided sodium.

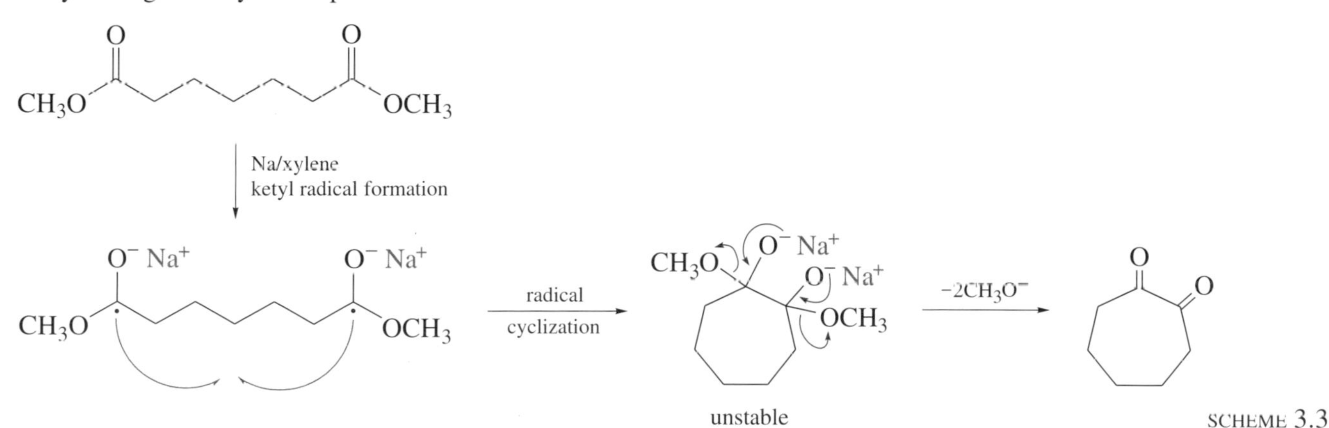


Once radical dimerization is completed the two carbonyl groups re-form with loss of two molecules of methoxide ion.

However, as we have seen, ketones themselves are susceptible to reduction under these conditions; treating a ketone with sodium metal, for example, leads ultimately to reduction to the alcohol. Since a 1,2-diketone is more reactive towards reduction than a normal ketone, the acyloin reaction does not stop here (see Scheme 3.4).

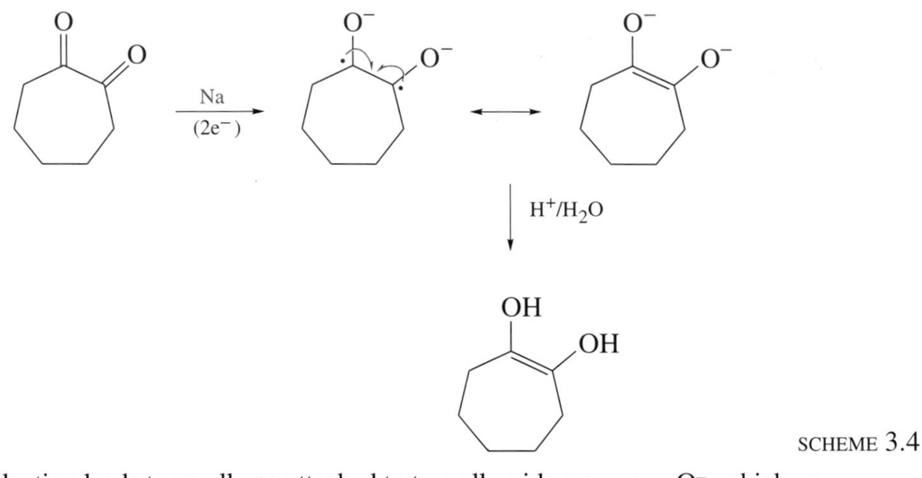


SCHEME 3.4

Reduction leads to an alkene attached to two alkoxide groups, $-\mathrm{O}^-$, which on addition of dilute acid gives the corresponding diol. However, this is not a stable configuration of atoms, and there is a rearrangement of one of the alcohol hydrogen atoms and the double bond, to generate a carbonyl compound. This phenomenon, which is known as *enol–keto tautomerism*, is very important in organic chemistry, but is beyond the scope of this Book. Thus, we obtain the α-hydroxyketone **3.6**, the ultimate product of the acyloin condensation:

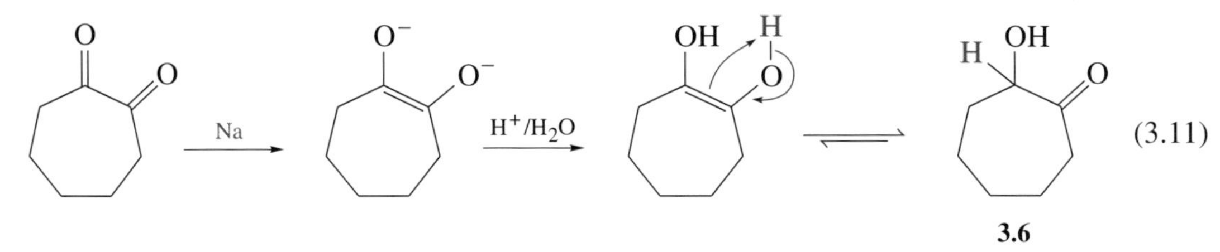


(3.11)

So the overall transformation is the intramolecular condensation of two ester groups to yield a cyclic α-hydroxyketone. The reaction need not be intramolecular: it will also take place intermolecularly between two ester molecules.

### QUESTION 3.2

Draw out the mechanism and predict the product of the following reaction:

$$2 \; \text{(branched alkyl)}\mathrm{COOCH_2CH_3} \xrightarrow[\text{(ii) } \mathrm{H^+/H_2O}]{\text{(i) Na}} \qquad (3.12)$$

We have seen several examples of radical combination reactions, but can you see any obvious problems associated with this type of reaction? As we suggested at the start of this Section, radical–radical reactions are relatively uncommon. One reason is that radicals are very reactive species, and tend to react with the first thing that they meet: they are relatively unselective. Although the energetics of the process may favour combination with another radical, they are more likely to collide with a molecule of a different type first, whether that be solvent or parent molecule. Since radicals react quickly with other species, the concentration of radicals in solutions tends to remain low.

So radical–radical combinations are very unlikely unless we do something to increase the likelihood of their meeting — such as tethering the two radicals as in the pinacol reaction.

As we continue our study of the chemistry of radicals, we shall see that radical combinations are a very small subset of the useful reactions of radicals.

## 3.2 Summary of Section 3

1 Radical–radical coupling reactions involve the pairing up of two radicals to give a neutral molecule.

2 Pinacol coupling is the divalent metal-promoted coupling of two aldehydes or ketones.

3 Pinacol coupling involves the coordination of two radical anions (or ketyl radicals) to the metal, followed by radical dimerization.

4 The use of titanium in a ketyl radical coupling reaction yields an alkene rather than a diol — the McMurry reaction.

5 The acyloin reaction is identical to the pinacol reaction but involves the coupling of two ester molecules to yield an α-hydroxyketone.

### QUESTION 3.3

Predict the products of the following two pinacol reactions:

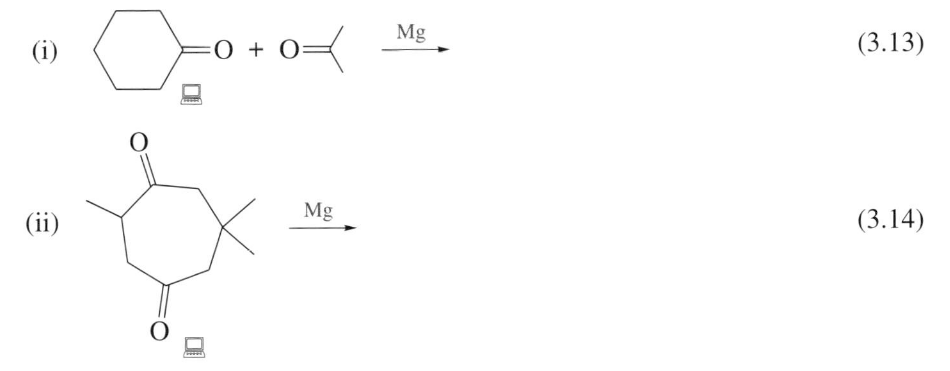


### QUESTION 3.4

Predict the product of the following reaction. What is the name of this reaction?

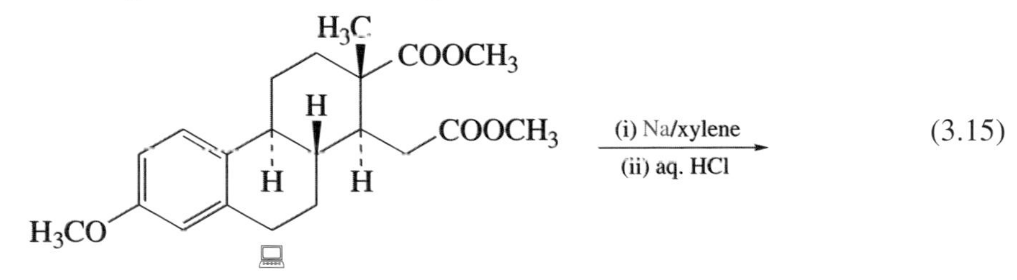

4

# RADICAL CHAIN REACTIONS

It is now time to turn our attention to the second class of radical reactions — **radical chain reactions**. This is certainly the largest class of radical reactions, and also by far the most useful type of reaction for the organic chemist.

All radical chain reactions consist of the same three steps. To learn more about these steps and what exactly they are, it is best to first consider an example of a radical chain reaction, to see what we can learn from it. Perhaps the simplest chain reaction to study is the reaction of alkanes with chlorine.

## 4.1 Radical halogenation of alkanes

Since alkanes possess their full complement of hydrogen atoms and have no multiple bonds or other functional groups, they are generally considered as being unreactive molecules. For example, hexane, $C_6H_{14}$, is a widely used inert organic solvent. However, there is one type of reaction readily undergone by alkanes, namely radical halogenation.

The overall reaction of methane with chlorine in the presence of light or high temperatures may be written:

$$CH_4 \xrightarrow[\text{light or heat}]{Cl_2} CH_3Cl + HCl \qquad (4.1)$$

methane

At first glance, this may appear to be a very simple reaction — a simple substitution of a hydrogen atom in methane for a chlorine atom, but in reality it is much more complex! In the absence of heat or light, no reaction will occur. Gaseous methane and chlorine will remain unreacted in a container for many hours and days in the dark or at temperatures below 100 °C, but only a relatively low number of photons of light are required to produce large amounts of the chlorinated product.

- Thinking back to Section 1.1, what do we know about the relative strengths of a carbon–hydrogen bond and a chlorine–chlorine bond?

- Table 1.1 shows that although the C—H bond is relatively strong and unreactive, a chlorine–chlorine bond is weak. It only takes a relatively small amount of energy to break the chlorine–chlorine bond.

### 1 Initiation stage

The bond dissociation energy, $D$(Cl—Cl), for a chlorine–chlorine bond is only 243 kJ mol$^{-1}$, compared with 435 kJ mol$^{-1}$ for the carbon–hydrogen bond of methane. Therefore, at comparatively low temperatures or under the influence of light, the chlorine molecule will dissociate homolytically and each atom takes one of the bonding electrons. The result is the generation of two highly reactive chlorine radicals. Since this step involves the generation of the radicals that subsequently drive the reaction, it is known as the **initiation stage**:

$$\text{Cl} : \text{Cl} \xrightarrow[\text{or heat}]{\text{light}} 2\text{Cl}^{\bullet} \qquad D(\text{Cl—Cl}) = 243\,\text{kJ}\,\text{mol}^{-1} \tag{4.2}$$

## 2 First step of the propagation stage

The reactive chlorine radical that we have generated is then capable of reacting with a molecule of methane (with a full octet of electrons), to generate hydrogen chloride (a neutral molecule also with a full octet) and a methyl radical.

$$\text{Cl}^{\bullet} \quad \text{H} : \text{CH}_3 \longrightarrow \text{HCl} + {}^{\bullet}\text{CH}_3 \tag{4.3}$$

methyl radical

In other words, a radical is reacting with a neutral molecule with a full octet of electrons to generate a new and different radical and another molecule with a full octet. There is no net loss or gain of radicals in this reaction, and so it is said to be the first part of the **propagation stage**. One feature of the propagation stage is that it permits continuation of the chain reaction, since radicals are not destroyed during a propagation step.

## 3 Second step of the propagation stage

We have now generated another highly reactive species — a methyl radical. This is capable of reacting with a number of species, but the most likely is the reaction with a chlorine molecule (containing the weakest bond in the system) in a second step of the propagation stage:

$$\text{Cl} : \text{Cl} \quad {}^{\bullet}\text{CH}_3 \longrightarrow \text{Cl—CH}_3 + \text{Cl}^{\bullet} \tag{4.4}$$

The product of this reaction — chloromethane — is the major product of the overall reaction sequence. It is important to notice that the other product of this second step of the propagation stage is a chlorine radical, which may now go through the whole propagation sequence again, starting with the first step of the propagation stage. From this we can see why radical reactions of this type are known as chain reactions: each reaction generates a new radical species which can further react, and is continually recycled.

However, radical chain reactions do not continue *ad infinitum*.

- What was the key reaction that radicals underwent in the pinacol reaction?

- Radical dimerization (that is, odd-electron species + odd-electron species to give an even-electron species).

## 4 Termination stages

During the pinacol reaction, we saw that two radicals could react with each other and dimerize to give a stable molecule; that is, there was a net loss of two radicals. However, we also said that radical–radical combinations were unlikely because radicals are so reactive that they normally collide with some other molecule first and react with this rather than with another radical. (This was overcome in the pinacol reaction because the two radicals were tethered together by the metal atom). That is not to say that radical–radical couplings do not happen; they are just unlikely. Inevitably, a few radicals will collide with other radicals in the reaction mixture and thus combine. This combination can occur between any two radicals in the mixture; radicals are not discriminating in their reactions. Examples of such couplings are shown below:

$$\mathrm{Cl^{\bullet}} + {}^{\bullet}\mathrm{CH_3} \longrightarrow \mathrm{CH_3Cl} \qquad (4.5)$$

$$\mathrm{H_3C^{\bullet}} + {}^{\bullet}\mathrm{CH_3} \longrightarrow \mathrm{H_3C{-}CH_3} \qquad (4.6)$$

$$\mathrm{Cl^{\bullet}} + {}^{\bullet}\mathrm{Cl} \longrightarrow \mathrm{Cl_2} \qquad (4.7)$$

As these reactions result in the formation of a stable molecule, they represent the end of the chain reaction sequence since no new radical is formed. Therefore, radical combination steps are part of the **termination stage**. Termination reactions are normally regarded as unfavourable because they stop the chain reaction proceeding and also lead to a wide range of byproducts.

As a general guideline, the propagation stage in a chain reaction such as the chlorination of methane will occur approximately $10^6$ times for each initiation reaction before a termination step will occur. Therefore, it is easy to see why only a few photons of light or a small input of energy is required to give a large extent of reaction.

In summary the three components of a radical chain reaction are:

- the initiation stage
- the propagation stage
- the termination stage

As we examine other radical reactions, it will be possible to identify these three stages of reaction in every reaction.

The real value and synthetic utility of these reactions is that they permit the synthesis of functional molecules (i.e. molecules with a functional group) from alkanes, which are an intrinsically unreactive or inert class of compound.

Before we look at some of the other details of this process, note that the mono chlorination products are not necessarily the only products from this reaction.

Consider the following reaction:

$$CH_4 + Cl_2 \xrightarrow{\text{light}(h\nu)} CH_3Cl \qquad (4.8)$$

If you look closely at the product, what else do you notice? The starting material, methane, had four equivalent hydrogen atoms. The product, chloromethane still has three hydrogens remaining. Therefore further reaction may take place:

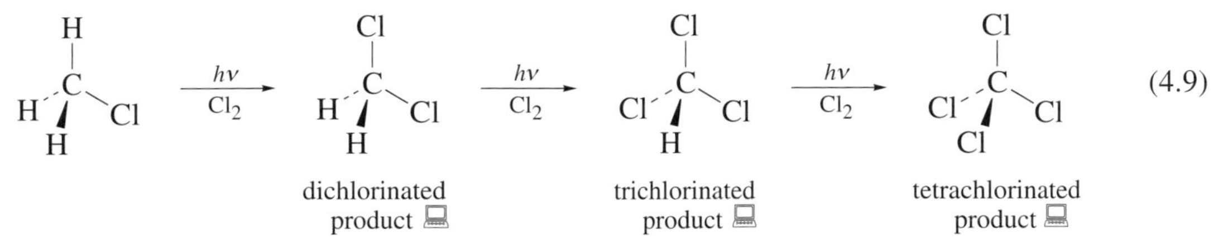


The next product formed is the dichlorinated product, which can then react further to form the tri- and tetrasubstituted products. In future, we shall normally assume that reaction conditions are carefully controlled to give only the monosubstituted product, unless otherwise stated.

### 4.1.1 Site selectivity

During our study of the chlorination reaction of methane, we have seen that the major product of the reaction is chloromethane. We obtain the same product, chloromethane, irrespective of which hydrogen atom we remove from the methane molecule; that is because the methane molecule possesses four equivalent hydrogen atoms.

Predict the product or products from the following reaction:

$$\text{propane} \xrightarrow[Cl_2]{h\nu} \qquad (4.10)$$

With alkanes, such as propane, the hydrogens are not all equivalent: there are six methyl hydrogens (primary sites) and two methylene ones (secondary sites). Therefore, on performing a radical chlorination reaction, we might expect two products, in the ratio 6 : 2

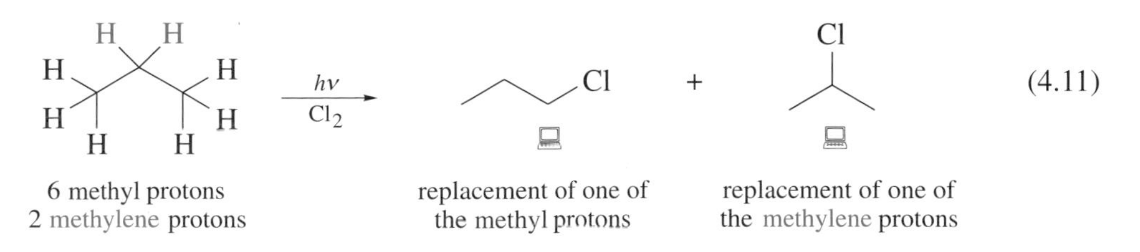


Although we do obtain these two products, they are not in the 6 : 2 ratio that we might expect:

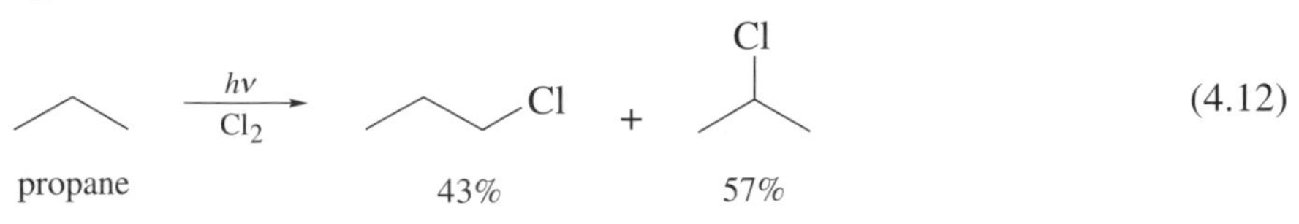

This is because there is a reactivity difference between hydrogens in primary, secondary and tertiary positions: they have a relative reactivity of 1 : 4 : 5, respectively. Thus, in propane, although there are only two secondary hydrogens as against six primary hydrogens, each secondary hydrogen is four times more likely to be removed. We can work out the theoretical yield based on these relative reactivities and the statistical factors as follows:

**Type of sites in** propane

| | SECONDARY | PRIMARY |
|---|---|---|
| number of equivalent hydrogens that could react to give the product shown | 2H | 6H |
| inherent relative reactivity of the site | 4 | 1 |
| include statistical factor (number of equivalent hydrogens) | $2 \times 4$ | $6 \times 1$ |
| relative abundance of product | 8 | 6 |
| theoretical yield | $\frac{8}{8+6} \times 100 = 57\%$ | $\frac{6}{8+6} \times 100 = 43\%$ |

The theoretical yield is identical to that determined experimentally!

Predict the four possible monochlorination products of Structure **4.1**.

**4.1**

The four possible products are shown as Structures **4.2–4.5** at the bottom of p. 146.

So, given that we know the number of each type of hydrogen and also their relative reactivities, it should be possible to predict the ratio of these four products that will be obtained from the reaction.

| **Type of site** | PRIMARY | TERTIARY | SECONDARY | PRIMARY |
|---|---|---|---|---|
| number of equivalent hydrogens that could react to give the product shown | 6H | 1H | 2H | 3H |
| inherent relative reactivity of the site | 1 | 5 | 4 | 1 |
| include statistical factor (number of equivalent hydrogens) | 6 × 1 | 1 × 5 | 2 × 4 | 3 × 1 |
| relative abundance of product | 6 | 5 | 8 | 3 |
| theoretical yield | $\frac{6 \times 100}{6+5+8+3}$ | $\frac{5 \times 100}{6+5+8+3}$ | $\frac{8 \times 100}{6+5+8+3}$ | $\frac{3 \times 100}{6+5+8+3}$ |
| | 27% | 23% | 36% | 14% |

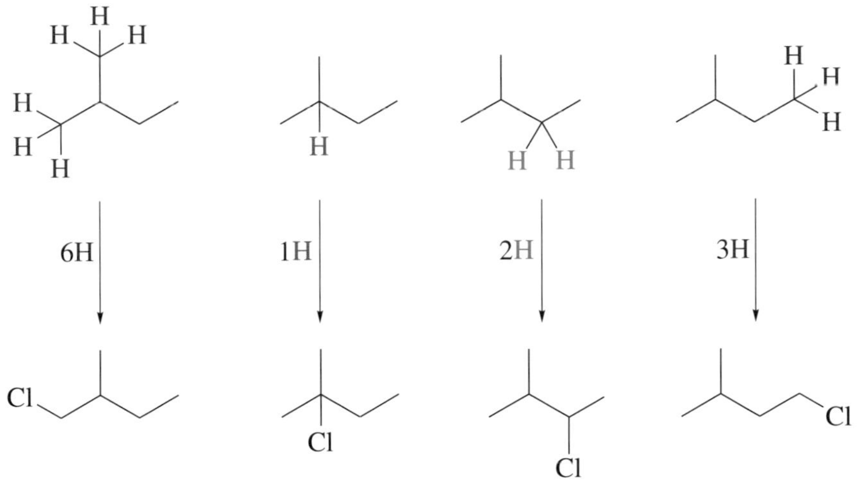

This is very close to the yields obtained experimentally.

There is an even larger difference between the reactivity of primary, secondary and tertiary sites in radical bromination; for example

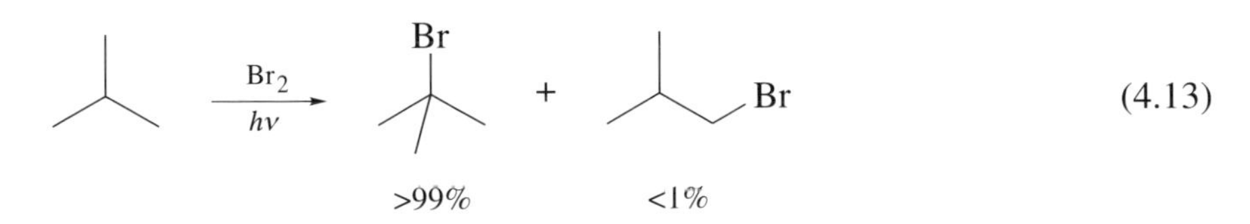


QUESTION 4.1

Predict the relative abundances of the products obtained from the radical monochlorination of Structure **4.6**.

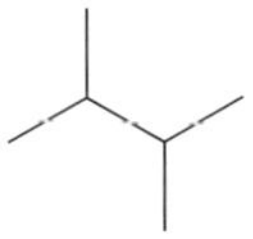

**4.6**

## 4.1.2 Halogenation of alkylbenzenes

A radical on a carbon atom adjacent to a benzene ring, known as a **benzylic radical** (**4.7**), is even more stable than a tertiary radical because of delocalization of the radical into the adjacent benzene ring;

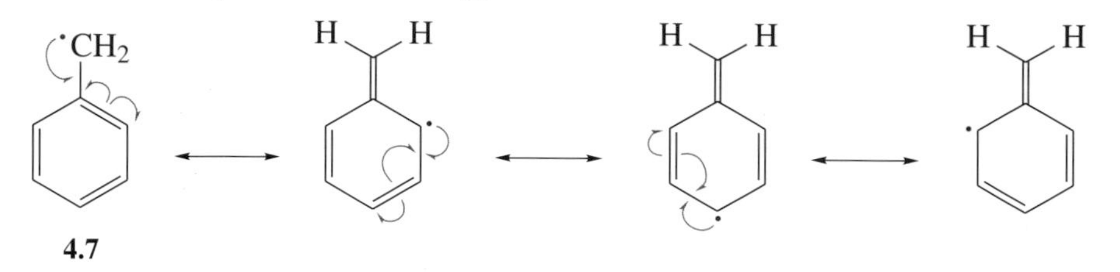


This explains why the benzyl radical is formed from toluene rather than the less-stable radical that would be formed by removal of a ring hydrogen. The extra stabilization of the benzylic radical can be exploited to achieve side-chain halogenation. The reaction of toluene with a halogen source in the presence of light yields benzyl bromide (**4.8**):

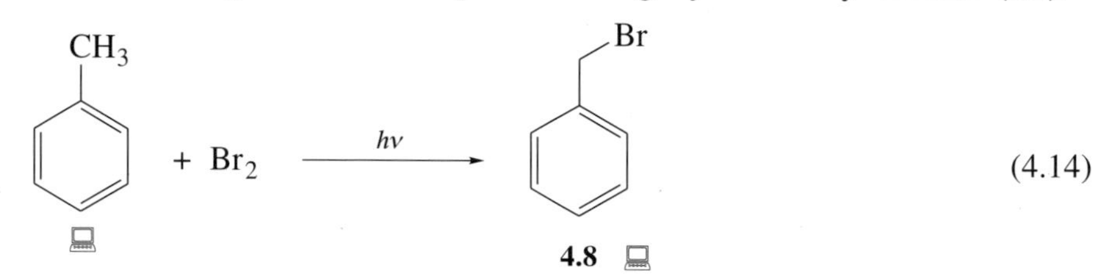


(4.14)

In fact, *N*-bromosuccinimide (**4.9**) is used to provide a low concentration of bromine in this reaction. In the presence of light (or a radical initiator), this bromine will undergo homolytic cleavage to give two bromine radicals:

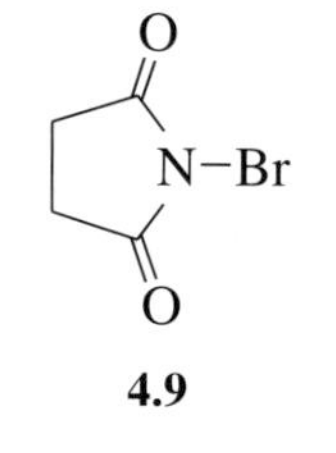


$$Br_2 \xrightarrow[\text{or light}]{\text{radical initiator}} 2Br^{\bullet} \qquad (4.15)$$

As with other radical chain mechanisms, two chain propagation steps then follow: the first is the abstraction of a benzylic hydrogen atom to form a benzylic radical and hydrogen bromide. The radical then reacts with another molecule of bromine to form benzyl bromide and a bromine radical, which then may go round the entire cycle again.

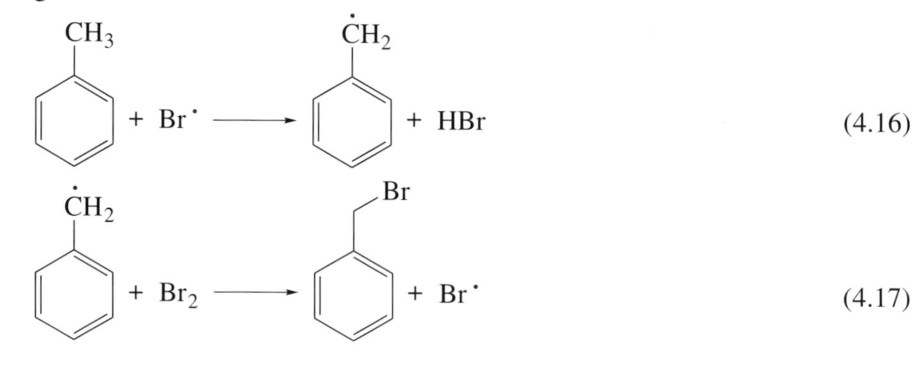


(4.16)

(4.17)

*Answer to question on p. 144*

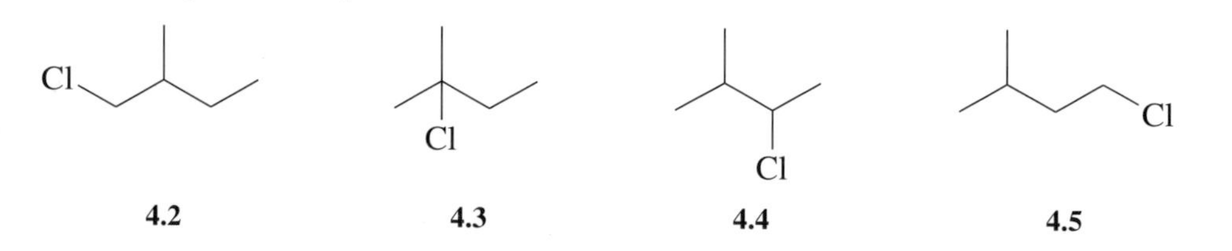

This type of reaction is typical of all alkylbenzenes; that is, under radical conditions, they will react preferentially through the benzyl radical rather than a normal alkyl radical.

Predict the product of the following reaction:

$Cl_2$, $h\nu$ (4.18)

The major product of the reaction is 1-chloro-1-phenylethane (**4.10**), formed via a benzylic radical intermediate.

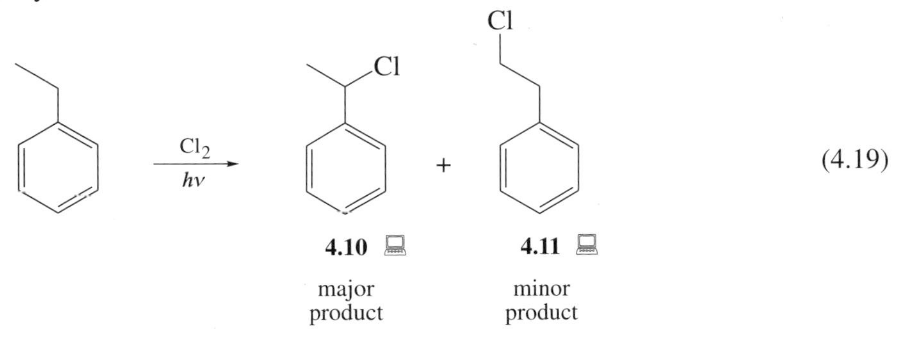


QUESTION 4.2

Write a mechanism for the formation of **4.10** and **4.11** in Reaction 4.19.

You should now know how to prepare both ring halogen-substituted alkylbenzenes using an electrophilic aromatic substitution* and side-chain substituted alkylbenzenes, simply by using the appropriate reaction conditions:

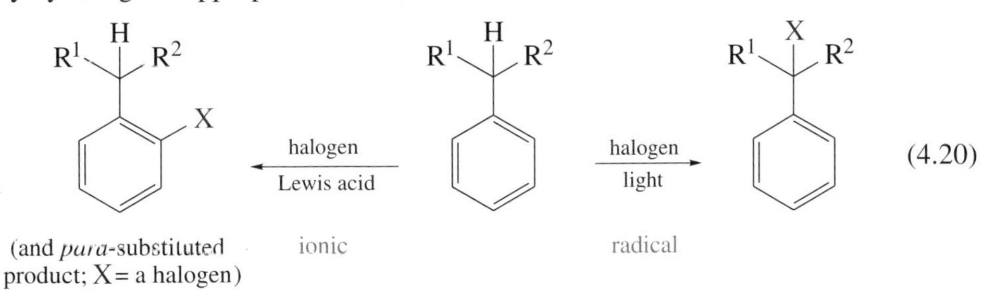


## 4.1.3 The stereochemistry of radical substitutions

One aspect of radical halogenation that we have not yet discussed is the stereochemistry of radical reactions; for example, would we expect the benzylic hydrogen in **4.12** to be substituted by a chlorine with retention of configuration, inversion or a mixture of both?

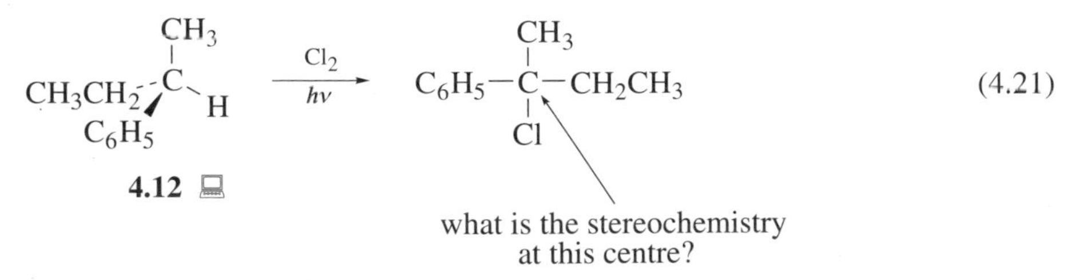


* Electrophilic aromatic substitution reactions are discussed in *Alkenes and Aromatics*[1].

The radical chlorination involves removal of the benzylic hydrogen by a chlorine radical to give a planar carbon-centred radical:

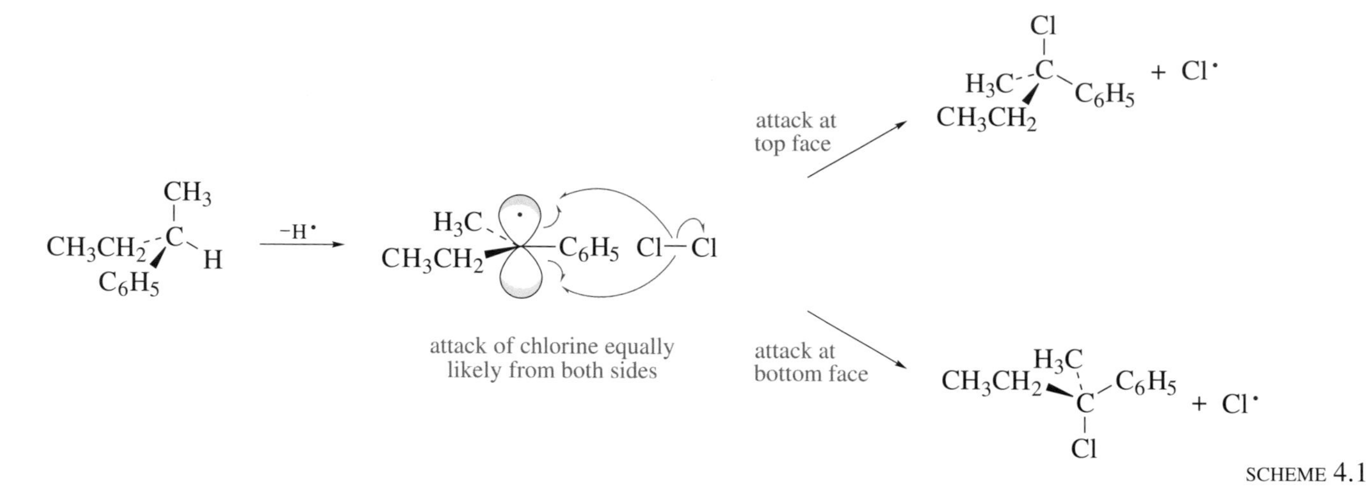


SCHEME 4.1

Just as the nucleophile may attack from either side of the carbocation in an $S_N1$ reaction, so the incoming chlorine atom may attack the radical carbon from either side of the planar system. Each side is equally likely, so, racemization occurs: both enantiomeric products result and in equal abundance.

## BOX 4.1 Radicals and health

Radicals and antioxidants are words and terms that are widely used in the scientific and popular press. The media tends to portray radicals as being bad and antioxidants as good. We need antioxidants because oxygen, $O_2$, is toxic to organisms. In fact, research has shown that it is not oxygen, $O_2$, that is directly toxic but rather a radical derived from oxygen, known as superoxide $O_2^{\bullet-}$, which has one electron more than an oxygen molecule, and can be designated as $^{-}O{-}O^{\bullet}$. It is now known that this is not the only toxic radical in the body; in fact, there are many very reactive oxygen- and nitrogen-derived radicals in the body. These can cause damage to DNA, lipids, proteins and many other biomolecules.

It is believed that exposure to environmental factors such as ultraviolet radiation in sunlight, smoke, ozone, radiation (such as from X-rays or oncology treatments), alcohol and even stress, cause an increase in the production of radicals in the body.

The human body has developed its own defence mechanism against radicals through a series of enzymes, the most famous being the superoxide dismutases (SODs). These are inadequate for preventing damage completely and so diet-derived antioxidants are also essential. Vitamin E (α-tocopherol) is known to be the most important, but is by no means the only diet-derived antioxidant. Others are believed to include vitamin C and β-carotenes and other carotenoids related to vitamin A. These all act by trapping the radicals and rendering them harmless. In conclusion, therefore, we must be careful to ensure that we obtain antioxidants in our diet.

We have now seen all the typical characteristics of a radical chain reaction. However, although useful, the radical halogenation of alkanes is not enough in itself to revolutionize organic chemistry. It was the discovery of a compound — tributyltin hydride — which was to turn radical chemistry into the powerful tool that it is today.

## 4.1.4 Summary of Section 4.1

1. Radical chain reactions comprise three stages: initiation, propagation and termination.
2. Although the reaction of a tertiary C—H is preferred over secondary and primary C—H in radical halogenation reactions, there is also an important statistical effect because of the differing number of sites of each type.
3. Radical halogenation is a useful method for the functionalization of alkyl side-chains in alkylbenzenes. This is due to the particularly stable benzylic radical that is formed first.
4. Radical intermediates tend to be planar in nature, which leads to racemization at chiral centres.

### QUESTION 4.3

Predict all the possible products for the following reaction and the relative ratios of each of these products.

(4.22)

### QUESTION 4.4

Suggest reagents for the following transformations:

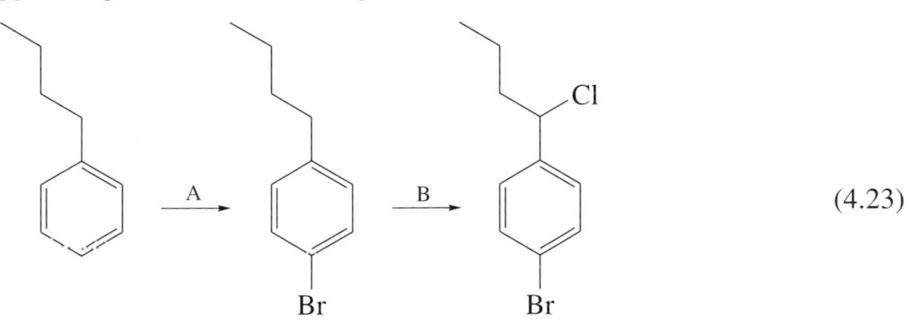


(4.23)

### QUESTION 4.5

Define the stereochemistry as *R* or *S* at the chiral centre in **4.13**. If this compound were irradiated in the presence of bromine, what do you think the major product would be, and what would be the resulting stereochemistry at the chiral centre.

$CH_3$, H, $C_6H_5$, C, $COOCH_3$

**4.13**

## 4.2 Tributyltin hydride

Tributyltin hydride (or $Bu_3SnH$, **4.14**, as it is often written) is the reagent most often used to conduct radical reactions, and is another example of an organometallic reagent).

tributyltin hydride ≡ $Bu_3Sn{-}H$ ≡ Sn–H

(Bu = $-CH_2CH_2CH_2CH_3$)

**4.14**

Although the tin–hydrogen bond in tributyltin hydride is weak, it is not weak enough to undergo homolysis on its own under most normal reaction conditions. Instead, we have to employ an **initiator**. The one most commonly employed is α,α′-azo-*bis-iso*butyronitrile (**AIBN**; see Table 1.1). Under fairly mild conditions (> 60 °C), AIBN undergoes thermal homolysis to give two nitrile-stabilized radicals, along with loss of nitrogen gas:

NC–N=N–CN ⟶ 2 CN + N≡N (4.24)

2-cyanopropyl radical

The 2-cyanopropyl radical is then able to abstract a hydrogen atom from the tributyltin hydride to generate a tin-centred radical:

$Bu_3Sn{-}H$ CN ⟶ $Bu_3Sn^{\bullet}$ + CN (4.25)

Let's now examine what can happen to this tin-centred radical.

- What do you think is happening in the following reaction?

Br $\xrightarrow[\text{AIBN}]{Bu_3SnH}$ H + $Bu_3SnBr$ (4.26)

- The bromine atom in the haloalkane is being replaced by a hydrogen atom.

- Where do you think that this hydrogen atom comes from?
- The only source of hydrogen in this reaction is from the tributyltin hydride.

- What is the byproduct of this reaction?
- It is tributyltin bromide.

So it would appear that the net overall result of this reaction is that *somehow* the hydrogen from the tin hydride and the bromine from the haloalkane have swapped places. We need to consider the mechanism by which this reaction occurs. We have already seen that the first step involves the decomposition of the initiator (AIBN), which in turn reacts with the tin hydride to generate tributyltin radicals.

- Once the Sn—H bond has been broken to generate tributyltin radicals, it will react with the bromoalkane. What do you think is the weakest bond in the bromoalkane (see Table 9.1 in the *Data Book* on the CD-ROM)?
- The carbon–bromine (C—Br) bond, whose bond enthalpy term is 290 kJ mol$^{-1}$ compared with 413 kJ mol$^{-1}$ for the C—H bond.

The tributyltin radical will react with the bromine in the starting material to form a stronger tin–bromine bond.

- Where do you think the electrons that form this new bond come from?
- One electron comes from the tributyltin radical. The other comes from the carbon–bromine bond, which is broken during the reaction.

The other electron of the pair from the C—Br bond moves onto the carbon atom of the C—Br bond to generate a carbon-centred radical, which is then able to react with another molecule of tributyltin hydride. This step provides the hydrogen atom that substitutes for the bromine in the bromoalkane, generating a new tributyltin radical, which is then able to go through the cycle again (and again, and again!)

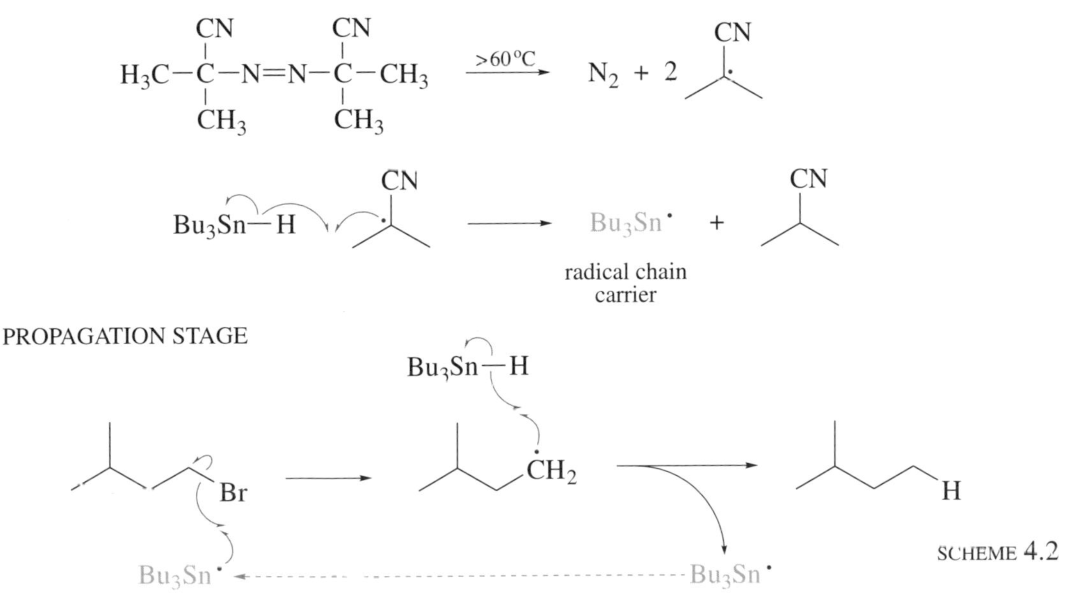


SCHEME 4.2

- How much initiator (AIBN) do you think would be required for a chain reaction like this?
- Very little! The AIBN is acting only as an initiator; it is the tributyltin hydride that is providing the hydrogen atom in the reaction. Therefore, only sufficient quantities of initiator are required to form a few radicals to start the chain process. In fact, only 0.02–0.05 molar equivalents (compared to starting material) of initiator are ever used in radical reactions.

- If the decomposition of AIBN occurs at temperatures above 60 °C (thereby starting the radical chain reaction), what type of solvent do you think radical reactions are normally performed in?

- Solvents that boil at temperatures above 60 °C and are inert to radicals (that is, have no reactive hydrogens or halogens which may participate in a reaction).

Typical solvents are benzene (b.t. 80 °C) and toluene (b.t. 110 °C), although dimethyl sulfoxide, **4.15** (b.t. 189 °C) is used in exceptional cases when a polar solvent is required.

$H_3C$ \ S=O / $H_3C$

**4.15**

- How much tributyltin hydride do you think would be required for a chain reaction like this?

- Each molecule of tributyltin hydride provides one hydrogen atom for reaction. Therefore we are going to need one molecule of tributyltin hydride for every one molecule of substrate. In fact, a slight excess (normally 1.2 molar equivalents) of tributyltin hydride is used. This allows for any other radical side reactions that may occur.

The overall reaction may be written as:

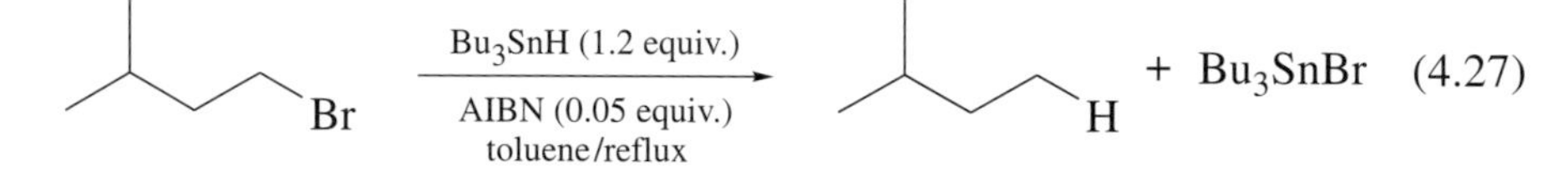


You will often encounter radical reactions written in this way. This reaction illustrates another important feature of tin-mediated radical reactions. Conventional radical chlorination of alkanes gave a range of products, although there was some preference for different hydrogens in the molecule. This does not happen with tin-mediated reactions, because a specific radical reaction occurs at just one site in the molecule; we say that this is a **selective reaction**.

## 4.2.1 Intermolecular radical reactions with tributyltin hydride

Despite the discovery of tributyltin hydride, things remained pretty static in the area of radical chemistry until the late 1970s. Chemists knew that they had a powerful methodology available both for the halogenation and de-halogenation of compounds, but that was all that was really expected of radicals.

We are going to use the following reaction to help us redefine the scope of radical reactions:

$C_6H_5$–CH=$CH_2$ $\xrightarrow[h\nu]{BrCCl_3}$ $C_6H_5$–CH(Br)–$CH_2$–$CCl_3$ (4.28)

- Using your knowledge of radical chemistry, suggest the first step in the mechanism of this reaction. Which is the weakest bond in the system (see Table 1.1)?

- This reaction, where light acts as the initiator, follows a pattern identical to those we have previously studied. The C—Br bond is the weakest. Therefore in the presence of light, it will cleave homolytically to give two radicals:

$$Br{-}CCl_3 \xrightarrow{h\nu} Br^{\bullet} + {}^{\bullet}CCl_3 \qquad (4.29)$$

The trichloromethane radical then adds to the least-hindered end of the double bond and thus generates a new carbon-centred radical, which is also a secondary and a benzylic radical:

$$C_6H_5CH{=}CH_2 \;\; {}^{\cdot}CCl_3 \longrightarrow C_6H_5\dot{C}HCH_2CCl_3 \qquad (4.30)$$

The final step is for this radical to abstract a bromine atom from another molecule of starting material and thus regenerate a trichloromethane radical to continue the chain reaction:

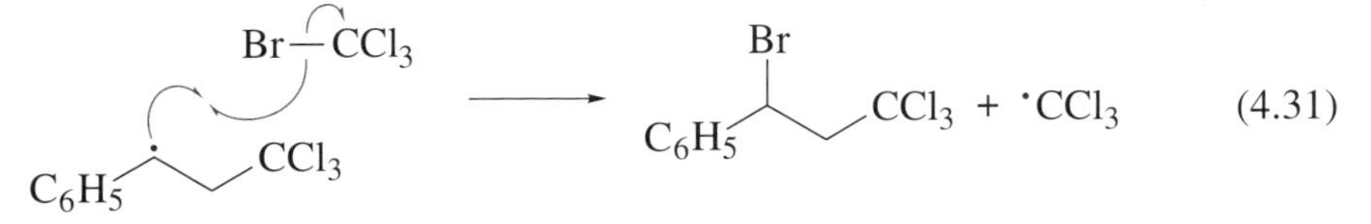


- What is the new feature about this radical reaction that we have not seen before?
- The formation of a new carbon–carbon bond via the reaction of an alkene with a radical. Overall this is an addition reaction.

In fact this is a unique reaction. If we picked any other bromoalkane, a horrendous mixture of products would be obtained. The reason that this reaction proceeds well is because the C—Br bond is particularly weak and the resultant trichloromethane radical is stabilized by the presence of three chlorine atoms. Where this extra stabilization is absent, reaction does not occur in such a controlled way.

From this reaction sprang the idea of combining known tributyltin hydride radical chemistry with the carbon–carbon bond-forming reaction described above to create a general and controllable method for the formation of carbon–carbon bonds. Such formation of carbon–carbon bonds is one of the main goals of organic chemists, as this permits the building and extension of carbon skeletons.

Consider the reaction between bromocyclohexane (or iodocyclohexane) and ethyl propenoate (acrylate) in the presence of tributyltin hydride and an initiator:

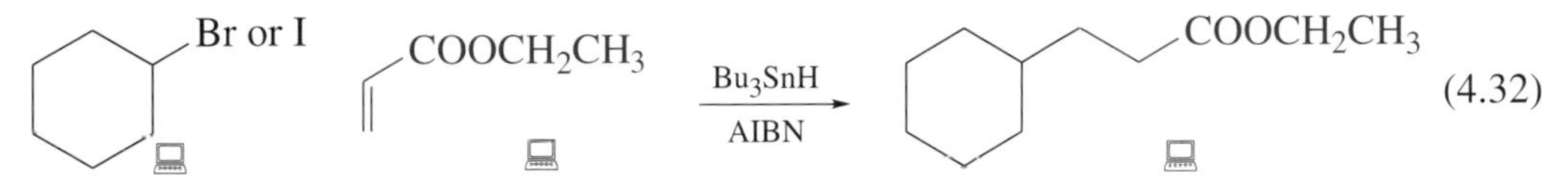


- What is the overall transformation that is occurring in this reaction?
- We have formed a new carbon–carbon bond via addition to an alkene group!

Let us consider the mechanism of the chain reaction that is taking place. As before, the first step is the generation of a tributyltin radical using AIBN as the initiator.

$$Bu_3SnH \xrightarrow[>60\,^{\circ}C/\text{ toluene}]{AIBN} Bu_3Sn^{\cdot} \qquad (4.33)$$

Once again, the tributyltin radical abstracts the halogen from the haloalkane to form a strong Sn—X bond and a cyclohexyl radical:

$$C_6H_{11}Br \;\; Bu_3Sn^{\cdot} \longrightarrow C_6H_{11}^{\cdot} + Bu_3SnBr \qquad (4.34)$$

- What stage of the radical chain reaction is this step?
- It is part of the propagation stage.

What follows next — and again this is something that we have seen before — is a second propagation step, whereby the cyclohexyl radical adds to the alkene (ethyl propenoate):

(4.35)

During this step, the alkene π bond is broken, and a new radical is generated, containing a new carbon–carbon σ bond.

- What factors help to stabilize this radical intermediate? (*Hint* Remember when we considered what types of group helped stabilize radicals.)
- The proximity of the electron-withdrawing carbonyl group allows delocalization of the single electron.

The third step in the propagation stage is the quenching of the carbon-centred radical with a hydrogen atom from tributyltin hydride, which, of course, generates another tin-centred radical. The entire cycle may then go around again.

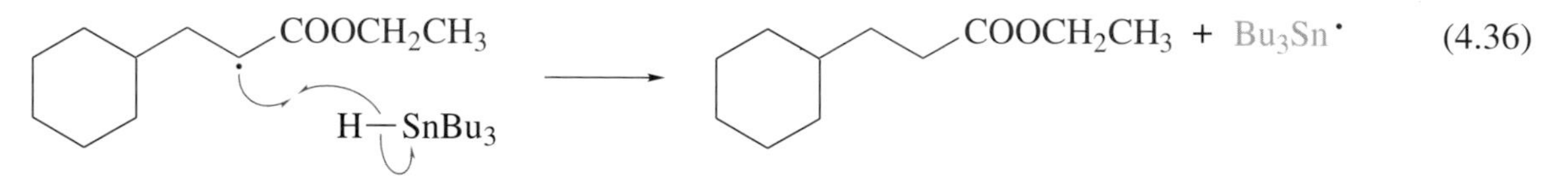


(4.36)

- What potential problems can you see with this reaction sequence? Apart from radical–radical coupling (termination) steps, are there any other potential side-reactions that could take place?
- There are many radicals in the reaction medium at the same time — for example, the 2-cyanopropyl radical (from AIBN), tributyltin radicals and cyclohexyl radicals, to mention but a few. What is to stop these radicals reacting together or with other molecules?

The reason why alternative products are not observed is because the radicals are controlled to selectively undergo only the desired reactions. To ensure the reactions give the desired products, there are three general requirements for a tributyltin radical chain reaction:

1. The starting haloalkane should contain a weak R—X bond (that is, a haloalkane with C—Br or C—I bonds, which are weaker than C—Cl or C—F bonds).
2. Slow addition of tributyltin hydride in a dilute solution is used.
3. An alkene (or other substrate) that is present in high concentration compared with that of tributyltin hydride, and that is preferably electrophilic in nature, is employed.

The synthesis of (–)-malyngolide (Reaction 4.37) by the famous German radical chemist Bernd Giese illustrates many of these key features very nicely. (In this reaction sequence, the wiggly line attached to the phenyl group means that the stereochemistry at this chiral centre is indeterminate.)

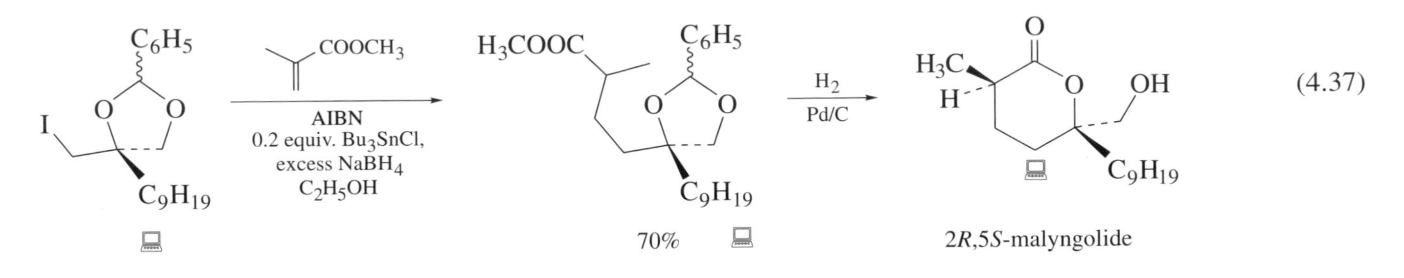


Giese used an intermolecular radical addition of an alkyl radical to an alkene as one of his key reactions, since it does not affect the stereochemistry at the adjacent centre and the desired product is isolated in 70% yield. This synthesis also introduces us to another method of radical generation. You may remember that we have stated several times that a stoichiometric amount of tributyltin hydride is required in radical reactions — one mole of tributyltin hydride for each halogen atom in the starting material. In this synthesis, however, an alternative method is employed. Only 0.2 molar equivalent of a tin compound is added to the reaction, and note that it is not tributyltin hydride which is used, but rather tributyltin chloride. An excess of a reducing agent, sodium borohydride, is added to the reaction; it serves to reduce the tributyltin chloride to tributyltin hydride and so at the start of the reaction we have 0.2 equivalent of tributyltin hydride. After the radical chain reaction, a tin–halogen bond is re-formed, followed by reduction with the sodium borohydride to regenerate tributyltin hydride; and so the cycle continues. In this way, we are effectively keeping the concentration of tributyltin hydride very low. The only drawback to this method is that the starting material must be able to tolerate the presence of an excess of sodium borohydride; since many reagents react with sodium borohydride, this has limited the use of this method.

## 4.2.2 Intramolecular radical reactions with tributyltin hydride

With the exception of the acyloin condensation in Section 3.1.2, every radical reaction that we have studied so far has involved the exchange of atoms or the joining together of *two* molecules. In some instances, these exchanges have been between identical molecules, and in others between different molecules. We are now going to look at one class of radical reaction which became a real growth area with the discovery of tributyltin hydride.

- We have already studied in some detail the reaction between a radical and an alkene. Predict the product of the following reaction:

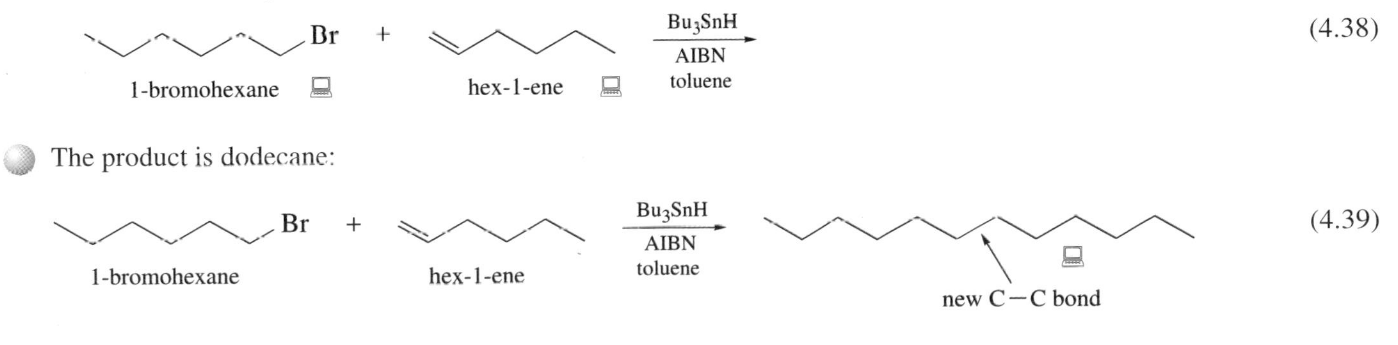


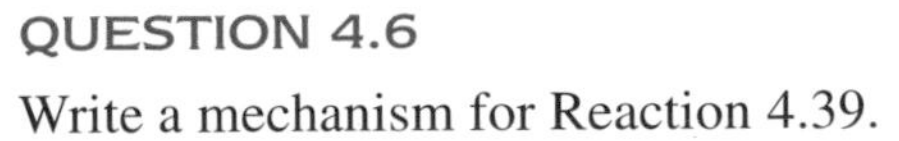

QUESTION 4.6

Write a mechanism for Reaction 4.39.

In this reaction we have formed a new carbon–carbon bond between the two molecules — specifically, between the carbon bearing the radical precursor (C—Br) and the terminal end of the alkene. What do you think might happen if the radical precursor and the alkene are part of the same molecule, as in 6-bromohex-1-ene (**4.16**)?

Br

**4.16**

There are two possibilities. One is polymerization, whereby the radical generated from one molecule of 6-bromohex-1-ene adds to the alkene part of another molecule, and so on to build up long chains. However, this is not observed.

Alternatively, the radical formed could attack the carbon–carbon double bond that is part of the same molecule in the propagation step:

INITIATION

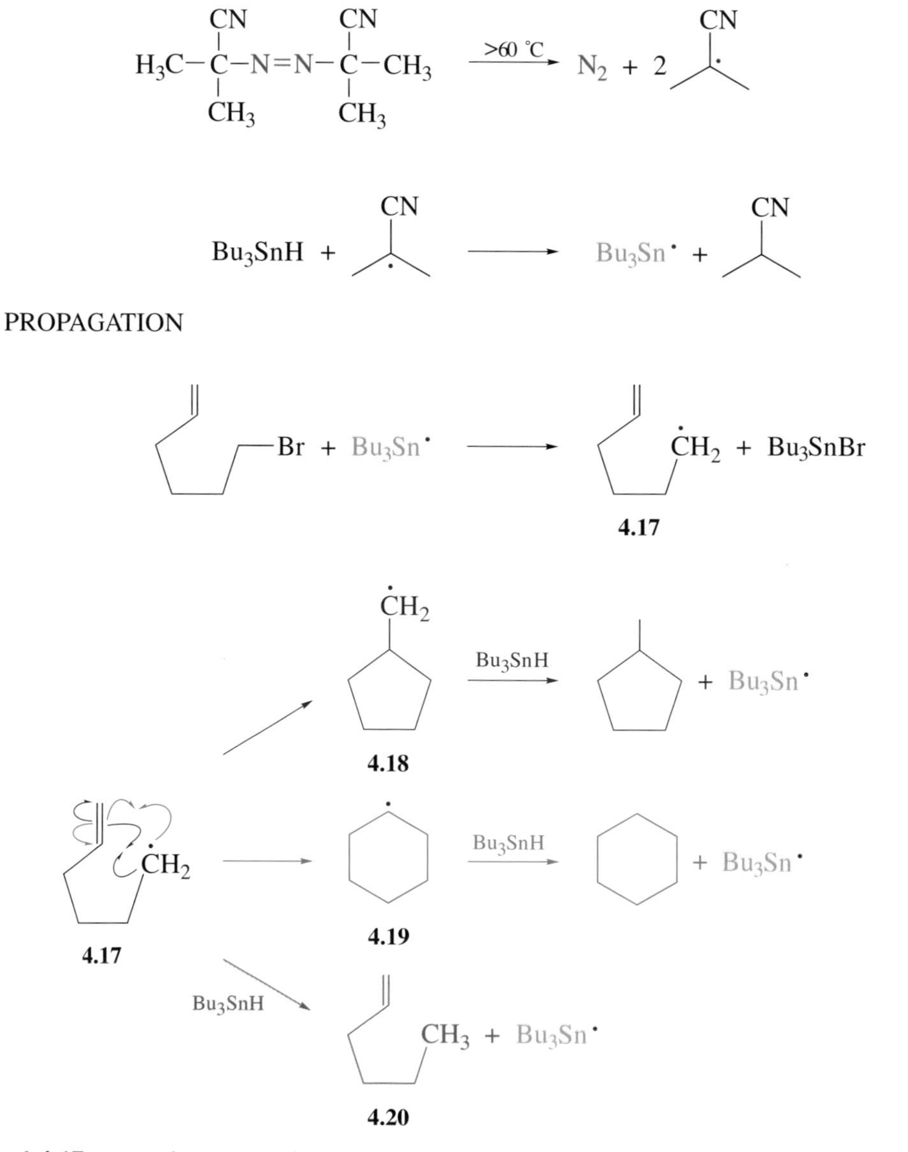


SCHEME 4.3

Radical **4.17** can undergo two alternative intramolecular cyclizations, giving radicals containing a five-membered ring (**4.18**) or a six-membered ring (**4.19**). Radical **4.17** can also react with tributyltin hydride to give the reduced product **4.20**. In fact, the first two reactions — the cyclizations — predominate, and a couple of empirical rules has been developed to help predict which of the cyclized products predominates.

1. Cyclizations are usually faster for the formation of five-membered rings than for any other ring size. For example, the cyclization of 6-bromohex-1-ene to give a five-membered ring occurs 20 times faster than does the cyclization of 7-bromohept-1-ene to give a six-membered ring on cyclization.
2. In cyclizations of any ring size, the cyclization that puts the carbon of one end of the original carbon–carbon double bond outside the newly formed ring is favoured; for example

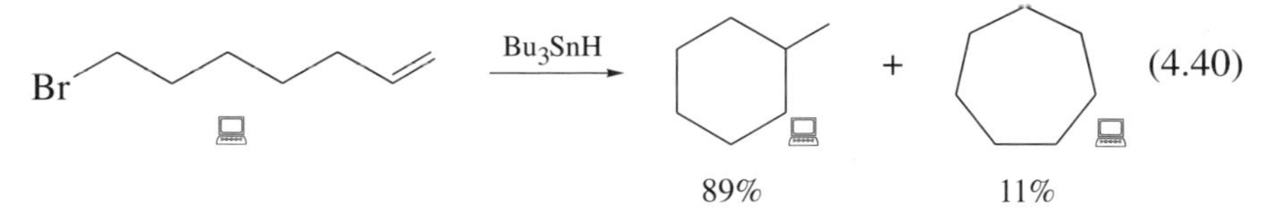


(4.40)

Formation of the five-membered ring with the carbon of one end of the original carbon–carbon double bond outside the ring is particularly favoured:

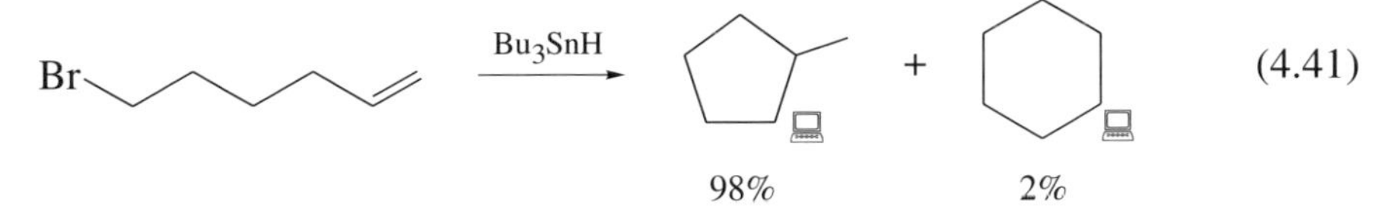


(4.41)

Formation of the five-membered ring is favoured because the geometry of the ring limits the way in which the radical can approach the alkene during its reaction. It is interesting that in these tin-mediated radical cyclization reactions, the results are contrary to what we might expect, since it is a *primary* radical that is formed in the cyclization, and it is the *most-hindered* site of the alkene which is attacked.

The ready formation of five-membered rings by radical cyclization has enabled chemists to use them in the synthesis of important natural products. We conclude this Section by looking at two such examples.

Although we have not considered them in any great detail, aryl radicals — that is, a radical on the carbon of an aromatic ring — may also be generated from aryl halides using tributyltin hydride and AIBN. Thus, this provides another important way of forming carbon–carbon bonds to benzene rings; for example

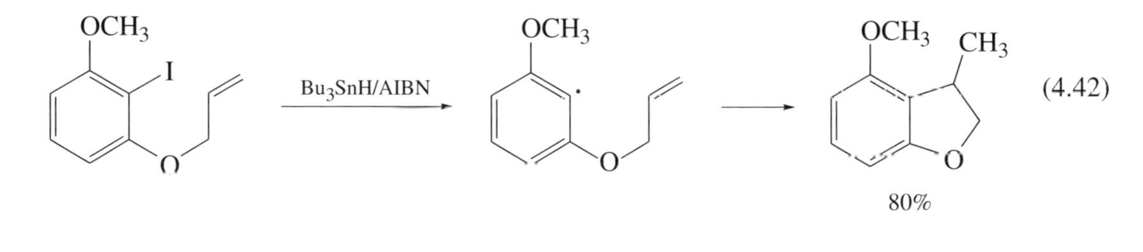


(4.42)

- Predict the product of the following reaction:

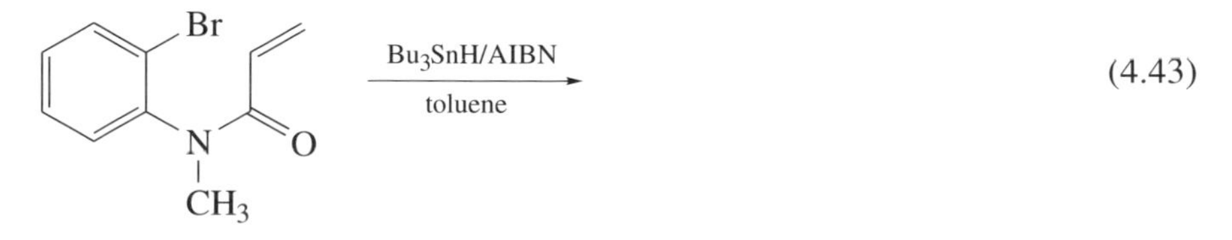


(4.43)

The product that you should have derived is **4.21**:

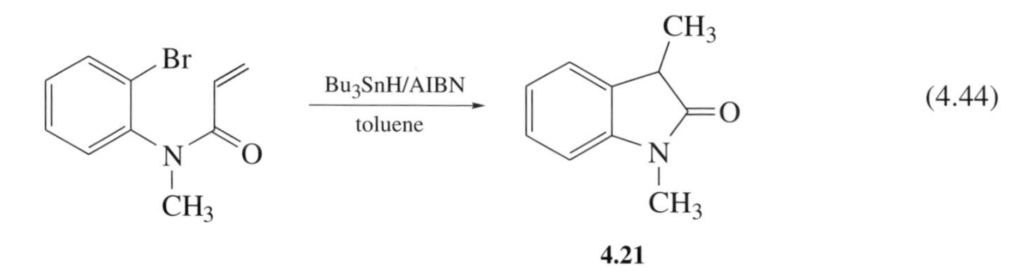


Our second example illustrates that it is possible to form more than one ring at the same time using a radical cyclization methodology. Dennis Curran and his coworkers at the University of Pittsburgh were faced with the challenge of synthesising the molecule hirsutene (**4.22**). This is one of a class of 'tricyclopentanoid' natural products, and possesses significant antibiotic and antitumour properties. One of the key structural features of this molecule is the three fused five-membered rings. He chose a substituted cyclopentene as the core central ring in the starting material (**4.23**), and decided to perform two radical cyclizations to create the other two rings. The mechanism is shown in Scheme 4.4.

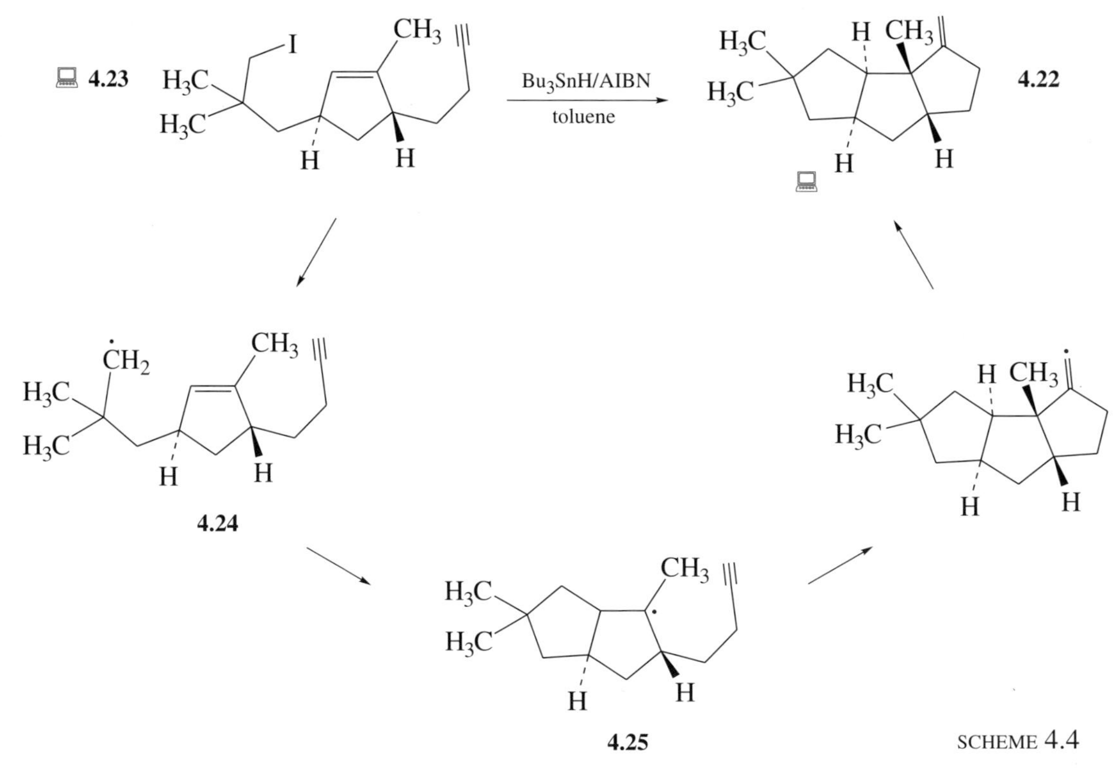


SCHEME 4.4

Treatment of **4.23** with tributyltin hydride generated an alkyl radical (**4.24**) in the normal way, and this in turn underwent a cyclization on to the double bond of the cyclopentene ring to give a new five-membered ring. However, the resultant radical (**4.25**) was not quenched by tributyltin hydride, but rather underwent a second cyclization on to the triple bond on the other side of the cyclopentene ring, again putting one end of the multiple bond outside the ring. It was the resulting vinyl radical that was then quenched by tributyltin hydride. Therefore, the overall transformation that Curran achieved was the synthesis of two five-membered rings via the same radical chain reaction.

## 4.2.3 Summary of Section 4.2

1. Tributyltin hydride is an important source of radicals in synthetic radical reactions.
2. Tributyltin hydride is normally used in conjunction with AIBN (α,α′-azo-*bis*-*iso*butyronitrile) as the initiator in a high boiling solvent such as benzene or toluene.
3. Carbon radicals undergo addition reaction with alkenes.
4. Typical reaction conditions for an intermolecular radical addition reaction are a low concentration of tributyltin hydride and a high concentration of the alkene.
5. During radical cyclizations, five-membered rings are always formed in preference to other ring sizes.
6. Cyclization that puts the carbon of one end of the original carbon–carbon multiple bond outside the newly formed ring is favoured.
7. Many variants may be introduced in radical cyclizations, such as using alkyl and aryl radicals.

### QUESTION 4.7

Predict the product of the following reaction:

O Si Br

$\xrightarrow[\text{toluene}]{Bu_3SnH/AIBN}$ (4.45)

### QUESTION 4.8

Predict the product of the following reaction. Give a full mechanism.

### QUESTION 4.9

Predict the products of the following reactions.

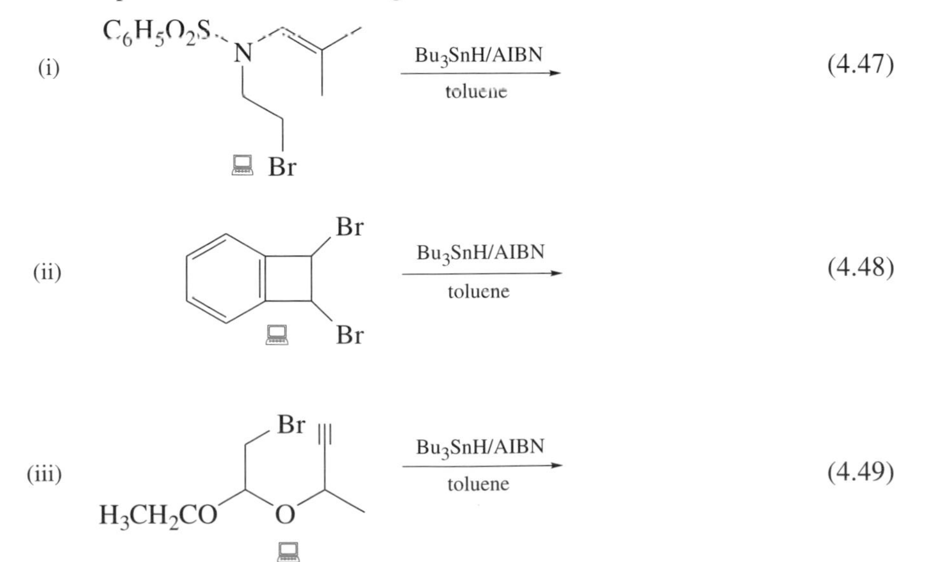

# RADICAL FRAGMENTATION REACTIONS

# 5

The final class of radical reaction (which we shall not consider in very much detail) is that of **radical fragmentation reactions**. In these reactions, a radical fragments (or literally 'falls to pieces') to form a new radical and a stable molecule. Usually, the new molecule is an alkene, making the overall process a simple elimination reaction. The following reaction represents a radical fragmentation in its simplest form:

Y + Y$^{\bullet}$ (5.1)

The key to this type of reaction is the ability of the group Y to be able to accept and stabilize the radical. The most commonly employed Y groups are trialkyltins and the thiophenyl group (**5.1**).

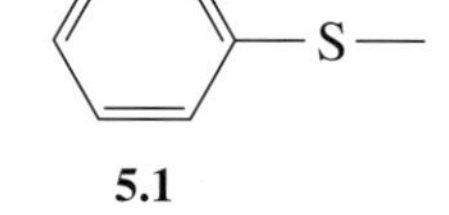


**5.1**

Write a mechanism for the following fragmentation reaction:

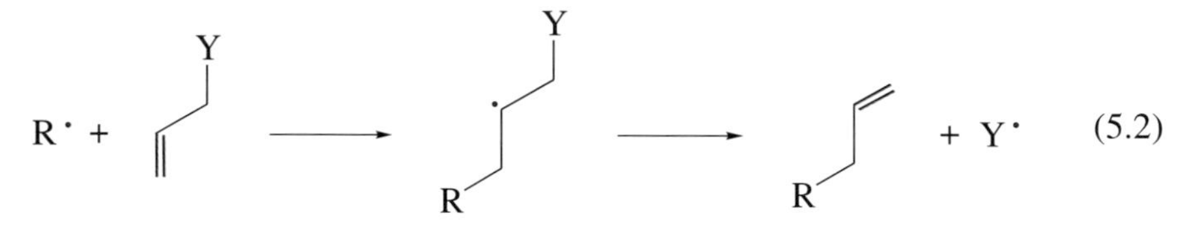


The full mechanism for the fragmentation is:

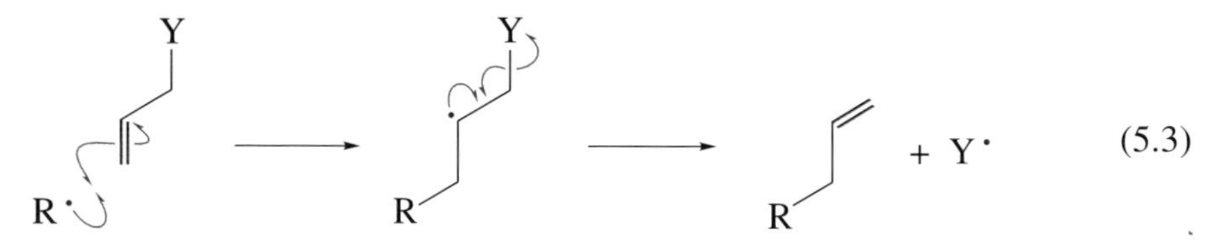


A typical example of a synthetically useful radical fragmentation reaction is shown in Scheme 5.1 in which the reactants are shown in red and the products in blue.

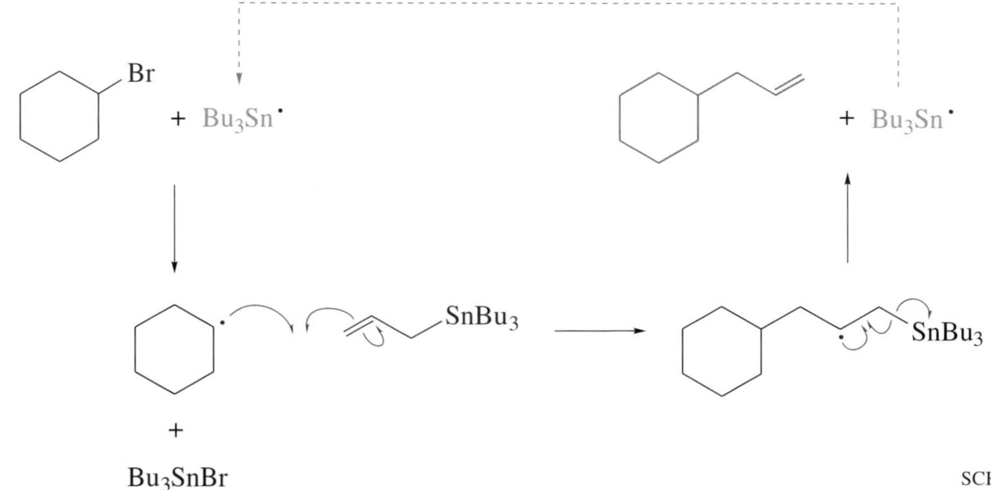


SCHEME 5.1

The tributyltin radical reacts with the bromocyclohexane to form a cyclohexyl radical. This performs a radical addition with the organotin reagent. However, on this occasion, the reaction does not terminate here: a fragmentation reaction then occurs to eliminate the tributyltin radical, and form a terminal alkene.

- What is the advantage of using an organotin compound as the starting material in this reaction?

- The radical that is formed as a result of the fragmentation process is the tributyltin radical, which may then proceed around the cycle again.

The American chemist Dale Boger has employed this approach in the synthesis of the antitumour and antibiotic agent *N*-(phenylsulfonyl)-CI (**5.3**) and related compounds. He formed one of the key intermediates (**5.2**) in the synthesis by performing a radical cyclization reaction involving elimination of a thiophenyl radical:

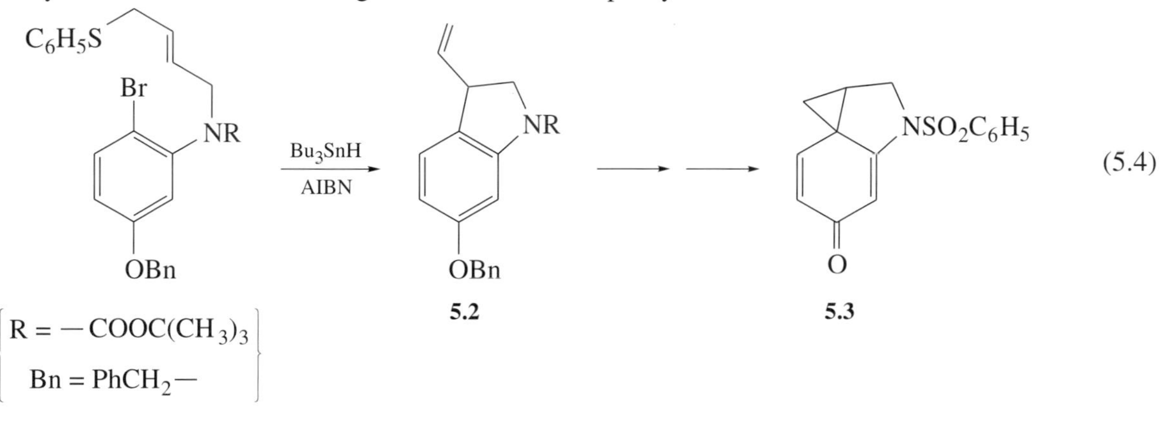


**QUESTION 5.1**

Suggest a reaction mechanism for the radical cyclization and fragmentation step in Reaction 5.4.

## 5.1 Summary of Section 5

1. Radical fragmentation reactions involve the cleavage or breaking up of a radical, to give a new smaller molecule plus a new smaller radical.
2. Typical radicals that are lost via fragmentation are the tributyltin and thiophenyl radicals.
3. Normally a double bond is formed as the result of a fragmentation reaction.

# 6 AN APPLICATION

Finally, to bring all the various threads of radical chemistry together and to illustrate the power of aryl radicals, cyclization methodologies and fragmentation reactions, we are going to look at the synthesis of the analgesic alkaloid* (±)-morphine (**6.1**). The American radical chemist, Kathy Parker, used an aryl radical cyclization, followed by the fragmentation/elimination of a thiophenyl radical as the key two ring-forming steps in her total synthesis of (±)-morphine.

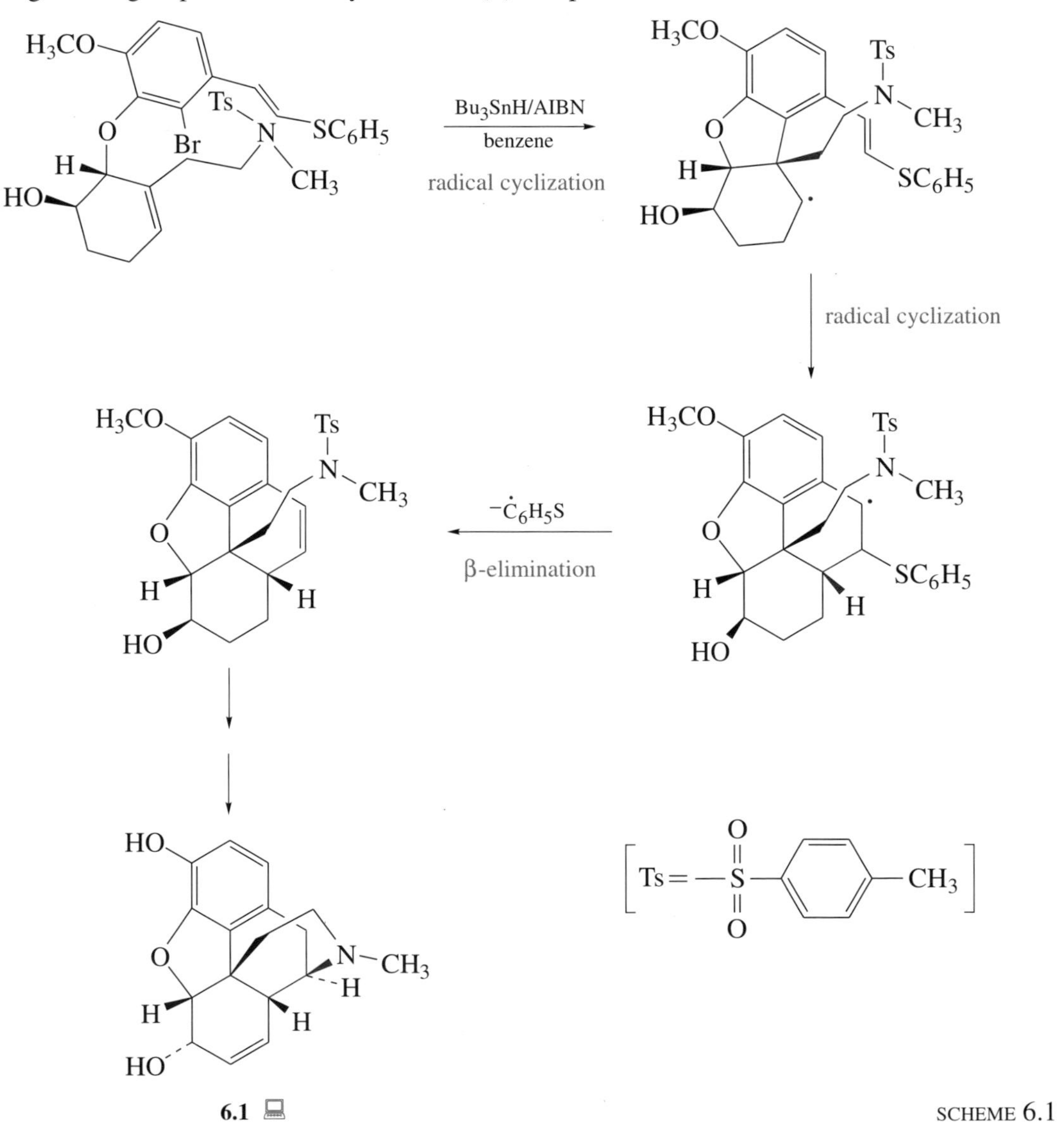


SCHEME 6.1

* An alkaloid is a basic nitrogen-containing natural product found in plant material. Alkaloids are characterized by their specific physiological action and toxicity. Many plants use them as a defence against herbivores, particularly insects. They are usually related to organic bases such as pyridine, quinoline, pyrrole, etc.

Treatment of the aryl bromide under the standard tributyltin hydride/AIBN conditions generates an aryl radical, which immediately undergoes radical cyclization at the cyclohexene double bond. However, rather than being quenched immediately by tributyltin hydride, the radical undergoes a second radical cyclization to the double bond in the side arm (the rigidity of the system forces it to form a six-membered ring rather than the usually more common five-membered ring). Finally, there is a fragmentation (β-elimination) reaction to eliminate a thiophenyl radical and establish a double bond in the newly formed ring. Thus, two new rings and a double bond have been produced in one radical reaction before the radical is finally quenched by the tributyltin hydride. The product could then be easily converted into the target morphine molecule.

## BOX 6.1 Morphine

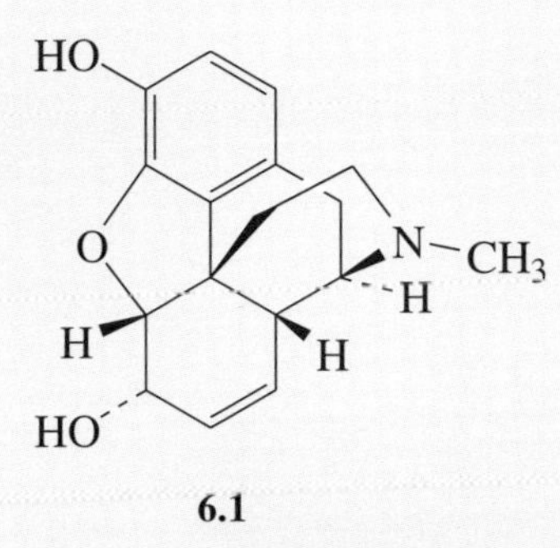


**6.1**

In 1805 the first alkaloid was isolated from a plant — the poppy. Its discoverer was a young German pharmacist called F. W. A. Sertürner, and the alkaloid was morphine (**6.1**), the principal constituent of opium.

Morphine takes it name from the Greek god of sleep and dreams, Morpheus, since the drug promotes drowsiness, although not necessarily culminating in sleep. Morphine is actually just one member of the opioid group of analgesics, which also includes heroin, codeine and methadone*. All are characterized as being powerful painkillers, but morphine is by far the safest, and is non-addictive in patients. It has found general clinical use as a powerful painkiller, particularly in chronically and terminally ill patients (e.g. in many forms of cancer).

**Figure 6.1**
Morphine poppies.

* Detection of morphine, heroin and codeine is discussed in the Forensic Science Case Study in *Separation, Purification and Identification*[2].

# 7 CONCLUSION

In Part 3 of this Book we have made an in-depth study of the chemistry and reactions of radicals. Hopefully, you will appreciate that rather than being the highly reactive and uncontrollable entity that they were considered to be at first, they are in fact highly controllable and useful reactive intermediates, which the synthetic chemist may now use and exploit in many applications.

# LEARNING OUTCOMES

When you have completed *Mechanism and Synthesis: Part 3— Radical reactions in organic synthesis*, you should be able to:

1 Recognize valid definitions of and use in a correct context, the terms, concepts and principles in the following Table (All Questions).

List of terms, concepts and principles introduced in *Radical reactions in organic synthesis*

| Term | Page number | Term | Page number |
|---|---|---|---|
| acyloin reaction | 137 | McMurry reaction | 137 |
| AIBN | 150 | pinacol coupling reaction | 132 |
| aprotic solvent | 132 | propagation stage | 141 |
| benzyl radical | 146 | protic solvent | 132 |
| bond homolysis | 123 | radical | 123 |
| captodative radicals | 127 | radical anion | 132 |
| fishhook curly arrows | 124 | radical chain reaction | 140 |
| heterolytic bond cleavage | 123 | radical fragmentation | 160 |
| homolytic bond dissociation energy | 125 | radical–radical coupling | 132 |
| initiation stage | 141 | selective reaction | 152 |
| initiator | 135 | steroid | 134 |
| intramolecular reaction | 135 | termination stage | 142 |
| ketyl radical | 132 | | |

2 Understand the difference between full (double-headed) and fishhook arrows when writing reaction mechanisms. Use fishhook (single-headed) arrows to represent the mechanism of radical reactions. (All Questions)

3 Recognize that different bonds have different bond energies. Understand the concept of initiators in radical reactions as compounds containing bonds that will cleave homolytically relatively easily, and the stability of the resultant radical (primary, secondary and tertiary) and captodative radicals. (Questions 1.1–1.4)

4 Recognize and predict the products of the pinacol, McMurry and acyloin reactions, and give reaction mechanisms for each of these reactions. (Questions 3.1–3.4)

5 Recognize radical chain reactions and be able to identify initiation, propagation and termination stages in a variety of radical chain reactions. You should also be able to draw fishhook arrow mechanisms for each stage. (Questions 4.2, 4.6–4.9 and 5.1)

6 Predict the product(s) of radical halogenation reactions of alkanes, and predict their relative abundances. Discuss the factors controlling these reactions, such as the differences in ionic and radical halogenation of alkylbenzenes and the stereochemical outcome of these reactions. Give a fishhook arrow mechanism for every step of the radical chain reaction. (Questions 4.1–4.5)

7 Be familiar with the chemistry of tributyltin hydride in radical reactions and recognize the importance of AIBN in initiating tin hydride radical reactions. (Questions 4.6–4.9)

8 Predict the products of both inter- and intramolecular radical chain reactions involving tributyltin hydride, and alkyl or aryl halides and alkenes. You should also be able to predict the different products possible from these reactions. (Questions 4.6–4.9)

9 Recognize that suitable leaving groups can lead to radical fragmentation reactions. (Question 5.1)

# QUESTIONS: ANSWERS AND COMMENTS

## QUESTION 1.1 *(Learning Outcome 3)*

Any combination of an electron-withdrawing group with an electron-donating group should lead to an extra-stable captodative radical; examples of these groups are

electron-donating groups: alkyl groups, —OR, $—NR_2$

electron-withdrawing groups: —CN, —COR, —COOR, $—NO_2$

Thus, we can combine $CH_3$ with CN; OR with COR; OR with $NO_2$; and so on.

## QUESTION 1.2 *(Learning Outcome 3)*

Only the last example in the list is a captodative radical. The methyl radical (i) has no stabilization at all, since it only has hydrogen attached. All the radicals (ii), (iii) and (iv) have some stabilization because they possess either an electron-donating ($OCH_3$) or electron-withdrawing group (CN or $CO_2CH_3$). However, only radical (v) benefits from the extra stabilization gained from having both an electron-donating group and an electron-withdrawing group attached.

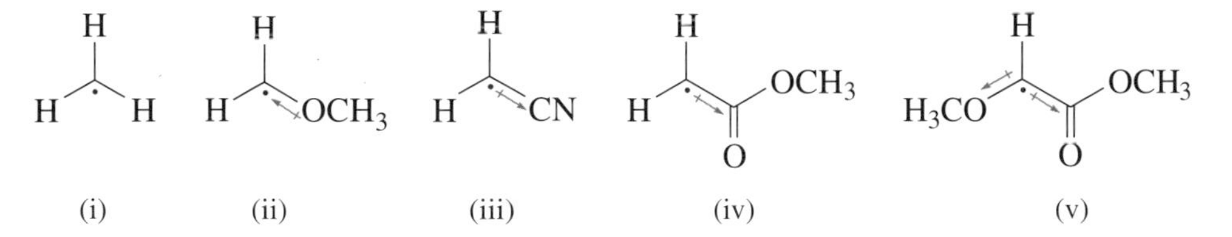


## QUESTION 1.3 *(Learning Outcome 3)*

From the values in Table 1.1, you should have arrived at the following order for bond strengths:

O=O > H—H > H—Cl > H—Br > Br—Br > F—F

strongest weakest

## QUESTION 1.4 *(Learning Outcome 3)*

From the representative values for similar bonds in Table 1.1, you would expect that the carbon–fluorine bond would be the strongest. Next is a primary C—H bond followed by a secondary C—H bond and a weak C—I bond. The weakest link is the O—O bond in the peroxide, which breaks very easily and hence explains why many peroxides are used as radical initiators.

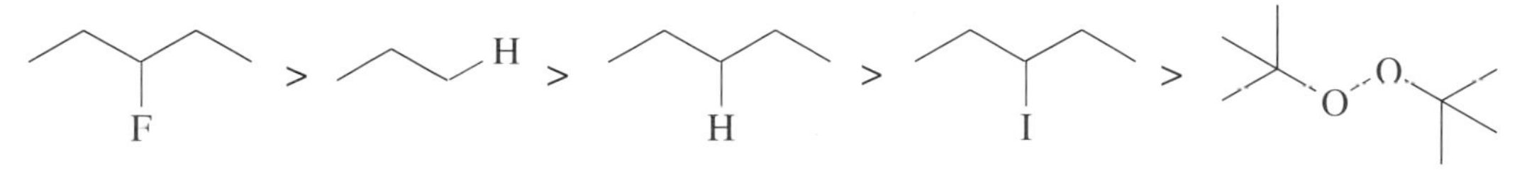

### QUESTION 3.1 (*Learning Outcome 4*)

The product of this reaction is the pinacol-coupled diol **Q.1**. This is a straightforward application of the pinacol reaction (Scheme Q.1).

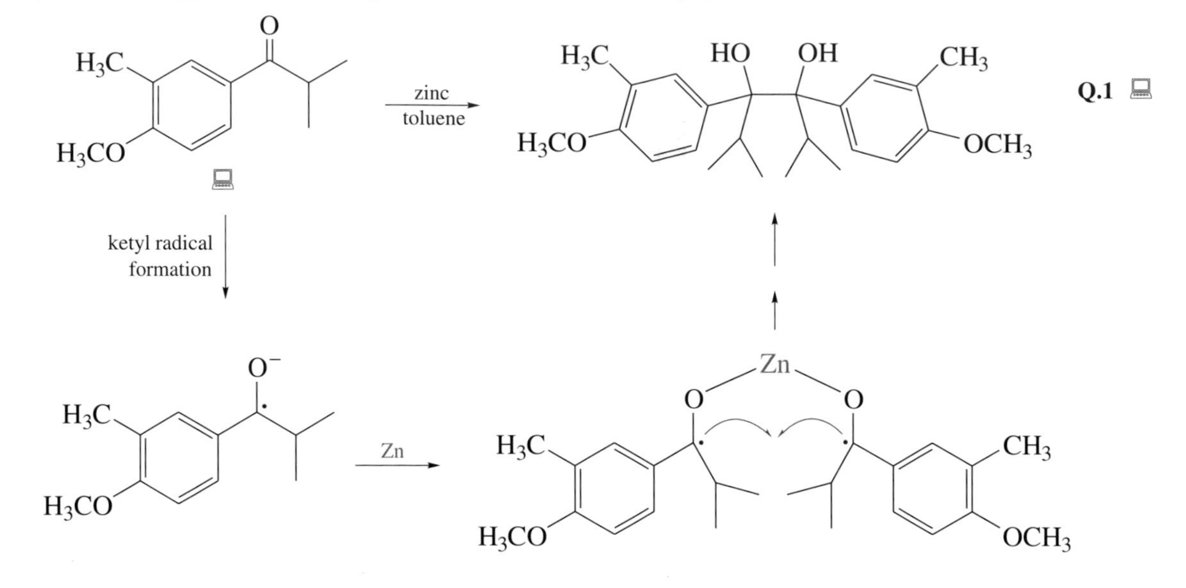


SCHEME Q.1

Ketyl radical generation occurs in the first step, which is followed by coordination to the metal (zinc) centre. Finally, radical dimerization followed by hydrolysis occurs to give the product.

### QUESTION 3.2 (*Learning Outcome 4*)

This is an example of the acyloin reaction, which is very similar in mechanism to pinacol coupling but involves the reaction of two esters instead of two aldehydes or ketones. The product and mechanism are shown in Scheme Q.2.

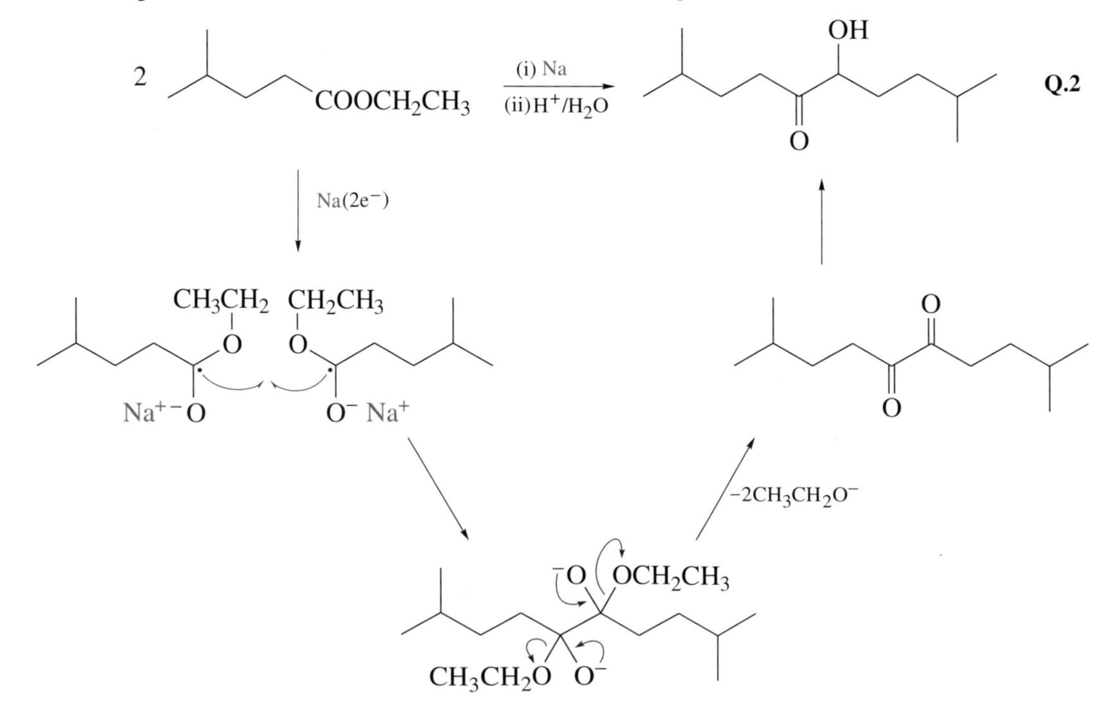


SCHEME Q.2

Ketyl radical formation and radical dimerization occur in the now familiar way, but this is followed by the loss of two ethoxide ions to give a 1,2-diketone. This is an unstable molecule under the reaction conditions, and is thus reduced to the α-hydroxyketone **Q.2**.

### QUESTION 3.3 (*Learning Outcome 4*)

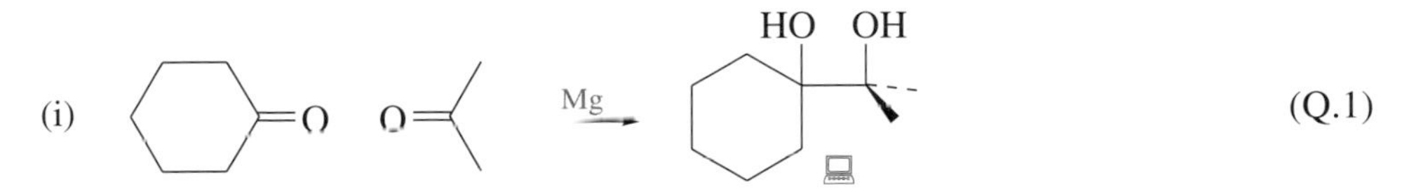


This is a straightforward application of the pinacol coupling reaction: we are employing two different ketones, but this has no effect on the outcome of the reaction! Ketyl radical formation from both ketones is followed by coordination to the magnesium and radical dimerization. Acidic work-up yields the product shown. In reactions such as this, it is inevitable that some self-dimerization will also occur, so low levels of two other products (**Q.3** and **Q.4**) are also seen.

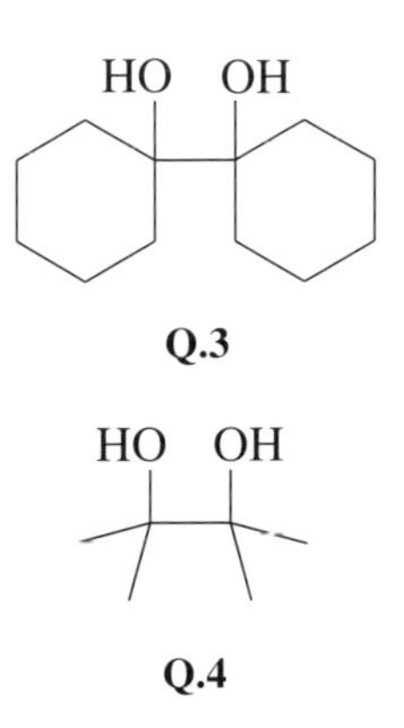


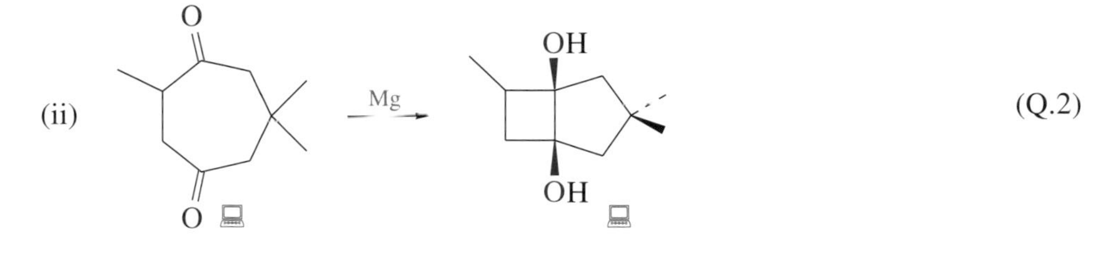


Although this is also a simple pinacol coupling, it may appear more complex as it is an intramolecular reaction, leading to the formation of two condensed rings. The two ketyl radicals form and dimerization occurs in the normal way. However, coordination to the metal is not so essential in this case, since the radicals formed are already in the same molecule. After work up, we obtain the product shown:

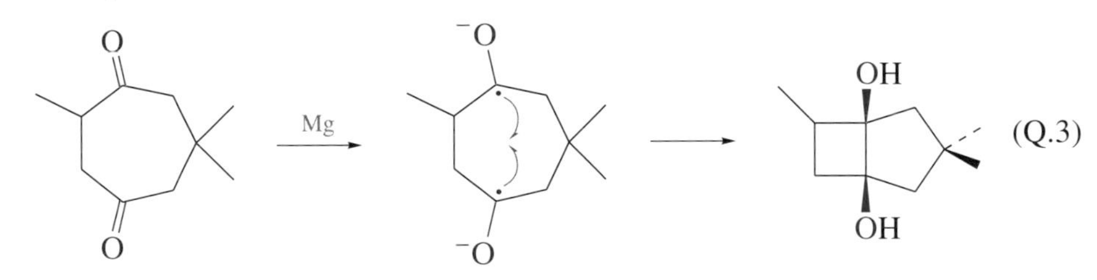

## QUESTION 3.4 (*Learning Outcome 4*)

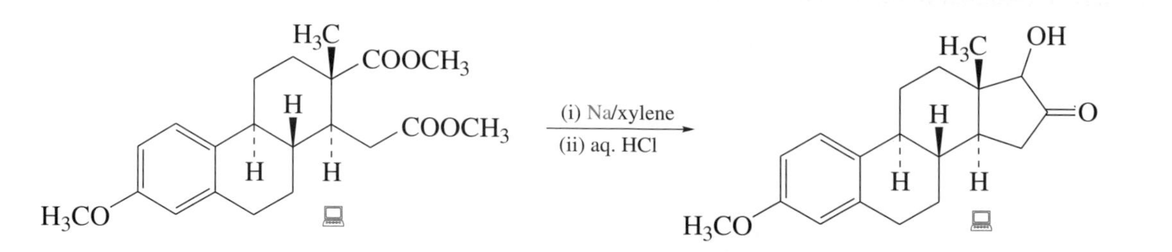


This is an application of the acyloin condensation, which is similar to the pinacol coupling reaction, but involves two esters instead of the aldehyde or ketone. Ketyl radical formation is followed by radical dimerization, and the loss of two $CH_3O^-$ ions to yield a 1,2-diketone. This functionality is even more susceptible to reduction under the reaction conditions, and after rearrangement gives the α-hydroxyketone.

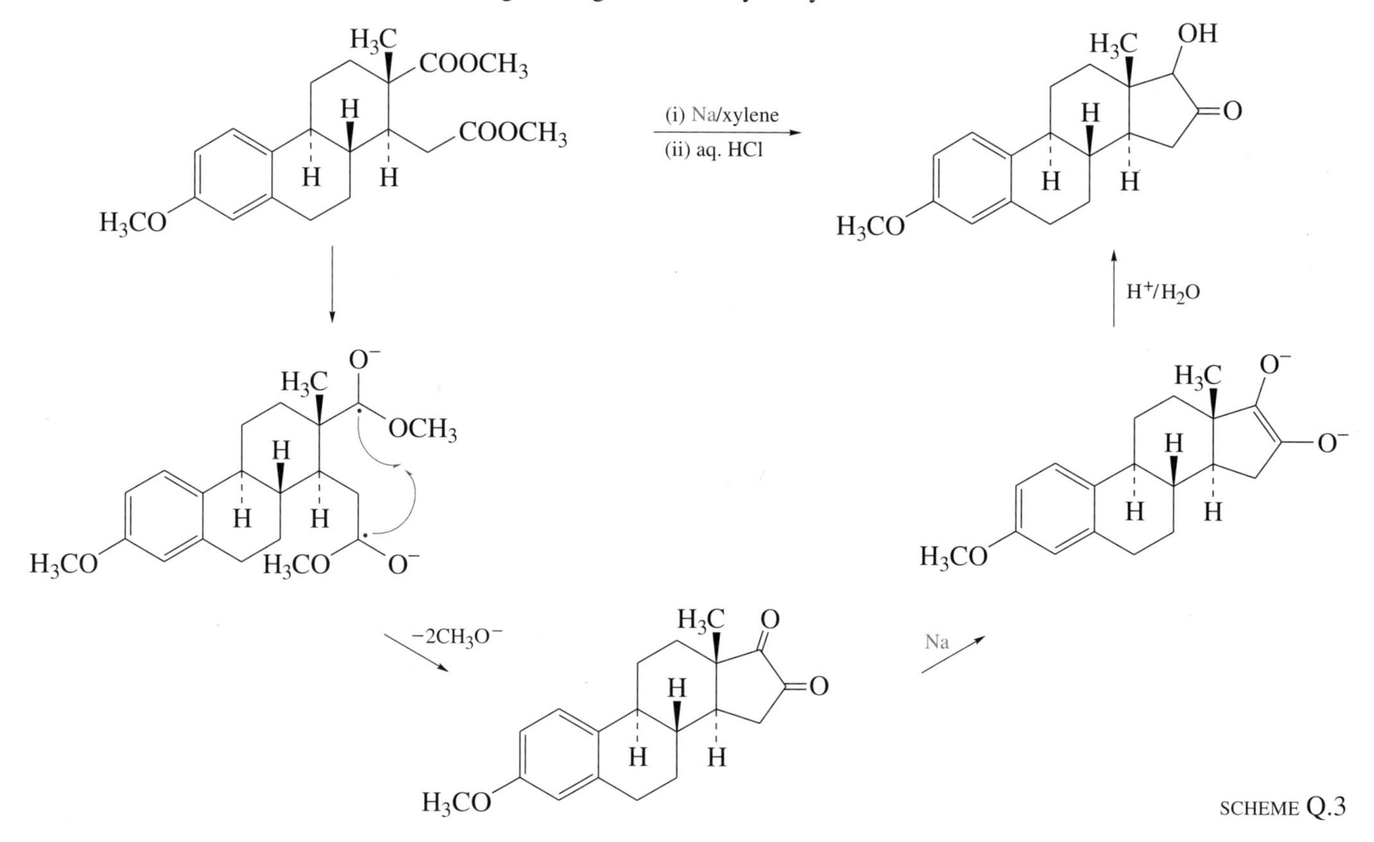


SCHEME Q.3

In the last step, addition of water can also lead to the alternative α-hydroxyketone **Q.5**:

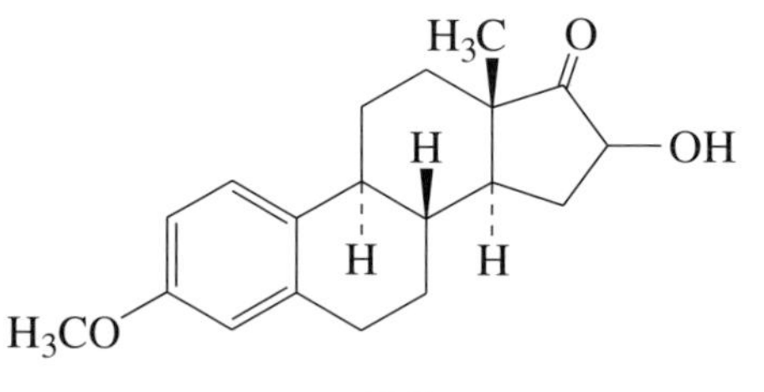


**Q.5**

QUESTION 4.1 (*Learning Outcome 6*)

The two possible products of monochlorination are shown in Reaction Q.5:

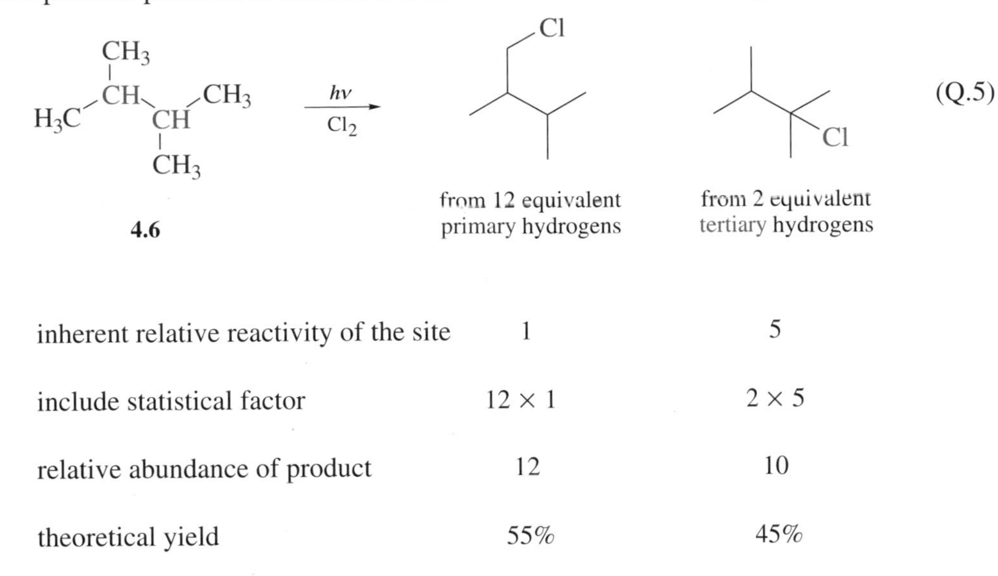


| | | |
|---|---|---|
| inherent relative reactivity of the site | 1 | 5 |
| include statistical factor | 12 × 1 | 2 × 5 |
| relative abundance of product | 12 | 10 |
| theoretical yield | 55% | 45% |

QUESTION 4.2 (*Learning Outcomes 5 and 6*)

Attack of a chlorine radical at $H_a$ gives a primary radical as the reaction intermediate; attack at $H_b$ gives a stabilized benzylic radical. (This is stabilized by delocalization around the aromatic ring). The benzylic radical is much favoured over the primary radical and so this is the major reaction pathway. Reaction of the benzylic radical with chlorine yields the major product; reaction of the primary radical with chlorine yields the minor product.

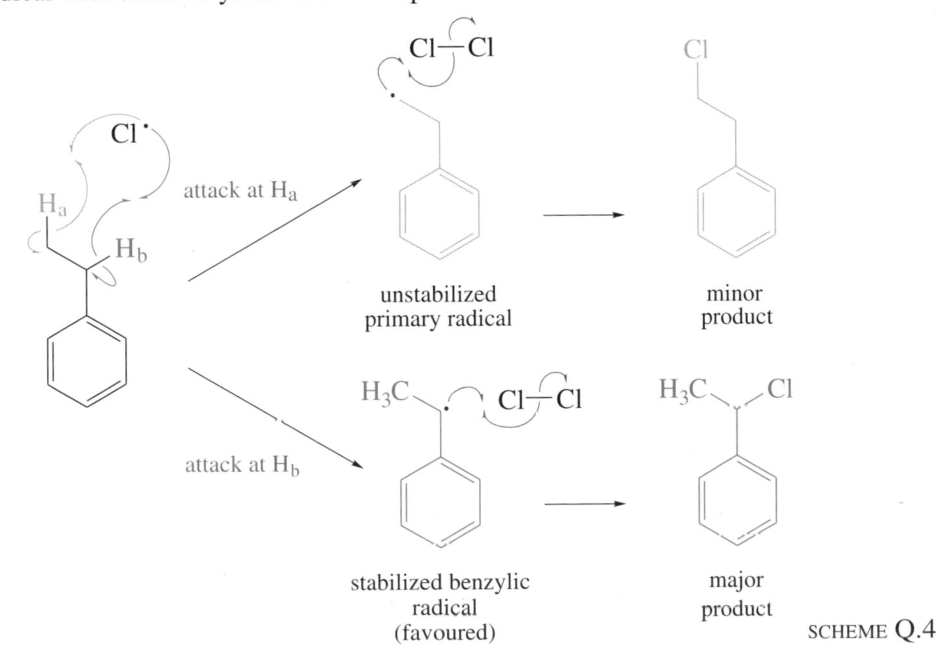


SCHEME Q.4

## QUESTION 4.3 (*Learning Outcome 6*)

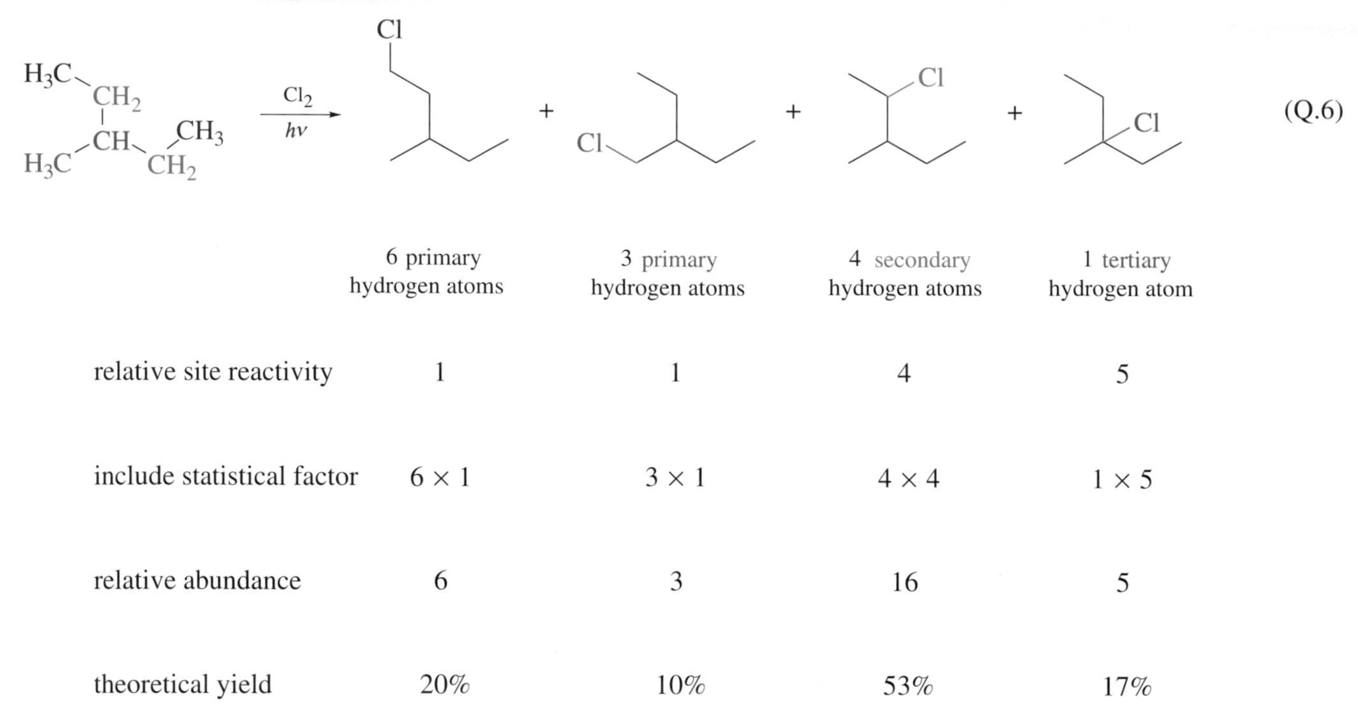


| | 6 primary hydrogen atoms | 3 primary hydrogen atoms | 4 secondary hydrogen atoms | 1 tertiary hydrogen atom |
|---|---|---|---|---|
| relative site reactivity | 1 | 1 | 4 | 5 |
| include statistical factor | $6 \times 1$ | $3 \times 1$ | $4 \times 4$ | $1 \times 5$ |
| relative abundance | 6 | 3 | 16 | 5 |
| theoretical yield | 20% | 10% | 53% | 17% |

## QUESTION 4.4 (*Learning Outcome 6*)

Halogenation of an aromatic ring will only take place under ionic conditions, and halogenation of the benzylic position only under radical conditions. Therefore the first reaction, the ring bromination of butylbenzene, must occur under Lewis acid-promoted conditions; for example, use bromine with $AlBr_3$. (This would probably lead to a mixture of *ortho* and *para* products.) The second part of the question asks for chlorination at the benzylic position. Substitution at this site will only take place under radical reaction conditions. Therefore chlorine with light as the initiator will effect side-chain halogenation. Thus, the reagents and conditions are as shown in Reaction Q.7:

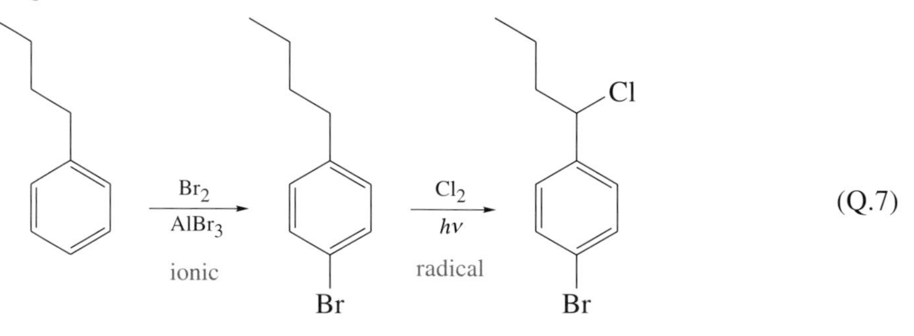


(Q.7)

## QUESTION 4.5 (Learning Outcome 6)

The configuration at the chiral centre is *R*. The priorities of the groups are shown in Structure Q.6:

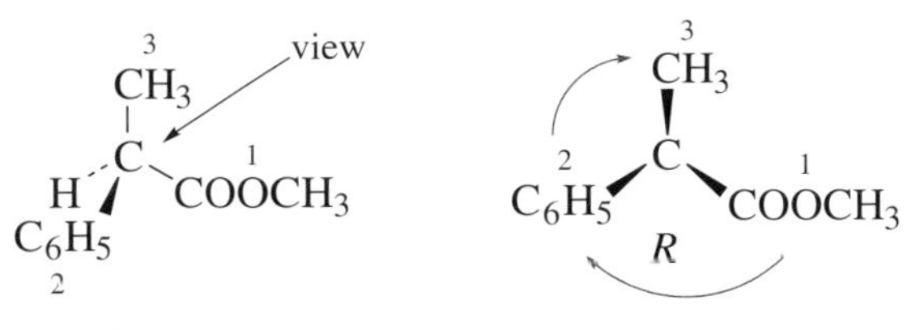


**Q.6**

The major product of bromination would arise from substitution at the benzylic position. As the intermediate radical formed is planar, the product is formed as a racemic mixture:

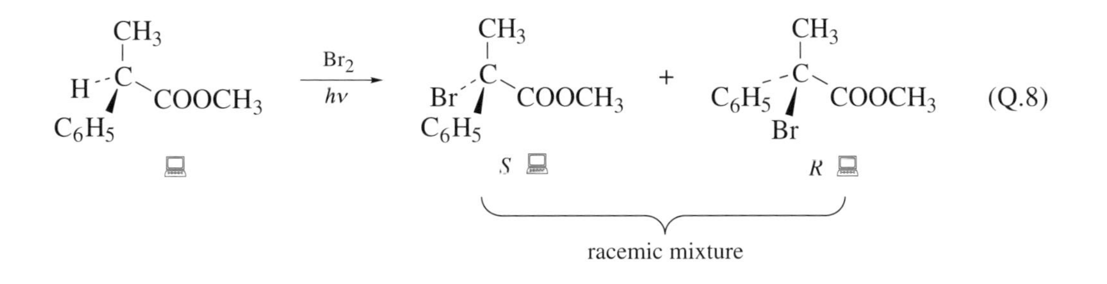


## QUESTION 4.6 (Learning Outcomes 5 and 7)

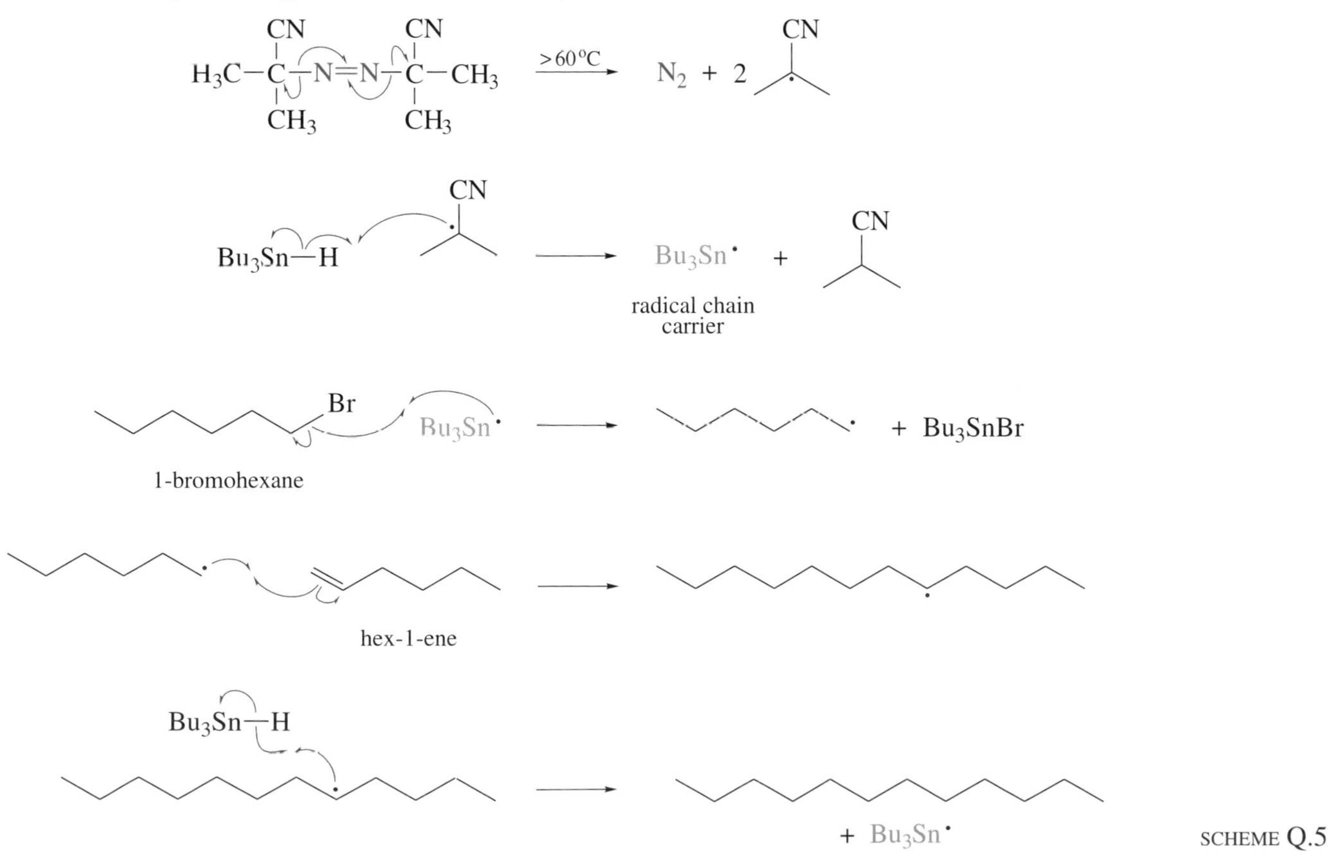


SCHEME Q.5

## QUESTION 4.7 (Learning Outcomes 5 and 7)

This question is primarily asking if you can understand how a radical-mediated cyclization reaction occurs. The fact that there are silicon and oxygen atoms in the chain (and thus the ring) is irrelevant.

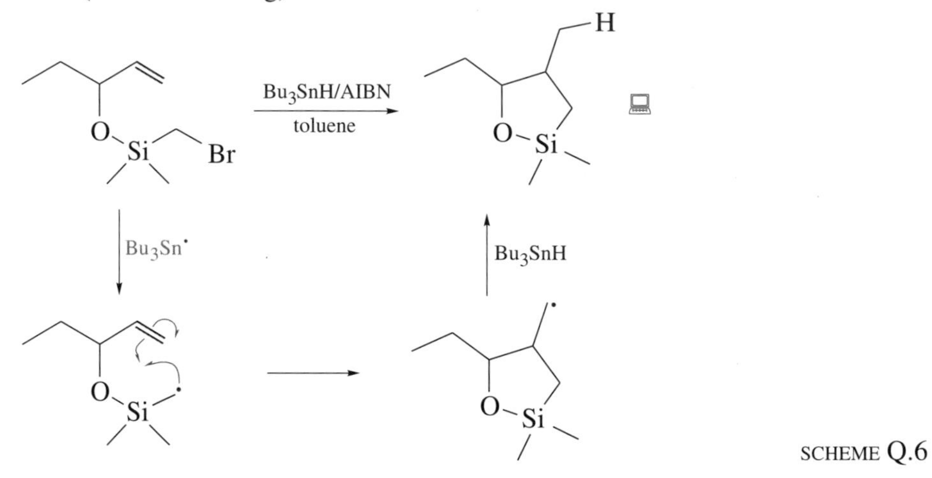


SCHEME Q.6

After chain initiation (from AIBN), the tin radical generated from the tributyltin hydride reacts with the bromine to form a strong Sn—Br bond and thus generates a carbon-centred radical. This radical then undergoes a cyclization, to give a five-membered ring. Reaction with tributyltin hydride then quenches the carbon radical and regenerates the tin radical to continue the chain reaction.

## QUESTION 4.8 (Learning Outcomes 5 and 7)

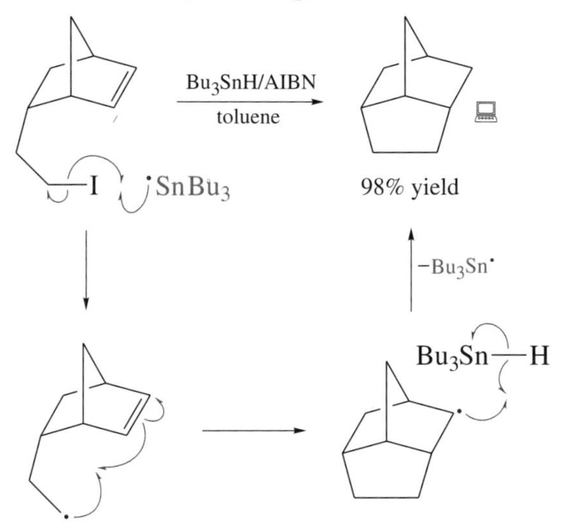


SCHEME Q.7

## QUESTION 4.9 (*Learning Outcomes 5 and 7*)

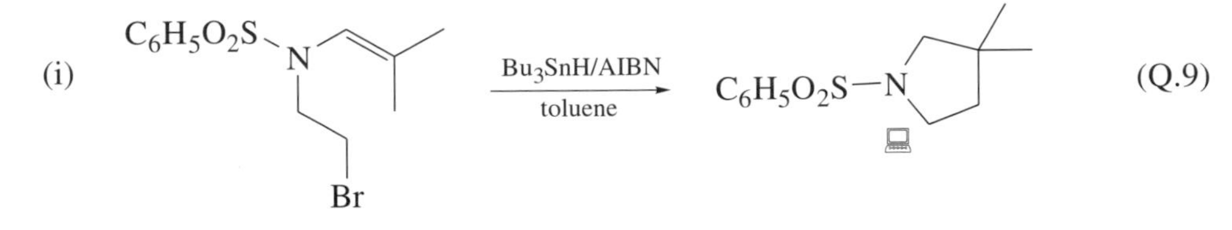


In this case, formation of the ring with the carbon–carbon bond outside the ring is disfavoured by the ring strain resulting from formation of a four-membered ring.

(ii) Since no other reactants are involved, the second example is a simple replacement of bromine with hydrogen.

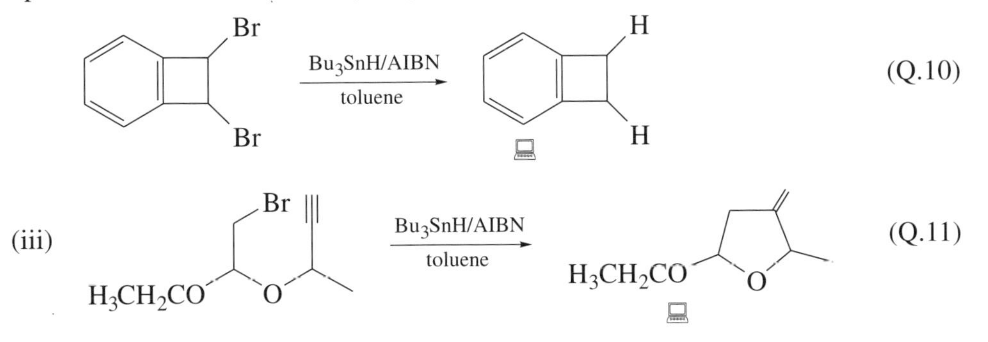


## QUESTION 5.1 (*Learning Outcome 3*)

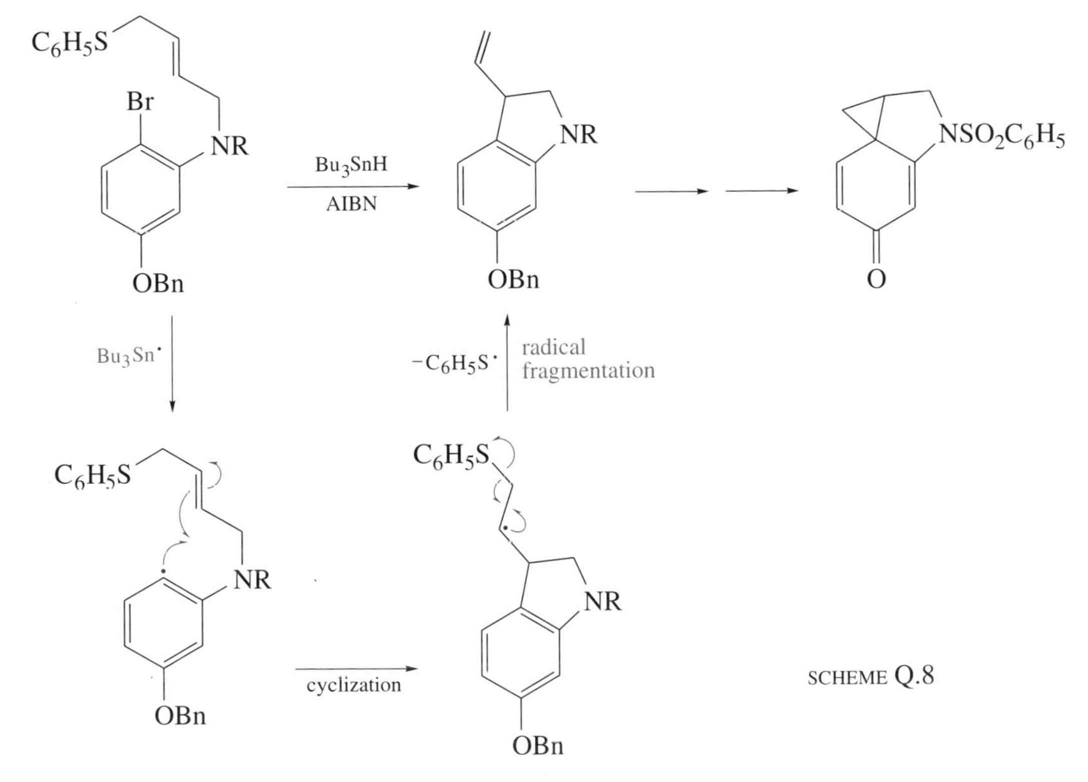


SCHEME Q.8

# FURTHER READING

1 P. G. Taylor and J. M. F. Gagan (eds.), *Alkenes and Aromatics*, The Open University and the Royal Society of Chemistry (2002).

2 L. E. Smart (ed), *Separation, Purification and Identification,* The Open University and the Royal Society of Chemistry (2002).

# ACKNOWLEDGEMENTS

Grateful acknowledgement is made to the following sources for permission to reproduce material in this book:

*Figure 1.2*: reproduced by permission from *Chemical & Engineering News*, **44** (41), October 1966; *Figure 1.4*: Paul Jaronski, University of Michigan Photo Services; *Figures 3.1 and 3.2*: Science Photo Library; *Figure 6.1*: Oxford Scientific Films, (1993).

Every effort has been made to trace all the copyright owners, but if any has been inadvertently overlooked, the publishers will be pleased to make the necessary arrangements at the first opportunity.

*Part 4*

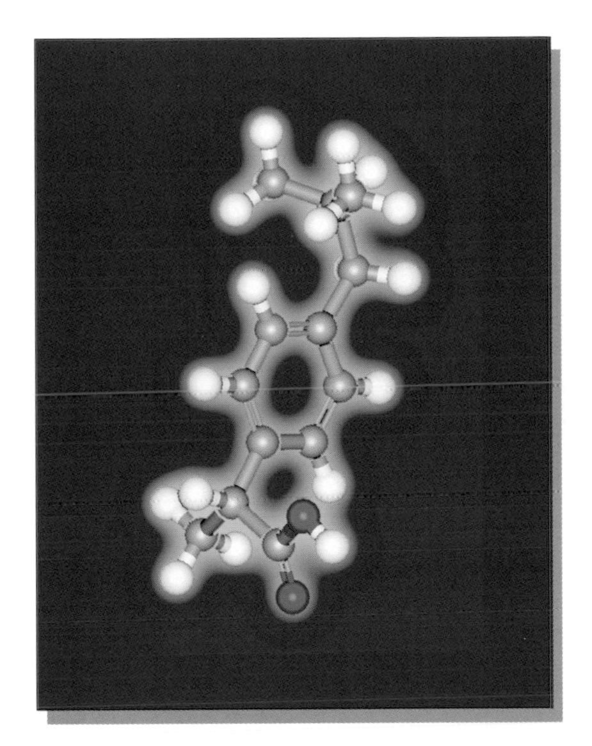

# Strategy and methodology in organic synthesis

edited by Peter Taylor

*based on*

Strategy and methodology in organic synthesis

*by Jim Iley, Ray Jones and John Coyle*

# SYNTHESIS IN ORGANIC CHEMISTRY

1

Much of organic chemistry is about learning the chemistry of functional groups, and the variety of organic reaction mechanisms that may be followed as one molecule is transformed into another. For example, in this Book you have seen how carbonyl compounds and organometallic compounds react with a variety of reagents, and how we perceive electrons to be redistributed during such reactions. By presenting organic chemistry to you using this mechanistic, functional group approach, we have attempted to provide you with a set of ground rules that can be used to understand, and perhaps predict, what the chemistry of an unknown compound might be.

However, we need to go further. Even though organic chemistry would still be exciting and important, it is not simply about single-stage transformations. In fact, it is true to say that much of the creativity of organic chemistry can be seen when a small organic molecule is converted to a much more complex organic molecule by a series of transformations. Herein lies the essence of what is usually meant by 'organic synthesis': firstly, one compound is converted into another compound; secondly, the starting material is more readily available than the product; thirdly, the sequence of reactions between starting material and product involves two or more separate stages*. Organic synthesis is often referred to as an 'artform', and the word 'elegant' is often used for syntheses that accomplish a difficult transformation in high yield with the minimum number of steps. It is into the realm of organic synthesis that we now travel.

Why do chemists embark on multi-step syntheses of organic compounds? There are many reasons. Often, it is to correlate the physical properties of a compound or material with its molecular structure. This is particularly important, for example, in the pharmaceutical and polymer industries, where precise physical properties are required. This leads to the need for precise molecular characteristics, for example shape, and the position of charges and functional groups, which in turn demands *careful molecular design* and synthesis.

Alternatively, a new synthesis may be attempted to provide a better, less wasteful and more economic route to important compounds. Sometimes, a synthesis of a compound is undertaken in order to confirm its structure unambiguously. It is not uncommon for a synthesis to be attempted simply because it will be extremely difficult, perhaps because the final compound has a complex structure. This demands great ingenuity on behalf of the chemist to draw up a synthetic plan that will achieve the goal. In some ways this is the chemist's equivalent of the mountaineer's 'Because it's there!'. It was rather

* These ideas are exemplified by the synthesis of pseudoephedrine in Part 3 of *Alkenes and Aromatics*[1].

well expressed by the American chemist R. B. Woodward (famous for many syntheses of natural products such as cortisone):

> Should I ask here explicitly why the chemist synthesises things? From the point of view of pure science, the question answers itself, and beyond that, there are obvious practical reasons for such activities, which are certain to become more compelling in the future. But I should like to mention here a basis for action more related to the spirit of man. The structure known, but not yet accessible by synthesis, is to the chemist what the unclimbed mountain, the uncharted sea, the untilled field and the unreached planet are to other men. The achievement of the objective in itself cannot but thrill all chemists, who even before they know the details of the journey, can apprehend from their own experience the joys and elations, the disappointments and false hopes, the obstacles overcome, the frustrations subdued, which they experienced who traversed a road to the goal. The unique challenge which chemical synthesis provides for the creative imagination and skilled hand ensures that it will endure as long as men write books, paint pictures, and fashion things which are beautiful, or practical, or both.
>
> ***R.B. Woodward*, Pointers and Pathways in Research *(1963)***

Whatever the reasons for the synthesis, organic chemists generally use a similar set of guiding principles in order to help them draw up a plan of the synthesis they wish to carry out. This Part of the Book attempts to set out those principles so that you too can devise suitable syntheses for a given molecule.

# 2 REQUIREMENTS FOR SYNTHESIS

Before we involve ourselves in detail with the processes that are used to develop a **synthetic plan or strategy** for making a compound, we shall start by identifying what, in general, are the requirements of synthesis.

Take a look at the following synthesis of the potential perfumery compound **2.1**.

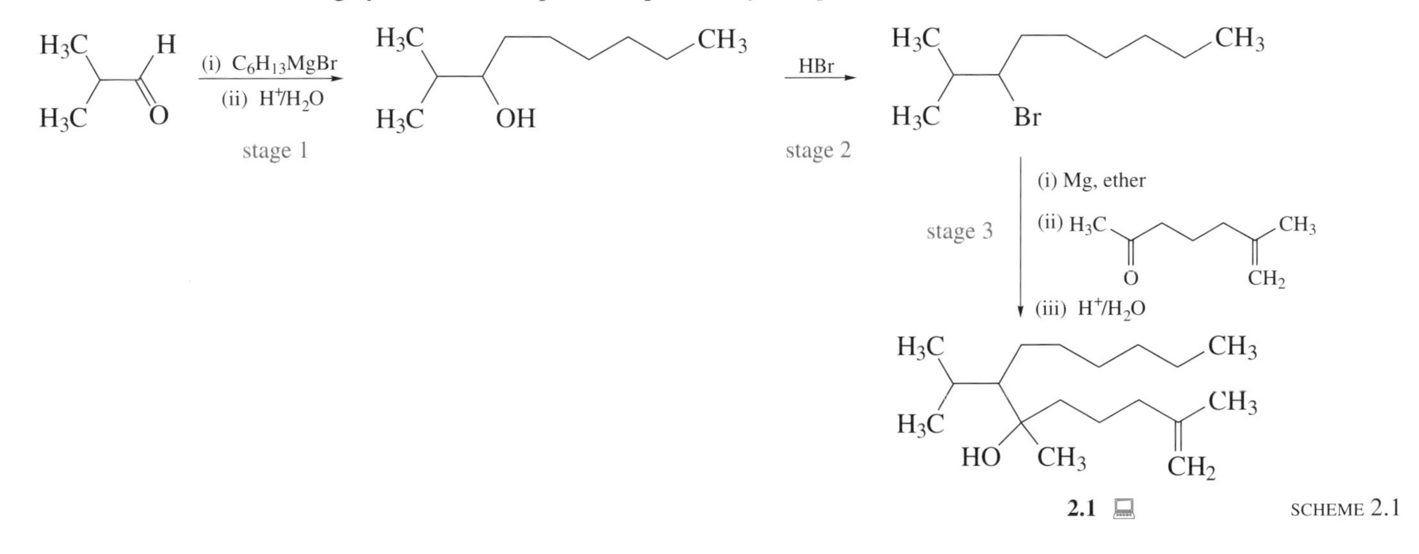


SCHEME 2.1

- What does this strategy achieve as the synthesis of the desired compound (the target molecule) progresses?
- The carbon framework of the target molecule is gradually built up.

> This is the first requirement of synthesis: the strategy should attempt to build up the **molecular framework** of the target molecule by using simpler compounds that contain fragments of the framework under scrutiny.

- Identify the stages and types of reaction that are used in this synthesis of compound **2.1** to assemble the framework of the target molecule.
- The molecular framework is assembled in the first and third stages of the synthesis, which both involve C—C bond-forming reactions. These reactions are accomplished using Grignard reagents.

In this synthesis, and indeed in all syntheses of organic compounds, **C—C bond-forming reactions** are central to the assembly of the appropriate molecular framework. It is perhaps not too surprising, then, that organometallic reagents, carbonyl compounds and haloalkanes assume a central importance in synthesis. As far as Part 4 is concerned, these C—C bond-forming reactions require one fragment to contain a nucleophilic carbon atom, and a second fragment to contain an electrophilic carbon atom. This idea will help us later in identifying appropriate strategies for synthesis.

$RCH_2$—MgX + C=O → $RCH_2$–C–OMgX (new carbon–carbon bond) (2.1)

nucleophilic carbon; electrophilic carbon

However, carbon–carbon bond-forming reactions are not the only type of reaction that enables a molecular framework to be assembled. Take a look at the synthesis of the potential anti-malarial compound **2.2**, shown in Scheme 2.2. (There are some reactions in this synthesis that you may not have met before, but don't worry about them. Just concentrate on the difference between the reactant and the product.)

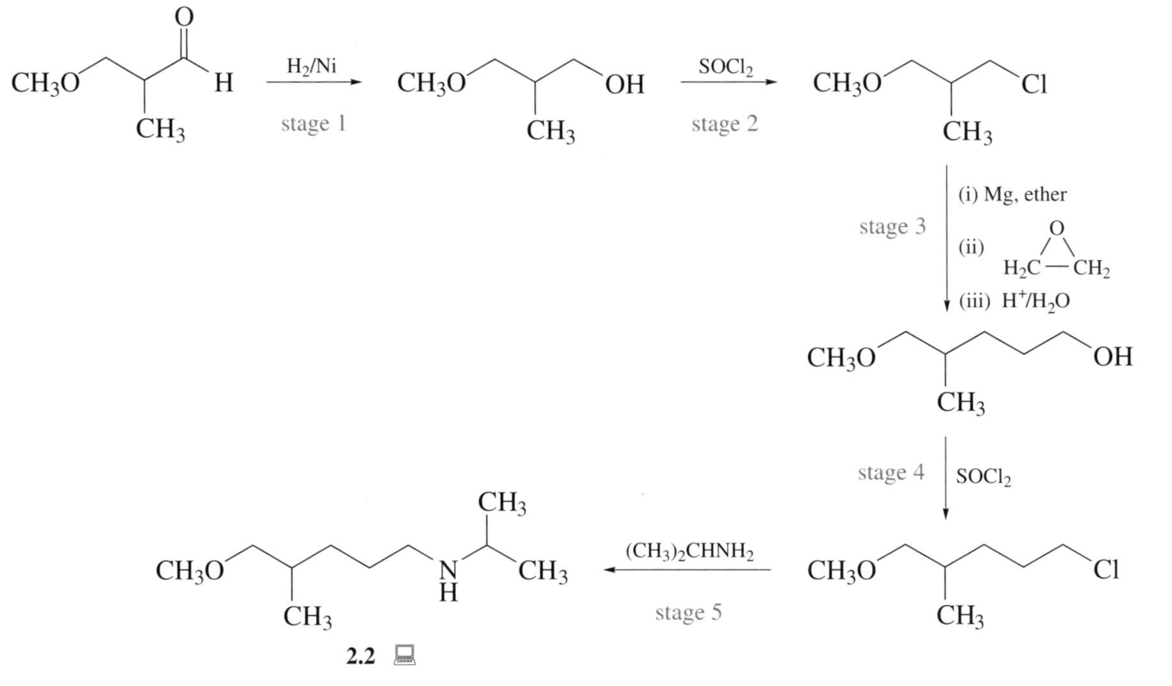


SCHEME 2.2

In this synthesis, the molecular framework is extended in stage 3, and completed in stage 5. Until now, you have met compounds that have a framework made up of carbon atoms only, but here is an example that demonstrates that a nitrogen atom can form part of the molecular framework.

- Identify the types of bond-forming reaction in these two stages.
- Stage 3 involves a carbon–carbon bond-forming reaction using a Grignard reagent; stage 5 involves the formation of a carbon–nitrogen bond by the $S_N$ reaction of an amine with a haloalkane.

This carbon–nitrogen bond-forming reaction is an example of a second type of bond-forming reaction (other than carbon–carbon bond formation, that is). This second type of reaction is classified as a **C—X bond-forming reaction**. These reactions are required whenever there is a heteroatom (that is, any atom other than C or H) *in the molecular framework of the target molecule*.

- In Scheme 2.2, what do we mean by target molecule?

- The target molecule is the molecule that is the overall objective of the synthetic plan, **2.2** in this case.

Essentially, the molecular framework is defined as the backbone of carbon atoms on to which the functional groups are attached. In the target molecule **2.2**, there is a nitrogen that joins two chains of carbons together, so the nitrogen is included in the molecular framework. Similarly, the oxygen joins a carbon atom to the framework, so is again part of the framework. As well as nitrogen and oxygen, the heteroatom could be sulfur, phosphorus or some other element that has a valency greater than 1. Univalent elements, such as chlorine and bromine, aren't included in this list because they can't extend the molecular framework: they can only be attached to one carbon. Similarly, if an oxygen or nitrogen is only attached to one carbon, as in C—OH or C—$NH_2$, then it is not included as part of the molecular framework because, once again, they do not extend it.

- Identify the molecular framework in the following target molecule.

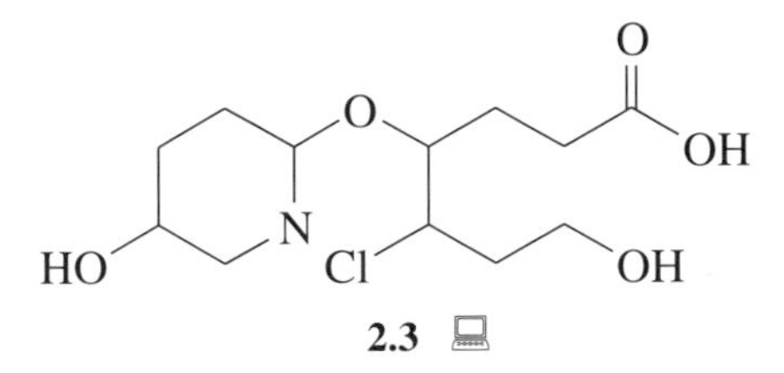


**2.3**

- The molecular framework is the carbon backbone (highlighted part of Structure **2.4**), including any heteroatoms that join the carbon atoms together:

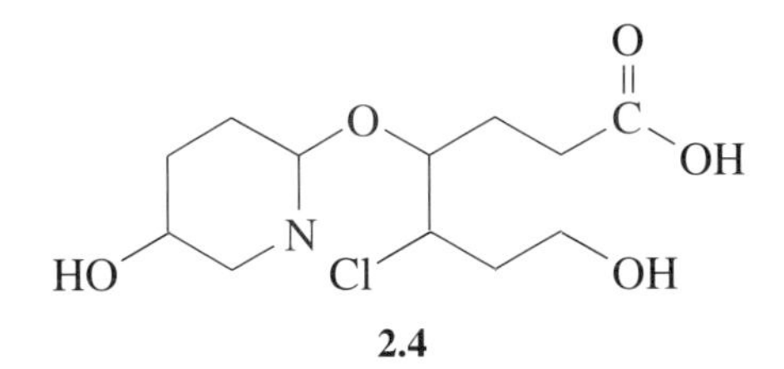


**2.4**

The two OH groups and the Cl are not part of the framework because they do not join carbon atoms together. We have included the carbon of the carboxylic acid in the framework because it is part of the chain of carbon atoms. This reflects the way we name carboxylic acids.

Returning to the synthesis of compound **2.1**, there seem to be other reactions that do not build up the molecular framework.

- What role does the second stage play in the synthesis?

- It converts the C—OH group into a C—Br group, so that a Grignard reagent can be made subsequently.

In any synthesis, there are likely to be many such reactions in which, in order to carry out a reaction that builds up the molecular framework, one functional group has to be converted into another, more appropriate, functional group. Such reactions are called **functional group interconversions (FGIs)**. FGIs only convert one functional group into another functional group. They never introduce additional *carbon atoms* into the molecular framework. One very important class of functional group interconversions is discussed in Box 2.1.

## BOX 2.1 Oxidation and reduction — important FGIs

In synthetic organic chemistry, oxidation and reduction usually have the following meanings:

*oxidation* — the addition of oxygen and/or the removal of hydrogen from a molecule;

*reduction* — the removal of oxygen and/or the addition of hydrogen to a molecule.

Thus, in terms of alcohols, aldehydes (or ketones) and acids we have:

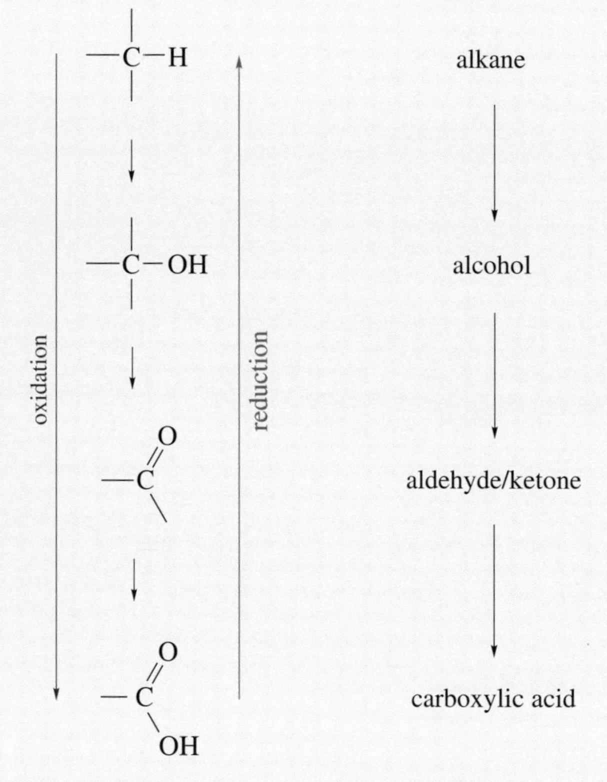


Going down the series, the number of bonds from the central carbon atom to an oxygen atom(s) increases, and the number to a hydrogen atom decreases. Thus, going down the series involves oxidation, whereas going up the series involves reduction.

A similar picture emerges with alkanes, alkenes and alkynes. Although there is no oxygen in hydrocarbons, as we go down the series the number of bonds to hydrogen decreases, which constitutes oxidation according to our use of the term. Going up the series, on the other hand, involves the addition of hydrogen — in other words, reduction.

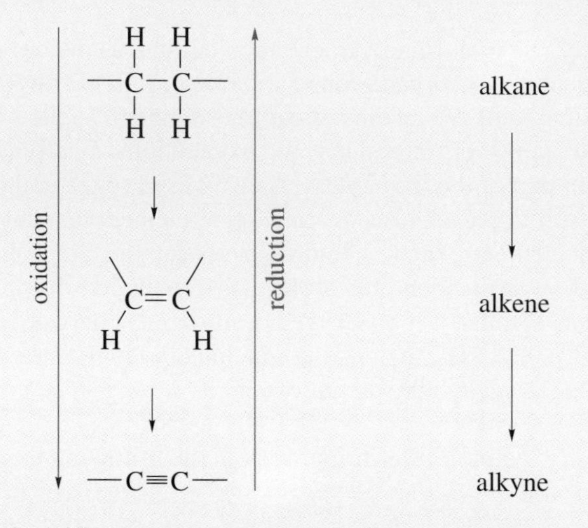


Specific reagents are required for each of these steps; for example, to oxidize an alcohol to a ketone (or aldehyde) without oxidizing the carbonyl further to give a carboxylic acid, $CrO_3$ in pyridine is needed:

$$RCH_2OH \xrightarrow[CrO_3]{\text{pyridine}} RCHO \qquad (2.2)$$

If potassium dichromate, $K_2Cr_2O_7$, in sulfuric acid is employed, the alcohol is oxidized completely, to give the carboxylic acid:

$$RCH_2OH \xrightarrow[H_2SO_4]{K_2Cr_2O_7} RCOOH \qquad (2.3)$$

The important thing to remember is that appropriate reagents can be used to interconvert these functional groups. For example, to reduce a carboxylic acid or a ketone/aldehyde to an alcohol, lithium aluminium hydride is used, as discussed in Part 1:

$$RCOOH \xrightarrow[\text{(ii) } H^+/H_2O]{\text{(i) } LiAlH_4} RCH_2OH \qquad (2.4)$$

$$RCHO \xrightarrow[\text{(ii) } H^+/H_2O]{\text{(i) } LiAlH_4} RCH_2OH \qquad (2.5)$$

To reduce an aldehye or a ketone to a hydrocarbon, the Clemmensen reduction would be suitable:

$$R^1{-}\overset{\displaystyle O}{\overset{\|}{C}}{-}R^2 \xrightarrow[HCl]{Zn/Hg} R^1{-}CH_2{-}R^2 \qquad (2.6)$$

Hydrazones provide a way of doing this under strongly basic conditions. The reaction is called the **Wolff–Kishner reduction** (discovered independently by Ludwig Wolff (1912) in Germany and N. M. Kishner (1911) in Russia), a reaction that usually requires high temperatures (ca 200 °C):

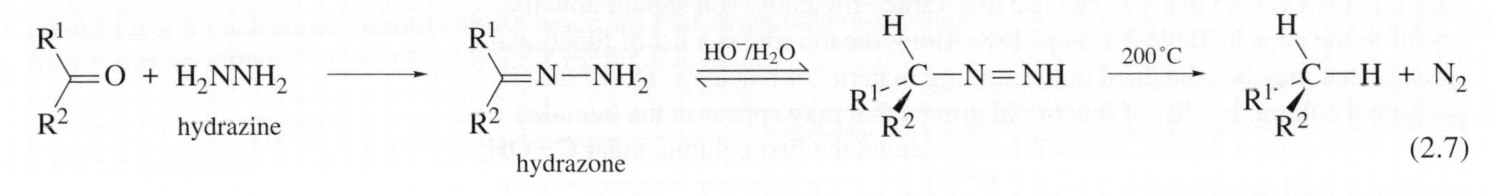

(2.7)

To reduce an alkyne to an alkane, hydrogen gas using a platinum or palladium catalyst is used, but to reduce it to an alkene, Lindlar's catalyst (palladium on barium sulfate poisoned with quinoline) is the reagent of choice:

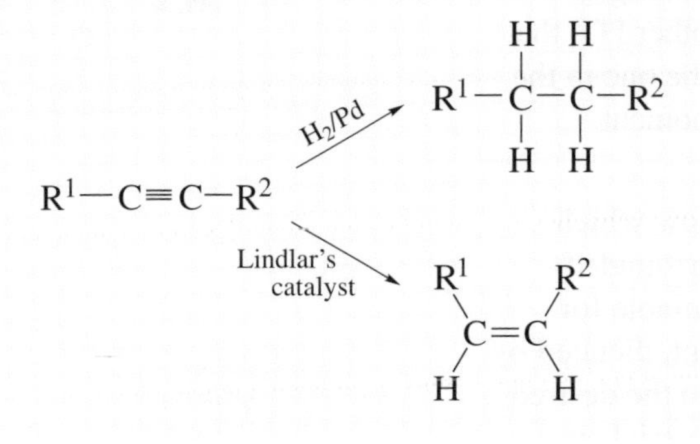


SCHEME 2.3

- Take another look at the synthesis of compound **2.2** in Scheme 2.2 (p. 182), and identify the FGIs.
- Stages 1 and 2 are FGIs, converting an aldehyde first into an alcohol and then into a chloroalkane; stage 4, another FGI, also converts an alcohol into a chloroalkane.

Notice that, at each of these three stages, the basic molecular framework remains the same: the purpose of the FGIs is to introduce an appropriate functional group for the next stage in the synthesis.

> This, then, is the second requirement of synthesis: that appropriate FGIs can be performed so that C—C or C—X bond-forming reactions can be carried out subsequently.

You may have noticed that the C—X bond formation in stage 5 of Scheme 2.2 involves the conversion of a chloroalkane into a secondary amine, so why don't we classify this as an FGI?

Well, although it does involve the conversion of one functional group into another, the main purpose is to build up the molecular framework by adding additional carbon atoms. It introduces carbon–heteroatom bonds that persist into the target molecule. So we don't classify it as an FGI, because it does more than just interconvert functional groups: it is classified as a C—X bond formation.

**Table 2.1** Functional group interconversions (FGIs) covered in this Book

| Functional group in product \ Functional group in starting material | C–OH | C–$NH_2$ | C–Cl (or C–Br) | C–C(=O)–C | C–C(=O)–H | C–C(=O)–OH | C–C(=O)–OR | C–C(=O)–NHR | C–C(=O)–Cl | C≡N | C=C | C≡C |
|---|---|---|---|---|---|---|---|---|---|---|---|---|
| C–OH | — | | | | | | | | | | | |
| C–$NH_2$ | | — | | | | | | | | | | |
| C–Cl (or C–Br) | 1 | | — | | | | | | | | | |
| C–C(=O)–C | | | | — | | | | | | | | |
| C–C(=O)–H | | | | | — | | | | | | | |
| C–C(=O)–OH | 2 | | | | | — | | | | B | | |
| C–C(=O)–OR | | | | | | | — | | | | | |
| C–C(=O)–NHR | | | | | | | | — | | | | |
| C–C(=O)–Cl | | | | | | A | | | — | | | |
| C≡N | | | | | | | | | | — | | |
| C=C | | | | | | | | | | | — | |
| C≡C | | | | | | | | | | | | — |
| –C–H (with bonds above and below the C) | | | | | | | | | | | | |

* The letters A and B in the Table refer to reactions that you have not met yet.

## *Notes to accompany Table 2.1*

| Table entry no. | Reaction | Table entry no. | Reaction |
|---|---|---|---|
| 1 | $ROH \xrightarrow{HCl\ (or\ HBr)} RCl\ (or\ RBr)$ | | |
| | or $ROH \xrightarrow{SOCl_2} RCl$ | | |
| 2 | $RCH_2OH \xrightarrow{K_2Cr_2O_7/H_2SO_4} RCOOH$ | | |

3

# PLANNING A SYNTHESIS

The starting point for any synthesis is a readily available starting material. But what is the starting point when we are *planning* a synthesis? Clearly it can't be any starting material(s) because we don't know what they ought to be, and indeed the purpose of the synthetic plan is to identify the appropriate starting materials. So, *we must start from the molecule that we ultimately want to make, the target molecule*. Our aim is to identify both an appropriate reaction that will form our target molecule, and also the reagents and a simpler organic molecule — called the **precursor** — which we use to achieve this reaction. Of course, if the precursor that has been identified is not commercially available, then we have to develop an appropriate way of making it. This process of examining each precursor in turn is continued until we arrive at a compound that *is* a readily available starting material.

Let's see how this works in practice by looking at a very simple example, phenylethanone (acetophenone, **3.1**). The question is 'How can this be synthesised from other compounds?'.

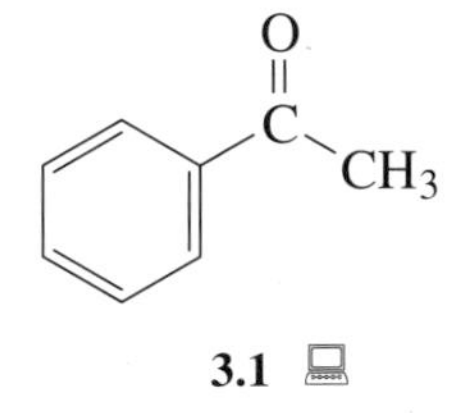


**3.1**

- Cite three or four methods for making phenylethanone.

- One method involves treating benzene with ethanoyl chloride (acetyl chloride) in the presence of aluminium trichloride (a Friedel–Crafts acylation):

$$C_6H_6 + CH_3COCl \xrightarrow{AlCl_3} C_6H_5COCH_3 \qquad (3.1)$$

Alternatively, the reaction between benzenecarbonitrile (benzonitrile) and methylmagnesium iodide, or that between ethanenitrile (acetonitrile) and phenylmagnesium bromide could be used:

$$\underset{\text{benzenecarbonitrile (benzonitrile)}}{C_6H_5C{\equiv}N} + CH_3MgI \longrightarrow \xrightarrow{H^+/H_2O} C_6H_5{-}\overset{O}{\overset{\|}{C}}{-}CH_3 \qquad (3.2)$$

$$\underset{\text{ethanenitrile (acetonitrile)}}{CH_3C{\equiv}N} + C_6H_5MgBr \longrightarrow \xrightarrow{H^+/H_2O} CH_3{-}\overset{O}{\overset{\|}{C}}{-}C_6H_5 \qquad (3.3)$$

Similarly, reaction of an acid chloride with an organocopper reagent could be used:

$$C_6H_5{-}\overset{O}{\overset{\|}{C}}{-}Cl + (CH_3)_2CuLi \longrightarrow C_6H_5{-}\overset{O}{\overset{\|}{C}}{-}CH_3 \qquad (3.4)$$

$$CH_3{-}\overset{O}{\overset{\|}{C}}{-}Cl + (C_6H_5)_2CuLi \longrightarrow CH_3{-}\overset{O}{\overset{\|}{C}}{-}C_6H_5 \qquad (3.5)$$

Another reaction involving an organometallic reagent that you may have included in your list is that between an organolithium reagent and a carboxylic acid:

$$C_6H_5COOH + 2CH_3Li \longrightarrow C_6H_5{-}\overset{\overset{\displaystyle O}{\|}}{C}{-}CH_3 \qquad (3.6)$$

$$CH_3COOH + 2C_6H_5Li \longrightarrow CH_3{-}\overset{\overset{\displaystyle O}{\|}}{C}{-}C_6H_5 \qquad (3.7)$$

Already we have quite a few routes to phenylethanone — and there are more! But for now we shall concentrate on those listed above.

- Look back at the reactions that we've identified that could yield phenylethanone. What characteristics of the target molecule are they all intended to produce?

- All the reactions are attempts both to build up the molecular framework and to form the ketone type of carbonyl functional group, **3.2**.

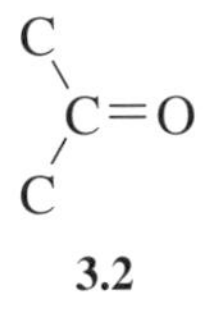

**3.2**

So, in choosing to use these reactions, we have implicitly chosen to focus our attention on the functional group. The reason for this is that it is at the functional groups that the chemical changes are carried out. In our examples, we have chosen to form the functional group in a particular way, by making a carbon–carbon bond in which one of the carbon atoms becomes the carbonyl carbon atom of the ketone group.

> This is true of all attempts to design synthetic pathways: *we focus our attention on the functional group(s).**

To form the new C—C bond we have decided to use a carbon nucleophile and a carbon electrophile.

- Identify the carbon nucleophiles and electrophiles in each of the potential syntheses of phenylethanone given above.

- The answer to this question is at the bottom of p. 192.

In adopting this approach we have chosen to think of the formation of the C—C bond in the following way:

$$\underset{\text{nucleophilic carbon}}{C^-} + \underset{\text{electrophilic carbon}}{C^+} \longrightarrow C{-}C \qquad (3.8)$$

In other words, a carbon–carbon bond is formed by linking together a carbon nucleophile and a carbon electrophile. The symbols $C^-$ and $C^+$ do not mean that the carbon atoms must carry negative and positive charges, respectively. Rather, they indicate that they are *sources* of nucleophilic, $C^-$, and electrophilic, $C^+$, carbon atoms. They also say nothing about the relative reactivity of the reagents; carbanions and carbocations are usually very reactive, but the reagents that are sources of $C^+$ and $C^-$ are not necessarily highly reactive. From now on we shall identify such sources of carbon nucleophiles and electrophiles by printing them in blue and green, respectively. Remember that these representations do not exist as compounds, but are just a symbolism that we use to help us plan a synthesis.

* This is referred to as the 'second golden rule' in *Alkenes and Aromatics*[1].

But what if we couldn't remember the above reactions? How then might we identify the types of reagent that we need? One way is to focus on the functional group, and *mentally* to break, or *disconnect*, the carbon–carbon bond that we wish to form; that is, in a 'thought experiment', we carry out the reverse process to bond formation.

$$C \wr C \longrightarrow C^- + C^+ \qquad (3.9)$$

In mentally breaking the bond heterolytically in this way (the orange wiggly line identifies the bond being broken), we end up with two *imaginary* carbon fragments, one of which must be nucleophilic, indicated by a minus (–) sign, and the other electrophilic, indicated by a plus (+) sign. The importance of this approach is that it enables us to identify the types of nucleophile and electrophile that we need to employ in our synthesis. This process of mentally breaking bonds is called **disconnection**. We use the symbol $\Longrightarrow$, rather than a conventional arrow $\longrightarrow$, to represent a disconnection process. *So* $\Longrightarrow$ *means we are working backwards in the planning of our synthesis, whereas* $\longrightarrow$ *refers to what is done in practice when we are making the target molecule. Put another way,* $\Longrightarrow$ *refers to a thought process — how we analyse the problem — whereas* $\longrightarrow$ *refers to what we can actually do in the laboratory!*

Of course, for any C—C bond, there are two potential pairs of carbon nucleophiles and carbon electrophiles that can be produced as a result of the disconnection process:

$$C \wr C \Longrightarrow C^- + C^+ \qquad (3.10)$$

$$C \wr C \Longrightarrow C^+ + C^- \qquad (3.11)$$

---

*Answer to question on p. 191*

| NUCLEOPHILES | ELECTROPHILES | REACTION |
|---|---|---|
| benzene | $H_3C-C(=O)-Cl$ | 3.1 |
| $CH_3MgI$ | $C_6H_5C{\equiv}N$ | 3.2 |
| $C_6H_5MgBr$ | $CH_3C{\equiv}N$ | 3.3 |
| $(CH_3)_2CuLi$ | $C_6H_5-C(=O)-Cl$ | 3.4 |
| $(C_6H_5)_2CuLi$ | $H_3C-C(=O)-Cl$ | 3.5 |
| $CH_3Li$ | $C_6H_5-C(=O)-OH$ | 3.6 |
| $C_6H_5Li$ | $H_3C-C(=O)-OH$ | 3.7 |

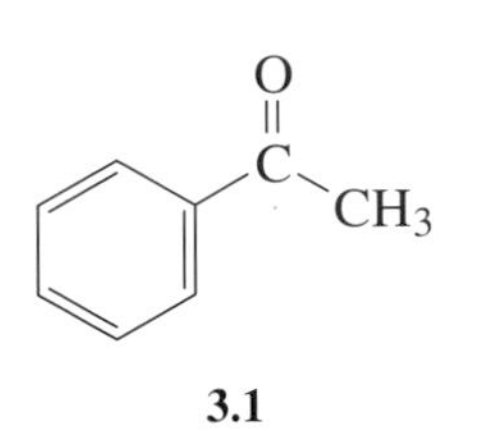

**3.1**

In the next few Sections we shall be studying the best way of making disconnections for a variety of functional groups, so that the most suitable nucleophiles and electrophiles can be identified. For now, let us see how such disconnections can be applied to our target molecule, phenylethanone (**3.1**).

- How many of the C—C bonds that are attached to the carbonyl group in phenylethanone can we disconnect?
- Both of them: $C—C_6H_5$ and C—$CH_3$.

As we've just mentioned, the C—$C_6H_5$ bond can be disconnected in two ways:

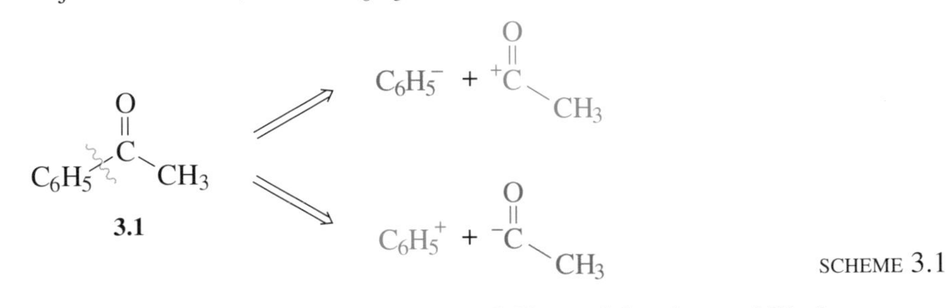

SCHEME 3.1

The first generates the nucleophilic fragment $C_6H_5^-$ and the electrophilic fragment $CH_3—\overset{+}{C}=O$; the second generates the electrophilic fragment $C_6H_5^+$ and the nucleophilic fragment $CH_3—\overset{-}{C}=O$.

> It cannot be stressed too strongly that *these fragments do not actually exist.* They are simply representations that identify $C_6H_5$ and $CH_3CO$ as nucleophilic or electrophilic.

Whenever the two groups at either end of a C—C bond are different, there are always these two ways of producing fragments. However, we can usually decide whether a fragment is likely to be nucleophilic or electrophilic, and the following Sections show how we do this.

- Look back to the list of nucleophiles and electrophiles that we identified for the reactions that can be used to synthesise phenylethanone. For those reactions in which the $C_6H_5—CO$ bond is formed, identify the nature of the $C_6H_5$ and $CH_3CO$ fragments involved.
- In every case, $C_6H_5$ is introduced as a nucleophile, and $CH_3CO$ as an electrophile. So, the fragments used are $C_6H_5^-$ and $CH_3—\overset{+}{C}=O$.

You may be wondering how the electrophilic reagent $CH_3—C\equiv N$ can be related to the $CH_3—\overset{+}{C}=O$ fragment, particularly as it does not contain a carbonyl group. This will be made clear in Section 5.6. So, of the two potential disconnections that can be imagined (Scheme 3.1), only one was adopted, that involving nucleophilic $C_6H_5^-$ and electrophilic $CH_3—\overset{+}{C}=O$ fragments. This is because $C_6H_5^-$ can be easily related to an organometallic compound (or benzene), and $CH_3—\overset{+}{C}=O$ to an appropriate carboxylic acid or carboxylic acid derivative, $CH_3COY$.

Nucleophilic and electrophilic fragments such as $C_6H_5^-$ or $CH_3{-}\overset{+}{C}{=}O$, are called **synthons**. As indicated above, synthons are purely imaginary entities, and have no real existence themselves. Rather, they represent the types of fragment — nucleophilic or electrophilic — that are required for a reaction to be carried out. So that there is no confusion between an imaginary nucleophilic synthon and a negatively charged carbanion, or between an imaginary electrophilic synthon and a positively charged carbocation, all synthons in this Book are printed in blue and green, respectively. The actual chemical entity that is used to provide a particular synthon is called a **reagent**. The reagent for $CH_3{-}\overset{+}{C}{=}O$ is a carboxylic acid (or derivative) $CH_3COY$, for example $CH_3COCl$. We know that the carbonyl carbon atom in $CH_3COY$ is electrophilic, and that Y is a good leaving group, capable of being replaced by a nucleophile.

Now let's turn our attention to the second carbon–carbon bond in the target molecule, the $C{-}CH_3$ bond.

- Write a disconnection of the $CH_3{-}CO$ bond in phenylethanone, and identify the potential pairs of synthons. From the list of nucleophiles and electrophiles identified earlier, which pair of synthons appears to be the more appropriate for this synthesis?

- The $CH_3{-}CO$ bond may be disconnected in two ways:

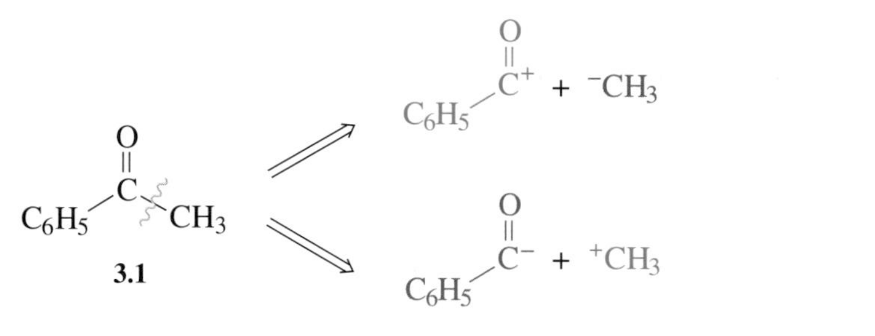


SCHEME 3.2

In each reaction so far identified that corresponds to this disconnection, the $CH_3$ group is introduced as a nucleophile, and $C_6H_5CO$ as an electrophile. So, the appropriate synthons are $^-CH_3$ and $C_6H_5{-}\overset{+}{C}{=}O$.

For both bond disconnections a pattern is emerging; the $R^1$ fragment ($C_6H_5$ or $CH_3$) is introduced as a nucleophile, and the $R^2CO$ fragment ($C_6H_5CO$ or $CH_3CO$) is introduced as an electrophile. Similar patterns can be discerned for all functional groups, as you will discover shortly. However, it should be apparent that by disconnecting the $C{-}C$ bond in order to generate synthons we were able to identify appropriate nucleophilic and electrophilic entities that could be related to known substrates and reagents. This is the purpose of the disconnection approach.

For molecules more complicated than phenylethanone, after each disconnection the organic precursor itself will probably need to be subjected to the disconnection approach, until a suitable starting material is identified. It is important that at each stage *a real molecule* is identified as the intermediate; don't try to disconnect synthons! As you progress through the rest of this Book, you will meet many examples of syntheses that require more than one disconnection step.

I for one will not conceal my hope, contrary though it may be to the often too narrowly utilitarian spirit of the day, that synthesis for its own sake will continue. There is excitement, adventure, and challenge, and there can be great art, in organic synthesis. These alone should be enough, and organic chemistry will be sadder when none of its practitioners is responsive to these stimuli.

***R. B. Woodward*, Perspectives, Organic Chemistry *(1956)***

## 3.1 Summary of Section 3

1. In planning a synthesis, we work backwards from the target molecule, in order to identify suitable starting materials.
2. In the target molecule, attention is focused on the functional group(s).
3. In order to identify suitable reactions that will enable a functional group to be introduced during the synthesis, we carry out a 'thought experiment', and mentally disconnect the bonds attached to the functional group.
4. Bond disconnection generates two imaginary fragments, one nucleophilic and one electrophilic, called synthons. Synthons are related to the reagents that are required for the synthesis.

### QUESTION 3.1

Carry out C—C bond disconnections at the functional groups in the following molecules, and identify all the types of synthon that can be produced.

(i) $C_6H_5CH(OH)CH_3$

(ii) $C_6H_{13}CH_2COOH$

(iii) $C_6H_{13}{-}C{\equiv}C{-}H$

### QUESTION 3.2

For the following synthesis of phenylethanenitrile (**3.2**), disconnect the target molecule to generate the pair of synthons that corresponds to the reagents used:

$$C_6H_5CH_2Br + K^{+}\,{}^{-}CN \longrightarrow \underset{\mathbf{3.2}}{C_6H_5CH_2CN} + K^{+}Br^{-} \qquad (3.16)$$

# SIMPLE DISCONNECTIONS: C—X BONDS

4

So far, you have been introduced to the general way in which organic chemists plan a synthesis; that is, by working backwards from the target molecule. This process of working backwards is called **retrosynthetic analysis**. The stages of a synthesis can be separated into those that modify functional groups (FGIs), and those that construct the molecular framework. In this and the following Section we shall concentrate on these construction stages. We plan these construction stages by identifying key disconnections to functional groups, and in these Sections we shall show you how to make these disconnections.

In Section 3 you learned that we focus on functional groups, but how do we spot a *good* disconnection? Well, in this and the following section we shall examine a selection of the simplest and most commonly used disconnections, and you will see how they relate to the chemical reactions that you should be familiar with. We must necessarily be selective because of the limited space available, but we shall cover all the principles of disconnection that you will need.

Construction steps in a synthesis — and therefore the places where we disconnect bonds in our retrosynthetic analysis — may involve C—X or C—C bond formation. We shall start by looking at C—X bond formation (where X is N or O). However, you should remember that in order to be classified as C—X bond formation the heteroatom must be part of the molecular framework of the target molecule, so that it serves as a link between carbon chains.

We have chosen to look at C—X, as opposed to C—C, bonds first because the best way to disconnect them is easily decided. For any C—X bond, it is possible to write down two sets of synthons that represent how it might be put together from a nucleophile and an electrophile.

- Suggest what these synthons might be.
- Using the symbolism introduced in Section 3, the two pairs are.

$$\text{C} \wr \text{X} \Longrightarrow \text{C}^+ + \text{X}^- \quad (4.1)$$

$$\text{C} \wr \text{X} \Longrightarrow \text{C}^- + \text{X}^+ \quad (4.2)$$

- Bearing in mind that X is oxygen or nitrogen, which of these pairs generally looks more promising?
- The atoms O and N are electronegative and familiar as nucleophiles, such as $^-$OH or $NH_3$, so the first pair, in which the C—X bond is disconnected to form $X^-$, will be much more useful.

In fact, the forward synthetic reaction here represents a heteroatom nucleophile attacking a carbon electrophile — a substitution reaction. When we disconnect C—O and C—N bonds:

$$\text{C} \wr \text{OR} \Longrightarrow \text{C}^+ + {}^-\text{OR} \quad (4.3)$$

$$\text{C} \wr \text{NR}^1\text{R}^2 \Longrightarrow \text{C}^+ + {}^-\text{NR}^1\text{R}^2 \quad (4.4)$$

we generate nucleophilic $RO^-$ and $R^1R^2N^-$ synthons. When we come to perform the reaction to construct the C—O and C—N bonds for our synthesis, we are just as likely to use an alcohol, ROH, and more likely an amine, $R^1R^2NH$, as we are to use the corresponding anions $RO^-$ and $R^1R^2N^-$. So, the nucleophilic reagents corresponding to the synthons identified by C—O and C—N bond disconnection are often the alcohols and amines, respectively.

$$RO^- \Longrightarrow ROH \quad (4.5)$$

$$R^1R^2N^- \Longrightarrow R^1R^2NH \quad (4.6)$$

Now, where do we find reagents for the carbon electrophiles? In nucleophilic substitution reactions we need a carbon atom that is positively polarized because it is attached to an electronegative element that can act as a leaving group. Useful compounds here are the haloalkanes ($R_3\overset{\delta+}{C}-\overset{\delta-}{X}$), familiar from $S_N1$ and $S_N2$ reactions*, and carboxylic acid derivatives, in particular acid chlorides, RCOCl.

- What type of compounds would be made using haloalkanes $R^1$—X as the electrophile, and simple amines, $R^2NH_2$ or alkoxide ions, $R^2O^-$, as the nucleophilic component?

- Reaction of simple amines, $R^2NH_2$ with haloalkanes will give an amine with two alkyl groups (assuming an $S_N2$ route):

$$R^2-\ddot{N}H_2 \;\; R^1-X \xrightarrow{-X^-} R^2-\overset{+}{N}H_2-R^1 \xrightarrow{-H^+} R^1R^2NH \quad (4.7)$$

Alkoxide ions, $RO^-$, will give ethers:

$$R^2-O^- \;\; R^1-X \longrightarrow R^2-O-R^1 + X^- \quad (4.8)$$

Let's now look at some types of molecule for which the carboxylic acid derivatives are appropriate.

## 4.1 Esters and amides

As our example, we shall consider C—X bonds where the electrophile needed to make them is the carboxylic acid derivative **4.1**. The target functional groups here are esters (**4.2**), and amides (**4.3**). We can deal with esters and amides together, as the disconnection is easily decided, and is the same for both types of compound. The bond we choose is always that between the oxygen or nitrogen atom and the carbon atom of the carbonyl group (**4.4**).

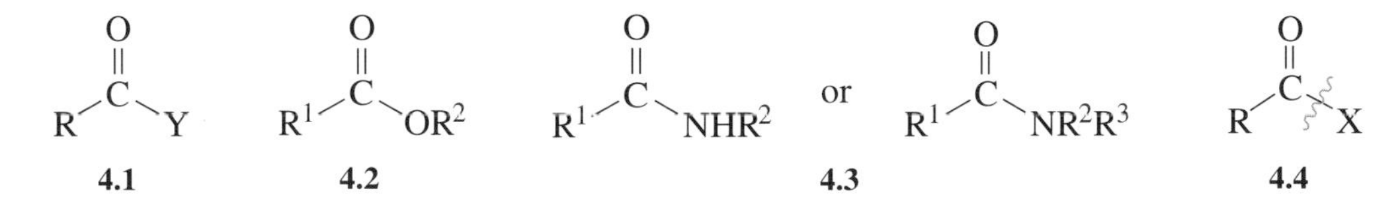


* The mechanisms of $S_N1$ and $S_N2$ reactions are discussed in *Chemical Kinetics and Mechanism*[2].

Thus, disconnection of the ester **4.5** (which is used as an insect repellent) suggests synthons **4.6** and **4.7** for its assembly, and the corresponding reagents are the alcohol **4.8** and the acid chloride **4.9**:

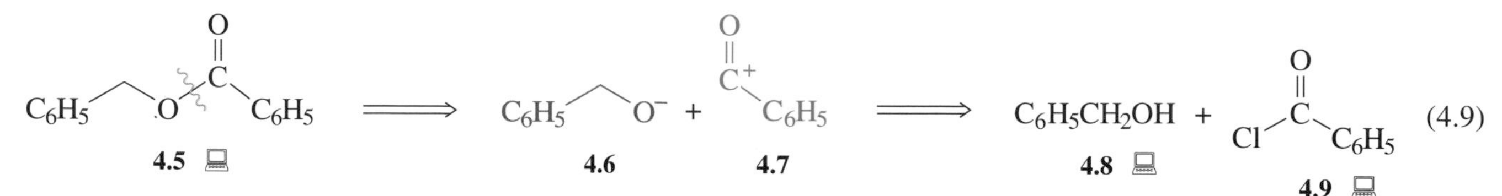


The forward, synthetic reaction that has been identified is now familiar; it is the formation of an ester from an alcohol and an acid chloride, often carried out in the presence of a base, such as pyridine. Acid chlorides, RCOCl, are the most commonly used electrophiles in such reactions, as they are the most reactive, although it is possible to use other acid derivatives, such as anhydrides, $(RCO)_2O$.

We use the same analysis for amides. Let's consider the amide **4.10**.

CH3 N CH3 O **4.10**

- Write down the synthons generated by the disconnection of **4.10**, and the reagents corresponding to those synthons.
- We choose to disconnect at the N—C=O bond. The nucleophilic synthon is **4.11**, supplied by reagent **4.12**, a cyclic secondary amine, piperidine. The electrophilic synthon is **4.13**, for which we can use the acid chloride **4.14**.

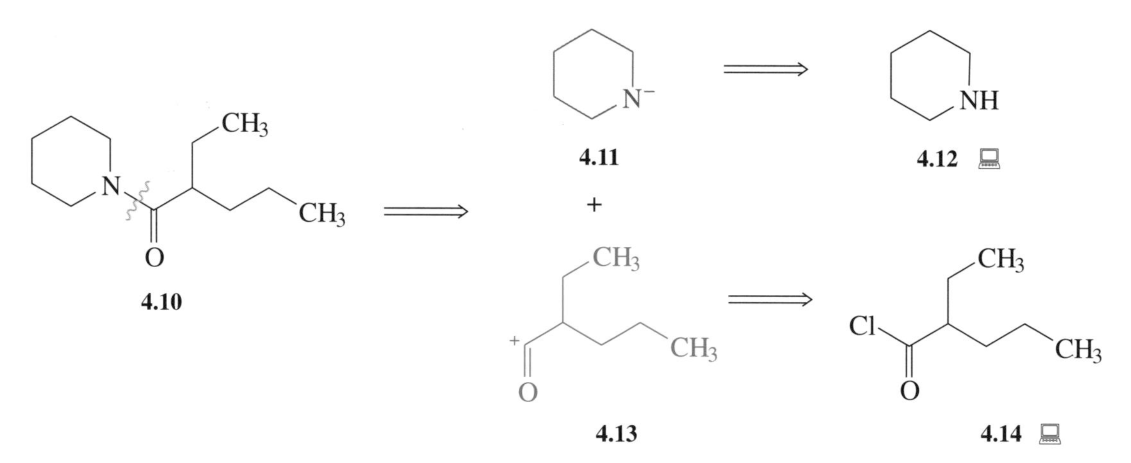

SCHEME 4.1

Again, the synthetic reaction that has been identified should be familiar: it is the formation of an amide from an acid chloride and an amine. The acid chlorides that we invoke in this type of synthesis may be commercially available, as is the case with benzoyl chloride (**4.9**), or they may themselves have to be prepared by an FGI, as is the case with **4.14**. The appropriate FGI to give **4.14** uses the carboxylic acid **4.15** and $PCl_5$ or $SOCl_2$, as discussed earlier. Our ultimate starting material is therefore the carboxylic acid **4.15**. We shall return to the synthesis of the acid **4.15** when we look at the disconnection of carboxylic acids.

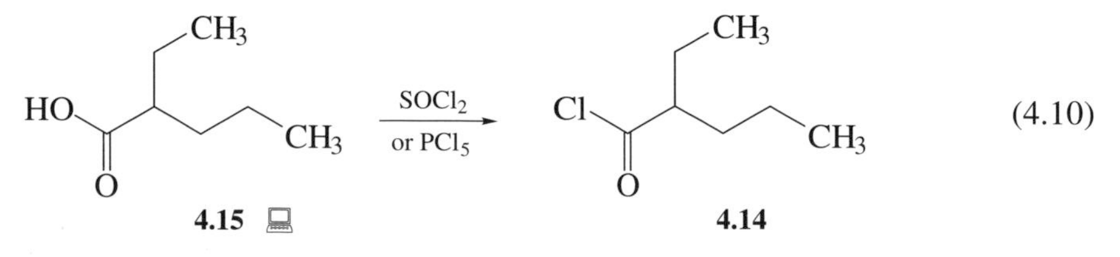

(4.10)

A general conclusion that we have illustrated is that *carboxylic acid derivatives, such as esters and amides, can always be disconnected to the corresponding carboxylic acid.* Interconversions between acid derivatives will be dealt with later when we consider the assembly of carboxylic acids.

## BOX 4.1 R. B. Woodward (1917–1979)

Robert Burns Woodward (Figure 4.1) was born on 10 April 1917 in Boston. He was the only child of Margaret and Arthur Woodward, and attended school in Quincy, a suburb of Boston. He was excited by chemistry from an early age, and in 1933 entered the Massachusetts Institute of Technology, but was excluded for lack of attention in 1934. He was allowed to re-enrol in 1935, and obtained his BSc degree in 1936 and his PhD in 1937. From then until 1963 he worked at Harvard, starting as a Postdoctoral Fellow, becoming Assistant Professor in 1944 and Full Professor in 1950. In 1963 he became director of the Woodward Research Institute at Basel. Woodward married Irja Pullman in 1938, and then Eudoxia Muller in 1946. He had three daughters and a son.

**Figure 4.1** R. B. Woodward 1917–1979.

Woodward dominated the field of synthetic organic chemistry for nearly half a century. He was always interested in synthesising very complicated natural products such as quinine, cholesterol, strychnine, reserpine, lysergic acid, chlorophyll and vitamin $B_{12}$ (Figure 4.2). As well as synthetic organic chemistry, Woodward developed analytical tools and rules for determining how reactions take place, based on the interactions of their orbitals. Woodward worked with more than 250 collaborators, most of whom have taken up academic positions all over the world. His intense devotion to his work is vividly illustrated by the fact that he named a synthetic steroid Christmasterol because it was first crystallized in his laboratory on Christmas Day!

He was awarded the Nobel Prize for Chemistry in 1965, for his outstanding achievements in the art of organic synthesis.

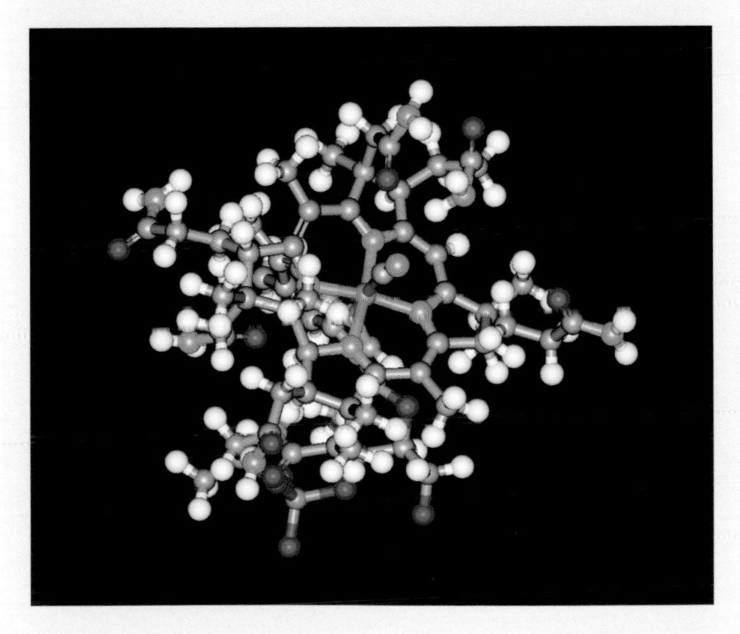

**Figure 4.2** Vitamin $B_{12}$ (cobalamin), synthesised by R. B. Woodward and A. Eschenmoser in 1973. More than 100 PhD students and postdoctoral workers contributed to the work, which took over ten years to complete.

The cobalt atom at the centre of the molecules is attached to five nitrogens (one from below) and a CN group from the top. At the bottom of the picture you can see an orange phosphorus, which is part of a phosphate group.

Vitamin $B_{12}$ is necessary for the formation of nucleic acids and red blood cells. Its sources in the diet are chiefly meat and milk.

## 4.2 Summary of Section 4

1. For molecules containing C—X bonds, where X = N or O, the best disconnection is generally the one that produces an $X^-$ synthon:

$$\mathrm{C{-}X} \Longrightarrow \mathrm{C^+} + \mathrm{X^-} \qquad (4.1)$$

2. For carboxylic acid derivatives, RCOX, the best disconnection is the one between the carbonyl carbon atom and X.

$$\mathrm{R{-}C({=}O){-}X} \Longrightarrow \mathrm{R{-}C^+({=}O)} + \mathrm{X^-} \qquad (4.11)$$

3. The reagents corresponding to $RO^-$ and $R^1R^2N^-$ synthons are often the alcohol, ROH, and secondary amine, $R^1R^2NH$.

4. The reagents corresponding to the $\mathrm{R{-}\overset{+}{C}{=}O}$ synthon are the acid chloride, RCOCl, or anhydride, $(RCO)_2O$, which can be made from the carboxylic acid itself, RCOOH.

### QUESTION 4.1

Chloramphenicol (**4.16**) is an antibiotic commonly used to treat conjunctivitis. Disconnect the appropriate bond (**4.17**) in chloramphenicol, decide on the appropriate synthons and identify reagents for its synthesis.

$$\mathrm{{-}C({=}O){-}X}$$

**4.17**

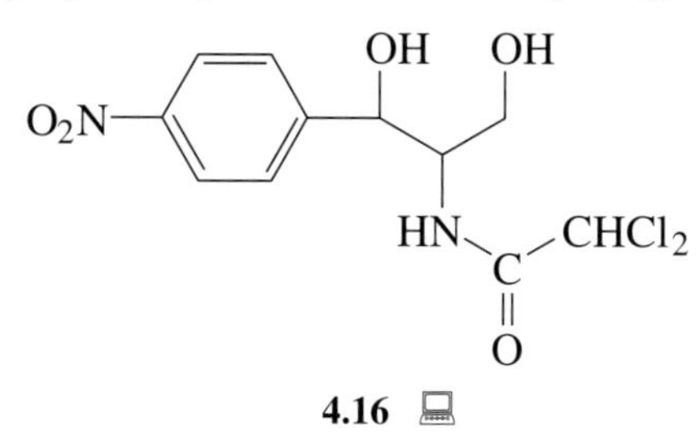

**4.16**

### QUESTION 4.2

Aspirin (**4.18**) is a well-known anti-inflammatory agent. Use the disconnection approach to identify reagents appropriate to its synthesis.

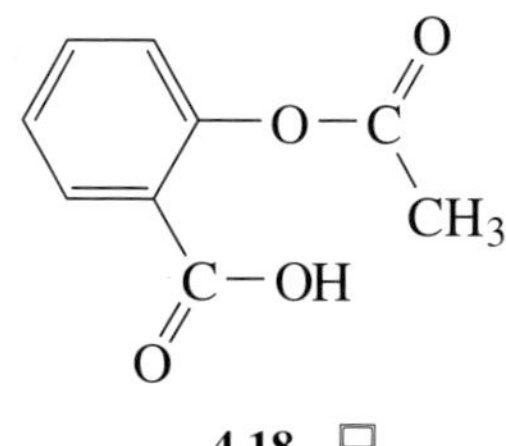

**4.18**

### QUESTION 4.3

Captopril (**4.19**) is a drug that has been designed to reduce blood pressure. Use the disconnection approach to identify reagents appropriate to its synthesis.

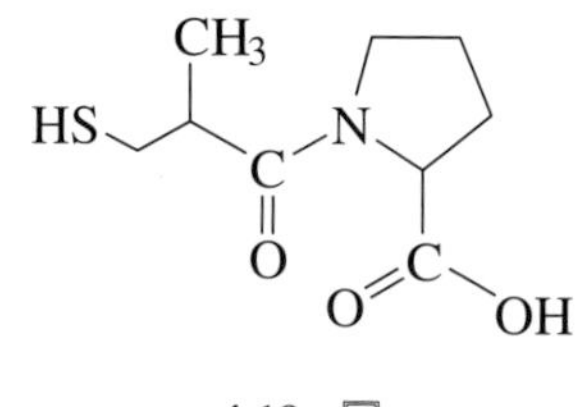

**4.19**

# SIMPLE DISCONNECTIONS: C—C BONDS

5

Now that we have looked at some examples of C—X bond formation, we can move on to look at the disconnection of C—C bonds. Once again, we shall focus on the functional groups, and this means that our survey can be ordered according to the types of functional group in the target molecules. If you think about all the various organic reactions that you may have met, you will soon conclude that the range of reactions that create C—C bonds is rather small. So, any disconnections in retrosynthetic analysis that we make will be limited by this range.

There is a good reason for the range of C—C bond-forming reactions being so limited. If we write down the disconnection of a C—C bond, it is obvious that we need both an electrophilic and a nucleophilic carbon synthon.

$$\mathrm{C{-}C} \Longrightarrow \mathrm{C^+ + C^-} \qquad (5.1)$$

- What reagents have we mentioned so far for the carbon electrophile?

- Haloalkanes, RX, and carboxylic acid derivatives, RCOY, are good carbon electrophiles. To this group we can add aldehydes (RCHO) and ketones ($R^1COR^2$), whose carbonyl carbon atom is electrophilic because it is attached to an electronegative oxygen atom.

Thus, we have no trouble finding electrophilic reagents to make C—C bonds, but what about the nucleophiles? Now we meet a limitation. You have met only a handful of types of carbon nucleophile. The most general are the Grignard reagents, RMgX, which contains a nucleophilic carbon because the R–Mg bond is polarized $\overset{\delta-}{\mathrm{R}}{-}\overset{\delta+}{\mathrm{Mg}}$ (recall the discussion of Grignard reagents in Part 2). Of course, other organometallic compounds, such as alkyllithium and lithium dialkylcopper reagents, are also sources of carbon nucleophiles. However, we shall generally illustrate examples using the Grignard reagent. All the organometallic reagents cited are made from the corresponding haloalkane. Aromatic (benzene) rings undergo substitution by electrophiles, so they must themselves be carbon nucleophiles. Two more specialized types of carbon nucleophile are the alkynides, $R{-}C{\equiv}C^- M^+$ (which are formed because alkynes are slightly acidic), and the cyanide anion, $^-CN$.

Now that we have identified the kinds of reagent available, let's move on to our selective survey of the major C—C bond disconnections.

## 5.1 Alcohols

We begin with alcohols for a very good reason. As will become apparent, the C—OH group is central to a great deal of retrosynthetic analysis. There are two reasons for this: (i) there is a reliable C—C bond disconnection that corresponds to known C—C bond-forming reactions used to make alcohols; (ii) alcohols are at the hub of a network of FGIs that allow access to other functional groups. But more of that later! First, we shall consider the general alcohol **5.1**. The C—C bonds closest to the functional group are those from the alcohol carbon atom itself.

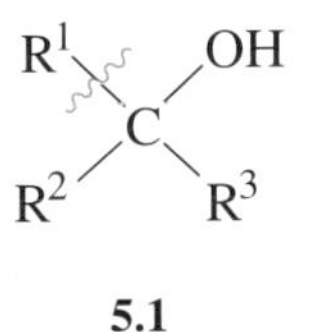


**5.1**

- Write down the two possible pairs of synthons derived from disconnection of the C—C bond indicated in **5.1**.

- One pair consists of the alkyl nucleophile **5.2** and the electrophile **5.3**; the other combination is the alkyl electrophile **5.4** and the electrophile **5.5**.

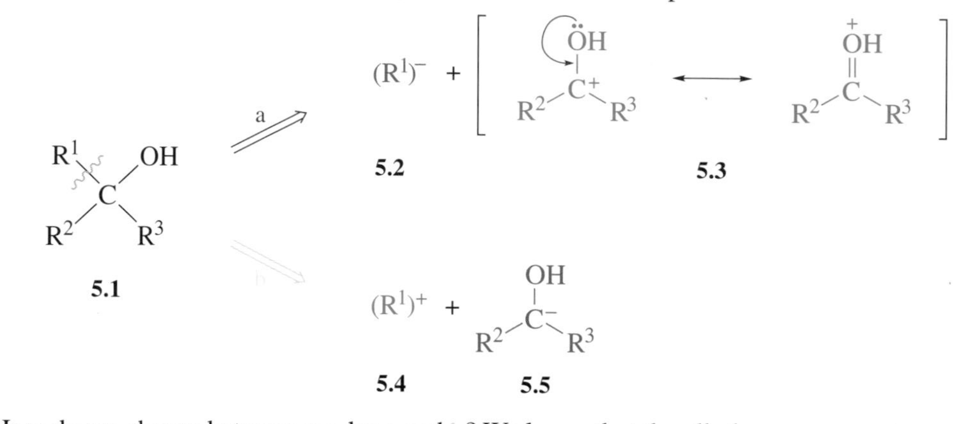


SCHEME 5.1

How do we choose between modes a and b? We know that the alkyl group can act as a nucleophile, **5.2**, or an electrophile, **5.4** (Grignard reagent or haloalkane, respectively), so the decision will be based on the oxygen-containing synthon. Referring to the alternative resonance form of **5.3** gives us a clue: the synthon corresponds to a protonated carbonyl compound (here, a ketone), so the reagent will be the carbonyl compound itself. The reaction we have identified by disconnection a is the addition of a Grignard reagent to an aldehyde or ketone, which is well known. It turns out that reagents for the synthon **5.5** are difficult to prepare, so that path a is our choice of disconnection. *As far as this Book is concerned, this illustrates how all alcohols should be analysed using retrosynthetic analysis; the synthon containing the alcohol oxygen atom is electrophilic.* In fact, so significant is this disconnection that we shall draw it out again, working back to the reagents.

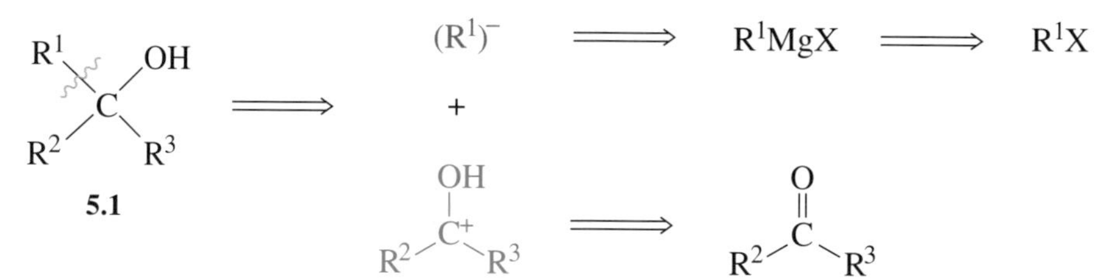


SCHEME 5.2

Almost any alcohol can be made by addition of an organometallic nucleophile (such as a Grignard reagent) to a carbonyl compound (such as an aldehyde or ketone). The Grignard reagent leads back further by an FGI to a haloalkane as the ultimate starting material.

Looking at Structure **5.1**, you may have noticed that we have a further element of choice in our disconnection. Any of the three groups $R^1$, $R^2$ or $R^3$ attached to the alcohol carbon atom could be used as the Grignard reagent, with the carbonyl compound carrying the other two groups.

- Suppose our target molecule is 1-phenylethanol (**5.6**). Draw the two possible sets of synthons and reagents that would be generated by the alcohol C—C bond disconnections a and b.

OH
b a
CH
$C_6H_5$ $CH_3$
**5.6**

- One approach, disconnection a, gives synthons corresponding to the reagents benzaldehyde and a methyl Grignard reagent, whereas the other, disconnection b, leads to a phenyl Grignard and ethanal (acetaldehyde).

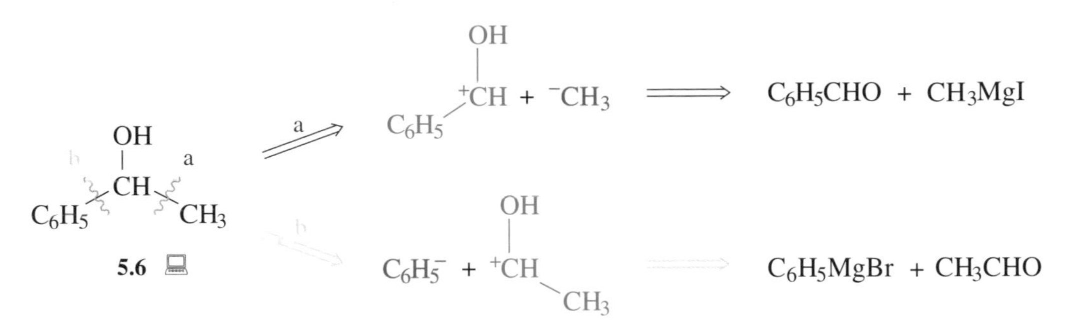


SCHEME 5.3

Both carbonyl compounds and both Grignard reagents (and the haloalkanes from which the latter are made) are available, and both approaches are expected to proceed in good yield and without difficulty. Sometimes, however, one possibility from the choice available may be preferred. Obvious cases would be where one carbonyl compound or one Grignard reagent (or the haloalkane used to make it) is not available, or is too expensive, or is known to be difficult to make or to handle. In other cases, the choice might be made on strategic grounds. The target alcohol **5.7**, for example, could be assembled a from the ketone **5.8** and methylmagnesium iodide, or b from the cyclohexyl Grignard reagent **5.9** and propanone (acetone). Disconnection a, however, merely chops off one carbon atom, and leaves a new target molecule **5.8**, which is almost as difficult to make as **5.7**. On the other hand, disconnection b breaks the molecule into two pieces that are more similar in size; indeed, it generates readily available starting materials, in propanone and bromocyclohexane. The latter approach thus produces a *simpler* synthetic route, which is one of the criteria for a good disconnection. So, path b is preferred as a synthesis of **5.7**.

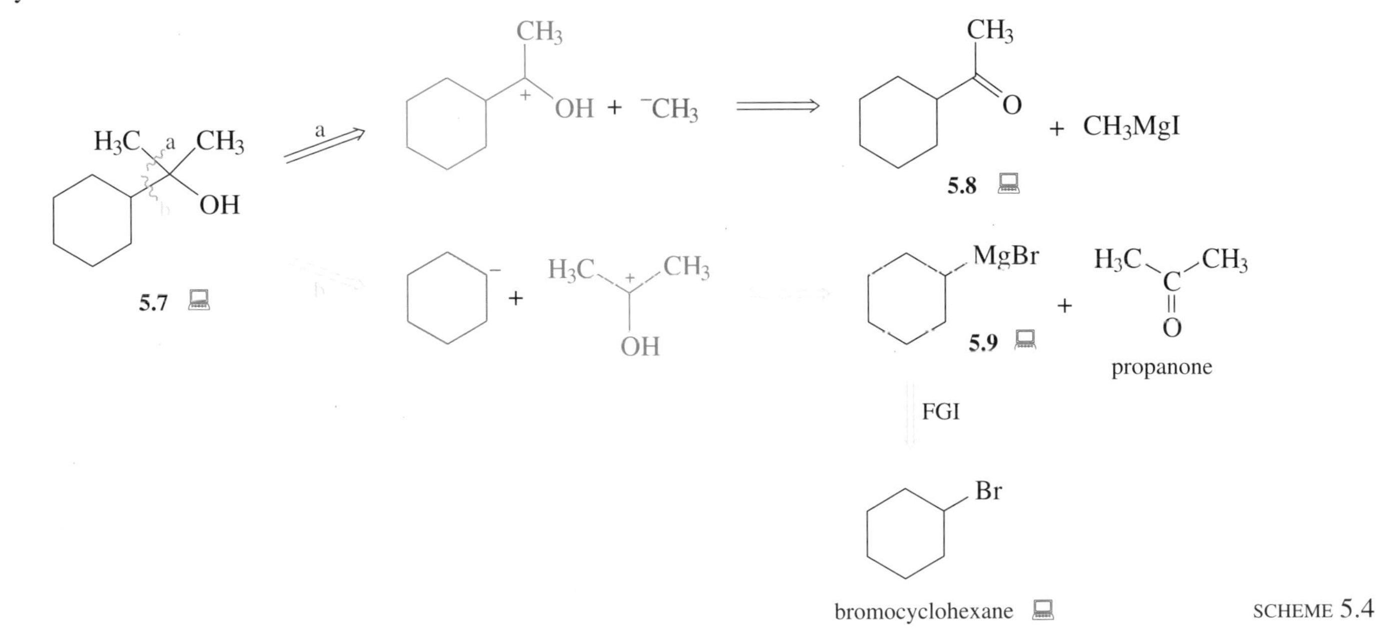


SCHEME 5.4

When two of the R groups attached to the alcohol carbon atom are identical, we can disconnect both bonds to these two R groups sequentially. For example, disconnection of the alcohol **5.10** leads back to the ketone **5.11** and the Grignard reagent, which, as we saw in Section 3, can be disconnected back to the ester.

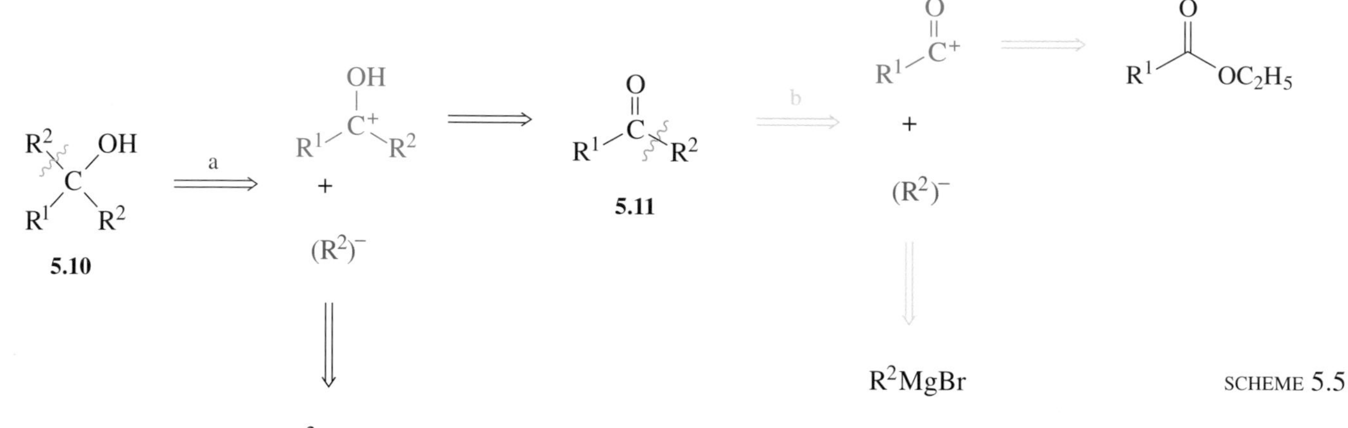


SCHEME 5.5

In the synthetic direction, the alcohol **5.10** can be made by reacting the ester with two mole equivalents of the Grignard. Since this is carried out in one step, we can simplify our retrosynthetic analysis by disconnecting both $R^2$ groups to give the doubly positive synthon **5.12**.

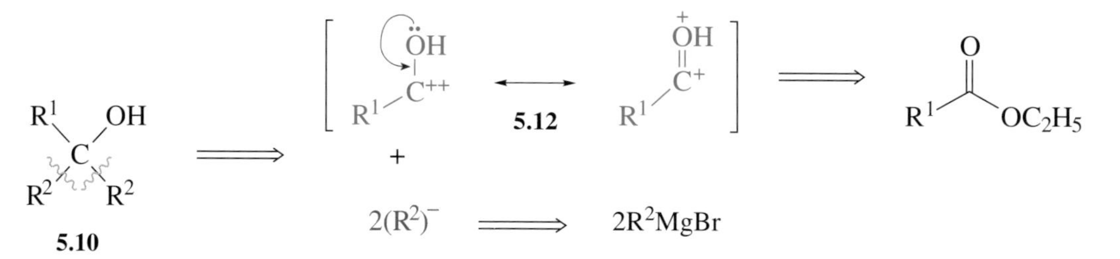


SCHEME 5.6

At first sight, the electrophilic synthon **5.12** may not look easy to translate into a reagent. However, $R^1\text{—}\overset{+}{C}\text{=}\overset{+}{O}H$, the other resonance form of synthon **5.12**, is similar to the synthon $R^1\text{—}\overset{+}{C}\text{=}O$, which as we saw earlier, corresponds to a carboxylic acid derivative **4.1**, such as an ester.

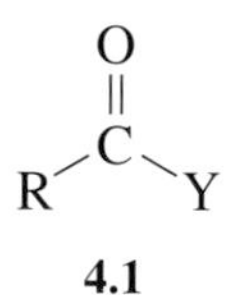


Consider the synthesis of the perfumery ester **5.13**, which is said to have 'a unique cloudy hyacinth lily-of-the-valley' smell. Our target molecule is an acid derivative (ester), so thinking back to the previous Section, we would first disconnect it to the alcohol **5.14** and ethanoic acid. Write down potential disconnections of the alcohol **5.14**, one involving the bond marked a, and the other involving both the bonds marked b, to identify the reagents required for its synthesis.

$C_6H_5$, $H_3C$, $CH_3$ (**5.13**) $\Longrightarrow$ $C_6H_5$, $H_3C$, $CH_3$, OH (**5.14**) + $CH_3COOH$ (5.2)

Disconnection a is a simple Grignard addition to a ketone, and leads to significant simplification of the target molecule. Alternatively, the double disconnection b can also be written, to leave an ester as starting material. In fact, syntheses involving both of these routes are successful.

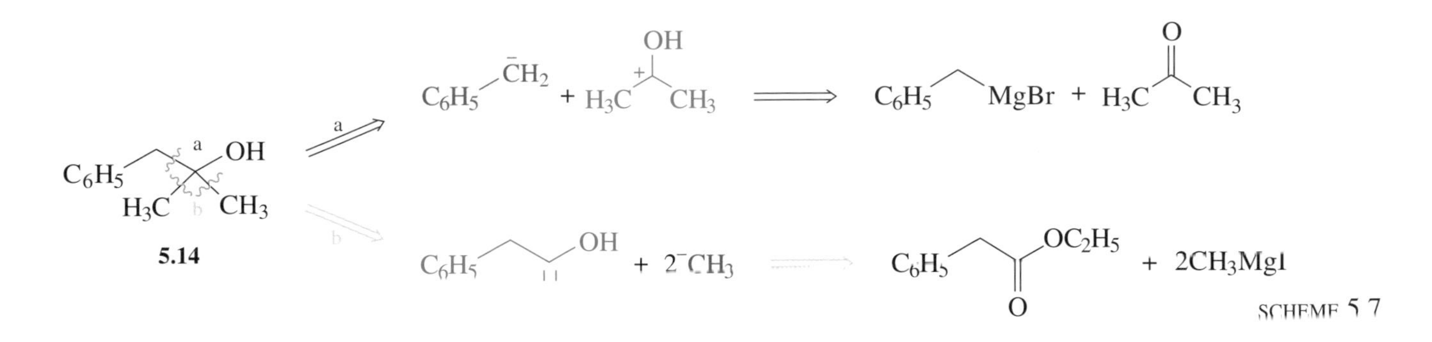

SCHEME 5.7

Alcohols are also formed when Grignard reagents react with oxiranes (epoxides):

$$R^2MgBr + H_2C\overset{O}{-}CH_2 \longrightarrow R^2CH_2CH_2O^- \xrightarrow{H^+/H_2O} R^2CH_2CH_2OH \quad (5.3)$$

In terms of the retrosynthetic analysis of alcohols, this corresponds to a C—C bond disconnection between the carbon atom next-but-one to the OH group, and the carbon atom next-but-two:

$$R^2 \wr CH_2-CH_2-OH \Longrightarrow (R^2)^- + {}^+CH_2-CH_2-OH \Longrightarrow H_2C\overset{O}{-}CH_2 \quad (5.4)$$

The synthon $^+CH_2—CH_2—OH$ corresponds to the oxirane reagent. This is one of the few cases where we do not disconnect bonds adjacent to the functional group. In this Book, such disconnections will only be used for retrosynthetic analyses of primary alcohols.

It is appropriate to explore our earlier assertion that the disconnection of alcohols is central to retrosynthetic analysis, and hence to organic synthesis. The reason lies in the fact that alcohols are at the centre of a network of FGIs, which convert them into numerous other functional groups. Look back at Table 2.1, at the FGIs involving alcohols. Some of the relevant ones are summarized in Scheme 5.8.

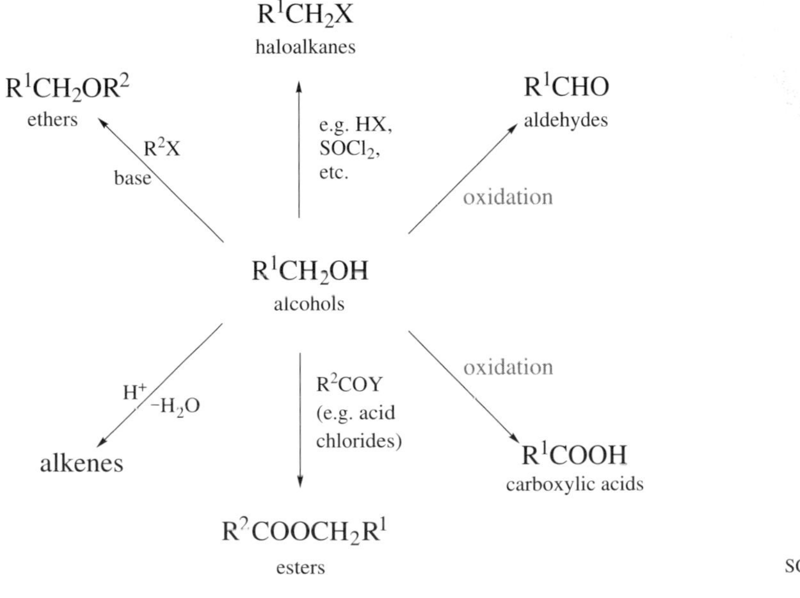

SCHEME 5.8

Alcohols can be transformed into haloalkanes or ethers by substitution reactions, into alkenes by elimination reactions* (directly by dehydration, or indirectly via haloalkanes), or into esters, by reaction with carboxylic acid derivatives such as acid chlorides or anhydrides. This last reaction type often turns out to be a C—O bond formation, building the molecular framework rather than as an FGI, as we saw in Section 4.1, and the same is true of ether formation.

We can also oxidize alcohols. The products here depend on the type of alcohol and the method used for oxidation. Primary alcohols, $R^1CH_2OH$, can be oxidized using milder methods (for example, the chromium trioxide–pyridine complex, $CrO_3$–pyridine) to aldehydes, $R^1CHO$, and under more vigorous conditions (for example, acidified aqueous potassium dichromate, $K_2Cr_2O_7$) to carboxylic acids, $R^1COOH$. Secondary alcohols, $R^1R^2CHOH$, on the other hand, are oxidized to ketones, $R^1R^2C{=}O$. However, oxidation of tertiary alcohols, $R^1R^2R^3COH$, breaks up the framework of the molecule. Don't worry about trying to remember all these details: they are summarized in the table of FGIs in the Appendix and the *Data Book* (available from the CD-ROM associated with this Book).

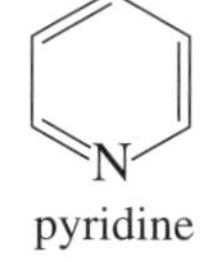


pyridine

We shall return to these elimination and oxidation reactions again when we examine suitable disconnections for alkenes, aldehydes and ketones, and acids. In many cases, the retrosynthetic analysis of these four functional groups is best approached first via an FGI to the corresponding alcohol, followed by disconnection of the C—C bonds adjacent to the alcohol. We did this earlier in this Section, in the analysis of the perfumery ester **5.13**, which was easily accomplished from the alcohol **5.14**. Figure 5.1 gives you an indication of the sort of ingredients used for preparing perfumery products.

Now we can go on to look more fully at the alkene functional group, which is available from alcohols by elimination.

**Figure 5.1** Rack of ingredients for developing perfumes.

* The mechanisms of substitution reactions and elimination reactions are discussed in *Chemical Kinetics and Mechanism*[2].

In the century that has passed, organic chemistry has literally placed a new Nature beside the old. And not only for the delectation and information of its devotees; the whole face and manner of society has been altered by its products. We are clothed, ornamented, and protected by forms of matter foreign to nature; we travel and are propelled in, on, and by them. Their conquest of our powerful insect enemies, their capacity to modify the soil and control its microscopic flora, their ability to purify and protect our water, have increased the habitable surface of the earth and multiplied our food supply; and the dramatic advances in synthetic medicinal chemistry comfort and maintain us, and create unparalleled social opportunities (and problems). We do not propose to examine this vast domain in detail, or to prognosticate the direction of its advance, in response to the need, desire, and fancy of man. We shall leave it that the evidence is overwhelming that the creative function of organic chemistry will continue to augment Nature, with great rewards, for mankind and the chemist in equal measure.

***R. B. Woodward,* Perspectives in Organic Chemistry *(1956)***

**STUDY NOTE**

All the key synthons described in this part, and the reagents to which they correspond, are summarized in Table 5.1 on p. 231 (this Table also appears in the *Data Book* for reference).

## 5.1.1 Summary of Section 5.1

1 Disconnection of the C—C bond involving the alcohol carbon atom is best carried out as follows:

$$\text{C} \wr \text{C—OH} \Longrightarrow \text{C}^- + {}^+\text{C—OH} \qquad (5.5)$$

2 The reagents corresponding to the nucleophilic synthon $C^-$ are organometallic reagents, such as RMgX.

3 The reagents corresponding to the electrophilic, oxygen-containing synthon $^+$C—OH are carbonyl compounds, such as aldehydes and ketones.

4 Any of the three groups attached to the alcohol carbon atom may be disconnected. Usually, the most appropriate disconnection involves breaking the molecule into two roughly equal-sized fragments.

5 When two of the groups attached to the alcohol carbon atom are identical, it is possible to disconnect both at the same time, to generate a doubly positive, oxygen-containing electrophilic synthon:

$$\text{R}^1\text{R}^1\text{C(OH)R}^2 \Longrightarrow 2(\text{R}^1)^- + \text{R}^2\overset{++}{\text{C}}\text{—OH} \qquad (5.6)$$

6 An ester is a suitable reagent that corresponds to this doubly positive synthon:

$$\text{R}^2\overset{++}{\text{C}}\text{—OH} \Longrightarrow \text{R}^2\text{C(=O)—OR}^3 \qquad (5.7)$$

7 Alcohols have a central role in organic synthesis, because they can be converted into many other functional groups.

### QUESTION 5.1

For the compounds **5.15–5.18**, consider all suitable bond disconnections in order to identify those that are the most effective, and relate the synthons so produced to the reagents required for the syntheses.

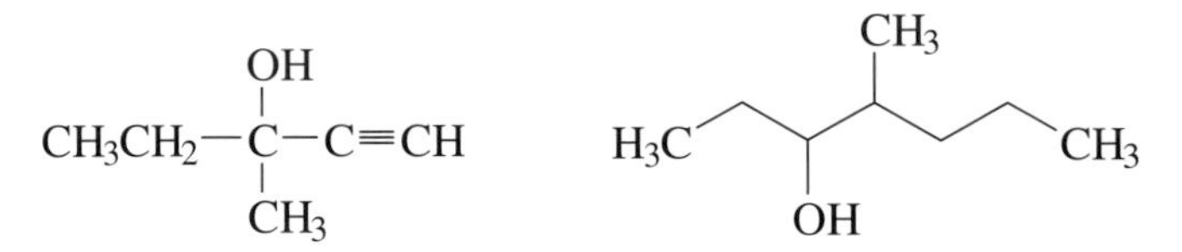


**5.15**, a tranquilizer **5.16**, a sex attractant of the elm bark beetle

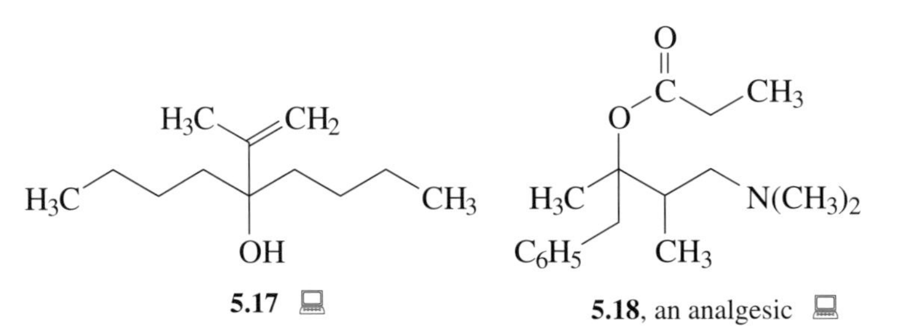


**5.17** **5.18**, an analgesic

## 5.2 Alkenes

In Section 5.1 you were introduced to the concept that an alkene can be related to an alcohol. Now we shall explore the retrosynthetic analysis of alkenes — and hence their synthesis — in more detail. The most important route for the preparation of alkenes is the use of β-elimination reactions*, in which HX is removed from a structure such as **5.19**, where X is a leaving group:

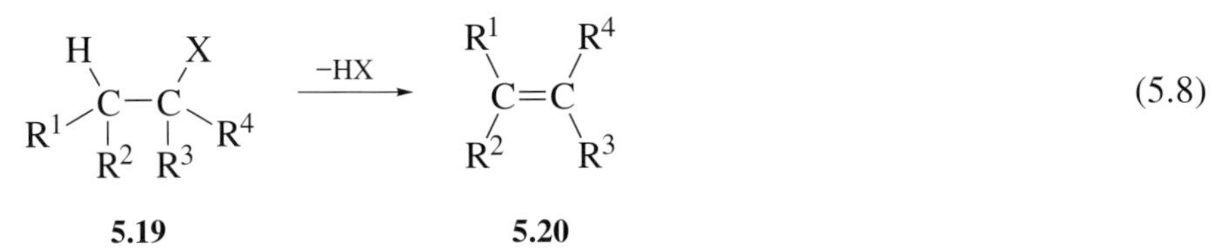


(5.8)

**5.19** **5.20**

You could argue that the conversion of a C—C bond to a C=C bond involves the formation of a new C—C bond, and that it is therefore not an FGI but a C—C bond-forming reaction. However, *our classification of C—C bond-forming reactions refers to reactions where previously unconnected carbon atoms are joined together.* β-eliminations do not therefore qualify as C—C bond-forming reactions, and are classed as FGIs.

The leaving groups used most frequently in synthesis are the hydroxyl group (OH), when the reaction is an acid-catalysed dehydration, or a halogen atom when the reaction is a base-promoted elimination of hydrogen halide. A haloalkane used for such an elimination would be made from the corresponding alcohol by an FGI

* β-elimination reactions are discussed in *Chemical Kinetics and Mechanism*[2].

(Scheme 5.8, p. 205). This means that *the disconnection of any alkene,* ***5.20****, can begin with an FGI to generate an alcohol,* ***5.21****, as the precursor.* Retrosynthetic analysis of this alcohol then follows in the way that we have seen in Section 5.1.

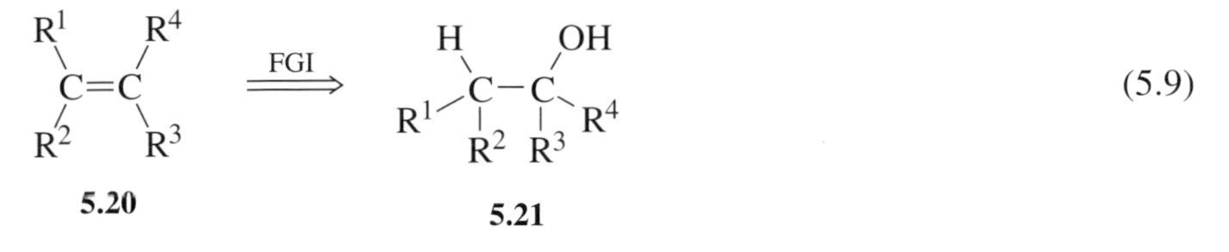


(5.9)

- The formation of an alkene using an FGI immediately gives us two possibilities. What are they?

- Working backwards, the (imaginary) addition of water across the target alkene, **5.20**, could be written in two ways, to generate two different alcohols. In the general case shown, alcohol **5.22** must join **5.21** as a potential precursor for the alkene **5.20**.

HO H
$R^1$ C—C $R^4$
$R^2$ $R^3$
**5.22**

The choice between the two ways of carrying out the FGI is made by looking at the alcohols themselves. We first perform a retrosynthetic analysis on each of them to see if one route is obviously better than the other. For the substituted cyclohexene target **5.23**, the two alcohols would be **5.24** and **5.25**, so let's compare them.

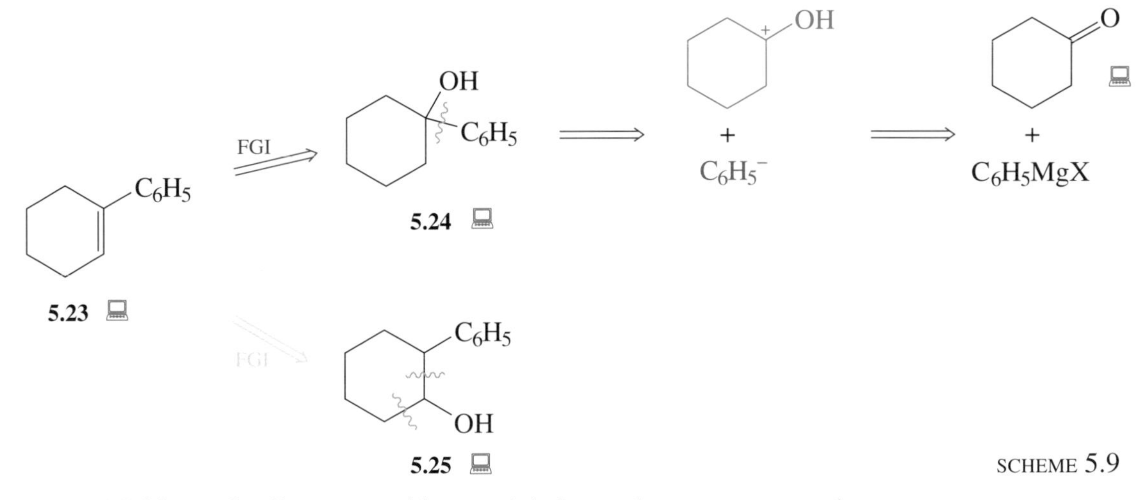


SCHEME 5.9

The alcohol **5.24** may be disconnected in a straightforward way, to suggest the reaction of cyclohexanone and phenyl Grignard reagent as a synthetic route. Alcohol **5.25**, on the other hand, has no helpful disconnection. If we try to use the normal approach to alcohols, the logical disconnections would require opening of the cyclohexane ring. That is, a cyclization reaction would be needed in the forward synthesis — something that we should try to avoid if possible.

Even if both precursors to a target alkene can be disconnected sensibly, there is a further question to answer. We need to look at the dehydration reaction that will be at the end of our synthesis, to see whether or not it will lead to isomers of the alkene as well as the required alkene itself. In the case of the alcohol **5.24**, this is not a problem; the symmetry of this molecule means that dehydration can give only the target alkene **5.23**, which is conjugated to the phenyl group as well. But what about other target molecules, say **5.26**? The two alcohol precursors in this instance are **5.27** and **5.28**:

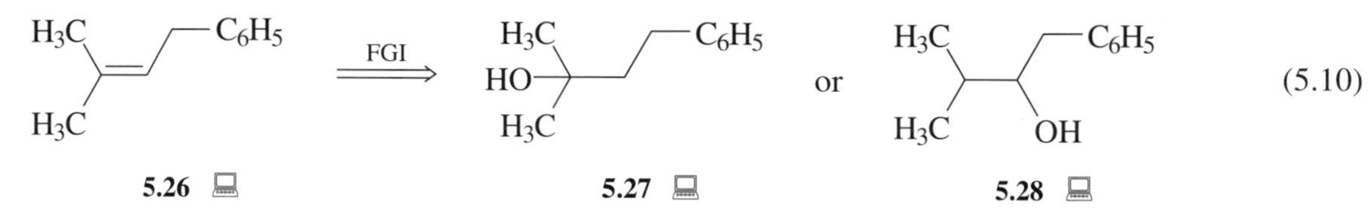


(5.10)

Write down all the possible dehydration products from these two alcohols. How do we assess which product will predominate from this type of β-elimination?

The alcohol **5.27** could give the alkene **5.29**, as well as the target alkene **5.26**; isomer **5.30** would be the alternative product from dehydration of alcohol **5.28**. These acid-catalysed eliminations would follow an El mechanism, and, according to the Saytzev rule, they should provide the more-stable (that is, the more-substituted) alkene **5.26** as the predominant product.

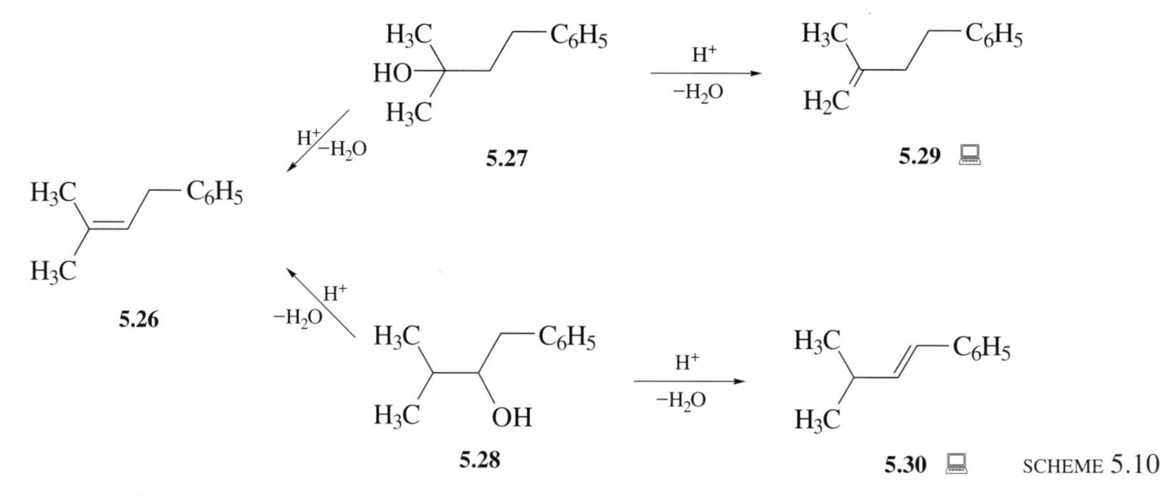


SCHEME 5.10

Dehydration of alcohol **5.27** would in fact give almost entirely the required target alkene, **5.26**, which is a trisubstituted alkene, and little or none of the disubstituted isomer, **5.29**. The alternative alcohol **5.28**, on the other hand, would give a little of the target alkene, **5.26**, but mainly the isomer, **5.30**, despite the fact that **5.26** is the more-substituted alkene. This is because conjugation of the alkene double bond with the phenyl group in **5.30** is a more powerful stabilizing factor, such that the conjugated disubstituted alkene predominates over the more substituted **5.26** as a product of the dehydration of alcohol **5.28**.

The alcohol **5.27** is therefore the preferred precursor to our target alkene **5.26**, since it gives predominantly the target molecule. The retrosynthetic analysis of **5.26** can now be completed (Scheme 5.11). To make the alcohol **5.27** we choose the simple ketone, propanone, as the electrophilic carbonyl precursor. The detailed synthesis of **5.26** is also illustrated, for comparison with the analysis.

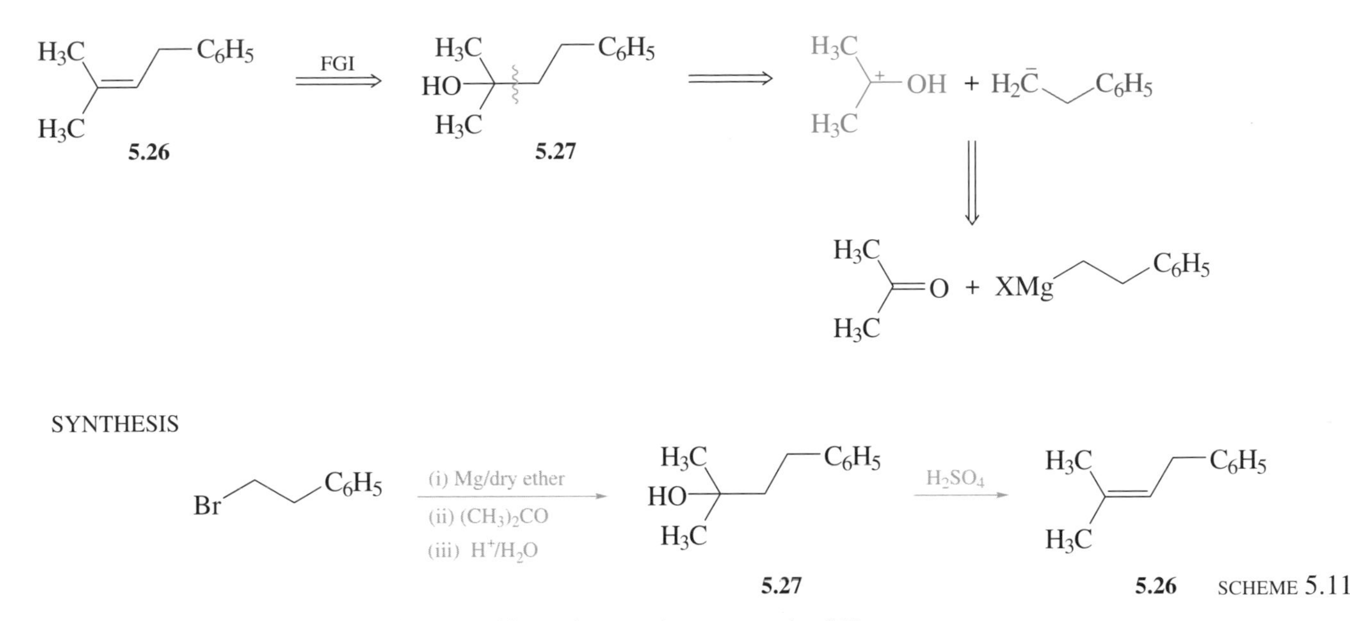


SCHEME 5.11

If you look at the analysis more carefully, you will see that we chose to put the OH group at a **branch point** of the carbon skeleton. It turns out that this is often a good thing to do.

> Even if there is nothing else to choose between the two alcohols, *having the OH group at a branch point will usually give simpler starting materials when the alcohol is disconnected.*

- There is still one property of some alkenes that we have not yet had to tackle in these retrosynthetic analyses. What is it?
- Some alkenes can exist as diastereoisomers, so we need to consider their stereochemistry.

This has not been an issue with the targets so far in this Section, namely **5.23** and **5.26**, since in these examples, diastereoisomers are not possible. But what about those cases in which diastereoisomers *are* possible, such as 1,2-disubstituted alkenes, **5.31**?

$R^1{-}CH{=}CH{-}R^2$

**5.31**

For these compounds, the configuration of the product of a dehydration reaction is usually controlled by the stability of the product alkene. The more-stable geometry predominates because the reaction occurs via an El mechanism. So the formation of 1,2-disubstituted alkenes using dehydration reactions will give more *E*-isomer, **5.32**, than *Z*-isomer, **5.33**. Often, however, mixtures of stereoisomers are obtained. We shall not go into any more detail on dehydration reactions here because, if the geometry of a target alkene is critical, it is actually better to avoid the dehydration approach, and to look for other ways to make it, which do not produce any stereochemical ambiguity.

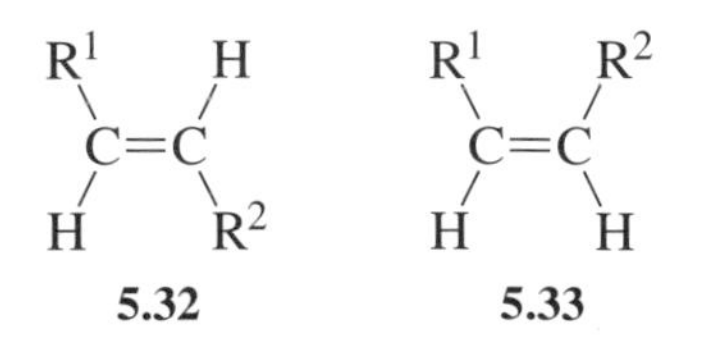

An alternative approach to making 1,2-disubstituted alkenes specifically, and one that *does* allow control over product stereochemistry, requires us to start the retrosynthetic analysis with an FGI. This time the FGI leads back to an alkyne, **5.34**, as precursor:

$$\underset{\textbf{5.31}}{R^1{-}HC{=}CH{-}R^2} \xRightarrow{\text{FGI}} \underset{\textbf{5.34}}{R^1{-}C{\equiv}C{-}R^2} \qquad (5.11)$$

What are the reagents that enable us to convert alkynes into alkenes in this synthetic direction?

One reagent is hydrogen gas and Lindlar's catalyst, which is a finely divided palladium catalyst 'poisoned' to prevent further reduction of the alkene product to the alkane:

$$\underset{\textbf{5.34}}{R^1{-}C{\equiv}C{-}R^2} \xrightarrow[\text{Lindlar's catalyst}]{H_2} \underset{\textbf{5.33}}{(R^1)(H)C{=}C(R^2)(H)} \qquad (5.12)$$

This procedure has a mechanism that ensures that the *Z*-alkene, **5.33**, is produced.

The *E*-isomer, **5.32**, is prepared in high yield by another method for this FGI, namely reduction by sodium metal in liquid ammonia:

$$\underset{\textbf{5.34}}{R^1{-}C{\equiv}C{-}R^2} \xrightarrow[\text{liquid } NH_3]{Na} \underset{\textbf{5.32}}{(R^1)(H)C{=}C(H)(R^2)} \qquad (5.13)$$

So you can see that, if the corresponding alkyne can be assembled, we can efficiently produce either isomer of a 1,2-disubstituted alkene as required. How, then, do we disconnect alkynes to plan their synthesis? That is the topic of Section 5.3, in which we shall also illustrate the use of alkynes in the synthesis of alkenes.

## 5.3 Alkynes

As we have indicated in the previous Section, alkynes are often valuable intermediates, especially in the synthesis of alkenes with defined stereochemistry. In fact, they are arguably more important as intermediates than they are in their own right!

Substituted alkynes are built up from smaller alkynes, and ultimately from ethyne (acetylene), which is readily available. We can see how this is done, by performing a retrosynthetic analysis on the generalized alkyne **5.34**. *The carbon–carbon bond we select to disconnect is a single bond attached directly to one of the alkyne carbon atoms.*

$$R^1 \wr C{\equiv}C{-}R^2$$

**5.34**

Think about the properties of alkynes. Which end of the disconnected bond in **5.34** bond will be better as the nucleophilic synthon?

The alkyne carbon end. Monosubstituted alkynes, RC≡CH, are slightly acidic, and are converted by strong base into nucleophilic alkynide anions.

This helps to define the way in which we shall disconnect the C—C bond to an alkyne carbon atom. One of the alkyl groups of **5.34** forms the electrophilic synthon $(R^1)^+$, usually through a haloalkane $R^1X$ as reagent, and the alkyne-containing fragment is the nucleophilic synthon:

$$\underset{\mathbf{5.34}}{R^1 \wr C{\equiv}C{-}R^2} \Longrightarrow (R^1)^+ + {}^-C{\equiv}CR^2 \Longrightarrow R^1X + HC{\equiv}C \wr R^2 \quad \mathbf{5.35}$$

$$\Downarrow$$

$$HC{\equiv}CH + R^2X \Longleftarrow HC{\equiv}C^- + (R^2)^+$$

SCHEME 5.12

The forward reaction is an $S_N2$ substitution of the haloalkane by a metal alkynide. The monosubstituted alkyne **5.35** itself may be disconnected in a similar fashion, as illustrated. So, to summarize our analysis of the disubstituted alkyne **5.34**, it can be formed from ethyne and two haloalkanes*. Remember that the real synthesis would involve two *successive* $S_N2$ substitutions of the appropriate haloalkanes by alkynide anions.

ANALYSIS

$$\underset{\mathbf{5.34}}{R^1 \wr C{\equiv}C \wr R^2} \Longrightarrow R^1X + HC{\equiv}CH + R^2X$$

SYNTHESIS

$$HC{\equiv}CH \xrightarrow[\text{(ii) } R^2X]{\text{(i) NaNH}_2} R^2{-}C{\equiv}C{-}H \xrightarrow[\text{(ii) } R^1X]{\text{(i) NaNH}_2} \underset{\mathbf{5.34}}{R^2{-}C{\equiv}C{-}R^1}$$

SCHEME 5.13

To see how this works in practice, let's look at the alkene **5.36**, which is a pheromone† (natural attractant) produced by the pea moth. The pheromone is used to trap the moths, to indicate to farmers exactly when to spray in order to eliminate this destructive pest. The target molecule is an unstructured ester, so we can immediately carry out a C—X disconnection and write the unsaturated alcohol **5.37** and ethanoic acid as precursors:

$$\underset{\mathbf{5.36}}{H_3C{-}CH{=}CH{-}(CH_2)_8CH_2O \wr \overset{O}{\overset{\|}{C}}{-}CH_3} \Longrightarrow \underset{\mathbf{5.37}}{H_3C{-}CH{=}CH{-}(CH_2)_8CH_2OH} + CH_3COOH \qquad (5.14)$$

The *E* geometry of the alkene is essential for the biological properties of the pheromone, and it must be produced selectively in a synthesis. We shall therefore use an FGI to generate the alkyne **5.38** as a precursor, and then disconnect the alkyne, as we have seen, to ethyne, and two halogen-containing compounds — in this case, iodomethane and the bromoalcohol **5.39**:

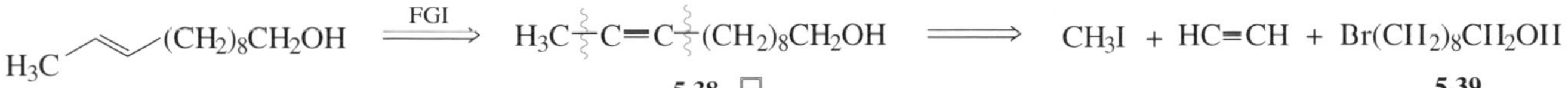

$$\underset{\mathbf{5.37}}{H_3C{-}CH{=}CH{-}(CH_2)_8CH_2OH} \overset{\text{FGI}}{\Longrightarrow} \underset{\mathbf{5.38}}{H_3C \wr C{\equiv}C \wr (CH_2)_8CH_2OH} \Longrightarrow CH_3I + HC{\equiv}CH + \underset{\mathbf{5.39}}{Br(CH_2)_8CH_2OH} \qquad (5.15)$$

* Note that in this reaction the alkynide ion is formed by treatment with sodamide, $NaNH_2$. Be sure not to confuse this reaction with that between an alkyne and sodium in liquid ammonia, which results in reduction to an alkene.

† Pheromones are discussed in Part 2 of *The Third Dimension*[3].

When the synthesis was carried out, the basic conditions needed to form the alkynide anion were found to cause side-reactions of the terminal alcohol **5.39**; a protecting group was therefore employed. As you will remember from Part 1 of this Book, a **protecting group** *temporarily* masks a functional group so that it remains unaffected while a reaction takes place at some other position in the molecule (you will meet this idea again in Section 7.4). An acetal protecting group was used here. We shall not go into detail here about this particular protecting group (abbreviated to THP). Suffice it to say that it is stable under basic conditions but may be removed using aqueous acidic conditions. The synthesis was carried out as shown in Scheme 5.14 (note that ethanoic anhydride was used to form the ester **5.36** from alcohol **5.37** in the final stage).

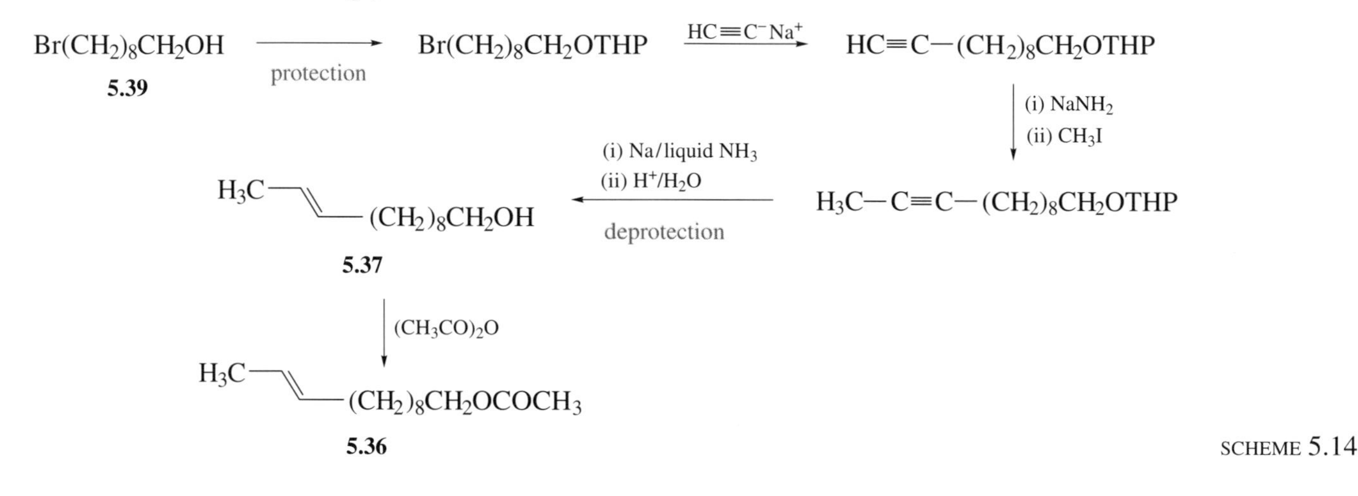


SCHEME 5.14

## BOX 5.1 Robert Bruce Merrifield

Robert Bruce Merrifield (Figure 5.2) was born in Fort Worth, Texas on 15 July, 1921, the only son of George E. and Lorene (Lucas) Merrifield. Later, they lived in several cities in California, and he attended a number of schools before joining Pasadena Junior College in 1939. Two years later, he transferred to the University of California at Los Angeles (UCLA). He obtained his PhD in 1949 and then joined the Rockefeller Institute for Medical Research. While there, he became interested in peptide (protein) synthesis, and developed a solid-phase method of protein synthesis in 1959, whereby a strand is grown outwards from the solid surface by adding the amino acids and forming amide bonds one at a time. The completed strand is then cleaved from the surface.

He married Libby Furlong in 1949 and has six children.

He was awarded the Nobel Prize for Chemistry in 1984 for his development of methodology for chemical synthesis on a solid matrix.

**Figure 5.2**
Robert Bruce Merrifield.

## 5.4 Summary of Sections 5.2 and 5.3

1. Alkenes may be derived from alcohols by dehydration.
2. One possibility for the retrosynthetic analysis of an alkene is therefore to carry out an FGI using an imaginary hydration of the double bond. This enables two alcohols to be identified as potential precursors to the alkene.
3. The alcohol that will give the target alkene as the major product by dehydration according to Saytzev's rule (or by being involved in conjugation, such as with a benzene ring) is the more appropriate.
4. All other factors being equal, the OH group is best located at a branch point of the carbon skeleton.
5. Diastereoisomerism is an important consideration for alkenes, and synthetic routes must take account of the diastereoisomer required. Dehydration reactions produce more *E*-isomer than *Z*-isomer.
6. An alternative analysis that takes account of the selective synthesis of one stereoisomer of 1,2-disubstituted alkenes, involves first carrying out an FGI between an alkene and the corresponding alkyne. The analysis then continues with the alkyne.
7. Alkynes may be converted into *Z*-alkenes using hydrogen gas and Lindlar's catalyst, or into *E*-alkenes using sodium metal in liquid ammonia.
8. Disconnection of an alkyne is best carried out at the C—C single bonds that are attached directly to the alkyne carbon atoms. The two synthons so produced are an electrophilic alkyl synthon and a nucleophilic alkyne synthon.

### QUESTION 5.2

Use retrosynthetic analysis to devise syntheses of the alkenes **5.40–5.42**, pointing out the synthons, precursors and reagents that are identified by such an analysis.

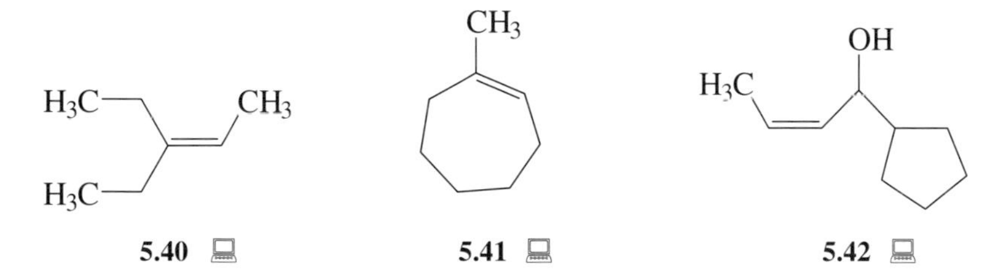


## 5.5 Aromatic compounds

The next group of compounds that we shall examine by retrosynthetic analysis are considered together because they have the benzene ring structural fragment in common, with at least one substituent attached to the ring via a C—C bond. The ready availability of benzene itself, and a number of simple, substituted derivatives, means that we very rarely contemplate making a benzene ring during a synthesis. We can therefore leave these rings intact in our retrosynthetic analyses.

For the purposes of retrosynthetic analysis, we focus on the benzene ring as our functional group, and choose to disconnect a C—C bond that is attached directly to the ring. The two pairs of potential synthons from such a disconnection are as follows, where Ar represents an aromatic ring (that is, a benzene or substituted benzene ring).

Ar⸾C — a ⇒ $Ar^- + C^+$

Ar⸾C — b ⇒ $Ar^+ + C^-$

SCHEME 5.15

- Which of these arrangements is more in keeping with the type of chemical reactivity we associate with a benzene ring?

- Aromatic rings are usually seen reacting as nucleophiles; their typical reaction is substitution (of hydrogen) by an electrophile*. Path a therefore looks to be the more logical way in which to disconnect C—C bonds to an aromatic ring.

Reagents corresponding to the aromatic nucleophilic synthon $Ar^-$ can be made in two ways. We can use benzene (the aromatic compound Ar—H) itself as a nucleophile, with the appropriate electrophile and a Lewis acid catalyst.

- Which electrophilic aromatic substitutions create new C—C bonds, and which reagents are used?

- The reactions in question are the Friedel–Crafts alkylation (i) and acylation (ii) reactions, and the electrophilic reagents are a haloalkane and an acid chloride, respectively:

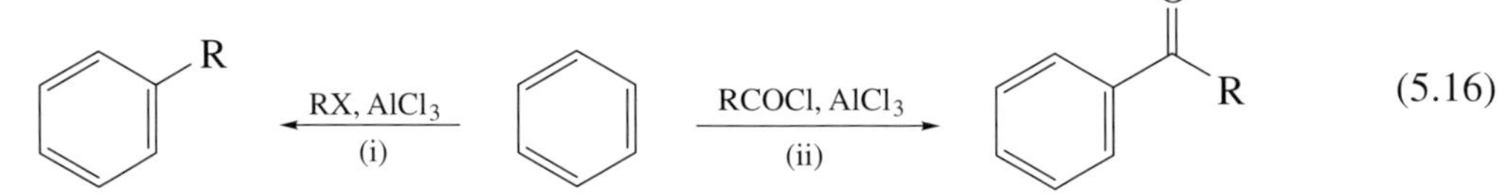


(5.16)

Alternatively, we can take the same approach as we did to generate an alkyl nucleophile — that is, to use a Grignard reagent ArMgX, prepared from a haloaromatic compound. These Grignard nucleophiles behave like any other Grignard reagent; for example, they react with aldehydes, ketones or esters to give alcohols (as in Section 5.1):

$$\mathrm{ArMgX + R^1COR^2 \longrightarrow \xrightarrow{H^+/H_2O} Ar{-}\underset{R^2}{\overset{R^1}{\underset{|}{\overset{|}{C}}}}{-}OH} \qquad (5.17)$$

Whether to use the direct Friedel–Crafts substitution, or the Grignard approach, is often a question of the availability of starting materials, such as the haloaromatic precursor for a Grignard reagent. You might think it would also depend on the exact functional group(s) required in the target molecule; the organometallic approach for example, usually produces alcohol products, whereas Friedel–Crafts reactions can provide hydrocarbon or ketone substituents on a benzene ring. We must always bear in mind, however, that these product functional groups can be interconverted if necessary, using appropriate oxidation or reduction FGIs. As the Grignard approach doesn't use any properties unique to benzene that haven't been discussed in Section 5.1, we won't discuss it any more here. You have already seen examples of the use of $C_6H_5MgX$, and more will appear later.

- The ketone **5.43** is a constituent of the fragrance of hawthorn blossom (Figure 5.3). Using a Friedel–Crafts type of C—C disconnection, suggest a suitable synthesis.

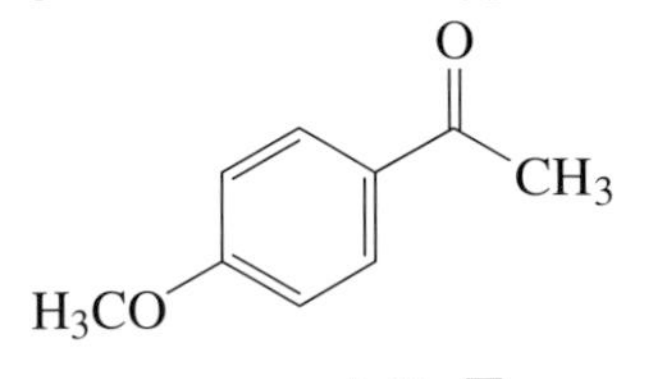


**5.43**

* Electrophilic aromatic substitution is discussed in Part 2 of *Alkenes and Aromatics*[1].

**Figure 5.3**
Hawthorn blossom.

Retrosynthetic analysis generates the synthons **5.44** and **5.45**, which can be translated into the acylation of methoxybenzene (readily available) with ethanoyl chloride, in the presence of $AlCl_3$. (This is an example of *para* disubstitution; *meta* and *ortho* substitution will be dealt with later.)

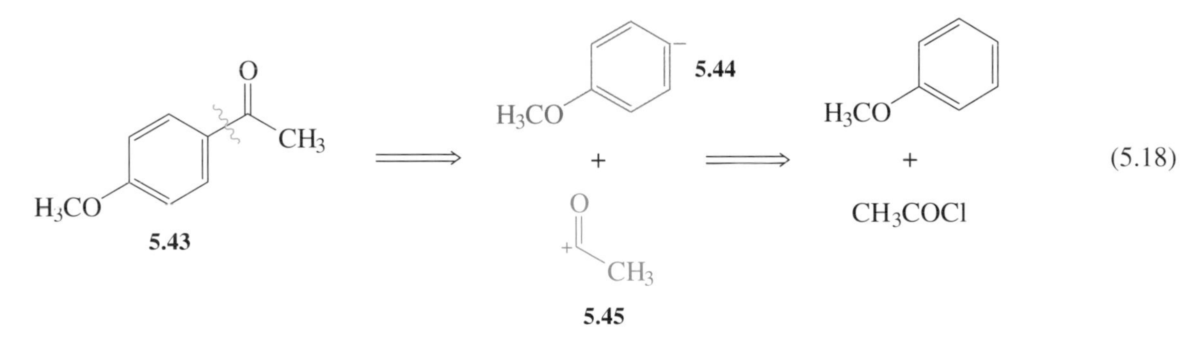


What if the target were the alkylbenzene **5.46**? By analogy with the ketone **5.43**, and thinking back to the Friedel–Crafts disconnection, we might propose reaction of methoxybenzene with chloroethane, in the presence of $AlCl_3$:

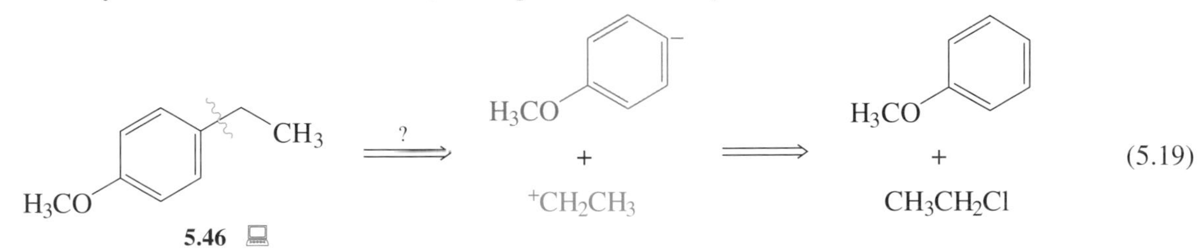


What problems can arise in Friedel–Crafts alkylation reactions?

The product is activated by the new alkyl substituent, and is therefore more reactive than the starting material, so that polyalkylation can be a problem. In general, there is also the possibility of the alkyl carbocation (formed from the reaction of the haloalkane with $AlCl_3$) undergoing rearrangement before substitution into the benzene ring. However, this is not a problem with the ethyl group in the synthesis of **5.46**.

If we wanted to make **5.46**, it would probably be best to avoid a Friedel–Crafts alkylation reaction, and instead use an acylation to make the ketone **5.43** (as shown in Reaction 5.18), followed by an FGI (in this case, a reduction). The Clemmensen reduction would be suitable. In retrosynthetic terms this means beginning our analysis of **5.46** with an FGI, to identify the ketone **5.43** as the new target.

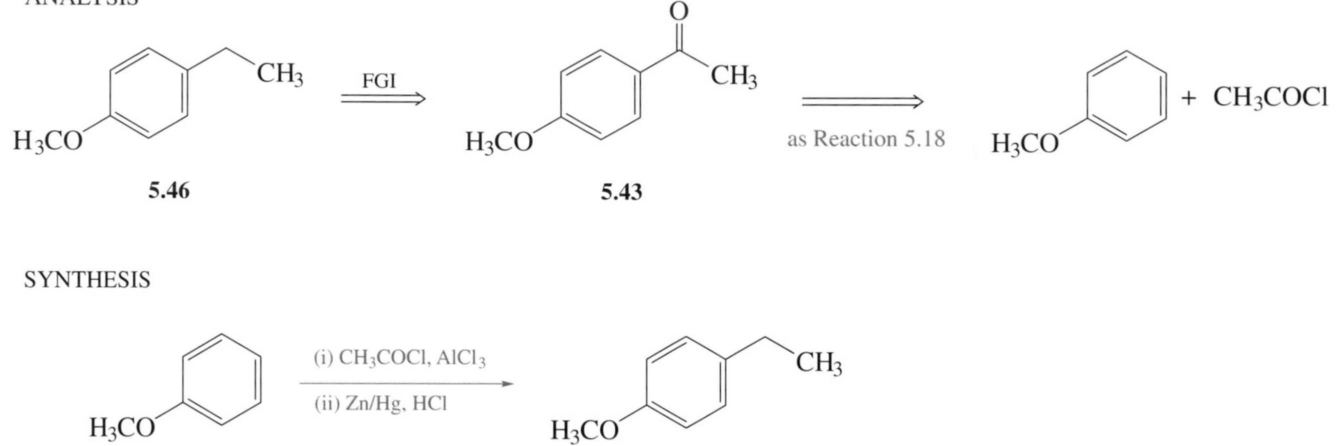


SCHEME 5.16

For the synthesis of a substituted aromatic compound, we must always remember that the mechanism of the reaction determines the position of substitution. In the previous examples, the *para*-substituted product **5.43** will be formed along with some *ortho*-isomer in the Friedel–Crafts acylation, since the methoxy group is an activating group and *ortho/para*-directing. *Ortho*- and *para*-substituted product isomers will often be formed together as mixtures, which means that subsequent separation* of isomers (by chromatography, crystallization or distillation, for example) will be needed. Normally in organic synthesis, reactions that generate mixtures are best avoided, but aromatic substitutions are usually so easy to carry out and have high yields, so that the problem of separation becomes acceptable. We just do the reaction on a big enough scale to make enough of the isomer that we want. (We can do this in the laboratory, but not on an industrial scale since it would be too wasteful!)

To illustrate this problem of the position of substitution in benzene rings, let's take as a further example the synthesis of the diaryl ketone **5.47**. The two pairs of reagents for the two possible Friedel–Crafts acylation reactions suggested by our proposed disconnections a and b are shown in Scheme 5.17. The choice between these routes should be made on the basis of mechanism. Route a is perfectly reasonable. The aromatic compound **5.48** is activated by the methoxy group, which is *ortho/para*-directing, so that **5.47** should certainly be formed as an acylation product, although it will need to be separated from the *para*-product (which may be formed preferentially). Disconnection b on the other hand, is not acceptable, as the nitro group in nitrobenzene is *meta*-directing, and a *para*-substitution is required. Worse still, the aromatic ring in nitrobenzene is too deactivated to react under Friedel–Crafts conditions! In general, *it is best to avoid electrophilic substitution reactions on deactivated benzene rings.*

Before we leave C—C bond formation to an aromatic ring, it is worth noting an alternative strategy that makes use of the electrophilic substitution reactions of benzene which form C—X bonds. A particularly useful sequence is nitration,

* Separation techniques are discussed in Part 1 of *Separation, Purification and Identification*[4].

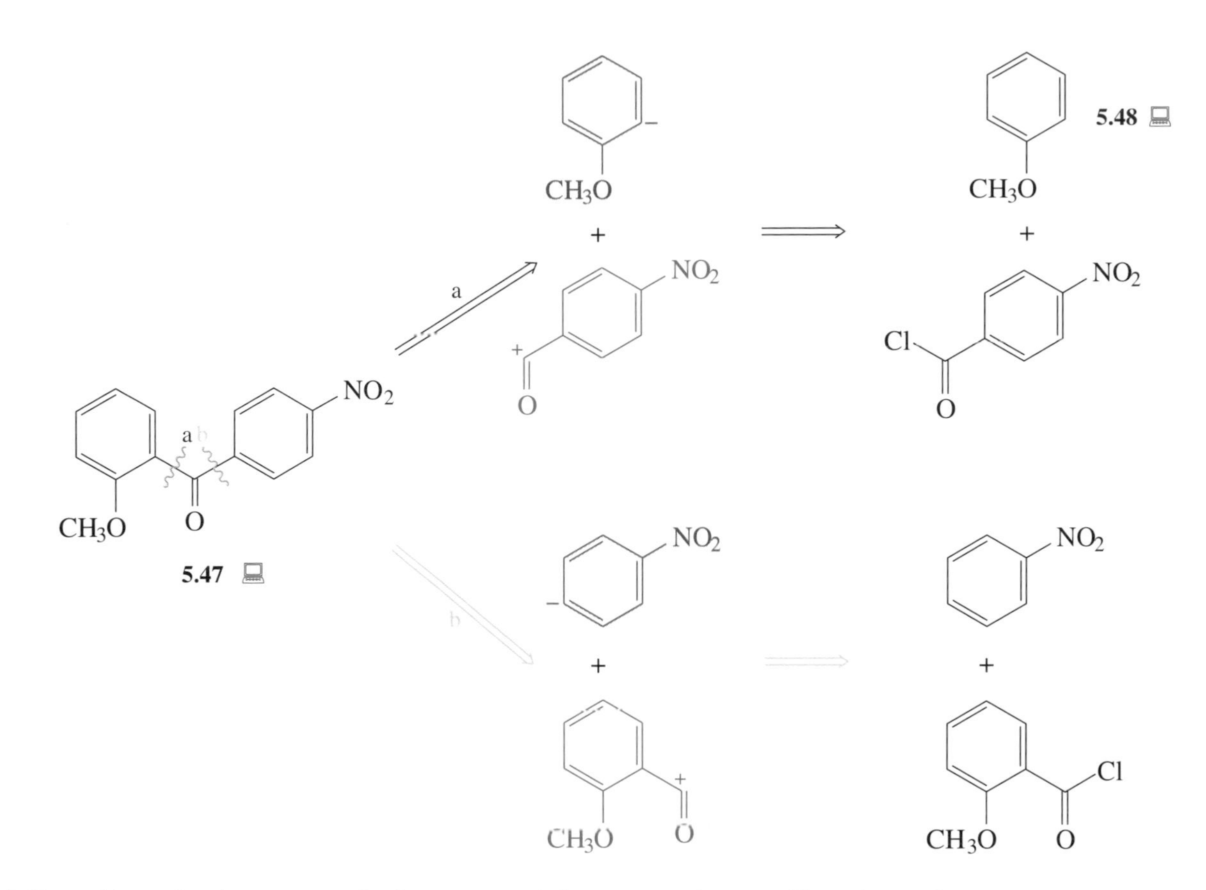


SCHEME 5.17

followed by reduction to an aminobenzene, and then conversion to a diazonium salt, $ArN_2^+ Cl^-$:

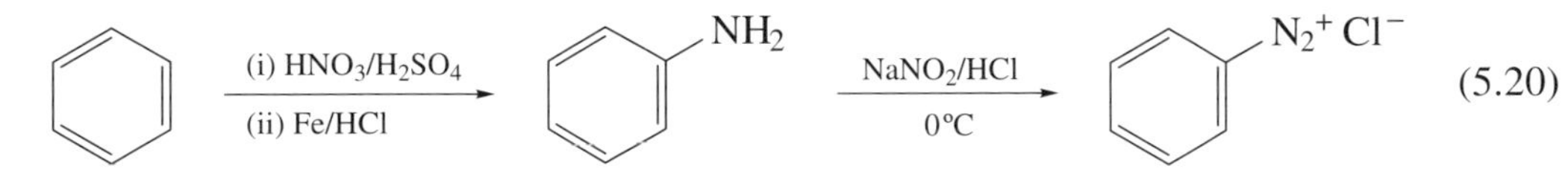


(5.20)

Diazonium salts undergo a variety of substitution reactions (see Figure 30.3 in the *Data Book* from the CD-ROM), such as the introduction of halogens, a hydroxyl group or a —CN group (of most significance for C—C bond formation):

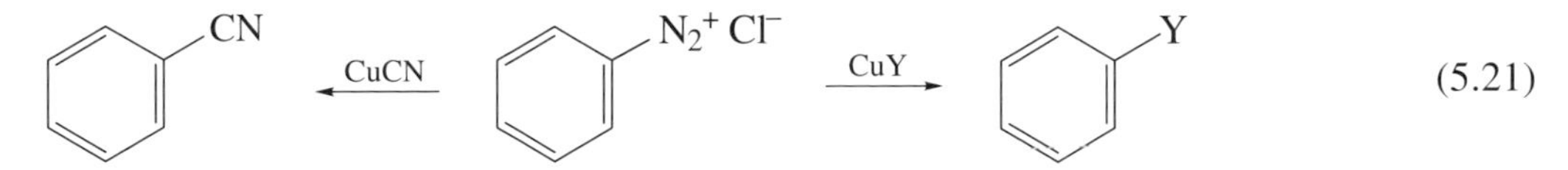


(5.21)

In the synthetic direction, therefore, we make a C—X (C—N) bond, and change it to a C—C bond. The nitrile group is particularly flexible, and can be changed by FGIs in various ways; for example, hydrolysis to an acid, and reduction to an amine, are possible, as shown in the table of FGIs in the Appendix. It may also be converted to a ketone using an organometallic reagent.

Retrosynthetic analysis of these aromatic nitriles is quite revealing as to the nature of the reagents involved. The cyanide anion is a nucleophile, so we are forced to invoke the electrophilic phenyl synthon **5.49**:

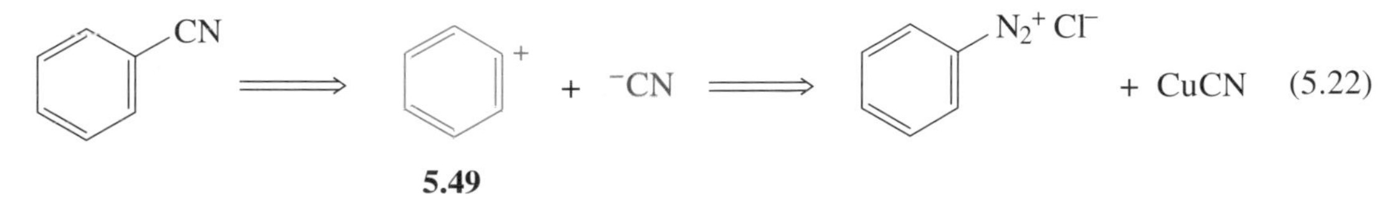


(5.22)

Aromatic diazonium salts thus act as the reagents for this unusual electrophilic synthon. This contrasts with our decision at the beginning of this Section that the nucleophilic $Ar^-$ synthon is compatible with the more usual reactivity of benzene rings.

### Box 5.2 Elias J. Corey

Elias Corey (Figure 5.4) was born in Methuen, Massachusetts, in July 1928 of Lebanese extraction. His father died when he was very young, and he and his brother and two sisters were brought up by their mother. He attended local schools and entered the Massachusetts Institute of Technology in July 1945. His favourite subject was mathematics, but he was converted to chemistry early in his studies. He obtained his PhD in 1950, and then moved to the University of Illinois at Urbana-Champaign, where he developed his interests in synthetic chemistry. He became Professor of Chemistry there in 1956. He moved to Harvard in 1959, where he used the basic concepts of retrosynthetic analysis to design the syntheses of many complex molecules. In September 1961, he married Claire Higham and had three children.

His research interests include the synthesis of complex, bioactive molecules, the logic of chemical synthesis, new methods of synthesis, reaction mechanisms, organometallic chemistry, bioorganic and enzyme chemistry, and the application of computers to organic chemical problems, especially to retrosynthetic analysis.

He was awarded the 1990 Nobel Prize for Chemistry for his development of the theory and methodology of organic synthesis.

**Figure 5.4** Elias J. Corey.

## 5.5.1 Summary of Section 5.5

1 Benzene rings are seldom made in synthesis, so, for the purposes of retrosynthetic analysis, we choose to disconnect a C—C bond attached directly to the benzene ring.

2 Generally, the most suitable disconnection is the one in which the aryl synthon is identified as being nucleophilic.

3 Suitable reagents for the nucleophilic aryl synthon ($Ar^-$) are the parent hydrocarbon (Ar—H) or an aryl Grignard reagent (Ar—MgX).

4 The most common reactions for constructing C—C bonds to aryl groups are those involving reaction between an aryl Grignard reagent and a carbonyl compound, or Friedel–Crafts alkylation and acylation reactions. Friedel–Crafts acylation is generally more appropriate than alkylation.

5 The most appropriate disconnection for Friedel–Crafts acylation is determined by the mechanism of the reaction, which governs the site of substitution.

6 An alternative way of constructing C—C bonds involving aromatic rings is nucleophilic attack of cyanide ion, $^-CN$, on an aromatic diazonium ion, which is itself a reagent for the electrophilic aryl synthon ($Ar^+$). The aryl cyanide so produced may be converted into a carboxylic acid or an amine, using FGIs, or a ketone using an organometallic reagent.

QUESTION 5.3

Use retrosynthetic analysis to devise syntheses of compounds **5.50** and **5.51**. Include appropriate synthons and reagents in your analysis.

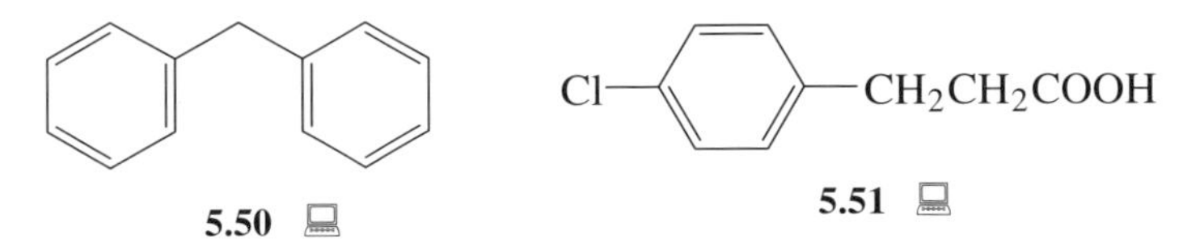


## 5.6 Aldehydes and ketones

Section 5.5 described how aromatic ketones can be disconnected and assembled via Friedel–Crafts acylation. In this Section we look at ways of synthesising aldehydes and ketones more generally.

Before considering any disconnection of C—C bonds in our retrosynthetic analysis of these carbonyl compounds, perhaps we should see if any simple precursors are available by FGI.

- What functional group covered already in Part 4 can be converted into an aldehyde or ketone by an FGI?

- The alcohol functional group: one easy approach to aldehydes and ketones is to consider them as oxidation products of alcohols.

$$R^1C(=O)R^2 \overset{\text{FGI}}{\Longrightarrow} R^1\text{—}CH(OH)\text{—}R^2 \text{ (also when } R^2 = H) \qquad (5.23)$$

Retrosynthetic analysis of a target aldehyde or ketone by this strategy involves, first, an FGI back to the alcohol, and then analysis of the new target alcohol by the methods we introduced in Section 5.1. At the end of a retrosynthetic analysis along these lines, we need to be sure that there are satisfactory methods available for the final oxidation FGI. In fact, several possibilities do exist, as you saw in Box 2.1 and the FGI table in the Appendix. A number of popular procedures are based on chromium reagents, and $CrO_3$–pyridine (Collins' reagent), or the more modern variant, pyridinium chlorochromate (PCC; $CrO_3$–pyridine–HCl), are widely used. These mild oxidizing reagents will selectively oxidize a primary alcohol, $RCH_2OH$, to an aldehyde, RCHO, or a secondary alcohol, $R^1R^2CHOH$, to a ketone, $R^1R^2CO$. The reagent $CrO_3$–$H_2SO_4$–propanone (Jones' reagent) is also used for the oxidation of secondary alcohols. You should add these reagents to your table of FGIs (Table 2.1). Remember that attempted oxidation of tertiary alcohols results in breaking the molecular framework.

- What potential problem arises in the oxidation of primary alcohols, $RCH_2OH$?

- The aldehydes, RCHO, produced are themselves able to undergo further oxidation to carboxylic acids, RCOOH.

Of course, the 'over-oxidation' of primary alcohols to carboxylic acids — for example, with acidified $K_2Cr_2O_7$ — may be useful as an approach to the acids, and we shall return to it in Section 5.7, but for aldehyde preparation it is obviously undesirable. Much research has gone into developing oxidizing agents that are sufficiently selective to prevent this over-oxidation; pyridinium chlorochromate was one result of this work, and there is now a choice of reliable methods.

Let's take a look at a retrosynthetic analysis of ketone **5.52**, starting with an FGI to an alcohol:

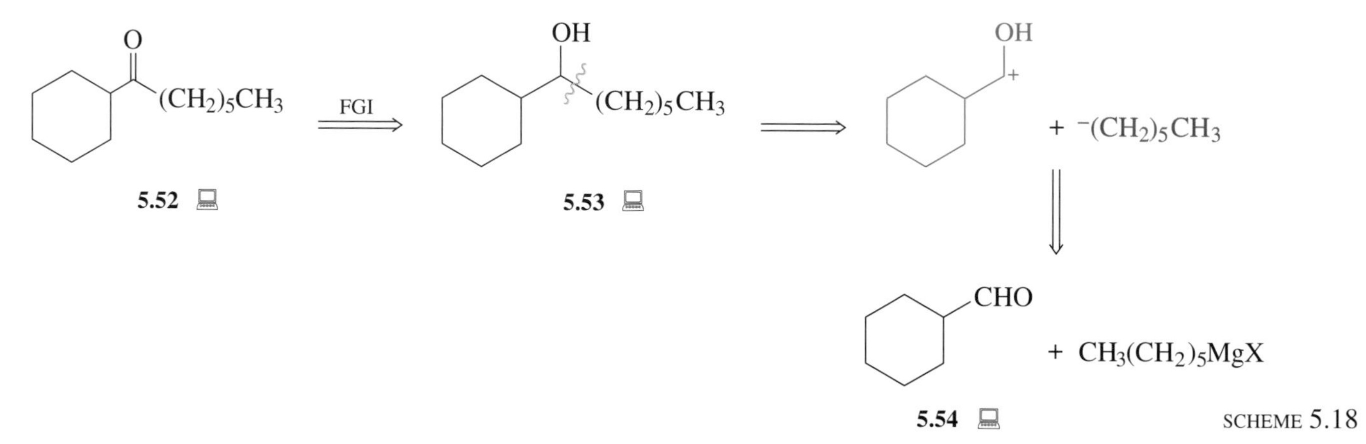


SCHEME 5.18

Notice how disconnection of the parent alcohol **5.53** leads back to another carbonyl compound as precursor, this time the aldehyde **5.54**.

- Can you use the same strategy to disconnect the aldehyde **5.54**?
- Yes, the parent alcohol will be **5.55**, and disconnection generates a cyclohexyl Grignard reagent (available from the corresponding haloalkane) and methanal, $H_2C{=}O$, as the starting materials:

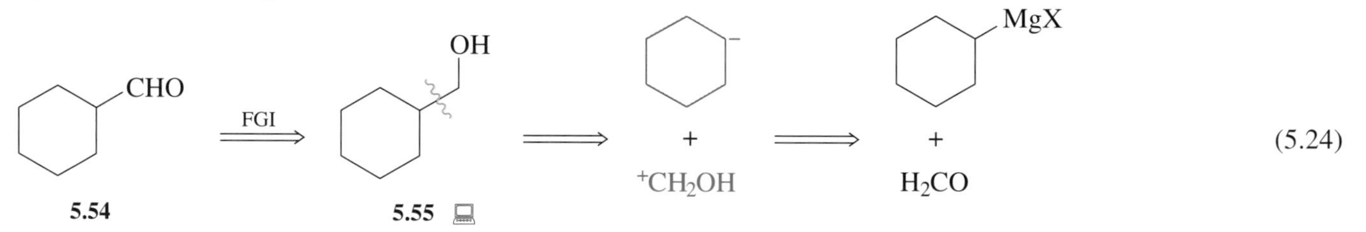


(5.24)

The full synthesis of ketone **5.52** reported in the chemical literature follows this analysis exactly; the oxidations were both performed using pyridinium chlorochromate (PCC):

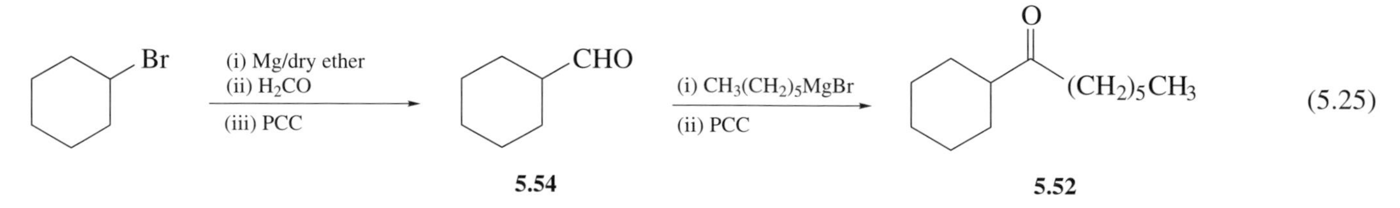


(5.25)

So far, we have looked at ketone synthesis as an extension of alcohol synthesis. Are there any direct ways of preparing ketones using C—C bond-forming reactions? To help us track down the type of reaction we need, let's remind ourselves of the disconnection of alcohols that we are using repeatedly. We disconnect a C—C bond to the alcohol carbon atom, to give an electrophile corresponding to a carbonyl compound, and a nucleophile corresponding to an organometallic reagent, such as a Grignard reagent:

$$R^1-\overset{\overset{\text{OH}}{|}}{\underset{\underset{R^2}{|}}{C}}\wr R^3 \Longrightarrow R^1-\overset{\overset{\text{OH}}{|}}{C^+}-R^2 + (R^3)^- \Longrightarrow R^1-\overset{\overset{\text{O}}{\|}}{C}-R^2 + R^3MgX \quad (5.26)$$

- Write down the corresponding disconnection for a ketone $R^1COR^2$ — that is, for a C—C bond to the carbonyl carbon atom (we looked at this in Section 3).
- The electrophile will be $R^1-\overset{+}{C}=O$, and the nucleophile will be $(R^2)^-$. Unless $R^2$ is aromatic (see Section 5.5), the obvious choices for reagents will apparently be a carboxylic acid derivative, $R^1COY$ (acid chloride or ester), and a Grignard reagent, $R^2MgX$:

$$R^1-\overset{\overset{\text{O}}{\|}}{C}\wr R^2 \Longrightarrow R^1-\overset{\overset{\text{O}}{\|}}{C^+} + (R^2)^- \Longrightarrow R^1-\overset{\overset{\text{O}}{\|}}{C}-Y + R^2MgX \quad (5.27)$$

So, we might reasonably suggest the reaction of a Grignard reagent with an acid chloride or ester to produce a ketone.

- What problem can you see with this?
- The ketone is certainly an initial product from these reactions, but, of course, ketones are themselves attacked by Grignard reagents, to produce alcohols; we used this route to make tertiary alcohols:

$$R^1-\overset{\overset{\text{O}}{\|}}{C}-Y \xrightarrow[\text{(ii) } H^+/H_2O]{\text{(i) } R^2MgX} R^1-\overset{\overset{\text{O}}{\|}}{C}-R^2 \xrightarrow[\text{(ii) } H^+/H_2O]{\text{(i) } R^2MgX} R^1-\overset{\overset{\text{OH}}{|}}{\underset{\underset{R^2}{|}}{C}}-R^2 \quad (5.28)$$

How can we stop the reaction at the ketone stage? It is not generally possible using Grignard reagents, so one obvious tactic is to choose an alternative nucleophilic organometallic reagent that is less reactive.

- What other organometallic nucleophiles have you met, and how do their reactivities compare?
- Organolithium, organosodium and organocopper reagents are alternative sources of organometallic nucleophiles. Organolithiums are more reactive than Grignard reagents, and organosodiums are too reactive to be generally useful. However, organocopper compounds seem to fit the bill.

In fact, lithium dialkylcopper reagents, $R_2CuLi$, are the usual choice; they react reasonably well with acid chlorides, but only slowly with ketones. Conveniently, the organocoppers are prepared from the organolithium reagents, RLi, and hence from haloalkanes, RX:

$$RX + 2Li \xrightarrow{\text{dry ether}} RLi + Li^+X^- \quad (5.29)$$

$$2RLi + CuX \xrightarrow{\text{dry ether}} R_2Cu^-Li^+ + Li^+X^- \quad (5.30)$$

The disconnection of ketones can now be completed properly with these new reagents:

$$R^1-\overset{\overset{\text{O}}{\|}}{C}\wr R^2 \Longrightarrow R^1-\overset{\overset{\text{O}}{\|}}{C^+} + (R^2)^- \Longrightarrow R^1-\overset{\overset{\text{O}}{\|}}{C}-Cl + (R^2)_2CuLi \quad (5.31)$$

How would you make the ant alarm pheromone **5.56** by the 'organocopper' approach, given that you had available the carboxylic acid **5.57**?

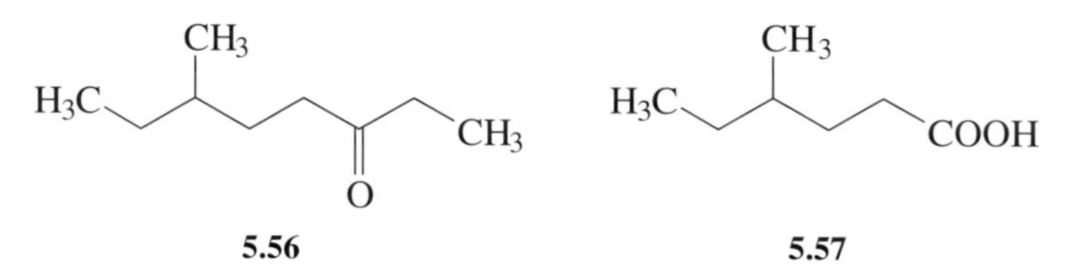


Retrosynthetic analysis of **5.56** suggests an acid chloride and lithium diethylcopper, $(CH_3CH_2)_2Cu^- Li^+$, as suitable reagents. The acid chloride can be made from the acid **5.57**, and the lithium dialkylcopper from bromoethane, by appropriate FGIs.

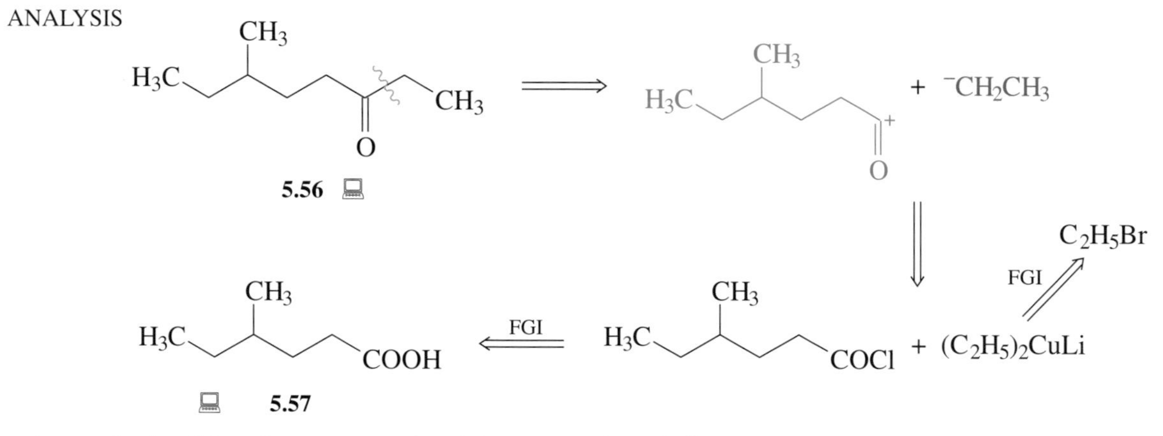


SCHEME 5.19

Thus, the synthesis of the pheromone, using this strategy is as follows:

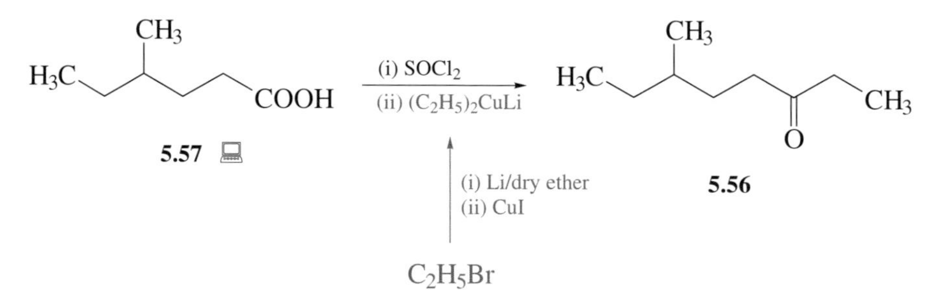


SCHEME 5.20

A different way of making ketones is to use a reagent for the electrophile, $R^1{-}\overset{+}{C}{=}O$, that does not form the ketone directly in the presence of a Grignard reagent or another organometallic reagent.

One of the two common solutions here is to use a carboxylic acid, $R^1COOH$, as the electrophile, and the reactive alkyllithium, $R^2Li$, as the nucleophile:

$$R^1{-}C({=}O){-}R^2 \Longrightarrow R^1{-}\overset{O}{\overset{\|}{C^+}} + (R^2)^- \Longrightarrow R^1COOH + 2R^2Li \qquad (5.32)$$

Two moles of $R^2Li$ are used, and a look at the mechanism will tell us why.

What is the first reaction of the organometallic reagent (reacting as a base) with the carboxylic acid?

A proton will be removed, to form the lithium carboxylate:

$$R^1COOH + R^2Li \longrightarrow R^1COO^-Li^+ + R^2H \qquad (5.33)$$

This doesn't look promising, but the negatively charged carboxylate ion is actually attacked by a second mole of the reactive alkyllithium to produce a dilithium salt. The ketone is not formed in the reaction mixture, but acid hydrolysis, which first destroys any excess alkyllithium, does give a ketone via the unstable ketone hydrate:

$$R^1C(=O)O^-Li^+ \xrightarrow{R^2Li} R^1C(O^-Li^+)(O^-Li^+)R^2 \xrightarrow{H^+/H_2O} R^1C(OH)_2R^2 \longrightarrow R^1C(=O)R^2 + H_2O \qquad (5.34)$$

ketone hydrate

The other solution to our ketone disconnection is to use the nitrile, $R^1CN$, as the electrophile, with a Grignard reagent:

$$R^1C(=O)R^2 \Longrightarrow R^1\overset{+}{C}{=}O + (R^2)^- \Longrightarrow R^1{-}C{\equiv}N + R^2MgX \qquad (5.35)$$

Again, consideration of the mechanism is needed to see how the nitrile $R^1CN$ is related to the synthon $R^1{-}\overset{+}{C}{=}O$. Nitriles are electrophilic at the carbon atom, because of the electronegative nitrogen atom.

- Write down the initial product of attack of a Grignard reagent on a nitrile.
- An addition reaction takes place, to give the magnesium salt of an imine:

$$R^1{-}\overset{\delta+}{C}{\equiv}\overset{\delta-}{N} \quad R^2{-}MgX \longrightarrow R^1C(=N^-\overset{+}{M}gX)R^2 \qquad (5.36)$$

A second mole of Grignard reagent doesn't add to the imine anion, because the latter is negatively charged and is therefore less reactive towards a nucleophilic organomagnesium reagent. The salt is stable under the reaction conditions, but on addition of dilute acid, any excess Grignard is destroyed and an imine is formed. This compound is unstable and rapidly hydrolyses to a ketone:

$$R^1C(=N^-\overset{+}{M}gX)R^2 \xrightarrow{H^+/H_2O} R^1C(=NH)R^2 \xrightarrow{H^+/H_2O} R^1C(=O)R^2 + NH_3 \qquad (5.37)$$

We can see one of these approaches to ketone synthesis in action in a synthesis of octan-3-one **5.58**, a compound that is said to contribute 'a remarkable freshness' to lavender oil.

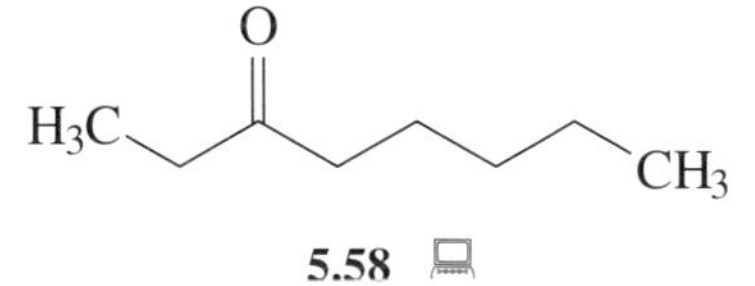


- Disconnection of the ketone towards the centre of the molecule will produce two fairly equal fragments. What are the reagents required for a 'nitrile-based' analysis of this compound?
- They will be propanenitrile and a pentyl Grignard reagent:

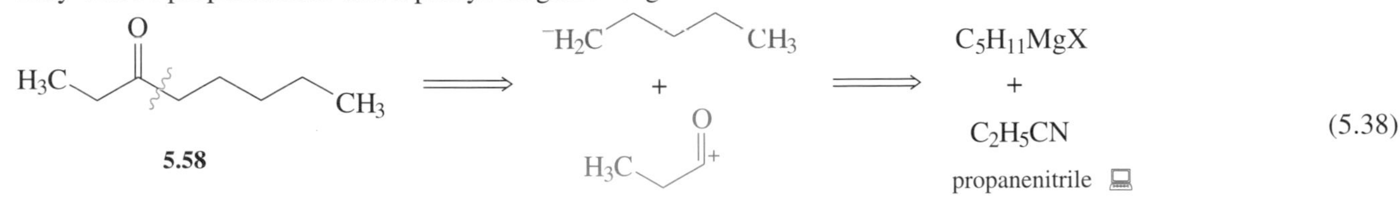


(5.38)

In fact, the synthesis was actually carried out in this way:

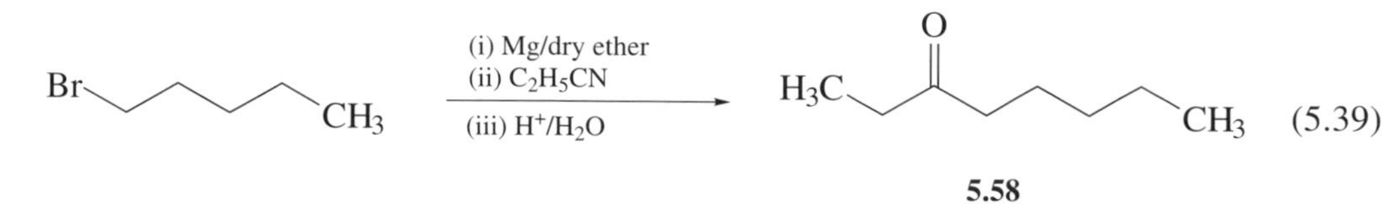


## BOX 5.3 Georg Wittig

Georg Wittig (Figure 5.5) was born in Berlin on 16 June 1897. Both his parents were artistic, and when he finished his secondary studies he had to choose between chemistry and art. He chose chemistry, but remained interested in the arts and was an excellent pianist; he even composed his own pieces.

In 1916, he started his chemical studies at Tübingen, but they were interrupted by the First World War. In 1919, he returned from captivity and resumed his studies at Marburg where he obtained his doctorate in 1926. In 1932, he became Head of Department at Braunschweig Technical College. At this time, National Socialism was growing in Gemany and since he did not agree with its doctrines his research suffered. In 1937, he moved to Freiburg to become Associate Professor. Then, in 1944, he became full Professor and Faculty Director at the Institute of Chemistry, Tübingen. Finally, in 1956, he moved to Heidelberg where he stayed until his retirement in 1967.

**Figure 5.5**
Georg Wittig (1897–1987).

His research interests covered stereochemistry, organometallic compounds and, in particular, organophosphorus compounds. He is remembered through the Wittig reaction, which is a reaction between an organophosphorus compound and a carbonyl compound that leads to an alkene. He also loved the outdoors, in particular the mountains, and he was a good skier.

He was awarded the Nobel Prize for Chemistry in 1979 (jointly with Herbert C. Brown, Box 7.2), for his development of the use of phosphorus-containing compounds into important reagents in organic chemistry.

## 5.6.1 Summary of Section 5.6

1 To devise syntheses of ketones and aldehydes, an FGI may be carried out to identify the corresponding alcohol as the target molecule.

2 There are mild synthetic methods for converting a primary alcohol into an aldehyde, or a secondary alcohol into a ketone: they are pyridine–$CrO_3$ or pyridine–$CrO_3$–HCl (PCC). Additionally, $CrO_3$–$H_2SO_4$–propanone is a system used for oxidizing secondary alcohols to ketones.

3 Aldehydes and ketones, $R^1R^2CO$, may be disconnected directly to generate an electrophilic $R^1{-}\overset{+}{C}{=}O$ synthon and a nucleophilic $(R^2)^-$ synthon.

4 The electrophilic–nucleophilic pairs of reagents that correspond to these synthons are (i) $R^1COCl/(R^2)_2CuLi$, (ii) $R^1COOH/R^2Li$, and (iii) $R^1CN/R^2MgX$.

### QUESTION 5.4

Concentrating solely on the carbonyl groups, use retrosynthetic analysis to devise syntheses for compounds **5.59–5.61**, identifying synthons and the corresponding reagents.

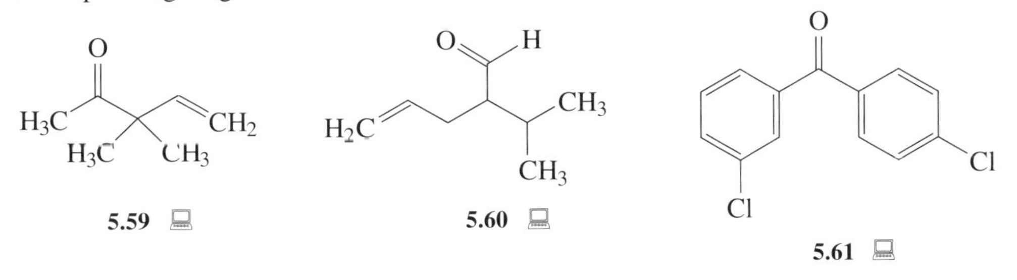


## 5.7 Carboxylic acids

Just as we can relate aldehydes and ketones to an alcohol by an oxidation FGI, we can similarly relate carboxylic acids, RCOOH, to an alcohol. Not surprisingly, some of the strategies here quite closely resemble those that we have seen for aldehydes and ketones. To make a carboxylic acid, the alcohol must be primary, $RCH_2OH$, and the oxidation goes via the aldehyde, RCHO. In fact, we have already mentioned a couple of times that we have to be careful not to over-oxidize if the aldehyde is our target molecule: acids are the end of the line for oxidation!

$$RCH_2OH \xrightarrow{\text{oxidation}} RCHO \xrightarrow{\text{oxidation}} RCOOH \qquad (5.40)$$

Various oxidizing agents have been used. The more reactive versions of the chromium-based oxidizing agents — for example, acidified $K_2Cr_2O_7$ — will do the job. But sometimes these strongly oxidizing conditions damage other functional groups in the molecule, so conditions and reagents with better selectivity have been developed. To make the acid **5.62**, for example, silver oxide (AgO) was used to oxidize the alcohol **5.63** (you should add this reagent to your FGI table):

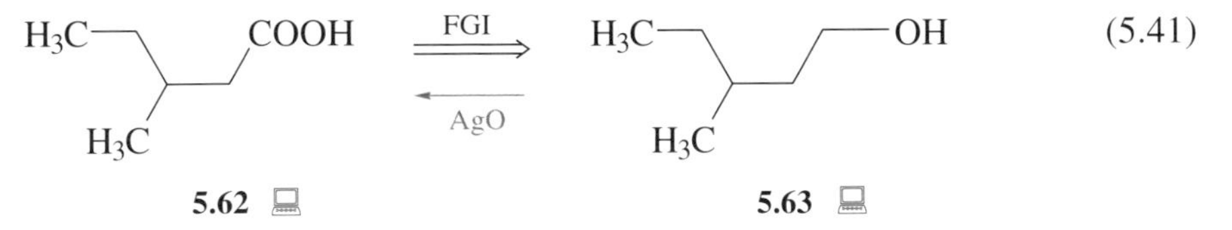


(5.41)

Thus, as we have seen earlier, one of the possibilities for the retrosynthetic analysis of a carboxylic acid is to begin with an FGI, and then to analyse the parent alcohol.

What about a 'direct' disconnection of carboxylic acids themselves? With the alcohols, and with the aldehydes and ketones, we were able to disconnect a C—C bond involving the functional group carbon atom. Perhaps we should try to do the same with carboxylic acids.

- Write down the two pairs of synthons generated by disconnection of the C—C bond to the carboxylate carbon atom of RCOOH.

- The two pairs will be as follows:

$$R \text{—} COOH \Longrightarrow R^- + {}^+COOH$$

$$R \text{—} COOH \Longrightarrow R^+ + {}^-COOH$$

SCHEME 5.21

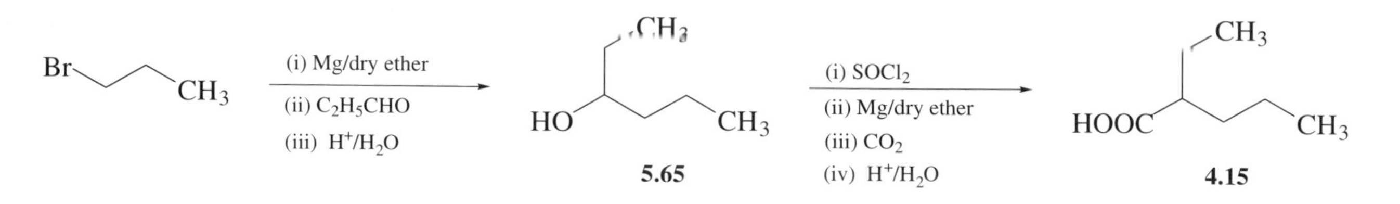


(i) $SOCl_2$

(ii) NH

CH3

N

CH3

O

**4.10**

SCHEME 5.25

Grignard reagents are sufficiently reactive towards $CO_2$ that even quite sterically hindered carboxylic acids can be made using this strategy. For instance, the acid **5.66**, which was required for a study of steric hindrance, was prepared by reaction of the Grignard reagent **5.67** with $CO_2$:

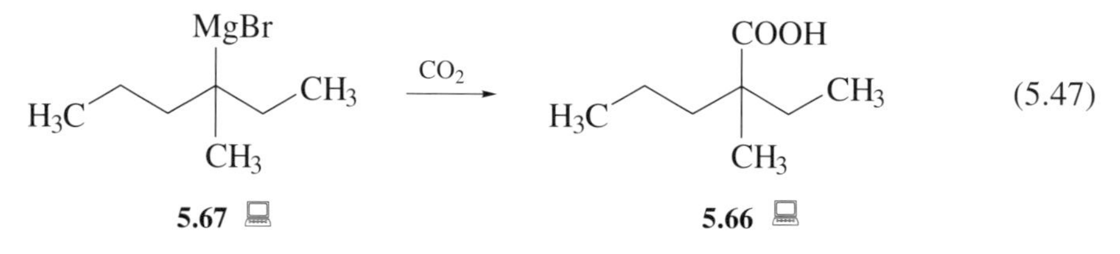


(5.47)

Before leaving carboxylic acids, we should say a word or two more about carboxylic acid derivatives, their preparation and interconversion. We have said before (Section 4.1) that such derivatives are always disconnected to the acid and the appropriate other component, by a C—X disconnection. The most important acid derivatives are the acid chlorides ($R^1COCl$), anhydrides ($R^1COOCOR^1$), esters ($R^1COOR^2$), and amides ($R^1CONHR^3$). They can be listed in order of reactivity, as shown in Scheme 5.26, from the most reactive ($R^1COCl$) to the least reactive ($R^1CONHR^3$). Compounds at the top of this list can be converted into those lower down simply by treatment with the appropriate nucleophile (see Appendix 2 in Part 1). Of course, all the carboxylic acid derivatives can be hydrolysed to the acid, and by treating the acid with $SOCl_2$ or $PCl_5$, to make the acid chloride, we can enter at the top of the list.

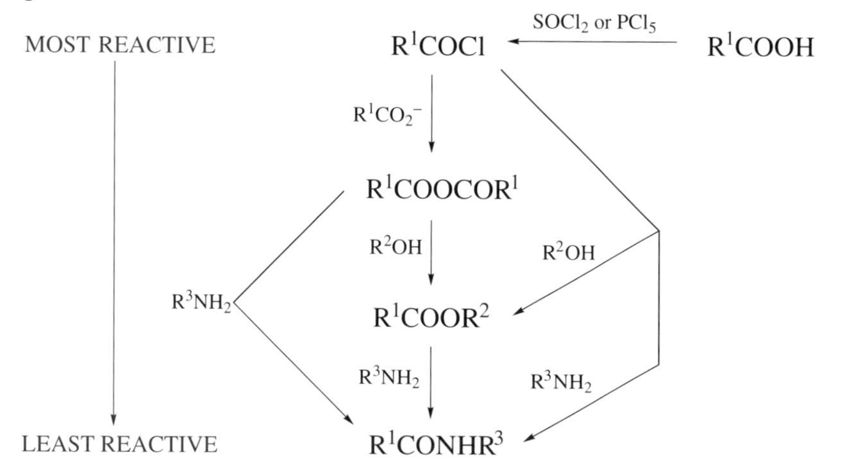


SCHEME 5.26

We have now completed our survey of the major C—X and C—C disconnections. To help you use retrosynthetic analysis, we have summarized in Table 5.1 the various synthons that we have introduced, and the reagents to which they correspond.

> Chemical synthesis always has some element of planning in it. But the planning should never be too rigid. Because, in fact, the specific objective which the synthetic chemist uses as the excuse for his activity is often not of special importance in the general sense; rather, the important things are those that he finds out in the course of attempting his objective.
>
> ***R. B. Woodward***

**Table 5.1** Some synthons and the corresponding commonly used reagents

| Synthon | Reagent | Comments |
|---|---|---|
| NUCLEOPHILIC | | |
| $RO^-$ | ROH, alcohols | also $Na^+\,{}^-OR$ or $K^+\,{}^-OR$ |
| $R^1R^2N^-$ | $R^1R^2NH$, amines | |
| $R^-$ | RMgX, Grignard reagents<br>RLi, organolithium reagents | also $R_2CuLi$ in reaction with acid chlorides |
| 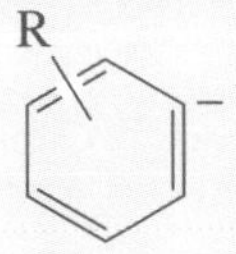 R-substituted phenyl anion | 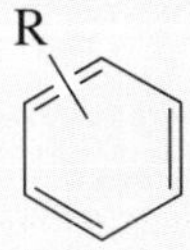 R-substituted benzene, (substituted) benzene | in Friedel–Crafts reactions;<br>also 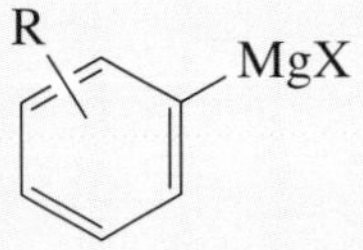 R-substituted phenyl MgX |
| $R{-}C{\equiv}C^-$ | $R{-}C{\equiv}CH$, alkynes | via alkynide anion |
| $^-C{\equiv}N$ | $M^+\,{}^-CN$, inorganic cyanides | |
| ELECTROPHILIC | | |
| $R^+$ | RX, haloalkanes | |
| $R{-}\overset{+}{C}{=}O$ | RCOY, acid derivatives;<br>for example, RCOCl, acid chlorides | also $(RCO)_2O$, acid anhydrides, and $RCOOR^1$, esters; all available from RCOOH by FGI |
| | RCOOH, carboxylic acids | in reaction with organolithium reagents |
| | RCN, nitriles | in reaction with Grignard reagents |
| $R{-}\overset{+}{C}H{-}OH$ | RCHO, aldehydes | |
| $R^1{-}\overset{OH}{\underset{+}{C}}{-}R^2$ | $R^1R^2C{=}O$, ketones | |
| $R^1{-}\overset{++}{C}{-}OH$ | $R^1(R^2O)C{=}O$, esters | in reaction with Grignard reagents |
| $^+COOH$ $(HO{-}\overset{+}{C}{=}O)$ | $CO_2$, carbon dioxide | |
| 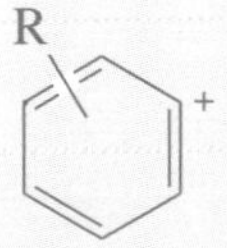 R-substituted phenyl cation | 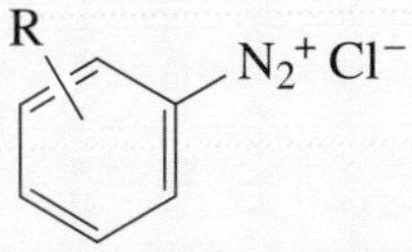 R-substituted phenyl $N_2^+\,Cl^-$, diazonium salts | in reaction with CuCN, etc. |
| 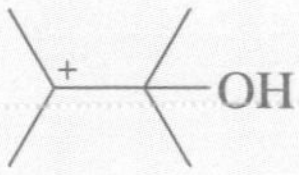 $>\overset{+}{C}{-}C<{-}OH$ | 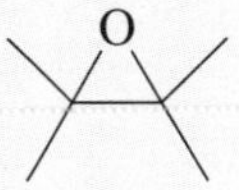 oxiranes (epoxides) | |

## 5.7.1 Summary of Section 5.7

1 Carboxylic acids may be related to a 'parent' alcohol. Retrosynthetic analysis can thus begin with an FGI between the acid and its corresponding alcohol.

2 Alternatively, direct disconnection of a carboxylic acid identifies a nucleophilic $R^-$ synthon and an electrophilic $HO{-}\overset{+}{C}{=}O$ synthon, for which the reagents are a Grignard reagent and $CO_2$, respectively.

## BOX 5.4 Radical reactions in retrosynthetic analysis

The approach we have taken to retrosynthetic analysis includes a number of radical reactions as FGIs — for example, the conversion of an alcohol or haloalkane to a hydrocarbon. However, when carrying out bond disconnections we have tacitly assumed an ionic mechanism. When working backwards, in our thought processes, we break bonds heterolytically to give positive and negative synthons. This is reasonable for C—X bond disconnections since the element X usually has a very different electronegativity from carbon, and thus bond heterolysis is most likely. However, should all carbon–carbon bond disconnections always lead to positive and negative carbon synthons? In Part 3 we saw a number of carbon–carbon bond-forming reactions that occur via radicals. For example,

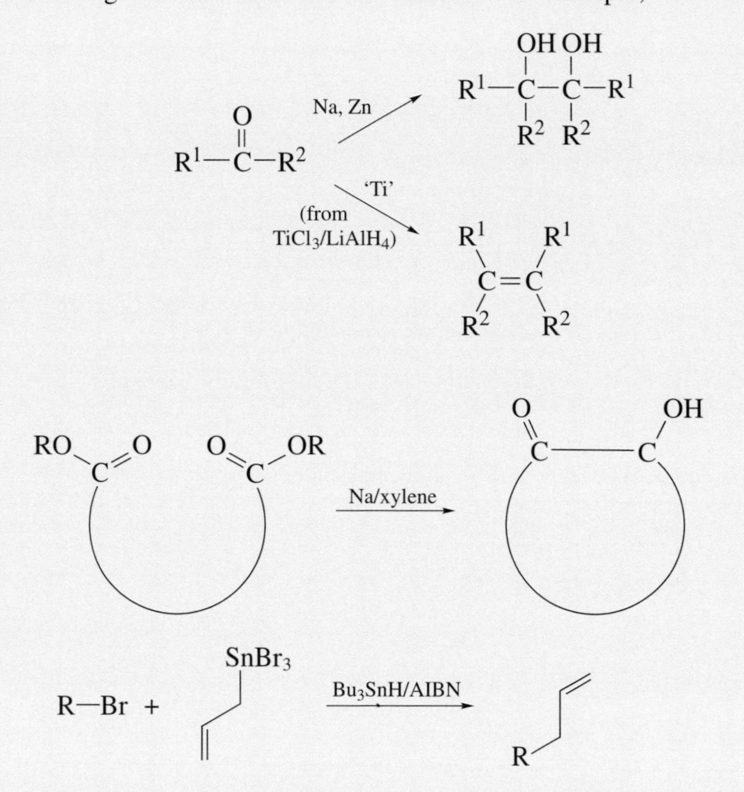


Our strategy for the retrosynthetic analysis of these products would certainly not suggest these reactions! Since most of the organic reactions you have met so far are ionic reactions, we shall not modify our strategy. However, you should watch out for target molecules that contain the specific substructures shown as products above. If they do, then you can use these radical reactions to disconnect the target molecule into new intermediates.

### QUESTION 5.5

Identify two strategies for the synthesis of each of the compounds **5.68** and **5.69**.

HOOC(C$_6$H$_5$)C=CH$_2$

**5.68**

$C_6H_5{-}C{\equiv}C{-}COOH$

**5.69**

# 6 CD-ROM ACTIVITY

In this Section we want to consolidate what you have learnt so far in this Part of the Book, and to illustrate how you can put these strategies together to devise syntheses for yourself. In particular we want to:

- reinforce your understanding of retrosynthetic analysis — especially the difference between working in the synthetic direction and the retrosynthetic direction;
- give you practice at looking at a target molecule, and spotting C—C and C—X bond disconnections and FGIs;
- understand what we mean by synthons — and remind you that they do not exist;
- give you practice at using the Tables in the Appendix, and thus identify reagents for particular conversions, and to convert synthons into reagents.

Before you start the CD-ROM exercise, look at the strategy given below for carrying out retrosynthetic analysis. You may not appreciate all the points listed, but if you use this strategy to solve the CD-ROM problems, some of the subtleties will become clearer.

### *A strategy for retrosynthetic analysis*

1. Work backwards from the target molecule, identifying intermediates one step at a time.
2. When working backwards, for each intermediate step, first look for C—X bond disconnections, and then for C—C disconnections.
3. Always try to disconnect the molecule roughly in half.
4. Always convert synthons into reagents before moving on to the next step.
5. For $-CH_2-X-CH_2-$ fragments, remember to look at disconnections on either side of X. For fragments of the general formula **6.1**, just disconnect the C—X bond next to the carbonyl carbon.
6. C—X disconnections produce electrophilic $C^+$ synthons, and nucleophilic $X^-$ synthons.
7. C—C disconnections are made at, or very near to, functional groups.
8. C—C disconnections can generate two pairs of synthons, with opposite patterns of polarities. Always assess which is the easier pair to convert to reagents.
9. Disconnections at benzene rings usually lead to $C_6H_5^-$ synthons.
10. In the absence of functional groups, look for disconnection at chain-branching points.
11. C—C bonds in rings are not usually disconnected. However, C—C bonds attaching fragments to rings are especially useful.
12. Look for symmetry in a molecule which will enable you to attach two identical fragments.
13. Disconnect the most-complex side-chain first.
14. Try to end up with alcohol or carbonyl groups at appropriate places which can be converted into other functional groups by FGIs.

$$-\overset{\overset{\displaystyle O}{\|}}{C}-X-CH_2-$$

**6.1**

## BOX 6.1 The problem of retrosynthetic analysis

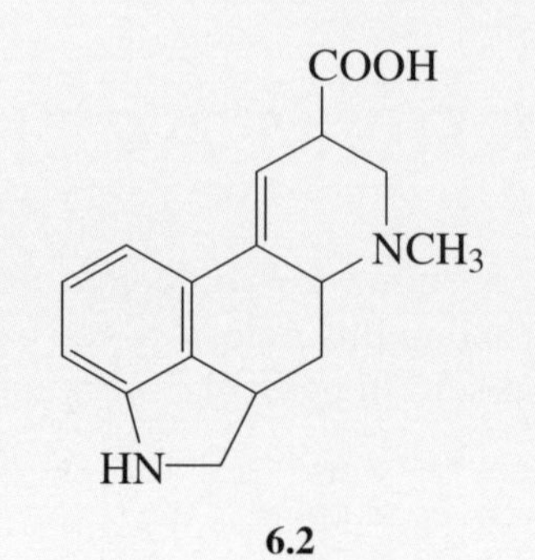


The real problem with retrosynthetic analysis is remembering which way you are working. For example, look at Scheme 6.1, which shows part of the retrosynthetic analysis of the hallucinogenic agent lysergic acid, **6.2**.

- Using the letters next to the retrosynthesis arrows, identify two C—C bond disconnections, one C—X bond disconnection and two functional group interconversions, indicating which two groups have been interconverted.

- The only C—C bond disconnections are steps a and e. The only C—X bond disconnection is step f. The remaining steps are FGIs (in the synthetic direction, b (hydroxy to chloro), c (carbonyl to hydroxy), d (hydroxy to alkene) and g (hydrogen to bromo).

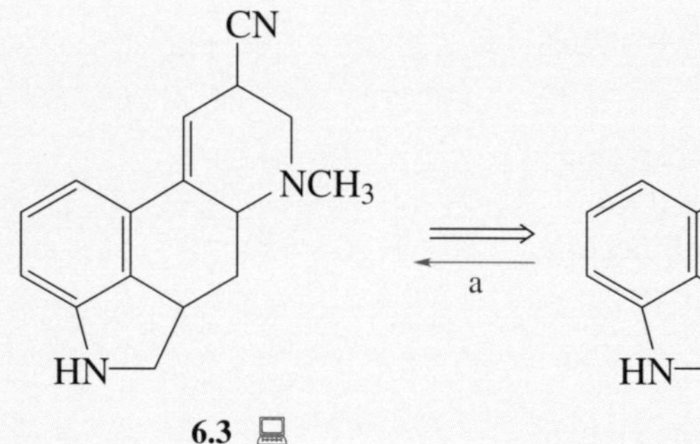


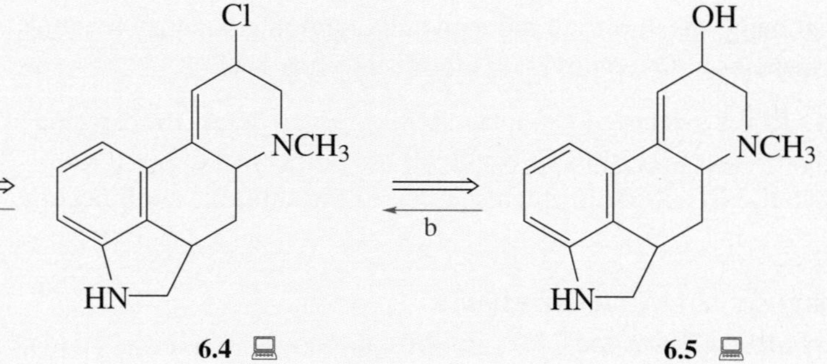


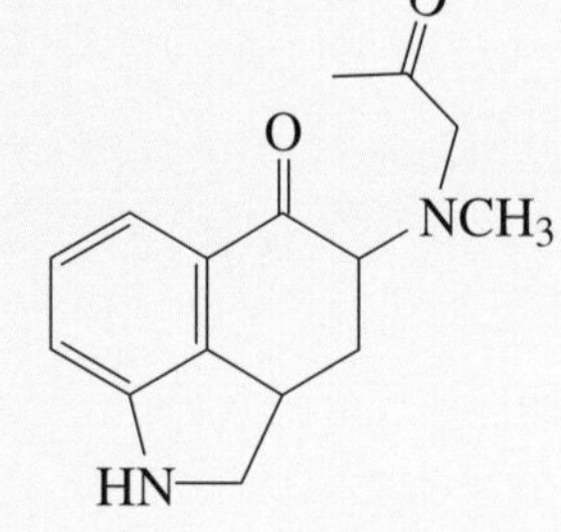


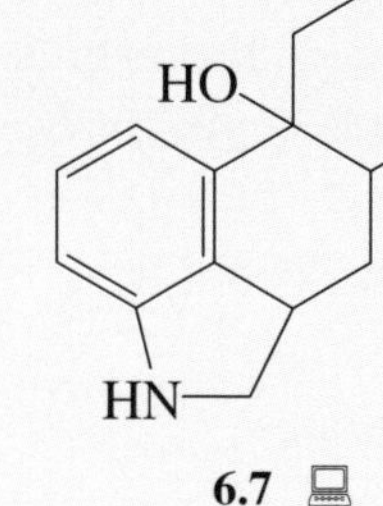


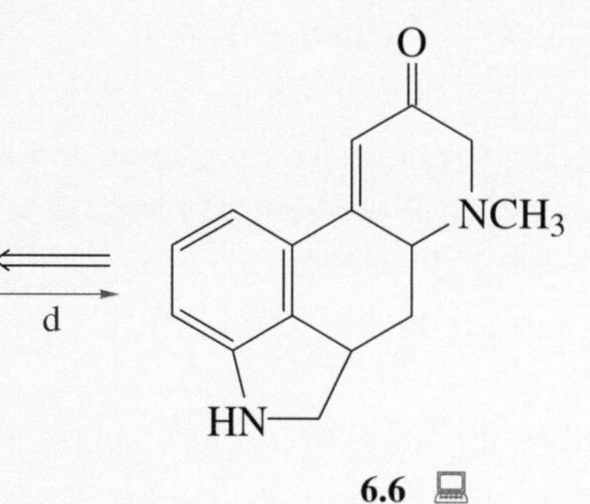


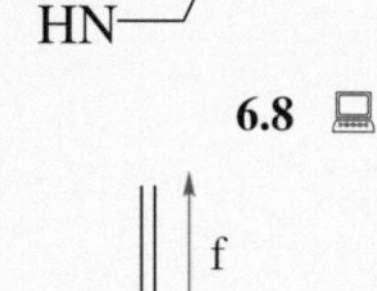


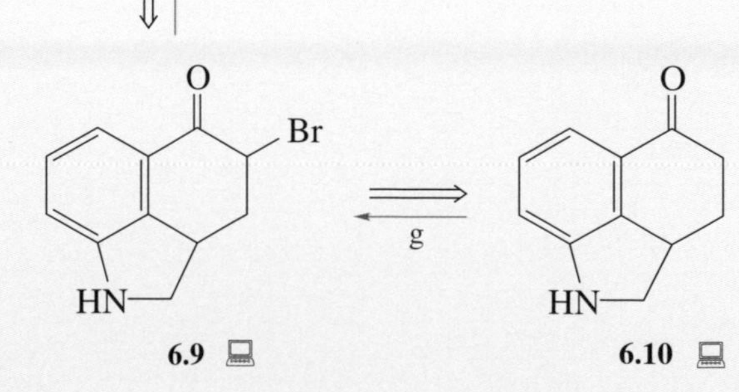


SCHEME 6.1

The most difficult part of this question is working out which way round is the synthetic direction and which way round is the retrosynthetic direction. The direction and type of arrows provide the clue. The target molecule is at the top left, and the retrosynthetic arrows point in a 'clockwise' direction; that is, **6.3** is made from **6.4**, which is made from **6.5**, etc. This means that the synthetic direction starts at the bottom right and works its way round from bottom to top; that is, **6.10** is converted into **6.9**, which is converted into **6.8**, and so on, until **6.3** is made. So step c involves a C—C bond disconnection, because in our thought processes, a C—C bond is broken in the retrosynthetic direction. In the synthetic direction, step e involves the formation of a carbon–carbon bond, which is what happens in the laboratory. Similarly, step c involves an FGI: in the synthetic direction, one functional group — a ketone — is converted into another — an alcohol — in a reduction reaction.

So if you are faced with such a question, the first thing to do is make sure you understand which is the retrosynthetic direction and which is the synthetic direction — that is, which compound is converted into which, if we were to carry this out in the laboratory.

## COMPUTER ACTIVITY 6.1 Retrosynthetic analysis

You should now attempt this CD-ROM activity. When you have finished it, you should return to the list of aims at the beginning of Section 6, just to make sure you feel confident in each of these areas; if not, have a go at some more problems on the CD-ROM.

It should take you about 1.5 hours to complete this activity.

# 7

# CONTROL IN SYNTHESIS

It may seem by now that, once we have devised our synthetic strategies, all we have to do is carry out each stage of our preferred route in order to obtain our target molecule. Unfortunately, this is seldom so — not because our chemistry is wrong, but because we have not checked our route to see if the reactions we plan to carry out will take place exactly as we want them to. For example, if the conditions of the reaction are not carefully controlled, oxidation of a primary alcohol can give both an aldehyde and a carboxylic acid. As an example of a different type of problem, let's assume that one step in a synthesis required us to carry out the following reaction:

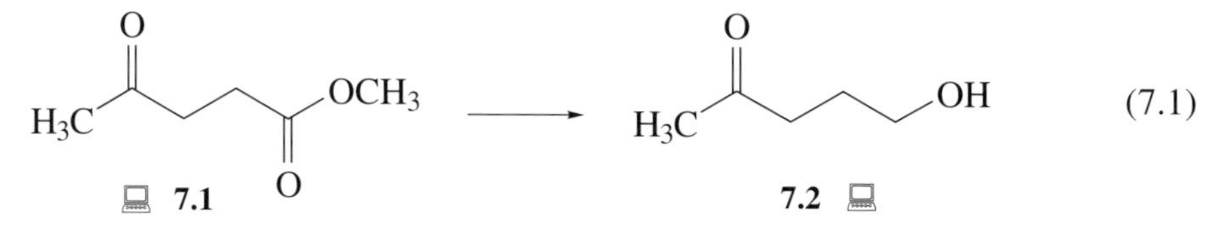


- What kind of reaction is this, and what type of reagent will it require?
- It is a reduction of an ester carbonyl group, requiring a hydride reducing agent.

You might be able to see a problem with this reaction if we were to carry it out using, say, lithium aluminium hydride. It is that the ketone carbonyl group in **7.1** would also react with the reducing agent, giving rise to the diol **7.3** and not the keto-alcohol **7.2**.

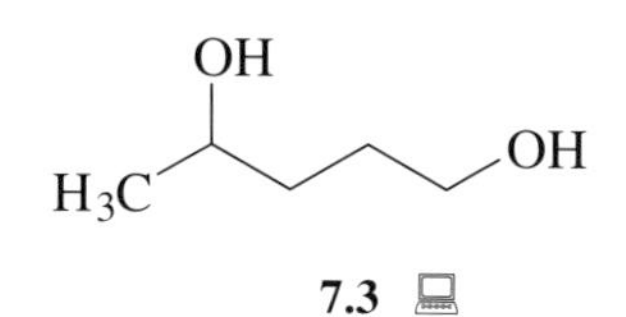


The problem boils down to one of **selectivity**. What we want is to carry out a reaction that selectively transforms the ester but not the ketone. (It turns out that this is very difficult to achieve directly in this case; you will find out how we can overcome this particular problem in Section 7.4.) The question of selectivity always arises when a molecule is capable of being attacked by a reagent to produce more than one product. In order to control an organic synthesis so that we produce the desired compound in high yield, we have to maximize the selectivity of the process that we want to employ. There are three types of selectivity that we need to consider: *chemoselectivity*, *regioselectivity* and *stereoselectivity*. We shall deal with each in turn.

## 7.1 Chemoselectivity

**Chemoselectivity** can be achieved when a molecule contains two functional groups of *different* reactivity. In such cases, *reaction can be preferentially effected at the more-reactive functional group*. Part 1 of this Book described how the ketone group is more reactive than an ester group. We can make use of this difference in reactivity to make compound **7.4**, by reduction of the keto-ester **7.1**. The nucleophilic hydride reagent, sodium borohydride, $NaBH_4$, is the reagent of choice here:

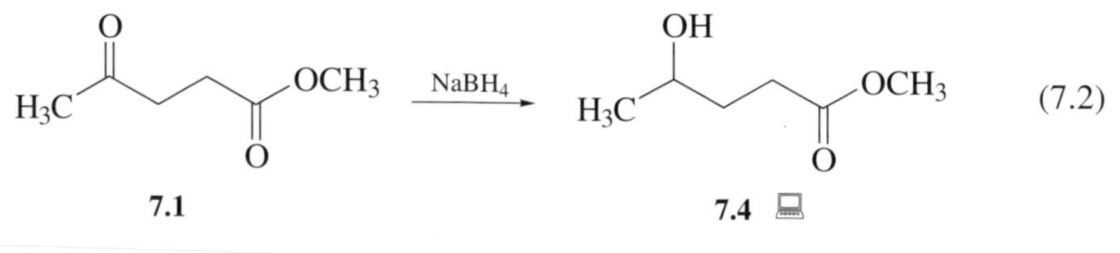

Sodium borohydride is similar to, but much less reactive than, lithium aluminium hydride. Unlike $LiAlH_4$, $NaBH_4$ reacts readily with ketones but not with esters. You should add this reagent to your table of FGIs (it's included in the Appendix). Another example is the oxidation of an alcohol adjacent to a double bond. Such compounds, for example **7.5**, are known as *allylic alcohols*. Both the alcohol and the alkene groups are susceptible to oxidation by aqueous potassium permanganate. $KMnO_4$ cleaves the alkene double bond to form ketones and/or carboxylic acids, depending on the alkene substitution pattern:

$H_3C$ $H$
$H_3C$ $CH_2OH$

**7.5**

$$R^1R^2C{=}CR^3R^4 \xrightarrow[H^+]{KMnO_4} R^1R^2C{=}O + O{=}CR^3R^4 \qquad (7.3)$$

To oxidize one of the functional groups selectively, an appropriate reagent needs to be chosen; manganese dioxide effects oxidation of the alcohol without attacking the carbon–carbon double bond:

$$H_2C{=}CH{-}CH_2OH \xrightarrow{MnO_2} H_2C{=}CH{-}CHO \qquad (7.4)$$

In this context, manganese dioxide is a **selective reagent** for the oxidation of an allylic alcohol to the corresponding unsaturated aldehyde. Again, you should include it in your table of FGIs (Table 2.1).

Sometimes, chemoselectivity can be achieved by adjusting reaction conditions such as concentration (or pressure for a gaseous reaction), temperature, solvent or catalyst.

## 7.2 Regioselectivity

Suppose we wanted to make 4,4-dimethylpent-2-ene, $(CH_3)_3C{-}CH{=}CH{-}CH_3$ (**7.6**). Two alcohols, **7.7** and **7.8**, are potential precursors, each giving rise to the target alkene by a dehydration reaction:

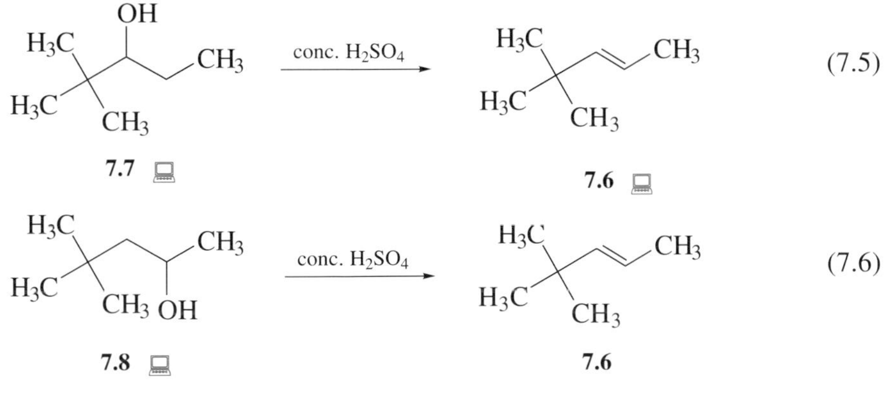


How do we choose which route to use?

How many alkene products can **7.7** and **7.8** form on dehydration?

**7.7** can produce only one alkene. Protonation of the alcohol leads to a carbocation adjacent to the dimethylethyl group, $(CH_3)_3C-$, and since there is only one set of β-protons, the double bond can only form in one position:

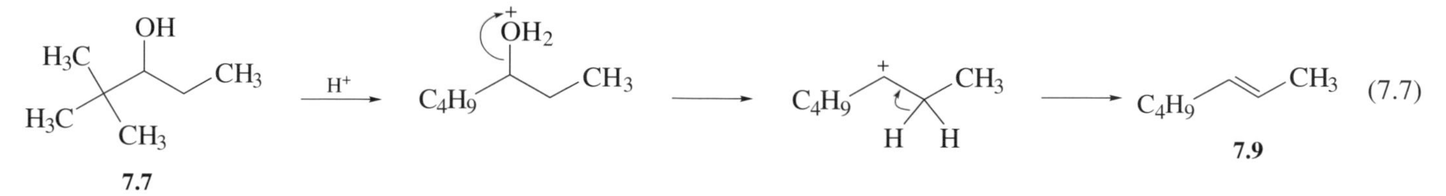


With **7.8**, there are two possible alkenes, **7.9** and **7.10**, formed by loss of a proton from either of the carbon atoms adjacent to the charge-bearing carbon in the carbocation. Saytzev's rule suggests that **7.9** would predominate. However, **7.10** may also be formed in appreciable amounts.

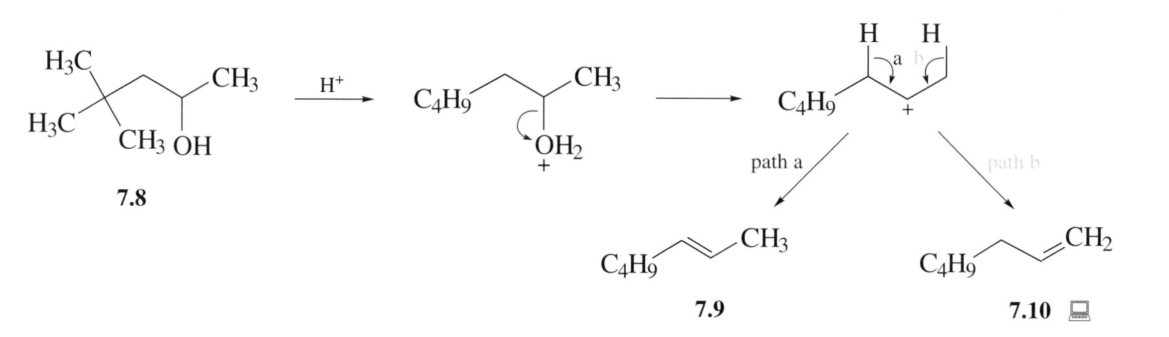


SCHEME 7.1

So, one reason for choosing **7.7** as the preferred precursor, rather than **7.8**, is the number of alkene products that **7.8** can produce. Dehydration of **7.8** is said to be less **regioselective** than dehydration of **7.7**, because reaction can occur at more than one position (region) around the functional group.

Dehydration is an elimination reaction, so it may not come as too much of a surprise to learn that the question of regioselectivity must also be considered for the reverse process, addition to an alkene.

If you wanted to make **7.11**, which alkene would you use to react with HBr, **7.12** or **7.13**?

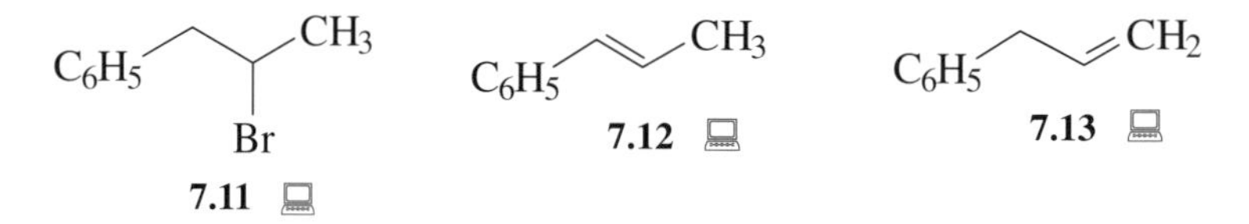


Addition of HBr to **7.12** would yield **7.14** preferentially, since the positive charge is stabilized by conjugation with the adjacent phenyl ring:

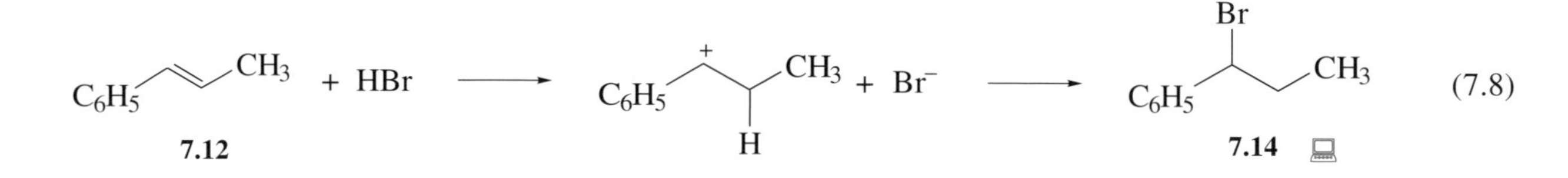

whereas alkene **7.13** would yield **7.11**, as required:

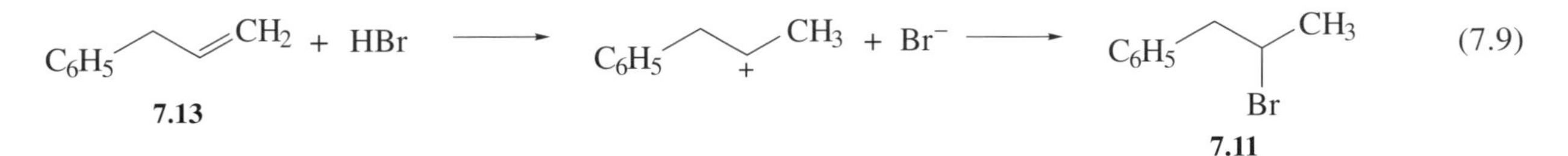


(7.9)

So, if our strategy involves the formation of a double bond by an elimination reaction, or an addition reaction of a double bond, we need to consider the regioselectivity of the process involved. That is, we must ask ourselves one of two questions:

1 For elimination, will the double bond be formed in the correct position?

2 For addition, will the substituents be added in the correct positions?

Another reaction for which regioselectivity is an important consideration is the electrophilic aromatic substitution of monosubstituted benzenes. For example, in the nitration of nitrobenzene there are three possible sites of attack (arrowed in **7.15**), but the major product is 1,3-dinitrobenzene; the reaction is regioselective.

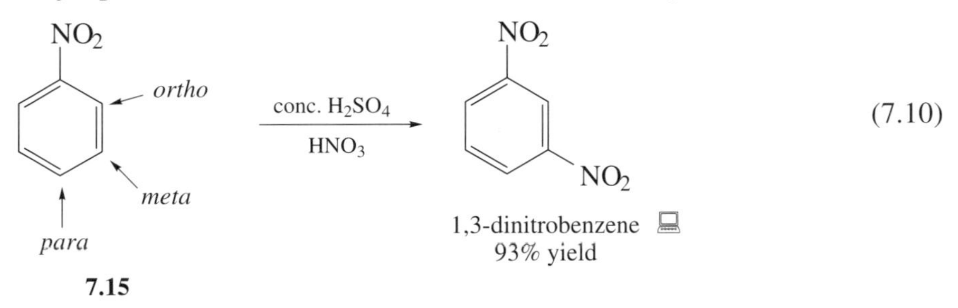


(7.10)

What factors do you think make this reaction regioselective?

The selectivity arises because of the electronic effect of the nitro group on the intermediates (more strictly, on the activated complexes) for the three reaction pathways for electrophilic substitution*.

When we use the regioselectivity of an aromatic substitution to control the synthesis of a disubstituted benzene derivative, we must ensure that the monosubstituted benzene precursor contains the appropriate substituent to 'direct' the incoming substituent into the correct position on the ring.

In the examples discussed so far, it has been electronic effects that have controlled regioselectivity. Steric effects are also able to affect the position of attack at a functional group; the attack of a nucleophile at an oxirane (see **7.16**) can, in principle, occur at either carbon atom in the ring.

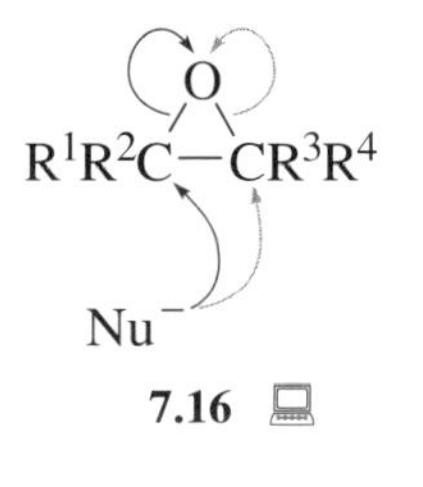


* Electrophilic aromatic substitution and the directing effects of substituents are discussed in Part 2 of *Alkenes and Aromatics*[1].

Pseudoephedrine, **7.18**, can be synthesised by such a route*. Attack of methylamine ($CH_3NH_2$) at the oxirane group in compound **7.17** occurs as shown, because the methyl group in the oxirane is smaller than the phenyl group: it is sterically easier for $CH_3NH_2$ to react at the less-hindered ring carbon atom.

$$\underset{\textbf{7.17}}{C_6H_5(H)C\overset{O}{\frown}C(H)CH_3} + CH_3NH_2 \longrightarrow \underset{\textbf{7.18}}{H{-}C(OH)(C_6H_5){-}C(H)(NHCH_3){-}CH_3} \qquad (7.11)$$

In Part 3 of this Book we discovered that some radical reactions were more selective than others. For example, the chlorination of propane gave 43% 1-chloropropane and 57% 2-chloropropane, which shows some regioselectivity, in that the secondary site is four times more reactive than the primary site. However, cyclization reactions to a five- or a six-membered ring were very regioselective. For example, in Scheme 7.2 the preference for one of the carbons of the double bond to end up outside the ring means the formation of a five-membered ring radical is 50 times faster than the cyclization to give a six-membered ring radical.

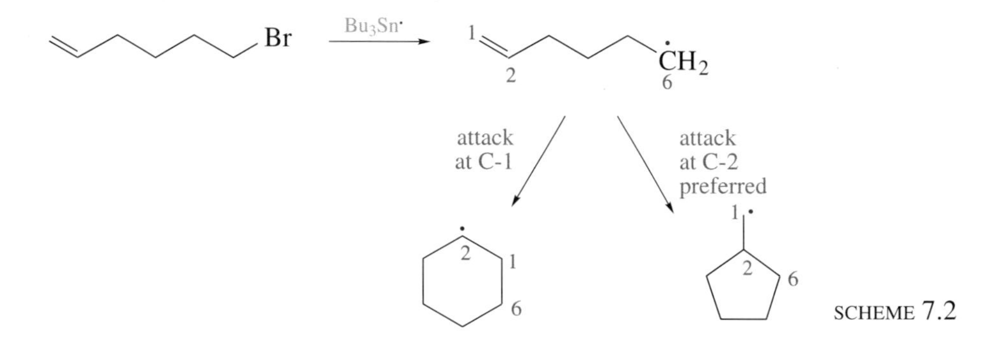


SCHEME 7.2

## 7.3 Stereoselectivity

Stereoselectivity is an important consideration if a reaction can potentially produce more than one stereoisomer, but only one of the stereoisomers is the product we want.

So for the synthesis of 1-phenylpropene from 1-phenylpropanol we need to consider which stereoisomer of the product we require, the *E* or the *Z*. In fact, such a synthesis yields a 95 : 5 mixture of the *E* and *Z* products (Scheme 7.3). We describe the dehydration of the alcohol precursor as being **stereoselective** for the *E* isomer. As we can separate the isomers by preparative GLC†, this approach is fine if we want *E*-1-phenylpropene. But what if we wanted *Z*-1-phenylpropene? The strategies we have formulated are clearly inappropriate, because they would be extremely wasteful. As discussed in Section 5.2, we must rethink our synthesis, and devise one that is stereoselective for our required isomer. Clearly, we need to incorporate in our strategy a reaction that generates a *Z* double bond.

* The synthesis of pseudoephedrine is discussed in Part 3 of *Alkenes and Aromatics*[1].

† Gas–liquid chromatography (GLC) is discussed in *Separation, Purification and Identification*[4].

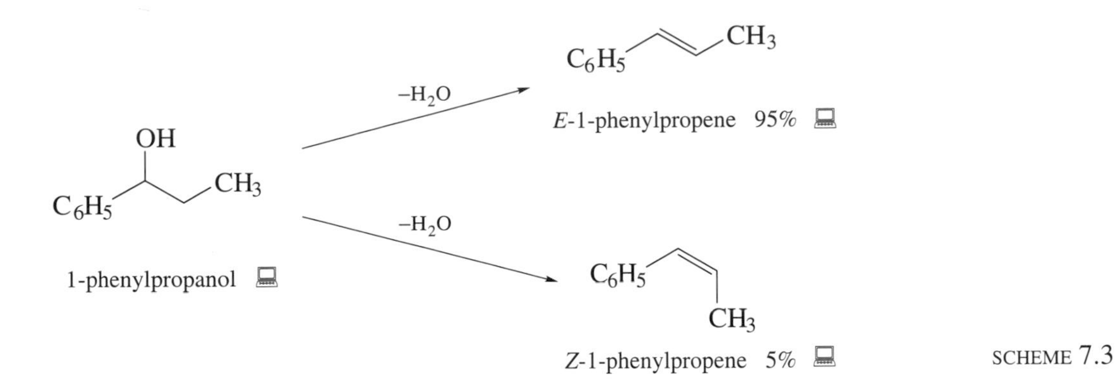


SCHEME 7.3

- Referring to your FGI table if necessary, identify a reaction suitable for forming a *Z* double bond.

- The hydrogenation of an alkyne using Lindlar's catalyst will produce a *Z* double bond:

$$R^1{-}C{\equiv}C{-}R^2 + H_2 \xrightarrow[\text{catalyst}]{\text{Lindlar's}} \underset{R^1\quad R^2}{\overset{H\quad H}{C{=}C}} \tag{7.12}$$

Such a reaction is stereoselective, because it preferentially yields one stereoisomer. So, a synthesis of *Z*-1-phenylpropene would necessitate a different strategy from that described earlier, and would involve 1-phenylpropyne:

$$\underset{C_6H_5\quad CH_3}{\overset{H\quad H}{C{=}C}} \overset{\text{FGI}}{\Longrightarrow} \underset{\text{1-phenylpropyne}}{C_6H_5{-}C{\equiv}C{-}CH_3} \Longrightarrow\Longrightarrow \text{appropriate starting matcrials} \tag{7.13}$$

Stereoselectivity is also a consideration in the synthesis of pseudoephedrine, **7.18**, which we looked at earlier. In particular, it is important to introduce the two chiral centres in such a way that they have the correct configuration. This is achieved by using the oxirane **7.17**, and by making use of the fact that an $S_N2$ substitution reaction inverts the stereochemistry at the carbon atom at which reaction occurs. So, the required methylamino group, $CH_3NH-$, is able to be introduced stereoselectively. Whenever a target molecule contains a chiral carbon atom, we must consider how the correct stereochemistry can be achieved using such stereoselective reactions. This reaction also demonstrates that some reactions can have more than one type of selectivity: it is regioselective as well as stereoselective because, as we saw in Rection 7.11, attack occurs at the less-hindered carbon atom

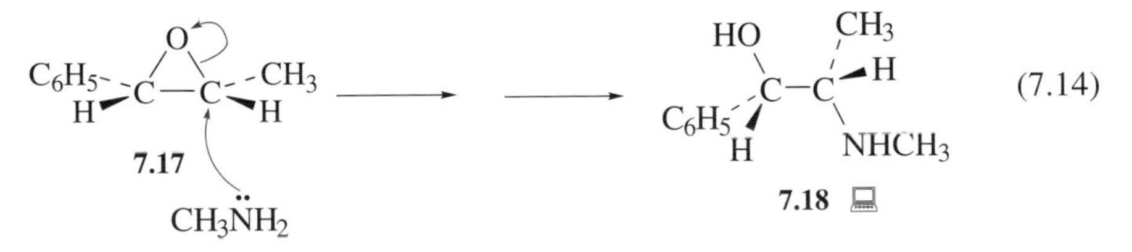

## Box 7.1 The Nobel Prize for Chemistry 2001

The 2001 Nobel Prize for Chemistry was awarded to three synthetic organic chemists for their work on stereoselective reactions. The Prize was awarded jointly to William S. Knowles and Ryoji Noyori 'for their work on chirally catalysed hydrogenation reactions', and K. Barry Sharpless 'for his work on chirally catalysed oxidation reactions'.

Knowles and Noyori's work involved using chiral catalysts to add hydrogen to just one face of a double bond so that just one enantiomer is formed, for example:

$R^1$, $CO_2H$, NHAc —($H_2$; chiral transition-metal catalyst)→ $R^1$, $CH_2$, $CO_2H$, H, NHAc (7.15)

The chiral catalyst ensures that the hydrogen is added from below the plane of the double bond to generate just one enantiomer of the amino acid.

Sharpless's work centred on the addition of oxygen to an alkene. For example, a chiral transition-metal catalyst was developed that delivered oxygen to just one face of a double bond to make just one enantiomer of an oxirane:

$R^3$, $R^2$, OH, $R^1$ —($(CH_3)_3CCOOH$; chiral transition-metal catalyst)→ O, $R^3$, $R^2$, $R^1$, OH (7.16)

William S. Knowles was born in 1917. In 1942, he obtained his PhD from Columbia University. He previously worked at the Monsanto Company, St. Louis, and is now retired.

Ryoji Noyori was born in 1938 in Kobe, Japan. In 1961 he obtained his BSc from Kyoto University followed by an MSc in 1963 and a PhD in 1967, also from Kyoto University. He is now at Nagoya University.

Barry Sharpless was born in 1941 in Philadelphia, USA. In 1963, he obtained a BA from Dartmouth College followed by a PhD in 1968 from Stanford University. He is now at the Scripps Research Institute, La Jolla, California.

# 7.4 Protecting groups

It is not always possible to choose the necessary reagent or reaction conditions to achieve the desired selectivity. Earlier, we introduced you to the idea that reduction of a ketone can be achieved by sodium borohydride in the presence of a less-reactive ester group. On the other hand, however, there is no direct way of reducing the ester group selectively to give a hydroxyketone because the ketone is always attacked first:

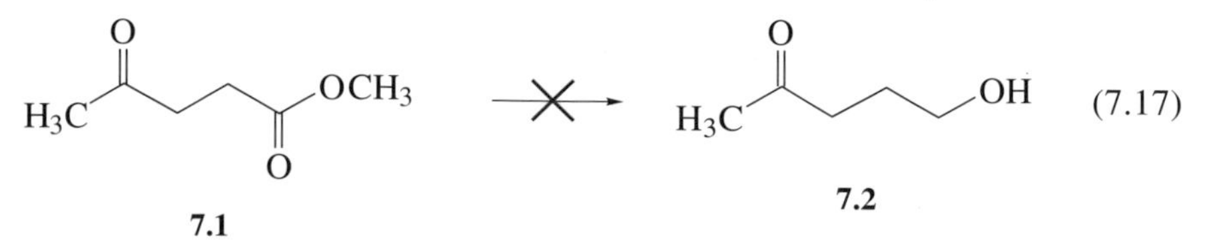

When this situation arises, use is made of a *protecting group*. The more reactive group (ketone in this example) is converted into an 'unreactive' group as a temporary measure, while the ester group is reduced. The ketone is regenerated afterwards. A possible sequence for the above example is as follows:

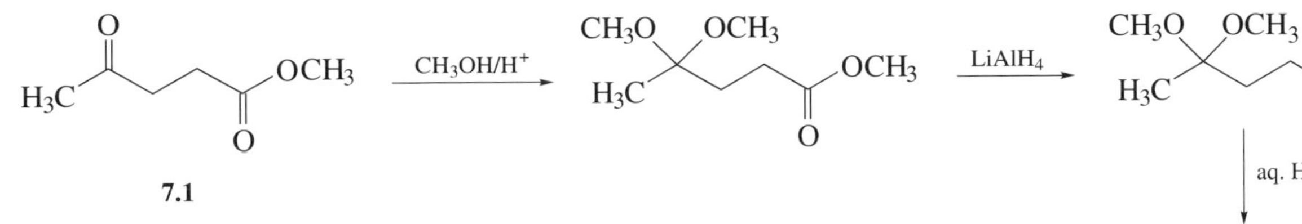


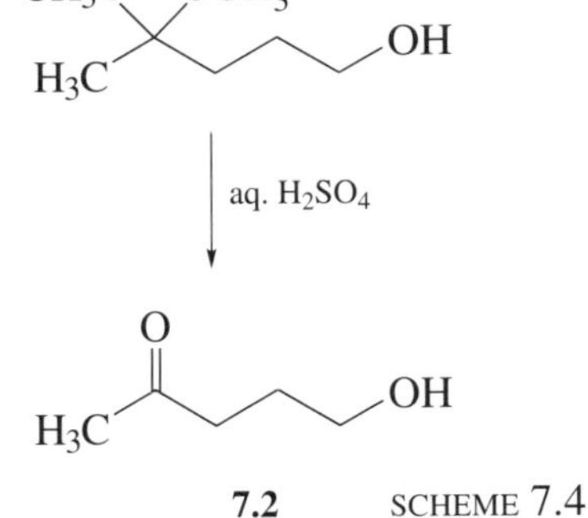


SCHEME 7.4

- What is the protecting group used in this reaction? How is it introduced, and how is it removed?

- The protecting group is an acetal (Part 1, Section 3.2.3). It is formed from the ketone group by reaction with methanol as a nucleophile (in the presence of an acid catalyst). The acetal is stable in the basic reducing conditions, and the ketone is regenerated by acid hydrolysis of the acetal in the final step.

A good protecting group has several desirable features. The group should be readily introduced by selective reaction in the initial stage of the reaction; it should be inert to the reagents used to carry out the desired transformation in the middle stage; and it should be easily removed (*deprotection*) in the final stage to regenerate the original functional group in high yield.

It is not always easy to find a sequence of reactions that fits all these specifications, and a further disadvantage is that two additional reaction stages are involved (**protection** and **deprotection**), with the associated implications for time, effort and lower yield. However, *protection is always necessary when we want to react a less-reactive functional group in the presence of a more-reactive functional group.*

Another situation in which *a protecting group is required is where a molecule contains two or more functional groups of the same type*. For example, propane-1,2,3-triol (glycerol) contains three hydroxy groups — two primary and one secondary. It would be difficult to acylate just one of these groups, because they all have similar reactivity. Even using one equivalent of acylating agent, a mixture of mono-, di- and triacyl derivatives will be formed:

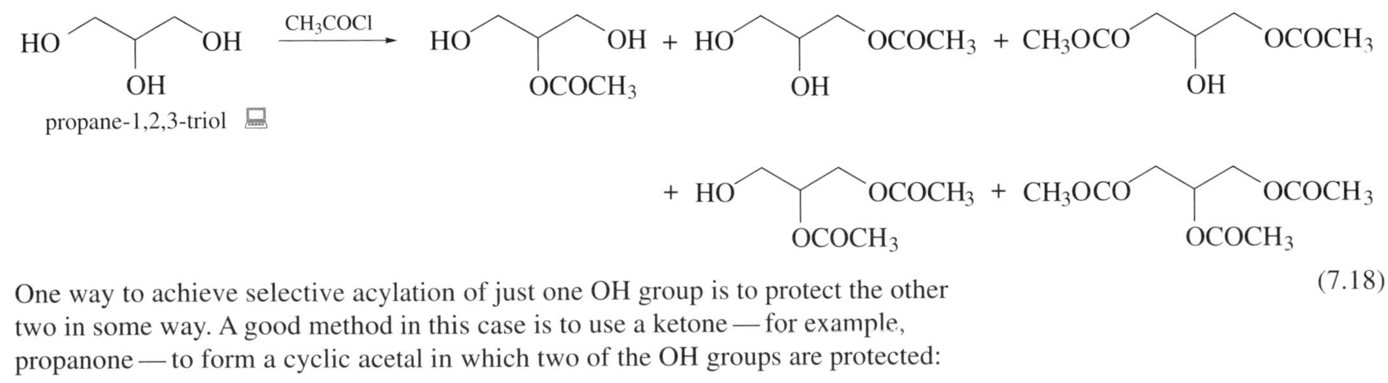


(7.18)

One way to achieve selective acylation of just one OH group is to protect the other two in some way. A good method in this case is to use a ketone — for example, propanone — to form a cyclic acetal in which two of the OH groups are protected:

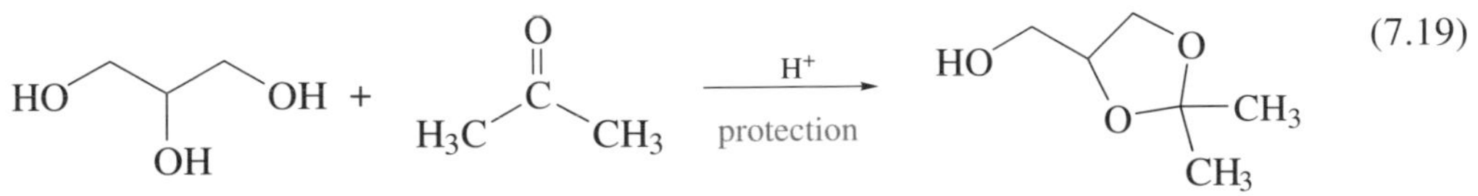


(7.19)

This enables a reaction at the free OH group to be carried out, and, following deprotection, the desired product is obtained:

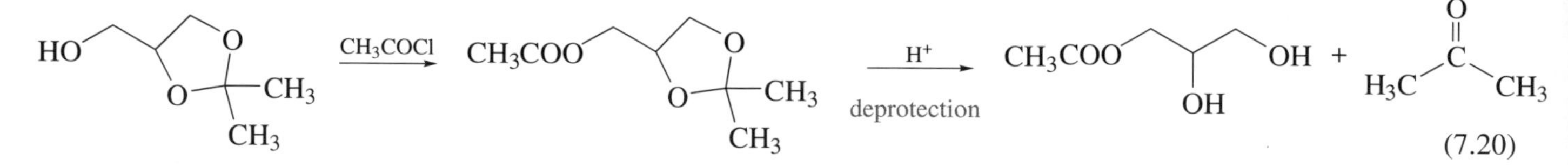

(7.20)

Protection of a functional group is also necessary when it is not required to undergo reaction, but is unstable under the reaction conditions. Such a case is the nitration of aniline. The action of nitric acid on aniline gives a complex mixture of products (in part, because the amine group is oxidized by the nitric acid), so the amine group must first be protected:

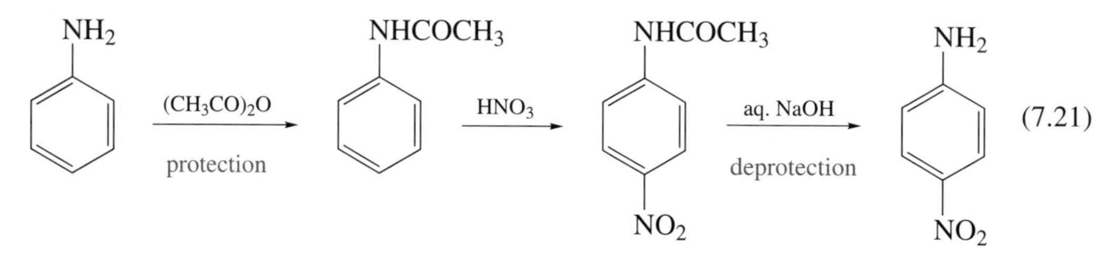

(7.21)

- How is the amine group protected in this reaction?
- It is converted into an amide, which is not oxidized by nitric acid.

- How is the protecting group removed?
- It is hydrolysed, to regenerate the amine group (and the sodium salt of ethanoic acid).

Another example of the protection of an otherwise reactive alcohol functional group was shown in Scheme 5.14, on p. 214.

## 7.5 Strategy and control

As you studied the Sections on selectivity and protecting groups, it should have become apparent that it is not sufficient to devise a synthetic strategy simply by using the disconnection approach. Of course, the approach is extremely useful in identifying potential routes, reactions and reagents. However, each proposed reaction must be checked to make sure that it will enable the desired product to be obtained in practice. It is best to check each step as it is identified, rather than devise a multi-stage synthesis only to find, on checking through, that one of the stages cannot be accomplished. It is better to abandon the planning for an unlikely strategy as soon as one can, rather than waste valuable effort.

## BOX 7.2 H. C. Brown

Herbert C. Brown (Figure 7.1) was born in London on May 22, 1912, of Ukranian parents (his grandfather's name had been anglicized to Brown). The family moved to Chicago in 1914. His father was a carpenter who opened a hardware store in Chicago. He attended local schools but on the death of his father in 1926, he left school to work in the shop. In 1929, he restarted school and went on to college in 1930 to study electrical engineering. He subsequently became fascinated with chemistry. However, the college closed in 1933, and so he went to night school, financing himself by working as a part-time shoe clerk. He then attended Wright Junior College, where he graduated in 1935. He subsequently went to the University of Chicago, receiving his BSc in 1936 and his PhD in 1938. He then moved to the University of Southern California and then Wayne State University in Detroit, where he became Associate Professor in 1946. In 1947, he moved to Purdue University to become Professor of Inorganic Chemistry. He remained at Purdue for the remainder of his career, becoming Emeritus Professor in 1978. His research interests include the study of steric effects and electrophilic aromatic substitution reactions, but it is for his work on organoborane chemistry that he is most famous.

**Figure 7.1**
H. C. Brown.

He was awarded the Nobel Prize for Chemistry in 1979 (jointly with Georg Wittig, Box 5.3), for the development of the use of boron-containing compounds as important reagents in organic synthesis.

# 7.6 Summary of Section 7

1 The need for selectivity arises when an organic compound can take part in a reaction to produce more than one product.

2 There are three types of selectivity: chemoselectivity, regioselectivity and stereoselectivity.

3 *Chemoselectivity* arises when a molecule contains two or more functional groups of different reactivity towards a particular reagent. The most-reactive functional group can be made to react preferentially.

4 *Regioselectivity* arises when a reaction can occur at more than one position around a functional group. Some examples are the formation of an alkene by dehydration of an alcohol, the addition of HBr to an alkene, electrophilic substitution of monosubstituted benzenes, and nucleophilic attack at an oxirane (epoxide).

5 *Stereoselectivity* arises when a reaction can produce more than one stereoisomer. Formation of a new chiral centre, and formation of an unsymmetrical alkene, are examples of processes for which stereoselectivity must be considered.

6 Protecting groups are required in the following circumstances:

(i) when you wish to react a less-reactive functional group in the presence of a more-reactive functional group;

(ii) when a molecule contains more than one functional group of the same type, but only one of them is required to react;

(iii) when a functional group is not required to undergo reaction but is unstable under the reaction conditions.

7 Protecting groups are used to convert a functional group into a different functional group that will not react under the conditions of the transformation that is being attempted. Protecting groups must be easily removable once the desired transformation has been successfully completed.

### QUESTION 7.1

Identify the types of selectivity involved in the following reactions.

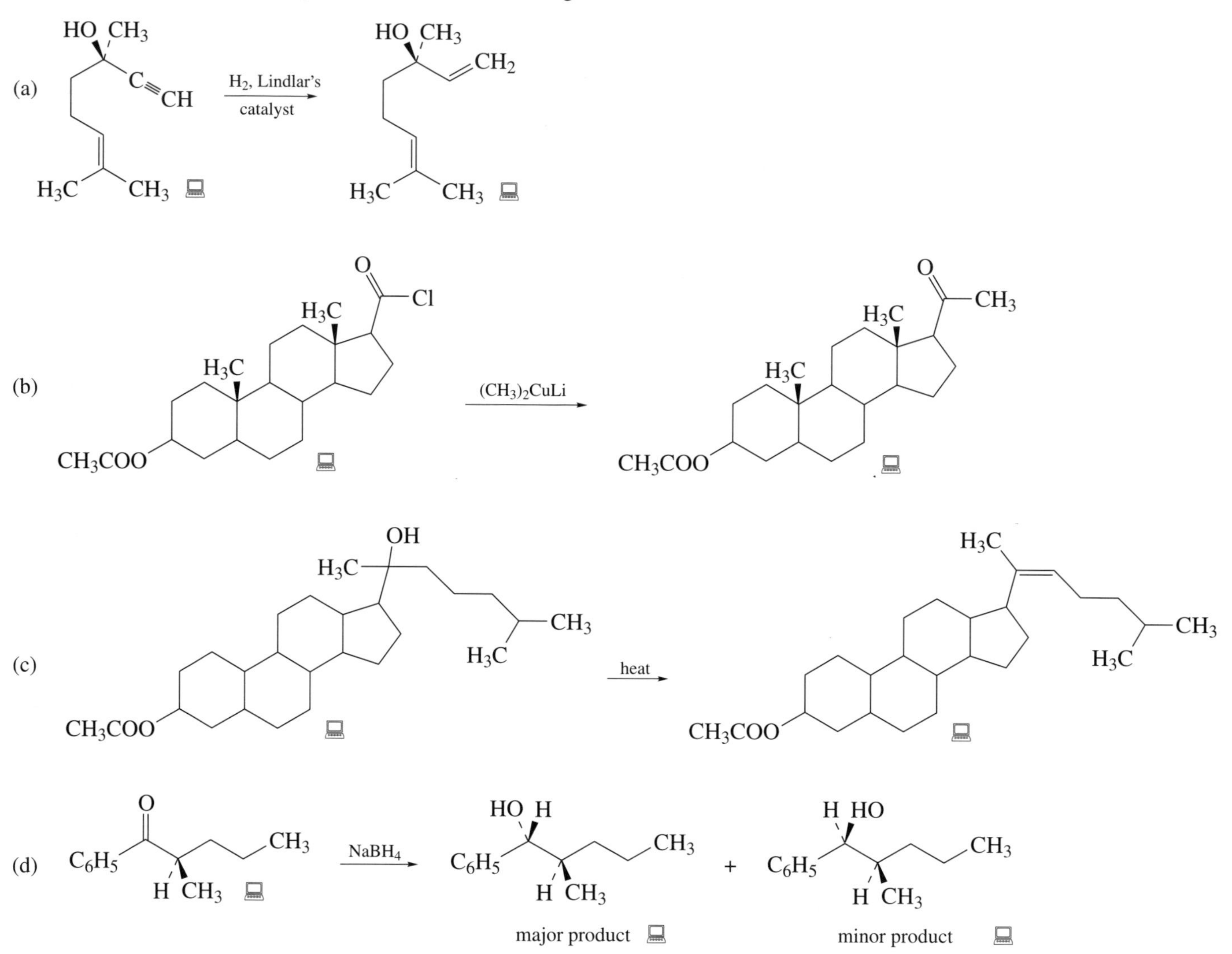

QUESTION 7.2

Describe the potential difficulties in carrying out the following conversion:

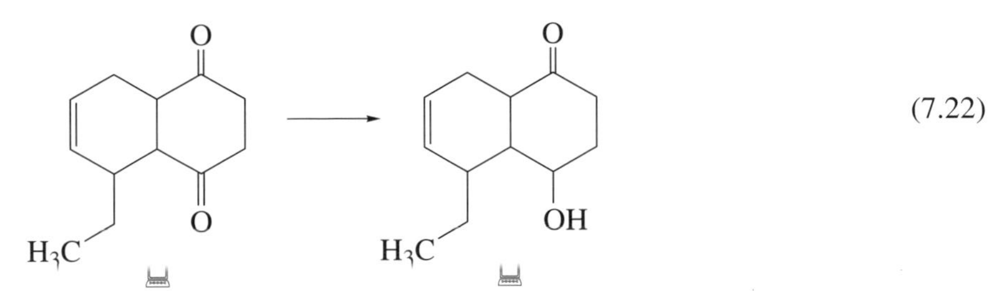


(7.22)

What must be done to overcome these problems?

QUESTION 7.3

Identify the potential selectivity problem associated with the following synthesis of *meta*-chloropropylbenzene, and suggest appropriate reagents to overcome it.

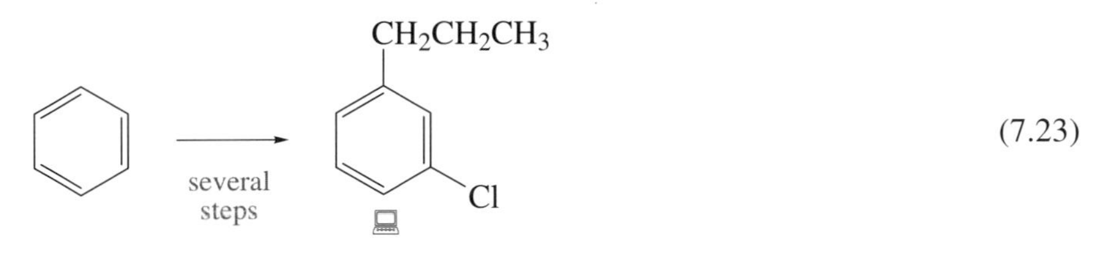


(7.23)

# FURTHER FACTORS AFFECTING THE CHOICE OF A SYNTHETIC ROUTE

8

Nearly all the disconnections and syntheses that we've studied so far have given rise to a number of possible synthetic routes to a particular target molecule. We are therefore led to the question: 'which route is the best one?'. There is often no simple answer, because several factors have to be balanced against one another, and what is best in one context may not be best in another. However, there are some obvious questions to ask about each route, such as 'what is the yield?', 'how long does it take?', and 'how much does it cost?', but the importance attached to each answer will depend on the nature of the project for which the synthesis is being designed.

## 8.1 Product yield

The yield* of a reaction is a measure of the chemical efficiency in converting starting material into product. It is normally calculated on a molar basis, and the amount of product formed is expressed as a percentage of the maximum that would be obtained if *all* the starting material were converted into product.

Try the following example to ensure you know how to calculate a percentage yield.

- After a hydrolysis of 20.0 g of benzyl chloride ($C_6H_5CH_2Cl$), 11.1 g of pure benzyl alcohol ($C_6H_5CH_2OH$) were isolated. What is the yield of this reaction?

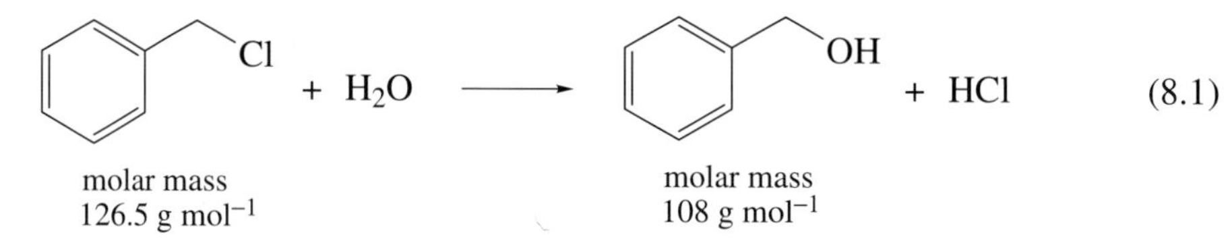


(8.1)

- Number of moles of $C_6H_5CH_2Cl$ = $\dfrac{20.0\,\text{g}}{126.5\,\text{g mol}^{-1}} = 0.158\,\text{mol}$.

  Thus, complete reaction would give 0.158 mol of $C_6H_5CH_2OH$.

  The *theoretical* yield of $C_6H_5CH_2OH$ is $0.158\,\text{mol} \times 108\,\text{g mol}^{-1} = 17.1\,\text{g}$.

  Thus, the percentage yield is given by:

$$\frac{11.1\text{g}}{17.1\text{g}} \times 100 = 65\%$$

- What assumption did you make in this calculation about the other reagent, water?
- It can be assumed that the water was in excess.

* The various ways of expressing yield are discussed in Part 3 of *Alkenes and Aromatics*[1].

In general, the yield is calculated from the starting material that is present in the *lowest molar amount*. This is the same as saying that when any reagent is used in excess, the excess is ignored in calculating the yield. The yield of a reaction can therefore vary from 0% to 100%, although the maximum is very rarely achieved, because some material is nearly always lost in the handling of chemicals. For example, when filtering the product, it is not usually possible to remove absolutely all of a precipitate from a filter paper and funnel. It is also important to realize that the yield refers to the product as obtained in the required state of purity, which means that it is measured *after* the necessary purification steps have been performed.

When a sequence of reactions is carried out, the *overall* yield (that is, the chemical efficiency in obtaining the final product from the first starting material) is obtained by multiplying together the fractional yields (the percentage yields expressed as fractions) for the individual stages, and expressing the result as a percentage.

- A four-stage synthesis is carried out in yields of 80%, 70%, 90% and 50% for the individual stages. What is the overall yield?

- The overall yield is

$$\frac{80}{100} \times \frac{70}{100} \times \frac{90}{100} \times \frac{50}{100} \times 100\% \approx 25\%$$

In a multi-stage synthesis the yields for the individual stages need to be quite high if the overall yield is to be reasonable.

Other things being equal, a route with fewer stages is preferred to one with more stages, because, if the mean yield per stage is the same, the overall yield diminishes as the number of stages increases. A route with fewer stages might also save time (though this is not necessarily so); it could present fewer losses during manipulation, for example, if each stage involves transfer to another reaction vessel. Finally, for a sequence that includes reactions that have not been carried out before, a route with fewer stages offers fewer possibilities for a reaction to go wrong!

A major consequence of the general decrease in overall yield as the number of stages increases is the preference for a convergent rather than a linear synthesis.

In a **linear synthesis** each intermediate is formed in sequence from the one before; for example,

$$\mathbf{A} \longrightarrow \mathbf{B} \longrightarrow \mathbf{C} \longrightarrow \mathbf{D} \longrightarrow \mathbf{E} \longrightarrow \mathbf{F}$$

In a **convergent synthesis** two or more intermediates are made independently before they react together; for example,

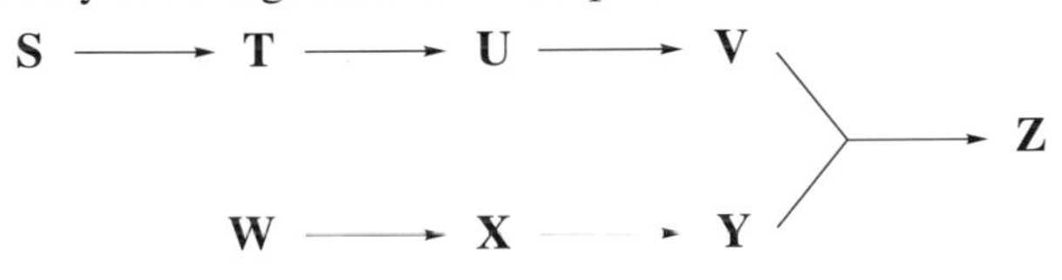


- If each stage of the linear synthesis from **A** to **F** can be carried out in 75% yield, what is the overall yield?

- The overall yield is

$$\frac{75}{100} \times \frac{75}{100} \times \frac{75}{100} \times \frac{75}{100} \times \frac{75}{100} \times 100 = 24\%$$

- If each stage of the convergent synthesis can be carried out in 75% yield, what are the overall yields of **V** and **Y**?
- **V** can be made from **S** in $\frac{75}{100} \times \frac{75}{100} \times \frac{75}{100} \times 100 = 42\%$ yield, and **Y** from **W** in 56% yield.

So, when we come to use **V** and **Y** to produce **Z**, we have more of **Y** than of **V**. That is, **Y** is in excess. So, we need only calculate our yield of **Z** based on the amount of **V** that we use. So the overall yield of **Z** is $\frac{42}{100} \times \frac{75}{100} \times 100 = 32\%$.

Now, it should be apparent from these examples of linear and convergent syntheses that **Z** is produced in higher yield than **F**, even though an extra stage is required in the synthesis of **Z**. It is a general rule that convergent syntheses give higher yields than linear syntheses, and are therefore preferred.

This preference for a convergent route means that it is usually more efficient to divide the synthesis up into compounds that constitute relatively large fragments of the target molecule, and then to join them together later in the sequence, rather than to carry out the synthesis in a single linear sequence. This concept was raised in Section 5.1, where you were introduced to the idea that it was better to make a disconnection in the middle of a framework.

## 8.2 Cost and time

For a synthetic route to be acceptable, the starting materials and reagents must be available in sufficient amounts at an acceptable price. As the goal in a research-orientated laboratory synthesis is to make enough of the target compound for the planned tests or measurements to be carried out, the cost of materials does not have the same high priority that it does for a commercial synthesis. Moreover, as the time of research chemists is considered valuable, the preferred route is the quickest one consistent with other factors. This may involve a decision to buy more expensive starting materials from a specialist supplier, rather than to buy cheaper materials from which they can be made, because the materials end up being expensive in time if you have to make them yourself.

As an example, consider two possible routes to 1-phenylpentan-l-ol, shown in Scheme 8.1. The relevant costs of chemicals, at 2001 prices, are listed in Table 8.1.

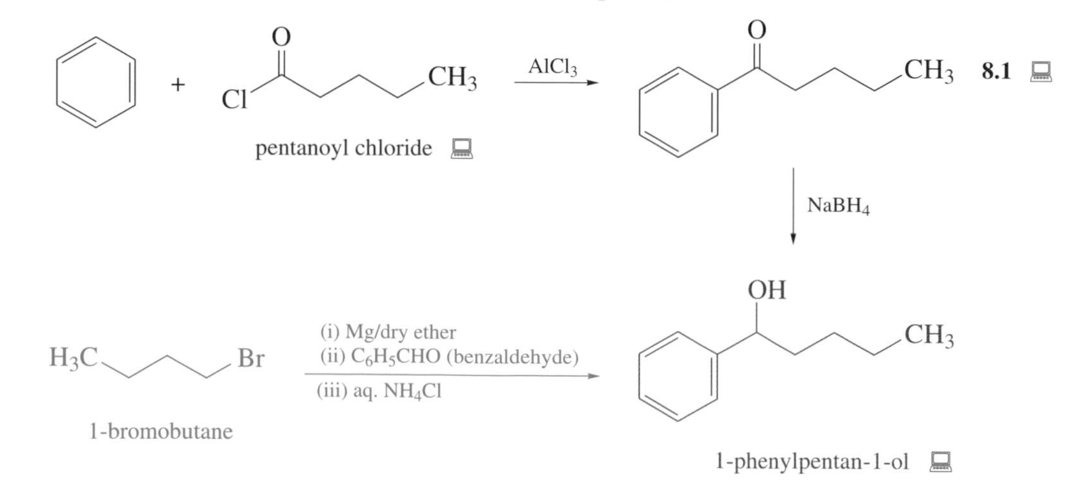


SCHEME 8.1

**Table 8.1** Prices of chemicals (in 2001) that can be used for the synthesis of 1-phenylpentan-l-ol

| | Pence per gram | Pence per mole |
|---|---|---|
| benzene | 2 | 156 |
| pentanoyl chloride | 15 | 1 830 |
| aluminium trichloride | 5.8 | 773 |
| sodium borohydride | 11.4 | 431 |
| 1-bromobutane | 1.7 | 238 |
| magnesium turnings | 2 | 49 |
| benzaldehyde | 0.77 | 82 |

- Study the two routes and Table 8.1 carefully. Which route would you choose for a laboratory synthesis of 1-phenylpentan-l-ol, and why?
- On the basis of cost, the Friedel–Crafts route is considerably more expensive because of the relatively high cost of pentanoyl chloride.

In the absence of information about yields, the Grignard route would probably be chosen, because it can all be carried out in one reaction vessel; in the other synthesis the ketone **8.1** would need to be isolated separately and purified before reaction with sodium borohydride.

## 8.3 Safety

In a chemical laboratory, safety factors should always be given their proper prominence. If a choice of routes is possible and one sequence involves an intermediate known to be particularly toxic or liable to explosion, this may be an overriding consideration, which would lead to the rejection of that route. In some cases there may be no alternative, and because, in general, chemists like to 'play safe', there may be a decision made not to proceed with the synthesis at all. These are extreme cases, but the point here is that all chemicals need to be handled with care, and some more so than others. When particularly hazardous (toxic, irritant, corrosive, explosive) materials are used, or when conditions of extreme temperature or pressure are required, special equipment and even buildings may be needed. This in itself may be a deciding factor if a particular laboratory does or does not have such equipment (for example, for high-pressure hydrogenation). The disposal of byproducts, waste solvents, etc., also needs to be taken into consideration, even on a laboratory scale, in the light of possible hazards.

Under the COSHH (Control of Substances Hazardous to Health) regulations, all these factors must be taken into account, so that every operation in the laboratory involving chemicals is assessed for the risk involved. Each risk assessment must be completed, written-up and independently authorized *before* the reaction is carried out. The aim is to minimize both hazards and accidents to the experimenter and other workers, and if necessary to abandon the proposed experiment if the danger is too great.

Another aspect of molecular complexity becomes apparent during the execution phase of synthetic research. For complex molecules even much-used standard reactions and reagents may fail, and new processes or options may have to be found. Also, it generally takes much time and effort to find appropriate reaction conditions. The time, effort, and expense required to reduce a synthetic plan to practice are generally greater than are needed for the conception of the plan. Although rigorous analysis of a complex synthetic problem is extremely demanding in terms of time and effort as well as chemical sophistication, it has become increasingly clear that such analysis produces superlative returns.

Molecular complexity can be used as an indicator of the frontiers of synthesis, since it often causes failures which expose gaps in existing methodology. The realization of such limitations can stimulate the discovery of new chemistry and new ways of thinking about synthesis.

***E. J. Corey*, The Logic of Chemical Synthesis (1989)**

## 8.4 Summary of Section 8

1 Convergent syntheses — that is, ones in which relatively large fragments of the target molecule are assembled independently and then linked together late in a synthetic strategy — are generally preferred to linear syntheses.

2 The yield of a chemical reaction is calculated from the starting material present in the lowest molar amount.

3 The overall yield of a linear synthesis is calculated by multiplying together the yields of the individual stages in the synthesis.

4 The overall yield of a convergent synthesis is calculated by treating it as a set of linear syntheses, and choosing the path that gives the *lowest* yield. For example, for the synthesis:

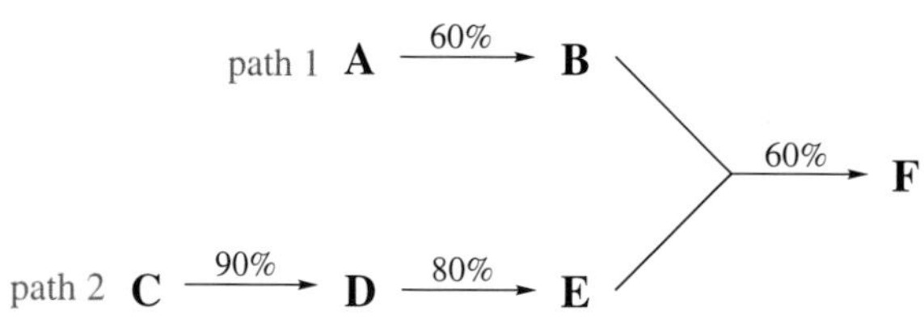


the yield of **B** from path 1 is 60%, and that of **E** from path 2 is 72%. Hence the overall yield of **F** is based on the yield of **B**, so is 36%.

5 Cost, time and safety are also factors that must be considered before embarking on any synthesis.

### QUESTION 8.1

Calculate the overall yields of the following two syntheses.

(a) **A** $\xrightarrow{97\%}$ **B** $\xrightarrow{77\%}$ **C** $\xrightarrow{80\%}$ **D** $\xrightarrow{85\%}$ **E** $\xrightarrow{85\%}$ **F**

(b) **G** $\xrightarrow{81\%}$ **H**

**I** $\xrightarrow{42\%}$ **J** $\xrightarrow{86\%}$ **K** $\xrightarrow{76\%}$ **L**

**H** + **L** $\xrightarrow{61\%}$ **M** $\xrightarrow{84\%}$ **N** $\xrightarrow{53\%}$ **O**

# 9 SYNTHESIS OF A DRUG

## 9.1 Introduction

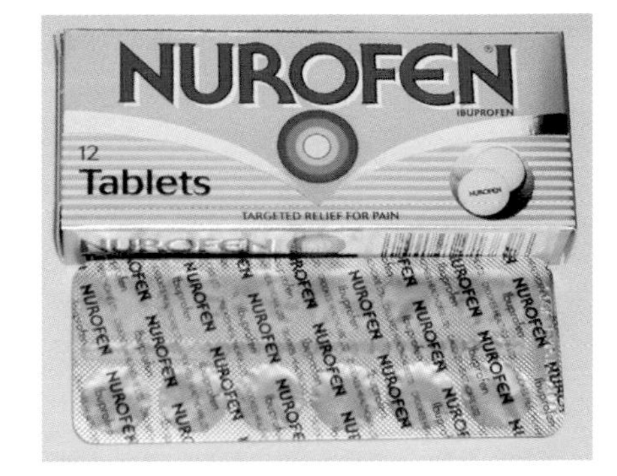


**Figure 9.1**
A commercial formulation containing ibuprofen.

Ibuprofen, **9.1**, is one of the UK's leading analgesics. It can be bought across the counter under trade names such as Nurofen (Figure 9.1). The active ingredient is 2-(4-(2-methylpropyl)phenyl)propanoic acid:

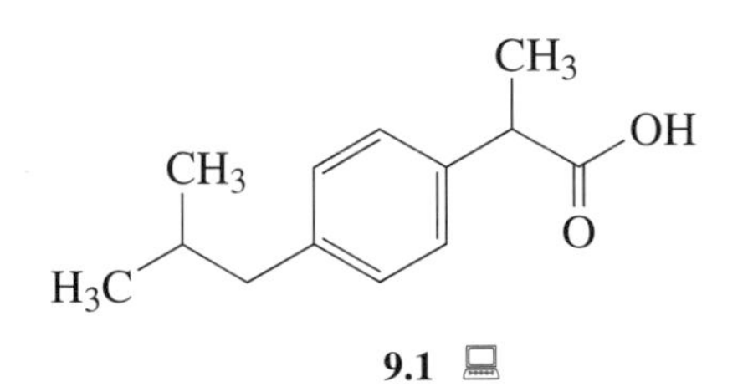


Ibuprofen was synthesised by the Boots Company in 1963*, and was first sold over the counter in 1983. It acts as an analgesic (pain killer), anti-pyretic (fever-reducing) and anti-inflammatory (reduces swelling and inflammation). It is used to treat a range of common painful conditions including dental pain, period pain, headache, migraine, gout, backache, rheumatic and muscular pains, and for reducing fever and the symptoms of colds and flu.

- Draw out the *R* enantiomer of ibuprofen.
- The chiral centre has a hydrogen, a methyl group, a carboxylic acid group and an aromatic group attached to it. In the *R* form, **9.2**, the priorities are ordered in a clockwise fashion, with the hydrogen pointing away.

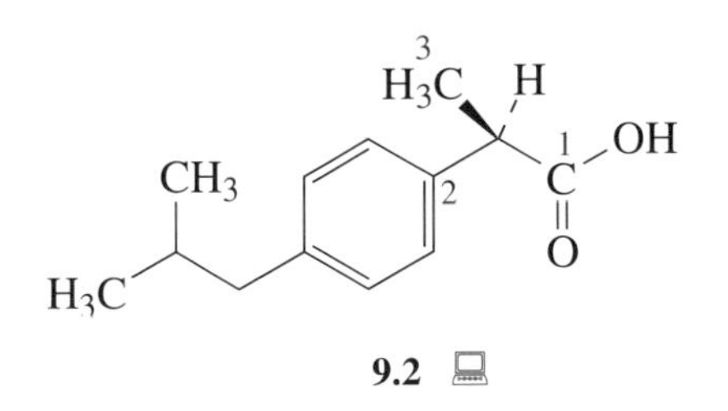


The active form is the *S* isomer, but the body converts the *R* enantiomer to the *S* enantiomer, so the commercial product can be sold as the racemic mixture.

---

**EXERCISE 9.1**

Based on the strategy that you developed in Section 6, use retrosynthetic analysis to devise a synthesis of a racemic mixture of ibuprofen. You need only go back as far as a monosubstituted benzene compound.

---

* The commercial synthesis of ibuprofen is considered in Part 3 of *Alkenes and Aromatics*[1].

# APPENDIX
# FUNCTIONAL GROUP INTERCONVERSIONS (FGIs)

**Table A.1** Summary of functional group interconversions

| Functional group in product \ Functional group in starting material | C–OH | C–$NH_2$ | C–Cl (or C–Br) | C<br>C=O<br>C | C<br>C=O<br>H | C<br>C=O<br>HO | C<br>C=O<br>RO | C<br>C=O<br>RHN | C<br>C=O<br>Cl | C≡N | C=C | C≡C | –C–H |
|---|---|---|---|---|---|---|---|---|---|---|---|---|---|
| C–OH | • | 9 | 14 | 20 | 20 | 23 | 23 | — | 23 | — | 36 | — | — |
| C–$NH_2$ | — | • | — | — | — | — | — | 29 | — | 34 | — | — | — |
| C–Cl (or C–Br) | 1 | 10 | • | — | — | — | — | — | — | — | 37 | 40 | 42 |
| C<br>C=O<br>C | 2 | — | — | • | — | — | — | — | — | — | 38 | — | — |
| C<br>C=O<br>H | 3 | — | — | — | • | — | — | — | — | — | — | — | — |
| C<br>C=O<br>HO | 4 | 11 | 15 | — | 22 | • | 27 | 30 | 31 | 35 | 39 | — | — |
| C<br>C=O<br>RO | 5 | — | — | — | — | 24 | • | — | 32 | — | — | — | — |
| C<br>C=O<br>RHN | — | 12 | — | — | — | 25 | 28 | • | 33 | — | — | — | — |
| C<br>C=O<br>Cl | — | — | — | — | — | 26 | — | — | • | — | — | — | — |
| C≡N | 6 | 13 | 16 | — | — | — | — | — | — | • | — | — | — |
| C=C | 7 | — | 17 | — | — | — | — | — | — | — | • | 41 | — |
| C≡C | — | — | 18 | — | — | — | — | — | — | — | — | • | — |
| –C–H | 8 | — | 19 | 21 | 21 | — | — | — | — | — | — | — | • |

**Table A.2** Functional group interconversions

| Table entry no. | Reaction |
| --- | --- |
| 1 | $ROH \xrightarrow[\text{(or HBr)}]{HCl} RCl$ (or RBr) |
| | $ROH \xrightarrow{SOCl_2} RCl$ |
| 2 | $R^1R^2CH{-}OH \xrightarrow[H_2SO_4]{K_2Cr_2O_7} R^1R^2C{=}O$ |
| | $R^1R^2CH{-}OH \xrightarrow{CrO_3/\text{pyridine } or\ CrO_3/\text{pyridine}/HCl \text{ (PCC) } or\ CrO_3/H_2SO_4/\text{propanone}} R^1R^2C{=}O$ |
| 3 | $RCH_2OH \xrightarrow[H_2SO_4]{K_2Cr_2O_7\ (or\ Na_2Cr_2O_7)} RCHO$ |
| | (care needs to be exercised so as to avoid further oxidation of the aldehyde product to the corresponding carboxylic acid; see entries 4 and 22) |
| | $RCH_2OH \xrightarrow[or\ CrO_3/\text{pyridine}/HCl]{CrO_3/\text{pyridine}} RCHO$ |
| | $RCH_2OH \xrightarrow{MnO_2} RCHO$ |
| 4 | $RCH_2OH \xrightarrow[H_2SO_4]{K_2Cr_2O_7\ (or\ Na_2Cr_2O_7)} RCOOH$ |
| | $RCH_2OH \xrightarrow{AgO} RCOOH$ |
| 5 | $R^1OH \xrightarrow[\text{conc. } H_2SO_4]{R^2COOH} R^1O{-}C({=}O){-}R^2$ |
| | $R^1OH \xrightarrow{R^2C(=O)OC(=O)R^2} R^1O{-}C({=}O){-}R^2$ |
| | $R^1OH \xrightarrow{R^2C(=O)Cl} R^1O{-}C({=}O){-}R^2$ |
| 6 | $R^1OH \xrightarrow{TsCl} ROTs \xrightarrow{KCN} R{-}C{\equiv}N$ |
| | (TsCl = *para*-toluenesulfonyl chloride, or tosyl chloride) |
| 7 | $RCH_2CH_2OH \xrightarrow{\text{conc. } H_2SO_4} R{-}CH{=}CH_2$ |
| 8 | $R{-}OH \xrightarrow[\text{(ii) } Bu^t_3SnH/AIBN]{\text{(i) } Cl{-}C(=S){-}R^1} R{-}H$    $Bu^t = (CH_3)_3C{-}$; AIBN = αα′-azo-*bis*-*iso*butyronitrile |

| Table entry no. | Reaction |
|---|---|
| 9 | $ArNH_2 \xrightarrow[\text{(ii) } H_2O]{\text{(i) HCl/NaNO}_2\text{/0 °C}} ArOH$ <br> (only for aromatic *primary* amines; that is, where Ar is an aryl group) |
| 10 | $ArNH_2 \xrightarrow[\text{(ii) CuCl}]{\text{(i) HCl/NaNO}_2\text{/0 °C}} ArCl$ <br> (only for aromatic *primary* amines; that is, where Ar is an aryl group) |
| 11 | $ArNH_2 \xrightarrow[\text{(ii) CuCN (iii) } H^+/H_2O]{\text{(i) HCl/NaNO}_2\text{/0 °C}} ArCOOH$ <br> (only for aromatic *primary* amines; that is, where Ar is an aryl group) |
| 12 | $R^1NH_2 \xrightarrow[\text{(ii) >200 °C}]{\text{(i) } R^2CO_2H} R^1HN{-}C({=}O){-}R^2$ <br> $R^1NH_2 \xrightarrow{R^2{-}C({=}O){-}OR^3} R^1HN{-}C({=}O){-}R^2$ <br> $R^1NH_2 \xrightarrow{R^2{-}C({=}O){-}Cl} R^1HN{-}C({=}O){-}R^2$ <br> $R^1NH_2 \xrightarrow{R^2{-}C({=}O){-}O{-}C({=}O){-}R^2} R^1HN{-}C({=}O){-}R^2$ |
| 13 | $ArNH_2 \xrightarrow[\text{(ii) CuCN}]{\text{(i) HCl/NaNO}_2\text{/0 °C}} Ar{-}C{\equiv}N$ <br> (only for aromatic *primary* amines — that is, where Ar is an aryl group) |
| 14 | $RCl \text{ (or Br)} \xrightarrow{H_2O/HO^-} ROH$ |
| 15 | $RCl \xrightarrow[\text{(ii) } CO_2 \text{ (iii) } H^+/H_2O]{\text{(i) Mg/ether}} RCOOH$ |
| 16 | $RBr \xrightarrow{NaCN} R{-}C{\equiv}N$ |
| 17 | $RCH_2CH_2Cl \text{ (or Br)} \xrightarrow{NaOBu^t} R{-}CH{=}CH_2$ $\quad Bu^t = (CH_3)_3C{-}$ <br> (elimination competes with substitution, and is preferred if a sterically hindered base like $NaOBu^t$ is used) <br> $R{-}CH(Br){-}CH_2Br \xrightarrow[or\ I^-]{Zn} R{-}CH{=}CH_2$ <br> (only for 1,2-disubstituted bromoalkanes) |

| Table entry no. | Reaction |
|---|---|
| 18 | $RCH_2CH_2Cl \xrightarrow{NaNH_2} R{-}C{\equiv}C{-}H$ <br> $RCHClCH_2Cl \xrightarrow{NaNH_2} R{-}C{\equiv}C{-}H$ |
| 19 | $R{-}Br \xrightarrow[AIBN]{Bu_3SnH} R{-}H$ $\quad$ $Bu = CH_3CH_2CH_2CH_2{-}$ |
| 20 | $R^1R^2C{=}O \xrightarrow[(ii)\ H^+/H_2O]{(i)\ LiAlH_4\ or\ NaBH_4} R^1R^2CHOH$ <br> (for aldehydes and ketones; aldehydes react faster than ketones, and $LiAlH_4$ is more reactive than $NaBH_4$) |
| 21 | $R^1R^2C{=}O \xrightarrow{Zn/HCl} R^1{-}CH_2{-}R^2$ <br> $R^1R^2C{=}O \xrightarrow[(ii)\ 200°C]{(i)\ H_2NNH_2} R^1{-}CH_2{-}R^2$ |
| 22 | $R(H)C{=}O \xrightarrow[H_2SO_4]{K_2Cr_2O_7} R(HO)C{=}O$ |
| 23 | $R(X)C{=}O \xrightarrow[(ii)\ H^+/H_2O]{(i)\ LiAlH_4} RCH_2OH$ <br> ($X = OH, OR, OCOR, Cl$) |
| 24 | $R^1(HO)C{=}O \xrightarrow[conc.\ H_2SO_4]{R^2OH} R^1(R^2O)C{=}O$ <br> $R^1(^-O)C{=}O \xrightarrow{R^2Br} R^1(R^2O)C{=}O$ |
| 25 | $R^1(HO)C{=}O \xrightarrow[(ii)\ heat, >200\ °C]{(i)\ R^2NH_2} R^1(R^2HN)C{=}O$ |
| 26 | $R(HO)C{=}O \xrightarrow[or\ SOCl_2]{PCl_5} R(Cl)C{=}O$ |

| Table entry no. | Reaction |
| --- | --- |
| 27 | $R^1(R^2O)C{=}O \xrightarrow[\textit{or}\ HO^-/H_2O]{\textit{either}\ H^+/H_2O} R^1(HO)C{=}O + R^2OH$ |
| 28 | $R^1(R^2O)C{=}O \xrightarrow{R^3NH_2} R^1(R^3HN)C{=}O + R^2OH$ |
| 29 | $R(H_2N)C{=}O \xrightarrow{LiAlH_4} RCH_2NH_2$ |
| 30 | $R^1(R^2HN)C{=}O \xrightarrow{H^+/H_2O} R^1(HO)C{=}O + R^2\overset{+}{N}H_3$ |
| 31 | $R(Cl)C{=}O \xrightarrow{H_2O} R(HO)C{=}O$ |
| 32 | $R^1(Cl)C{=}O \xrightarrow{R^2OH} R^1(R^2O)C{=}O$ |
| 33 | $R^1(Cl)C{=}O \xrightarrow{R^2NH_2} R^1(R^2HN)C{=}O$ |
| 34 | $R{-}C{\equiv}N + 2H_2 \xrightarrow[\text{e.g. Pd}]{\text{catalyst}} RCH_2NH_2$ |
| 35 | $R{-}C{\equiv}N \xrightarrow[\textit{or}\ H_2O/NaOH]{\textit{either}\ H_2O/H_2SO_4} RCOOH$ |
| 36 | $R{-}CH{=}CH_2 \xrightarrow{H^+/H_2O} R{-}CH(OH){-}CH_3$ (this is the Markovnikov product) <br> $R{-}CH{=}CH_2 \xrightarrow[\text{(ii) } H_2O_2/NaOH]{\text{(i) 'BH}_3\text{'}} RCH_2CH_2OH$ (this is the anti-Markovnikov product) |

| Table entry no. | Reaction |
| --- | --- |
| 37 | $R{-}CH{=}CH_2 + HX \longrightarrow R{-}CH(X){-}CH_3$ (X = Cl, Br, I)<br>$R{-}CH{=}CH_2 + Br_2 \longrightarrow R{-}CH(Br){-}CH_2Br$ |
| 38 | $R^1R^2C{=}CR^3R^4 \xrightarrow{KMnO_4/H^+} R^1R^2C{=}O + R^3R^4C{=}O$ |
| 39 | $R^1{-}CH{=}CH{-}R^2 \xrightarrow{KMnO_4/H^+} R^1COOH + R^2COOH$ |
| 40 | $R{-}C{\equiv}C{-}H + 2HBr \longrightarrow RCBr_2CH_3$ |
| 41 | $R^1{-}C{\equiv}C{-}R^2 \xrightarrow[\text{catalyst}]{H_2/\text{Lindlar's}} (R^1)(H)C{=}C(R^2)(H)$<br>(the product has *Z* stereochemistry)<br>$R^1{-}C{\equiv}C{-}R^2 \xrightarrow[\text{liquid } NH_3]{Na} (R^1)(H)C{=}C(H)(R^2)$<br>(the product has *E* stereochemistry) |
| 42 | $\geq C{-}H \xrightarrow[h\nu]{Br_2} \geq C{-}Br$<br>(not always very selective!) |

# LEARNING OUTCOMES

Now that you have completed *Mechanism and Synthesis — Part 4: Strategy and methodology in organic synthesis*, you should be able to:

1 Recognize valid definitions of, and use in a correct context, the terms, concepts and principles listed in the following Table. (All Questions)

List of scientific terms, concepts and principles introduced in *Strategy and methodology in organic synthesis*

| Term | Page number | Term | Page number | Term | Page number |
|---|---|---|---|---|---|
| branch point | 211 | linear synthesis | 249 | retrosynthetic analysis | 196 |
| C—C bond-forming reaction | 181 | molecular framework | 181 | selective reagent | 237 |
| C—X bond-forming reaction | 182 | precursor | 190 | selectivity | 236 |
| chemoselectivity | 236 | protecting group | 214 | stereoselectivity | 240 |
| convergent synthesis | 249 | protection | 243 | synthetic plan or strategy | 181 |
| deprotection | 243 | reagent | 194 | synthon | 194 |
| disconnection | 192 | regioselectivity | 238 | Wolff–Kishner reduction | 185 |
| functional group interconversion (FGI) | 183 | | | | |

2 Identify reactions in a synthesis as C—C or C—X bond-forming reactions, or as FGIs. (Question 2.1)

3 Carry out the disconnection of a C—X or C—C bond in a target molecule, to identify potential nucleophilic and electrophilic synthons. (Questions 3.1, 3.2, 4.1, 4.2, 4.3, 5.1, 5.2, 5.3, 5.4 and 5.5; Exercise 9.1)

4 Analyse, using the disconnection approach, a given reaction to a target molecule, identify the synthons, and relate the synthons to the reagents used in the reaction. (Question 3.2; Exercise 9.1)

5 (i) Make suitable FGIs and bond disconnections for compounds containing the following functional groups: esters, amides, alcohols, alkenes, alkynes, aromatics, aldehydes, ketones, nitriles and carboxylic acids; and (ii) relate the synthons identified by such bond disconnections to suitable reagents. (Questions 3.1, 3.2, 4.1, 4.2, 4.3, 5.1, 5.2, 5.3, 5.4 and 5.5; Exercise 9.1)

6 Distinguish between chemoselectivity, regioselectivity and stereoselectivity, and identify when each type of selectivity needs to be considered. (Question 7.1; Exercise 9.1)

7 For given examples, choose appropriate reactions or reagents to produce the desired product selectively. (Question 7.3; Exercise 9.1)

8 Identify when functional groups need to be protected, choose suitable protecting groups for given functional groups, and identify the use of protecting groups in given syntheses. (Question 7.2)

9 Calculate the overall yields for linear and convergent syntheses. (Question 8.1)

# QUESTIONS: ANSWERS AND COMMENTS

### QUESTION 2.1 (Learning Outcome 2)

Stage 1 involves the addition of $H_2$ to the double bond of an alkene; it is an FGI.

Stage 2 converts the ester into an amide functional group, it is also an FGI. This is *not* a C—X bond-forming reaction, because the nitrogen atom *is not part of the target molecule.*

Stage 3 involves the formation of an aldehyde from an amide. Since the carbon frameworks of the two molecules are the same, this too is an FGI.

Stage 4 involves the introduction of the remaining carbon atoms of the juvabione framework, so this reaction extends the molecular framework. It involves a C—C bond-forming reaction between a nucleophilic Grignard reagent and the electrophilic carbon atom of the aldehyde group.

### QUESTION 3.1 (Learning Outcome 3)

(i) Either the $C-C_6H_5$ or the $C-CH_3$ bonds may be disconnected. Taking the $C-C_6H_5$ bond first, we get two pairs of synthons:

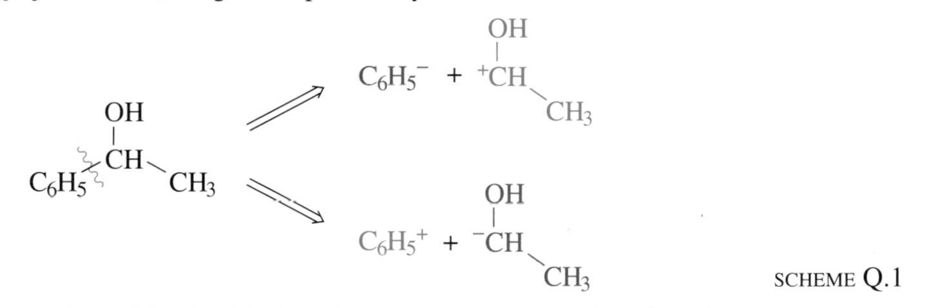

SCHEME Q.1

Disconnection of the $C-CH_3$ bond also generates two pairs of synthons.

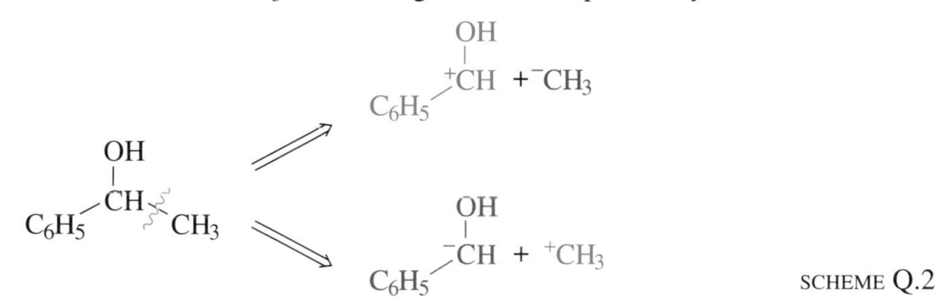

SCHEME Q.2

(ii) Disconnection of the $CH_2-COOH$ bond generates the following pairs of synthons:

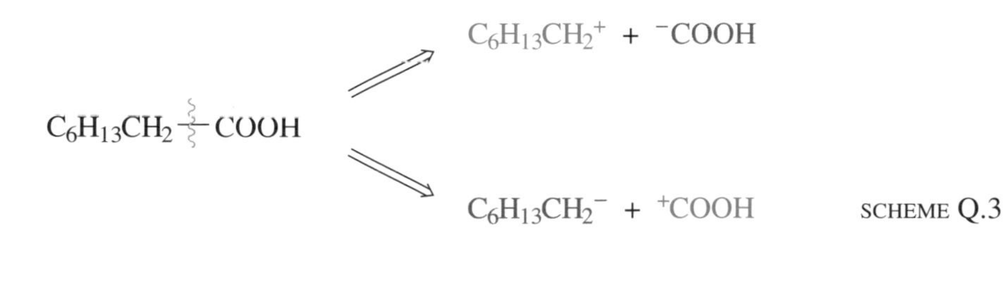

SCHEME Q.3

(iii) Disconnection of the $C{-}C_6H_{13}$ bond generates the following pairs of synthons:

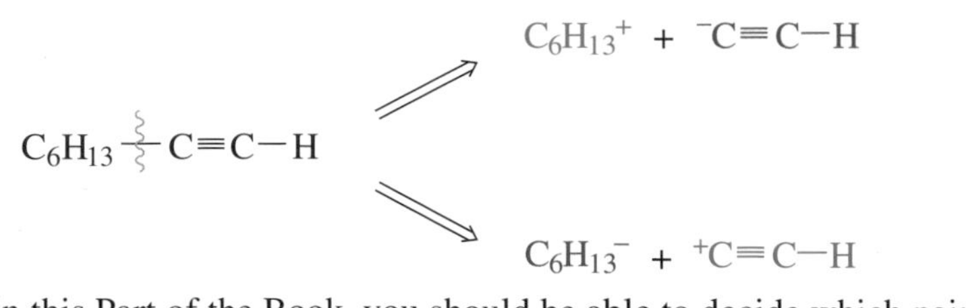


SCHEME Q.4

Later in this Part of the Book, you should be able to decide which pair of synthons is preferred in these disconnections.

## QUESTION 3.2 *(Learning Outcomes 3 and 4)*

Disconnection of the $-CH_2{-}C{\equiv}N$ bond generates two pairs of synthons.

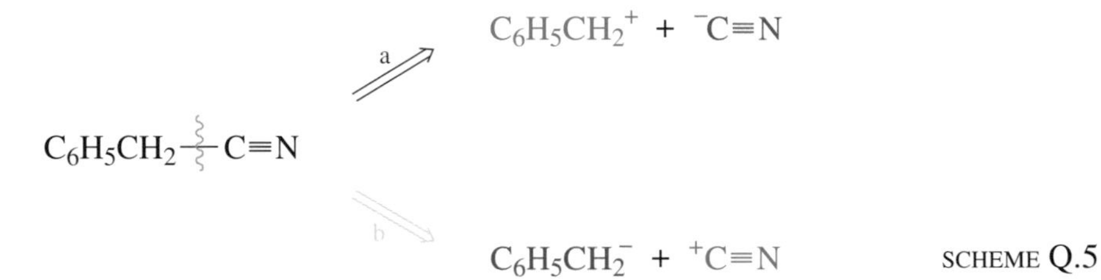


SCHEME Q.5

In the first pair of synthons, the $C_6H_5CH_2^+$ fragment is identified as electrophilic, and $^-C{\equiv}N$ as nucleophilic, whereas the reverse is true for the second pair of synthons. Looking at the reagents, it is clear that $C_6H_5CH_2Br$ acts as an electrophile because it undergoes displacement of the bromine atom by the nucleophilic cyanide ion. So, it is the first pair of synthons, $C_6H_5CH_2^+$ and $^-C{\equiv}N$, that corresponds to the reagents used.

## QUESTION 4.1 *(Learning Outcomes 3 and 5)*

Chloroamphenicol contains the amide functional group, and it is most appropriate to disconnect the C—N bond between the nitrogen atom and the carbonyl carbon atom.

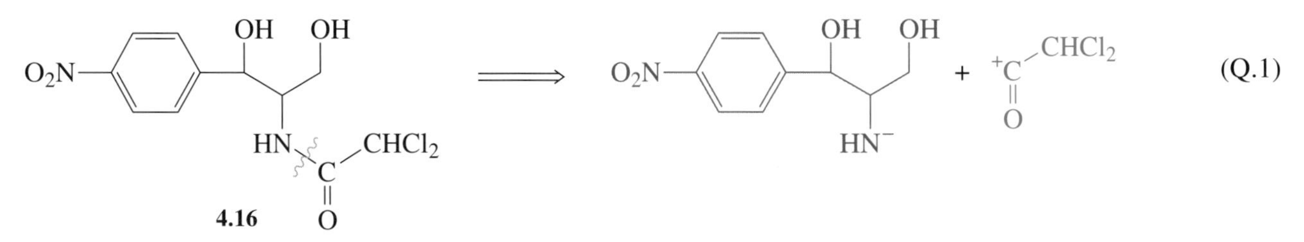


The appropriate reagents are the corresponding amine and acid chloride (and ultimately the carboxylic acid).

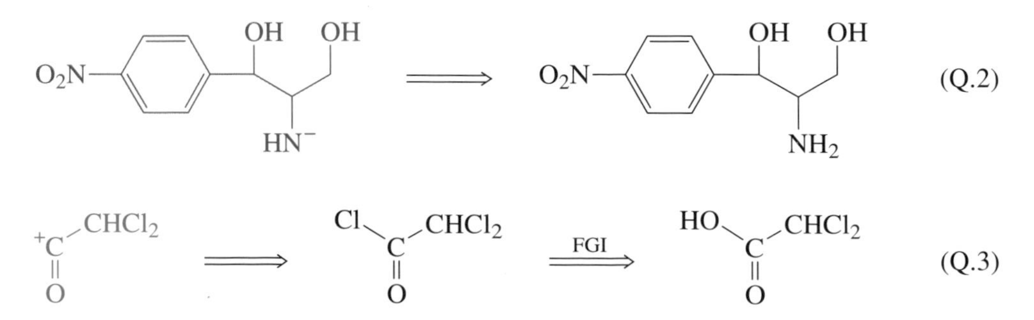

## QUESTION 4.2 *(Learning Outcomes 3 and 5)*

Aspirin contains the ester functional group, which is disconnected as follows:

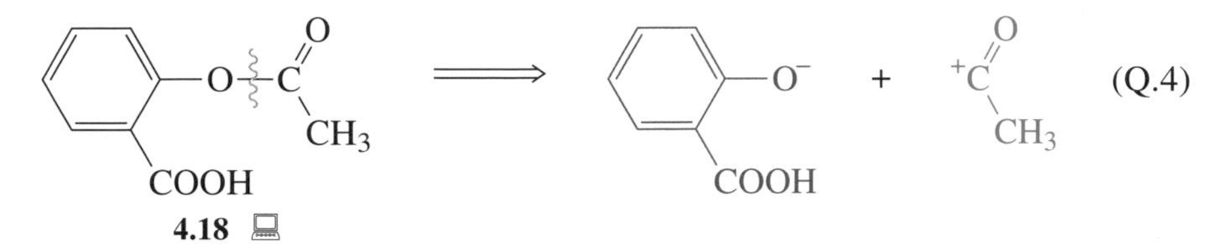


The reagents corresponding to these synthons are the phenol **Q.1** and ethanoyl chloride (or ethanoic acid). In fact, ethanoic anhydride is usually used in this synthesis.

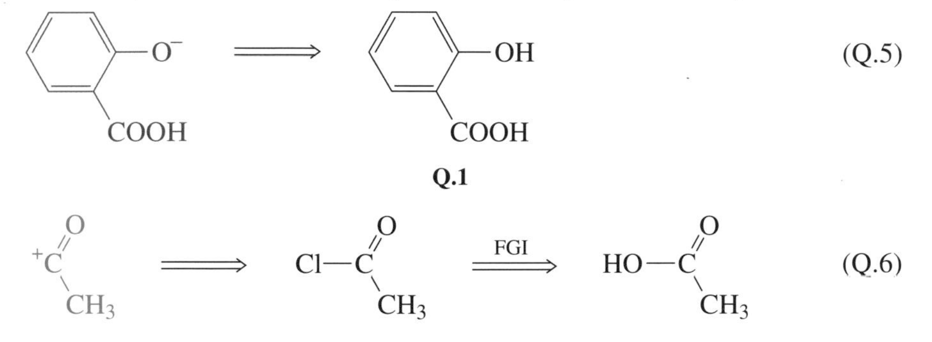


## QUESTION 4.3 *(Learning Outcomes 3 and 5)*

Captopril contains the amide functional group, which is disconnected as follows:

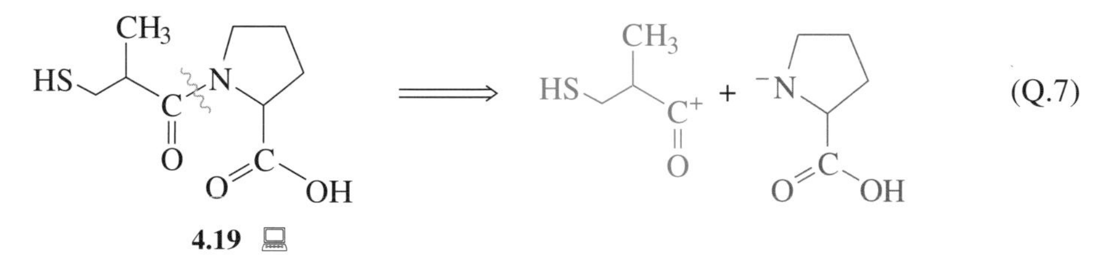


The appropriate reagents for these synthons are the corresponding carboxylic acid and amine:

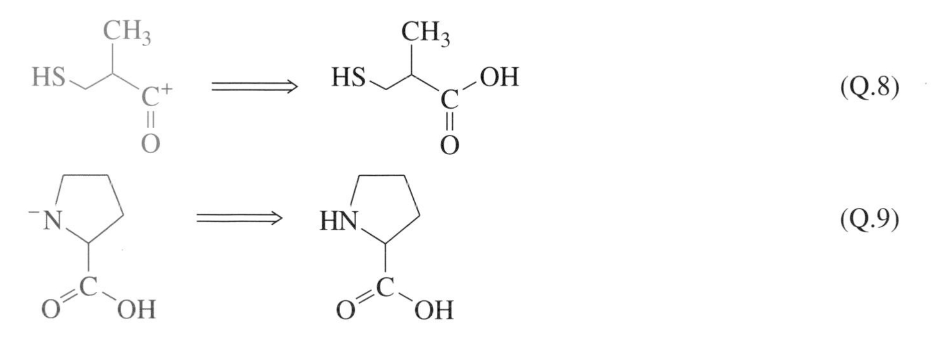

## QUESTION 5.1 *(Learning Outcomes 3 and 5)*

Compound **5.15** may be disconnected at any of the three C—C bonds involving the alcohol carbon atom. These disconnections generate three pairs of synthons, and hence the appropriate ketones and organometallic reagents required for the synthesis of **5.15** can be identified.

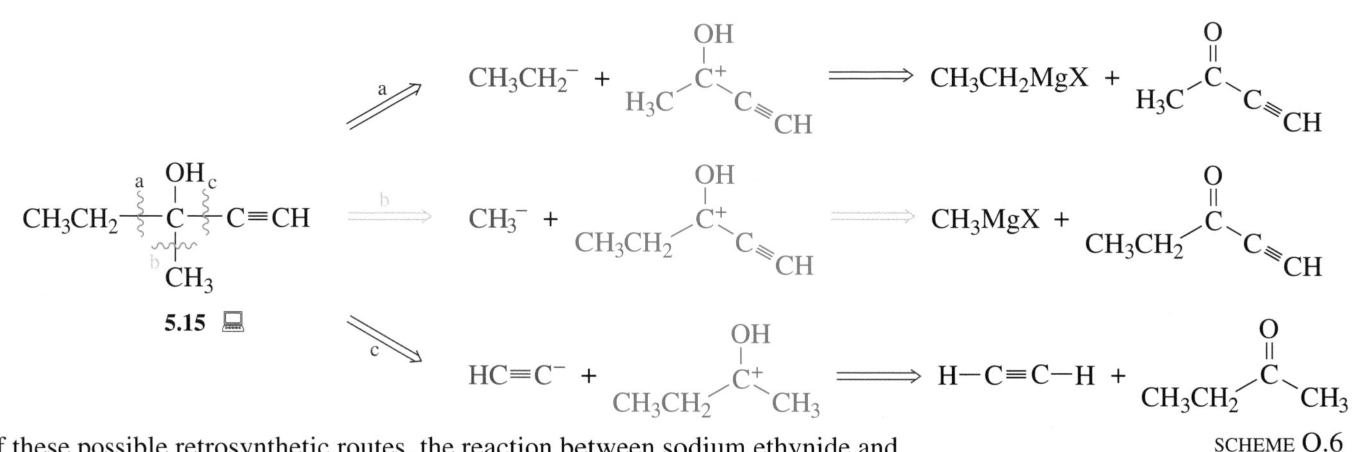


[Of these possible retrosynthetic routes, the reaction between sodium ethynide and butanone (disconnection c) uses the most readily available starting materials.]

SCHEME Q.6

Compound **5.16** may be disconnected at either of the two C—C bonds, a and b, that involve the alcohol carbon atom. Of the two, disconnection b is preferred because it breaks the molecule into two fragments that are more similar in size, and results in greater simplification. Thus, propanal and a pent-2-ylmagnesium halide are suitable reagents.

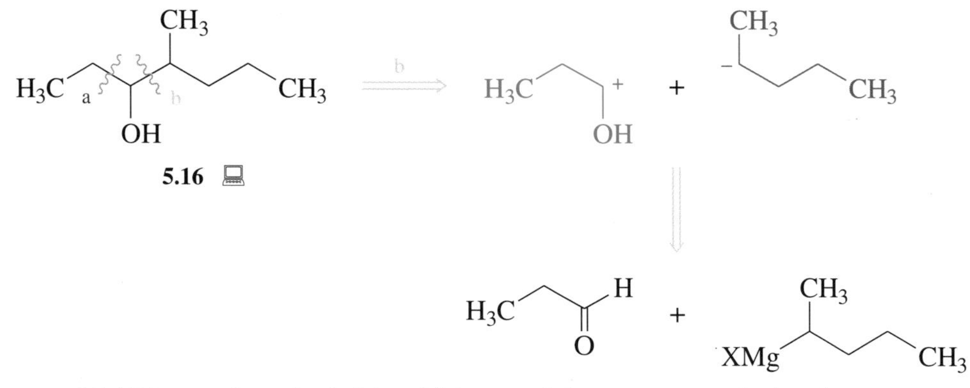


SCHEME Q.7

Compound **5.17** is a tertiary alcohol in which two of the groups attached to the alcohol carbon atom are identical. If we disconnect both these groups at the same time, we can generate synthons that identify an appropriate Grignard reagent as the carbon nucleophile, and an ester as the carbon electrophile:

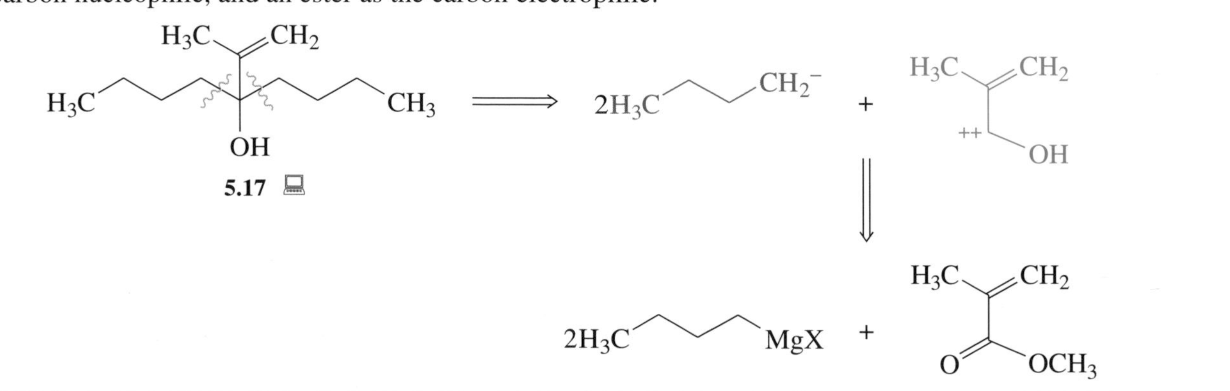


SCHEME Q.8

[This ester, methyl 2-methylprop-2-enoate (methyl methacrylate), is readily available: it is the monomer used for making the polymer Perspex!]

Compound **5.18** is an ester, which must first be converted into the corresponding alcohol. This is easily done by invoking a C—O bond disconnection.

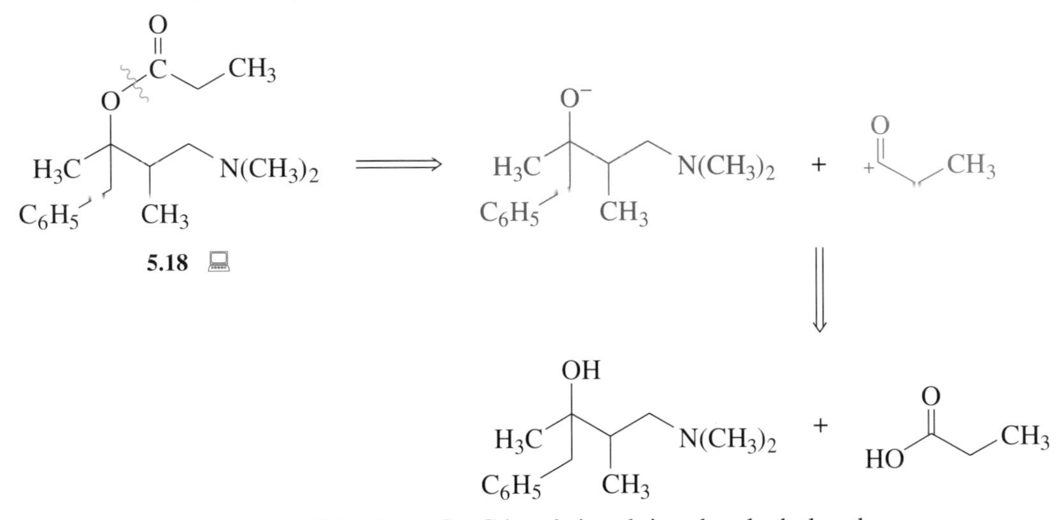


SCHEME Q.9

Now we may disconnect any of the three C—C bonds involving the alcohol carbon atom, as shown in Scheme Q.10. This enables us to identify the appropriate ketones and Grignard reagents required for the synthesis.

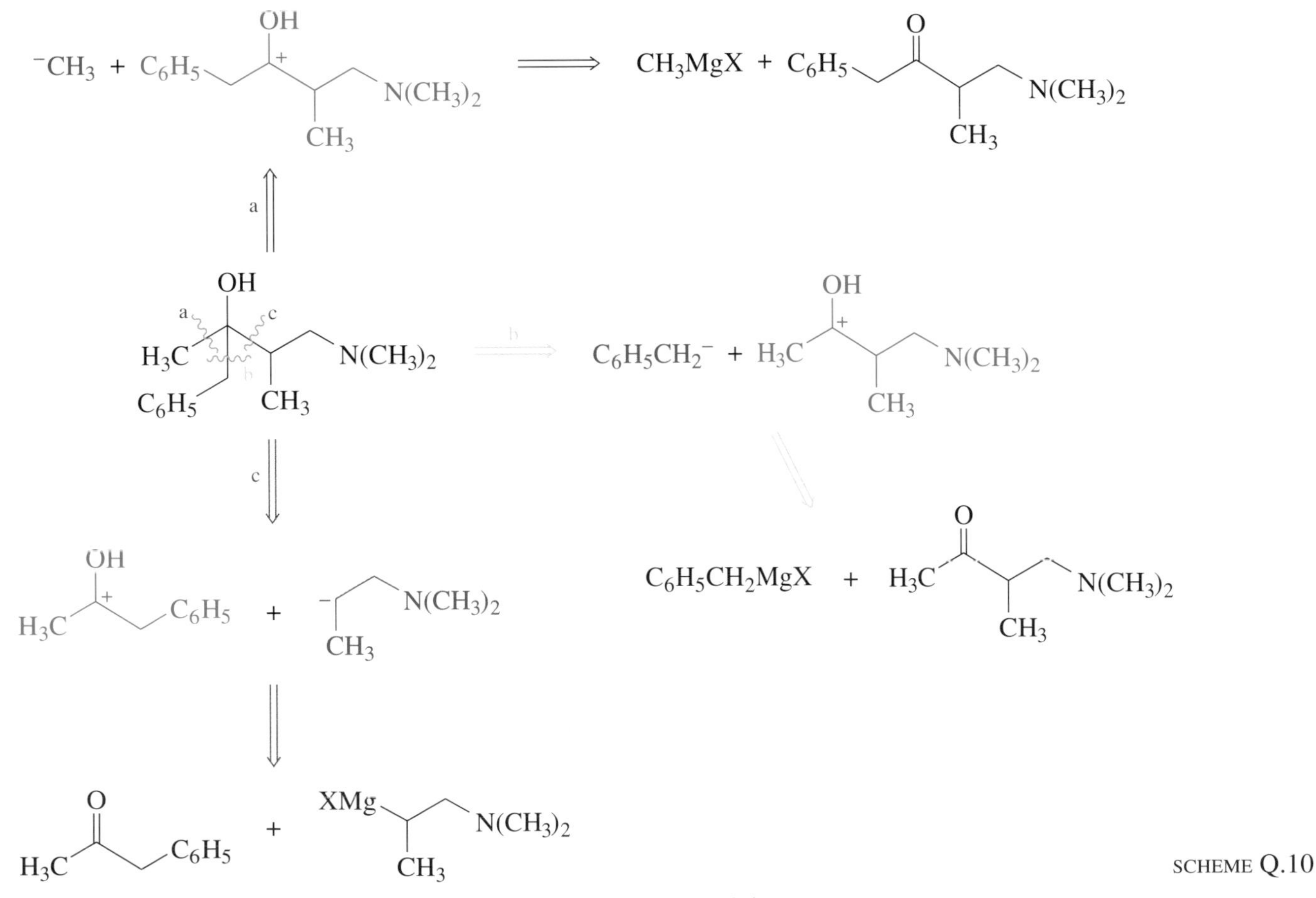


SCHEME Q.10

Routes a and b give quite complex amino ketones as the starting materials, so route c, using a simple haloamine and an available ketone would be preferred.

## QUESTION 5.2 (Learning Outcomes 3 and 5)

Compound **5.40** is a trisubstituted alkene with a branch point, so an appropriate analysis would involve first an FGI in which an OH group is placed at the branch point:

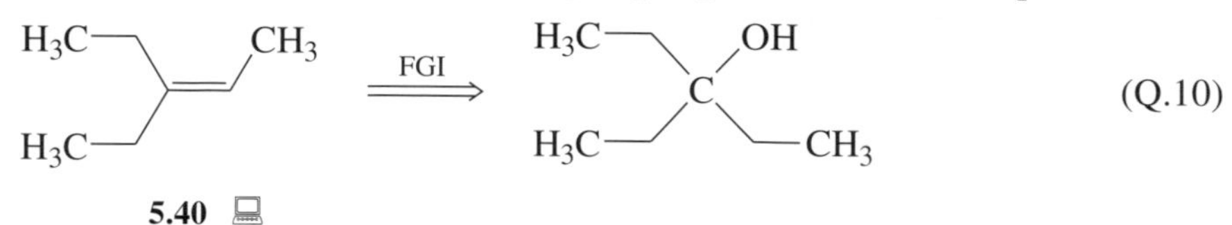


Since the alcohol carbon atom is attached to three identical groups, we can choose to disconnect one or two of these groups. This identifies a ketone or an ester as the electrophile, and a Grignard reagent as the organometallic nucleophile:

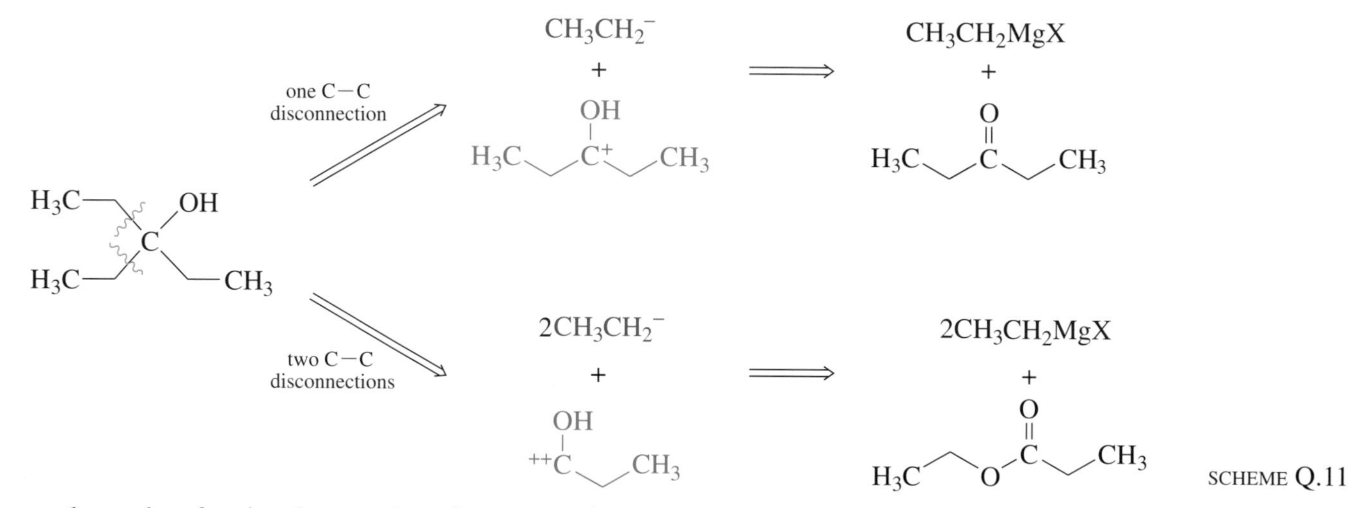


The syntheses therefore involve reaction of an appropriate amount of a Grignard reagent with either a ketone or an ester, followed by acid-catalysed dehydration of the intermediate alcohol to give the trisubstituted alkene.

Compound **5.41** is a cyclic, branched alkene, so, again, the first step in the analysis is to carry out an FGI, hydrating the double bond and putting the OH group at the branch point:

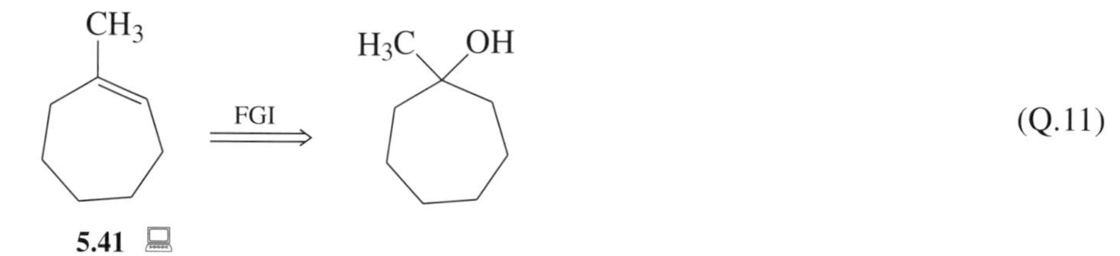


Disconnection of the alcohol now involves breaking the C—C bond to the methyl group rather than one of the ring C—C bonds. This disconnection identifies a methyl Grignard reagent and cycloheptanone as the required reagents:

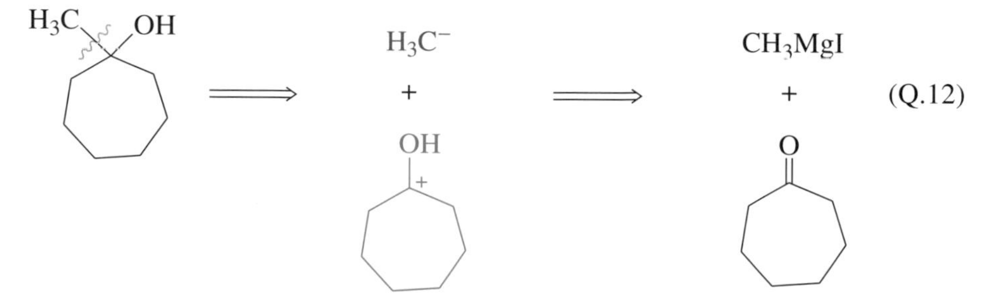

The synthesis is carried out by treating the methyl Grignard reagent with cycloheptanone, followed by acid-catalysed dehydration to form the trisubstituted alkene, according to Saytzev's rule.

Compound **5.42** is a *Z*-1,2-disubstituted alkene. It is best considered as being derived from an alkyne by stereoselective reduction. So our analysis should start by using an FGI to convert the alkene into the corresponding alkyne:

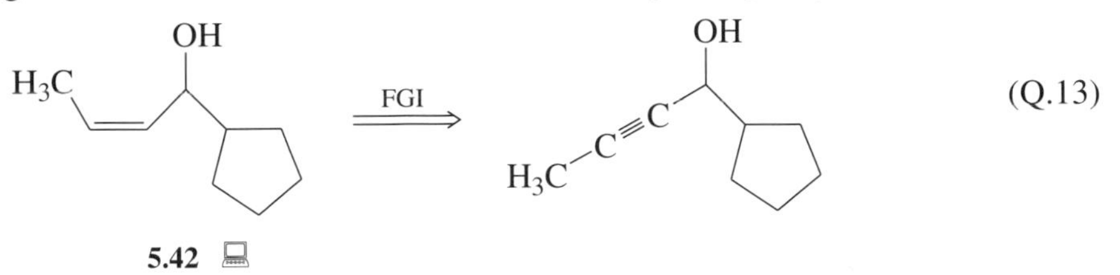


Disconnection of the alkyne then follows at the adjacent C—C single bond:

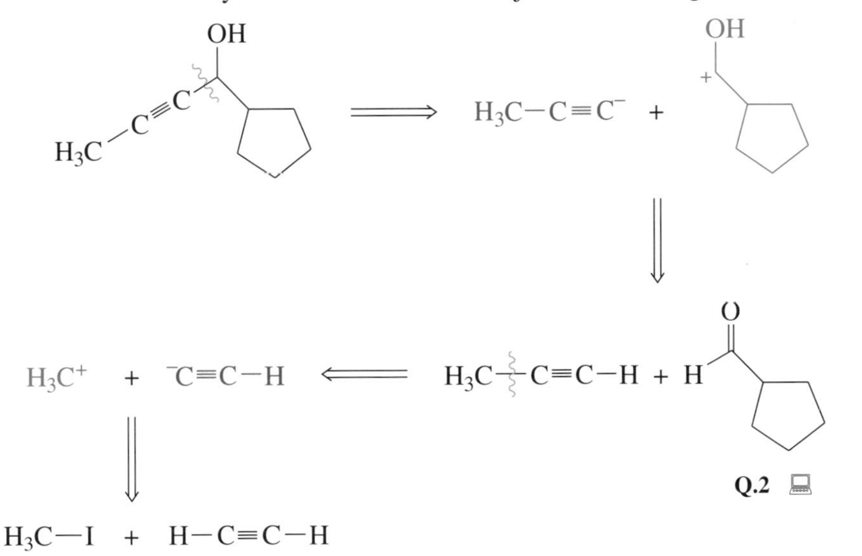


SCHEME Q.12

(Note that in Scheme Q.12 we show the first alkyne disconnection at the C—COH bond in order to obtain better-balanced synthons than if we placed it at the C—$CH_3$ bond.)

Initial disconnection of the C—C bond involving the alcohol carbon atom identifies propyne and the aldehyde **Q.2** as appropriate starting materials (the reaction between an alkynide anion and an aldehyde or ketone was dealt with in Part 1 of this Book). Even though propyne is readily available, further analysis leads ultimately to ethyne and iodomethane.

### QUESTION 5.3 (Learning Outcomes 3 and 5)

Compound **5.50** may be disconnected directly, in which case the Friedel–Crafts alkylation of benzene is identified as a suitable synthesis:

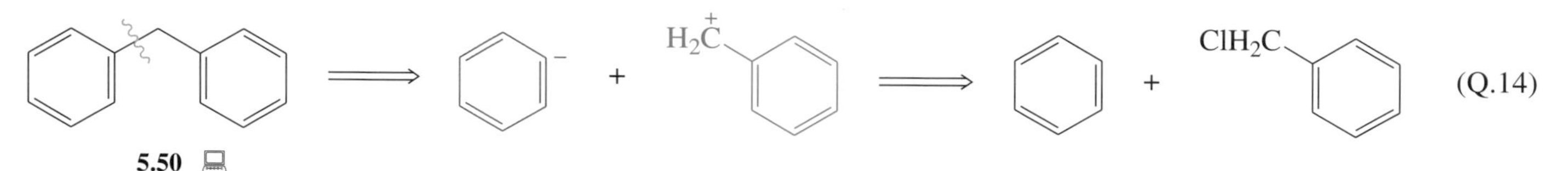


Alternatively, to avoid the possibility of polyalkylation, an FGI can be envisaged to identify the ketone **Q.3** as our new target compound. This analysis identifies Friedel–Crafts acylation of benzene as a potential route.

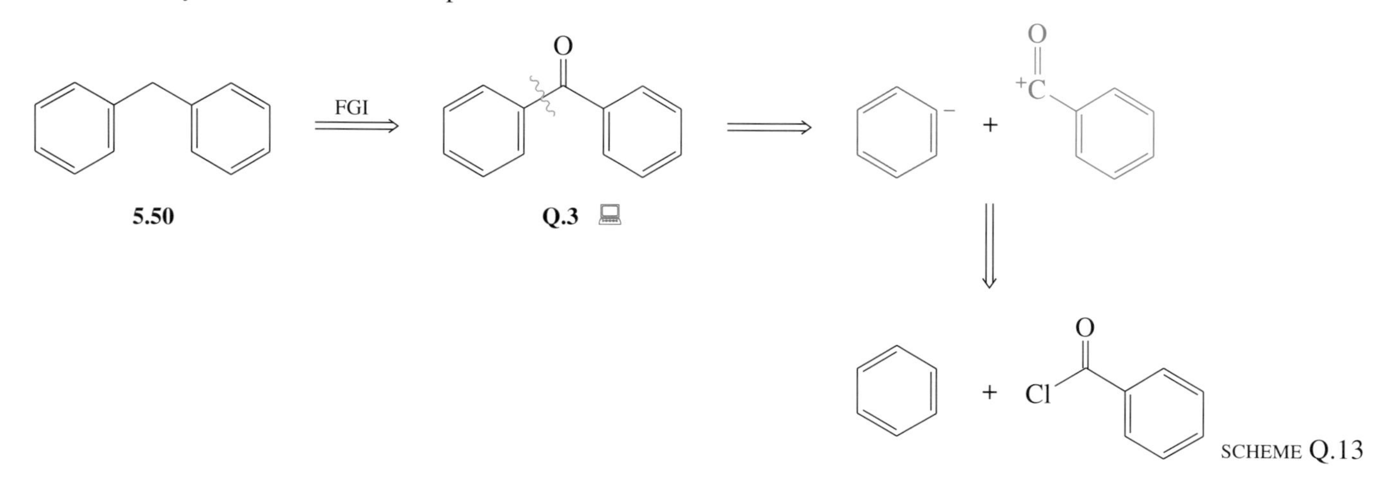


SCHEME Q.13

Compound **5.51** is best disconnected after carrying out an FGI to form a keto acid:

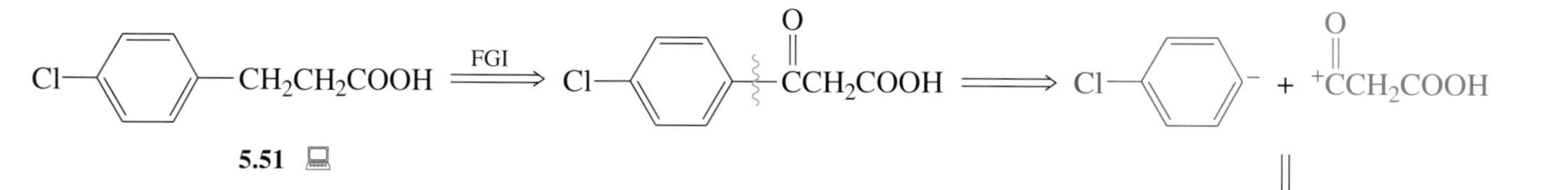


This approach identifies chlorobenzene and the acid chloride **Q.4** as potential reagents for a Friedel–Crafts acylation reaction. Actually, as indicated in Scheme Q.14, the diacid **Q.5** would probably be used instead of **Q.4**, because the latter is not readily available. Acylation will occur, to give a mixture of *ortho and para*-isomers, since chlorine is *ortho/para*-directing. However, some of the diacid **Q.5** may also react with two aromatic rings to give Ar—CO—$CH_2$—CO—Ar. Reduction of the ketone carbonyl group can be accomplished using the Clemmensen reduction.

$$\text{Cl—C}_6\text{H}_4 + \text{Cl—}\overset{O}{\overset{\|}{C}}\text{CH}_2\text{COOH} \;\; \textbf{Q.4}$$

FGI

$HOOCCH_2COOH$

**Q.5**

SCHEME Q.14

### QUESTION 5.4 (Learning Outcomes 3 and 5)

Compound **5.59** may be disconnected at a or b, but disconnection a simplifies the strategy more, so this is the disconnection of choice:

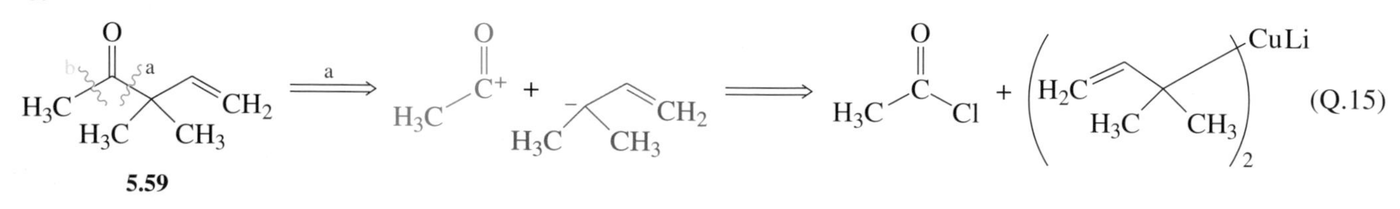

This strategy identifies a carbonyl electrophile, such as ethanoyl chloride, and an organometallic nucleophile, such as a lithium dialkylcopper compound. Alternatively, ethanenitrile, $CH_3CN$, and the appropriate alkyl Grignard reagent would be suitable reagents.

Compound **5.60** is an aldehyde, which is analysed by carrying out first an FGI to the corresponding alcohol, **Q.6**. Retrosynthetic analysis then identifies synthons that correspond to methanal and 5-methylhex-l-en-4-ylmagnesium bromide as appropriate reagents for the synthesis of **Q.6**. The Grignard reagent can be made from 4-bromo-5-methylhex-l-ene (**Q.7**). For the final synthesis step of **5.60,** oxidation of alcohol **Q.6** may be achieved using pyridine–$CrO_3$–HCl.

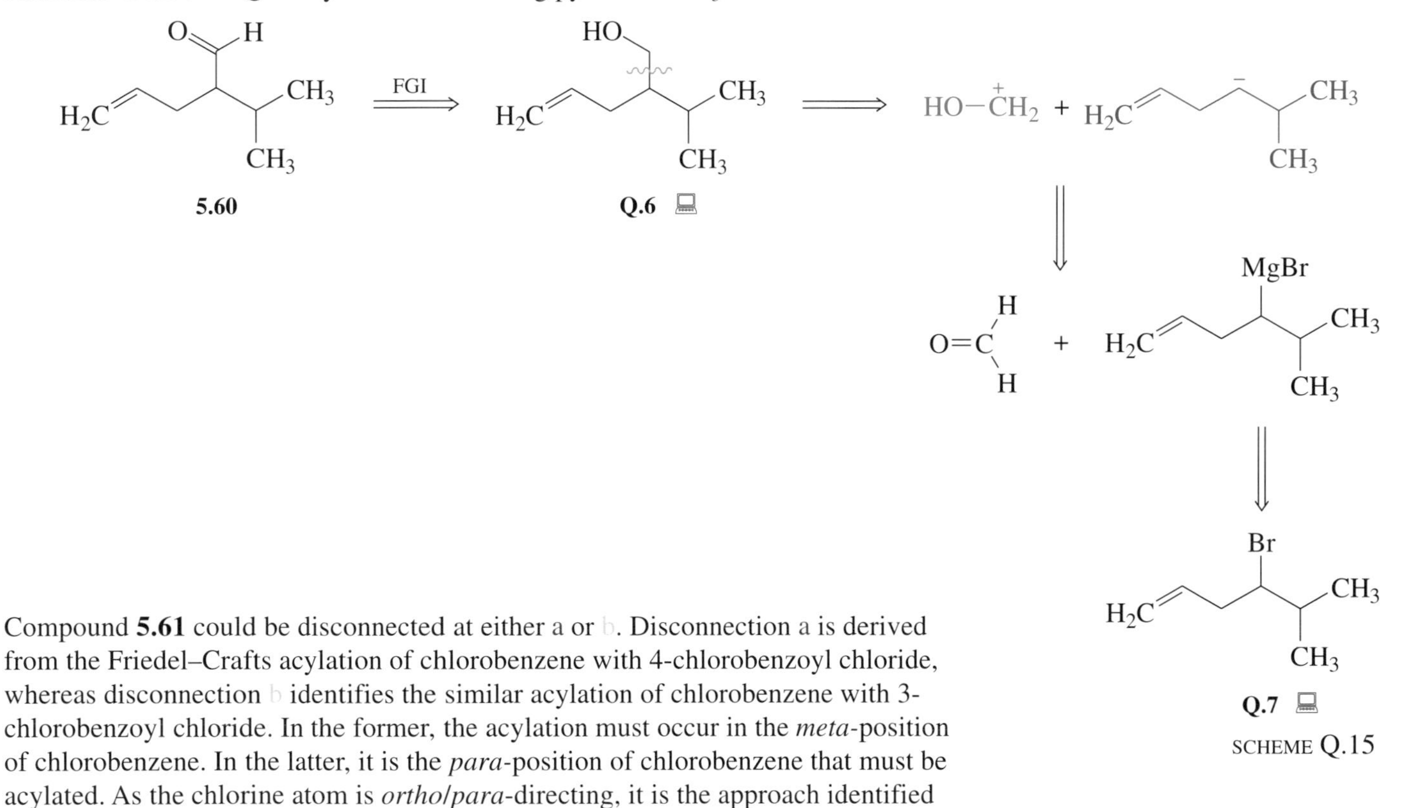


SCHEME Q.15

Compound **5.61** could be disconnected at either a or b. Disconnection a is derived from the Friedel–Crafts acylation of chlorobenzene with 4-chlorobenzoyl chloride, whereas disconnection b identifies the similar acylation of chlorobenzene with 3-chlorobenzoyl chloride. In the former, the acylation must occur in the *meta*-position of chlorobenzene. In the latter, it is the *para*-position of chlorobenzene that must be acylated. As the chlorine atom is *ortho*/*para*-directing, it is the approach identified by disconnection b that provides a suitable synthesis.

SCHEME Q.16

**QUESTION 5.5 (Learning Outcomes 3 and 5)**

Compound **5.68** can either be disconnected directly, or considered to be formed from an alcohol, in which case the COOH group must first undergo an FGI to a $CH_2OH$ group. Direct disconnection yields the synthons $^{+}COOH$ and $C_6H_5\bar{C}{=}CH_2$. The $^{+}COOH$ synthon is provided by $CO_2$, and the $C_6H_5\bar{C}{=}CH_2$ synthon is provided by the corresponding Grignard reagent, itself derived from 1-bromo-l-phenylethene.

$$\underset{\mathbf{5.68}}{(HOOC)(C_6H_5)C{=}CH_2} \Longrightarrow {}^{+}COOH + (C_6H_5)\bar{C}{=}CH_2$$

$$^{+}COOH \Longrightarrow CO_2$$

$$(C_6H_5)\bar{C}{=}CH_2 \Longrightarrow (BrMg)(C_6H_5)C{=}CH_2 \Longrightarrow (Br)(C_6H_5)C{=}CH_2$$

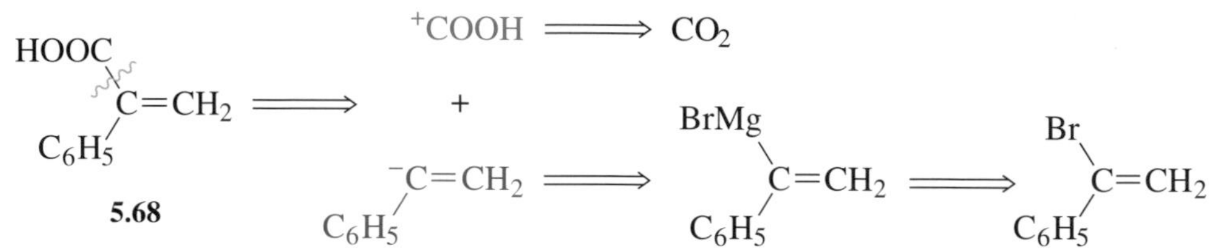

SCHEME Q.17

An initial FGI to an alcohol enables the same nucleophilic synthon (and hence reagents) as for direct disconnection to be identified, whereas the electrophilic synthon is $\overset{+}{C}H_2OH$, which is provided by the reagent methanal.

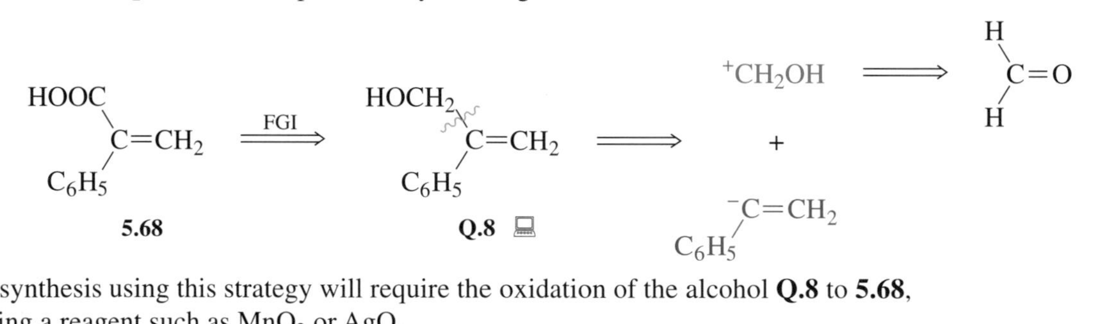


SCHEME Q.18

A synthesis using this strategy will require the oxidation of the alcohol **Q.8** to **5.68**, using a reagent such as $MnO_2$ or AgO.

Similar arguments can be applied to compound **5.69**, for which the disconnection strategies are:

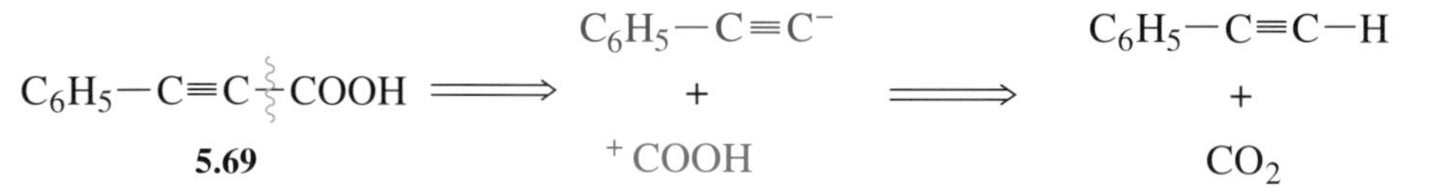

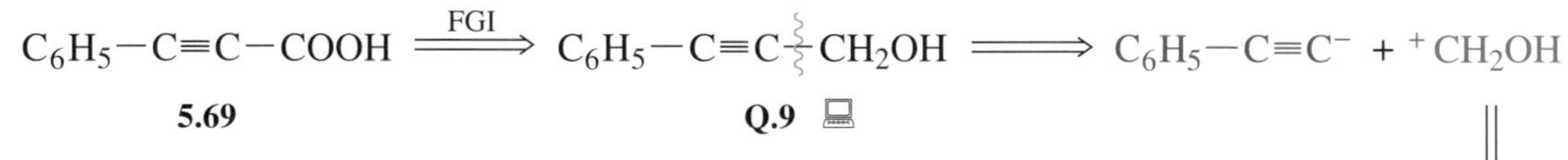

$$\underset{\mathbf{5.69}}{C_6H_5{-}C{\equiv}C{-}COOH} \Longrightarrow C_6H_5{-}C{\equiv}C^{-} + {}^{+}COOH \Longrightarrow C_6H_5{-}C{\equiv}C{-}H + CO_2$$

$$\underset{\mathbf{5.69}}{C_6H_5{-}C{\equiv}C{-}COOH} \xrightarrow{\text{FGI}} \underset{\mathbf{Q.9}}{C_6H_5{-}C{\equiv}C{-}CH_2OH} \Longrightarrow C_6H_5{-}C{\equiv}C^{-} + {}^{+}CH_2OH$$

$$^{+}CH_2OH \Longrightarrow H_2C{=}O$$

The forward synthesis requires the formation of a sodium alkynide from an alkyne using a strong base followed by its reaction with (a) $CO_2$ to form the acid directly, or (b) methanal to form the intermediate alcohol **Q.9**, which must then be oxidized to the target carboxylic acid **5.69**.

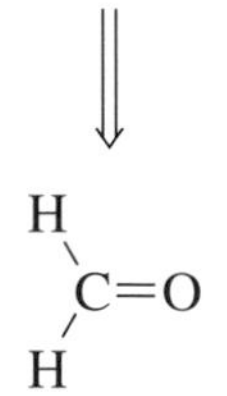

SCHEME Q.20

### QUESTION 7.1 (*Learning Outcome 6*)

(a) This reaction is chemoselective, because the alkyne triple bond is reduced to an alkene and no reaction occurs at the carbon–carbon double bond in the starting material. Since the new alkene group is monosubstituted, the question of stereoselectivity does not arise, because there is only one stereoisomer of a monosubstituted double bond.

(b) This is an example of a chemoselective reaction, in which the lithium dialkylcopper reagent reacts with the acid chloride in preference to the ester.

(c) This reaction exhibits both regioselectivity and stereoselectivity. The reaction is regioselective because, of the three positions in which the double bond can be formed, only one is observed:

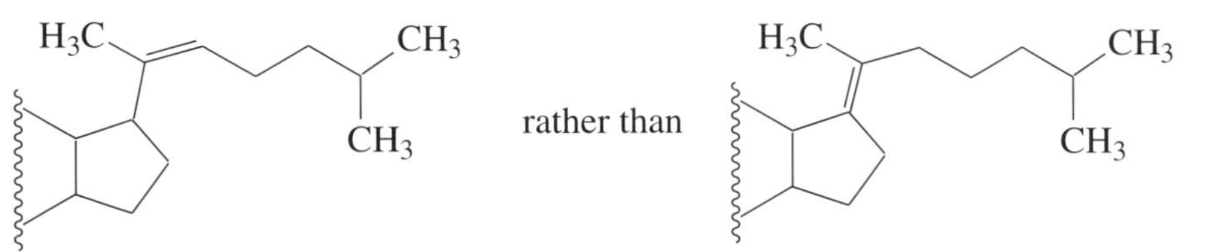


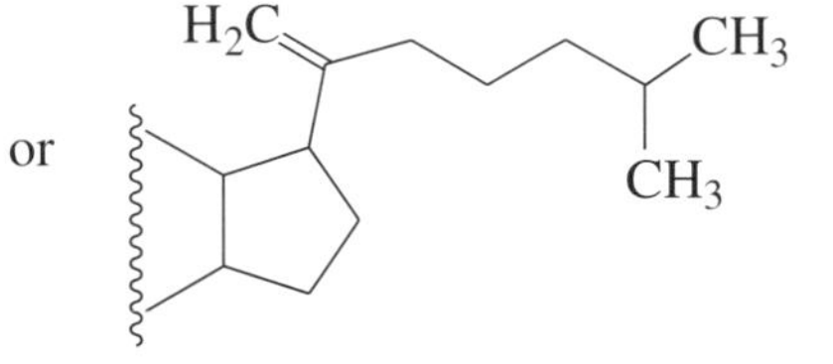


The reaction is stereoselective because the *Z*-diastereoisomer is produced in preference to the *E*-diastereoisomer.

(d) This reaction is stereoselective because a new chiral centre is introduced, with the *R*,*R*-isomer being formed preferentially.

### QUESTION 7.2 (*Learning Outcome 8*)

There are two potential problems here.

The conversion requires the reaction of just one of the two ketone carbonyl groups in the molecule. Since both will have similar reactivity, it will be difficult to carry this out directly as planned. The ketone that is not required to react will need to be protected in some way.

In fact, the way in which this problem is overcome is to protect the appropriate ketone group, by converting it into an acetal using ethane-1,2-diol:

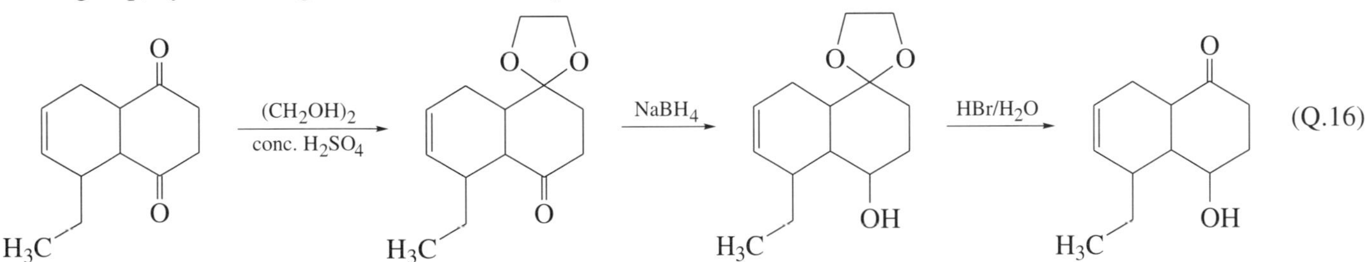


As drawn, the upper ketone group is more reactive than the lower one. This is because the ethyl substituent on the left-hand ring hinders attack at the lower ketone group, thereby forcing the acetal to be formed preferentially at the upper ketone. This is an example of introducing a protecting group regioselectively!

The second problem is reducing the ketone to the alcohol without reducing the carbon–carbon double bond. Table A.2 (p. 257) shows that $NaBH_4$ or $LiAlH_4$ are suitable.

## QUESTION 7.3 (Learning Outcome 2)

The synthesis requires that the two substituents are positioned *meta* to one another, so the strategy must address the regioselectivity of any proposed reactions. The problem resides in the known directing effects of the chlorine and propyl substituents, both of which are *ortho*/*para*-directing. The following strategies will therefore fail.

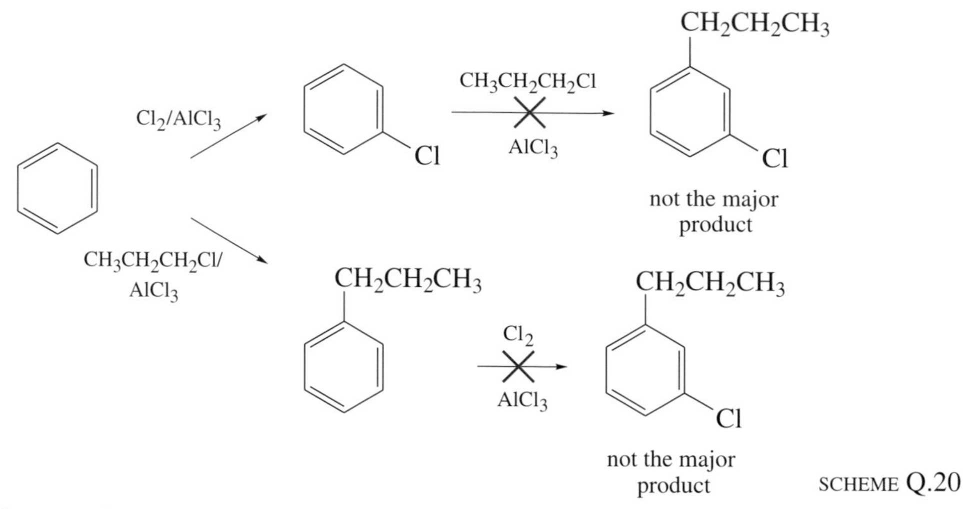


SCHEME Q.20

(Of course, introducing an alkyl group, using RCl and $AlCl_3$, is always fraught with problems because of the possibility of rearrangement of the alkyl group, as well as polyalkylation*.)

What we must do first is to introduce a *meta*-directing group that will direct the second substituent into the correct position, and which can then be transformed to the substituent that is required. Such a substituent is an acyl group, RCO—, and the following strategy will produce *meta*-chloropropylbenzene:

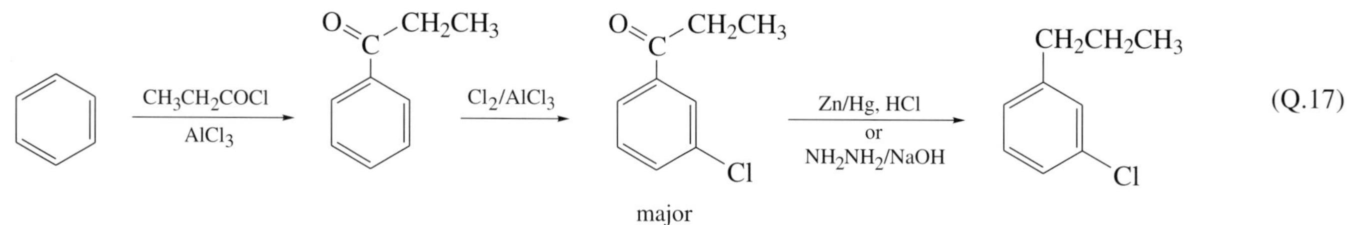


(Q.17)

This acylation strategy also circumvents the problems of rearrangement and polyalkylation.

## QUESTION 8.1 (Learning Outcome 9)

(a) The yield of this linear synthesis is:

$$\frac{97}{100} \times \frac{77}{100} \times \frac{80}{100} \times \frac{85}{100} \times \frac{85}{100} \times 100 = 43\%$$

(b) The yield of this convergent synthesis is 7.5%.

The yield of **H** is 81% and the yield of **L** is:

$$\frac{42}{100} \times \frac{86}{100} \times \frac{76}{100} = 27.5\%$$

Since this is lower than 81%, this is the value that is used for the remaining calculation.

$$27.5 \times \frac{61}{100} \times \frac{84}{100} \times \frac{53}{100} \times 100 = 7.5\%$$

* These issues are discussed in Part 2 of *Alkenes and Aromatics*[1].

# ANSWERS TO EXERCISES

## EXERCISE 2.1

Compare the Table you developed with Table A.1 in the Appendix. Don't worry if you missed some of the FGIs: this is not a memory test, but an exercise to help you understand how the Table works.

You may find, too, that some FGIs require more than one step. For example, to convert an alcohol into a cyano group requires the alcohol to be converted into a good leaving group using *para*-toluenesulfonyl chloride (TsCl), followed by $S_N2$ substitution with cyanide ion.

$$\text{R—CH}_2\text{—OH} \xrightarrow{\text{Cl—S(=O)}_2\text{—C}_6\text{H}_4\text{—CH}_3} \text{R—CH}_2\text{—O—S(=O)}_2\text{—C}_6\text{H}_4\text{—CH}_3 \xrightarrow[\text{–TsOH}]{\text{CN}^-} \text{R—CH}_2\text{—CN} \quad \text{(E.1)}$$

When you use the Appendix, you will notice that many transformations are shown as dashes. Most of these are FGIs that cannot be brought about easily, or are of little synthetic value. Some, however, are quite important, but are beyond the scope of this Book.

The incomplete Table 2.1 you were given in this Exercise also contains two entries indicated by letters. These entries refer to reactions that you may not have met yet, but which are important FGIs for synthesis. Entry B, conversion of a carboxylic acid into the more reactive acid chloride, is a reaction that is not discussed explicitly elsewhere. However, it is related to the conversion of an alcohol into a haloalkane (see entry 1), and is brought about by using thionyl chloride ($SOCl_2$) or phosphorus pentachloride ($PCl_5$):

$$\text{R—C(=O)—OH} \xrightarrow[\text{or PCl}_5]{\text{SOCl}_2} \text{R—C(=O)—Cl} \quad \text{(E.2)}$$

Entry B is the hydrolysis of a nitrile to a carboxylic acid. Again, you have not met this reaction, but all you need to know is that it is similar to amide and ester hydrolysis.

$$\text{R—C}\equiv\text{N} \xrightarrow{\text{H}_2\text{O/H}^+ \text{ or HO}^-} \text{R C(=O)—OH} \quad \text{(E.3)}$$

Even though you have probably not met these reactions before, and they will not be dealt with in detail elsewhere, we have included these reactions in the Table because they are very useful synthetic transformations.

The conversion of an alcohol into a ketone or an aldehyde is dealt with in more detail in Section 5, and the related entry, the conversion of an aldehyde into a carboxylic acid, is discussed in Section 5.

## EXERCISE 9.1 (Learning Outcomes 3, 4, 5, 6 and 7)

When I looked at the target molecule, I noted three features. There is one functional group, the carboxylic acid group, and there are two side-chains connected in a *para* arrangement to the benzene ring.

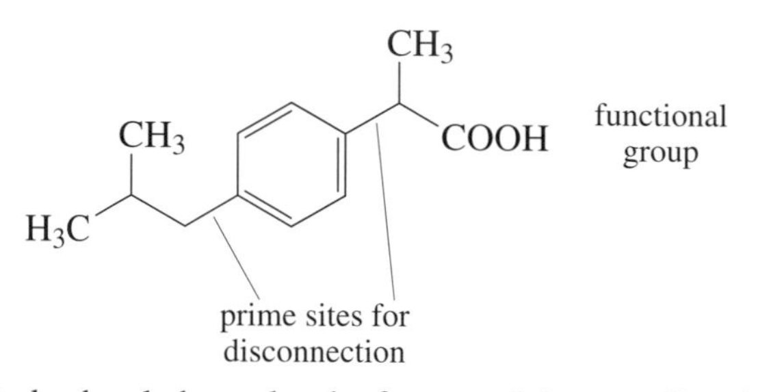


If I wanted to break the molecular framework into smaller pieces, I would focus on C—C disconnections to the aromatic ring. However, the molecule is not very big, so I might as well start with the carboxylic acid and work backwards from this seat of reactivity.

Section 5.7 provides the information I need for working backwards from a carboxylic acid. I could go back to the aldehyde or alcohol via an FGI, but that doesn't break the molecule up enough. A better strategy is to disconnect the C—C bond to the carboxylic acid to give two possible pairs of synthons.

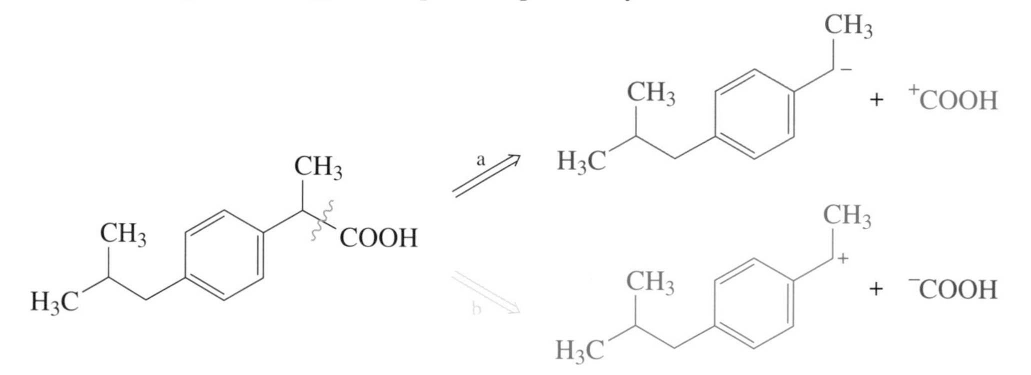


SCHEME E.1

Disconnection a leads to an $R^-$ synthon, so the most likely reagent is a Grignard reagent. Table 5.1 (p. 231) shows that the $^+COOH$ synthon corresponds to carbon dioxide. The Grignard reagent is prepared from the haloalkane, so the preparation in the synthetic direction is

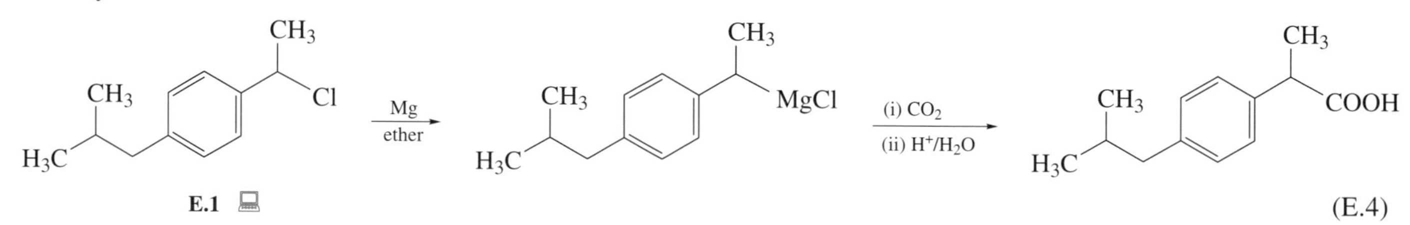


(E.4)

Disconnection b gives the $R^+$ synthon, which corresponds to a haloalkane, and the $^-COOH$ synthon. Table 5.1 shows that this synthon corresponds (after an FGI) to the cyanide ion. So, in the synthetic direction a possible preparation is

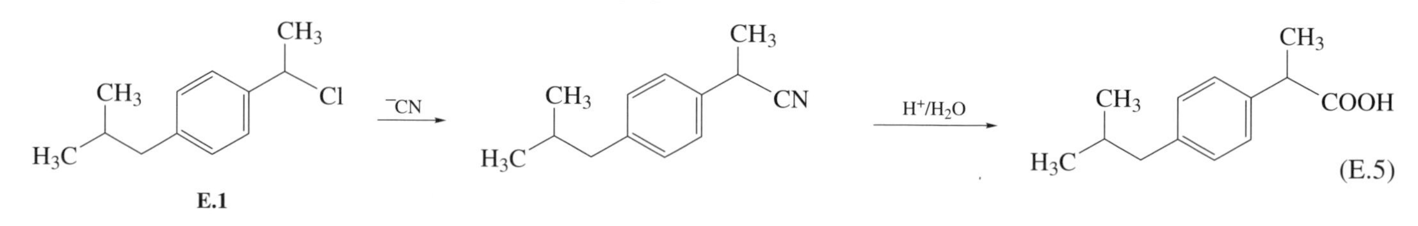


(E.5)

An $S_N2$ reaction of the nucleophilic cyanide ion with the haloalkane gives the nitrile, which can be hydrolysed to the carboxylic acid. I would favour the first approach, using a Grignard reagent and carbon dioxide, because of the health and safety concerns of using cyanide! Interestingly, both modes of disconnection lead back to the same chloroalkane, **E.1**, so this becomes my new objective. Our strategy developed in Section 6 suggests that it is always best to revert to alcohols and carbonyl compounds whenever possible, so an FGI leads us back to the alcohol **E.2**:

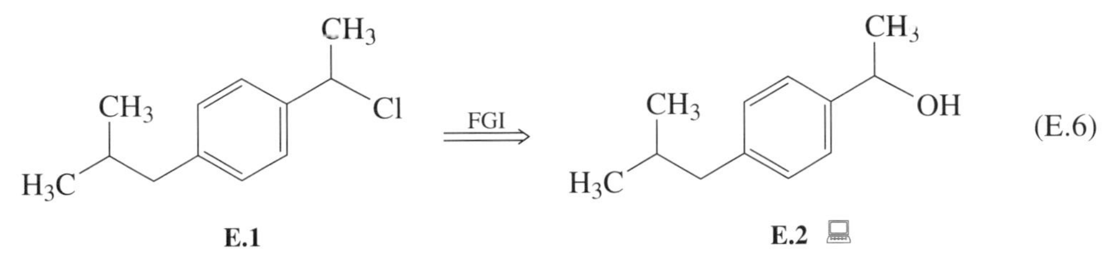


(E.6)

In the synthesis direction, the Appendix shows that this conversion can be carried out by reacting the alcohol with thionyl chloride, $SOCl_2$:

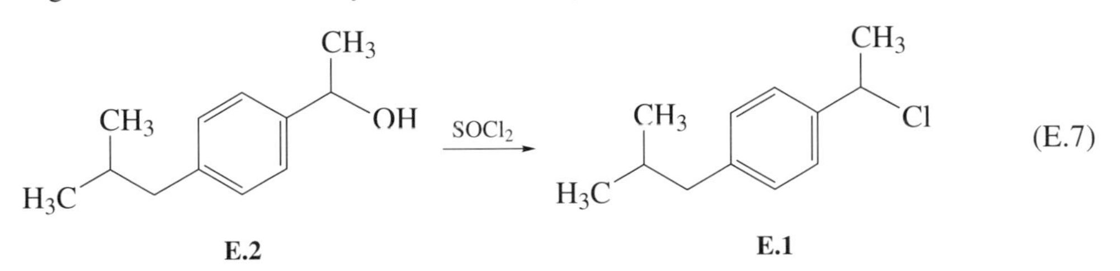


(E.7)

Section 5.1 discussed the retrosynthetic analysis of alcohols — via the reaction of Grignard reagents with carbonyl compounds. For this secondary alcohol, I can make two disconnections.

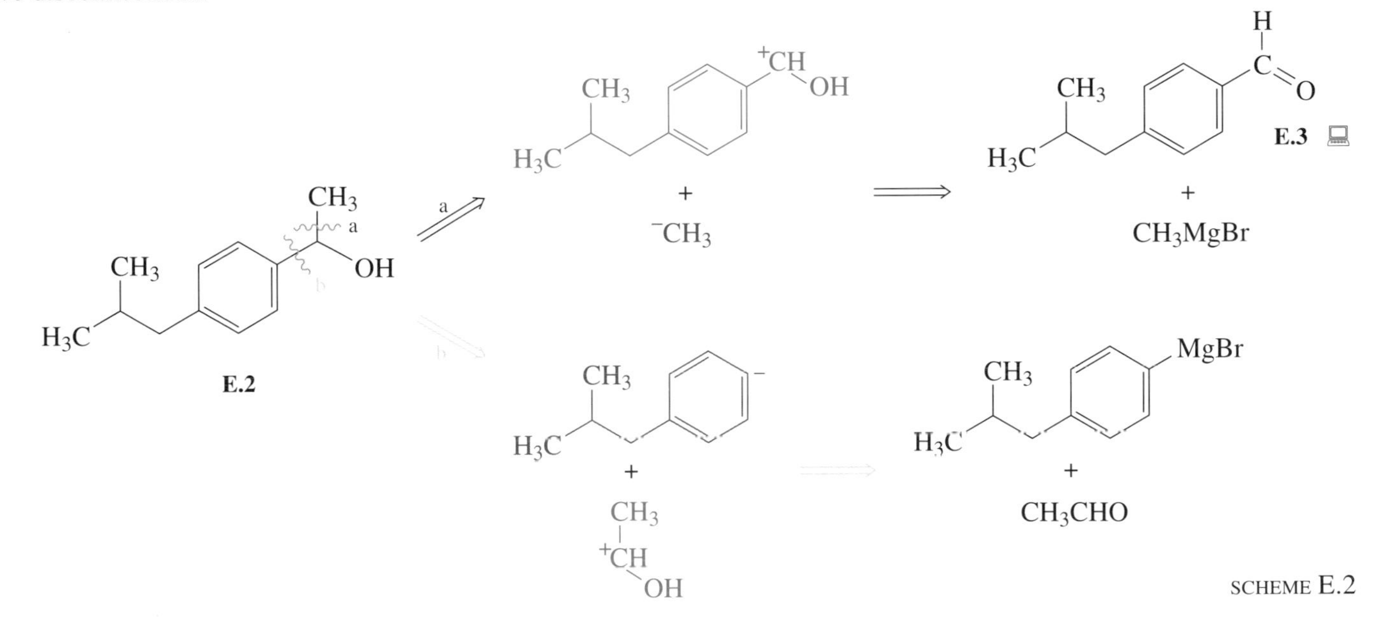


SCHEME E.2

Disconnection a leads to a methyl Grignard reagent and the substituted benzaldehyde **E.3**, whereas disconnection b leads to an aromatic Grignard reagent and ethanal. I prefer this latter disconnection because it breaks the molecule into larger fragments. Since Grignard reagents are generated from halo compounds, the reactions in the synthesis direction are as follows:

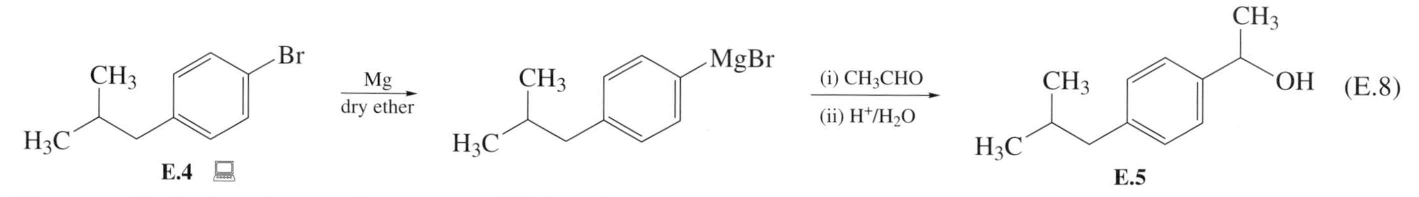


The *para*-disubstituted aromatic compound **E.4** can be made by disconnecting the bonds to the aromatic ring to give an $Ar^-$ synthon.

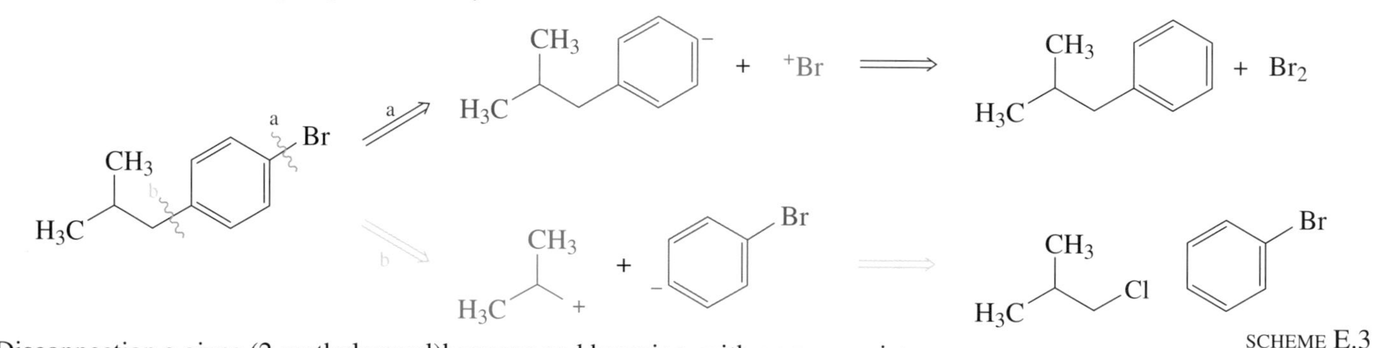


Disconnection a gives (2-methylpropyl)benzene and bromine, with an appropriate catalyst such as $FeBr_3$. Route b gives 1-chloro-2-methylpropane and bromobenzene. Both of these steps involve an electrophilic aromatic substitution in the forward direction.

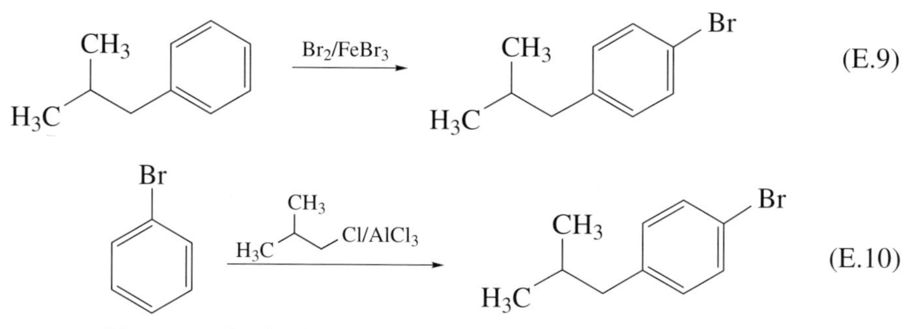


Whenever we add a second substitutent to a benzene ring using an electrophilic aromatic substitution, we must always check the expected regiochemistry based on the directing effect of the first substituent. Both bromine and alkyl groups are *ortho*/*para*-directing, and as a result of steric effects we might expect our desired *para*-products to predominate. However, the Friedel–Crafts alkylation can be complicated by polyalkylation and rearrangement of the haloalkane. The carbocation formed when aluminium trichloride reacts with 1-chloro-2-methylpropane will be a primary carbocation, and this can easily isomerize to a more stable tertiary carbocation:

$$H_3C-CH(CH_3)-\overset{+}{C}H_2 \rightleftharpoons H_3C-\overset{+}{C}(CH_3)-CH_3 \qquad \text{(E.11)}$$

So I shall discount this route. An alternative route is Friedel–Crafts acylation, followed by Clemmensen reduction. Since this adds further steps, my favoured route is the bromination of (2-methylpropyl)benzene, giving the overall sequence shown in Scheme E.4.

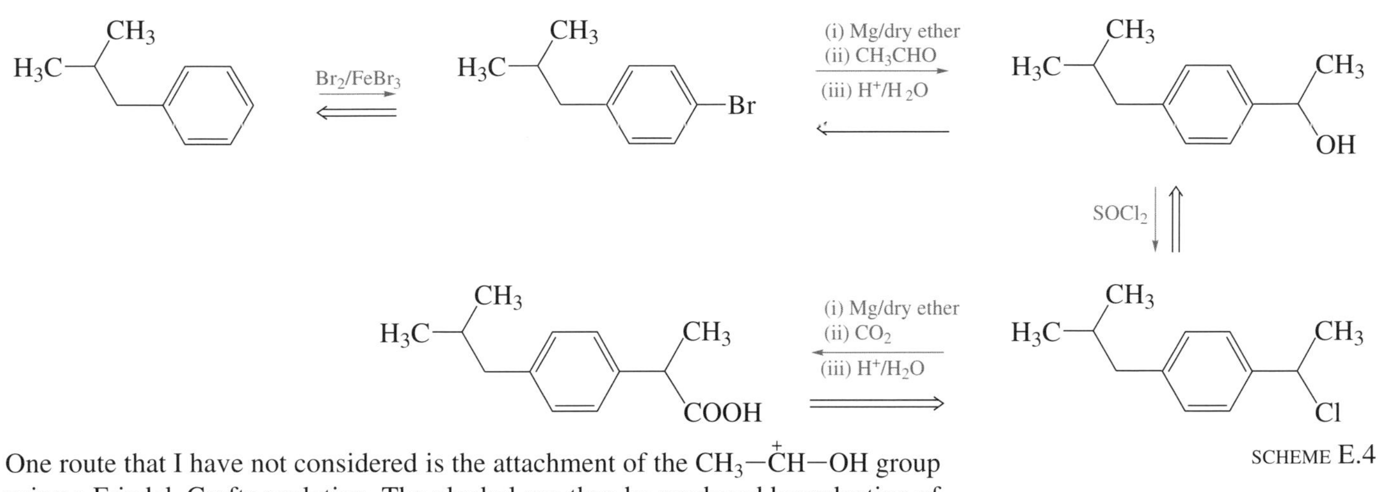


SCHEME E.4

One route that I have not considered is the attachment of the $CH_3{-}\overset{+}{C}H{-}OH$ group using a Friedel–Crafts acylation. The alcohol can then be produced by reduction of the corresponding ketone:

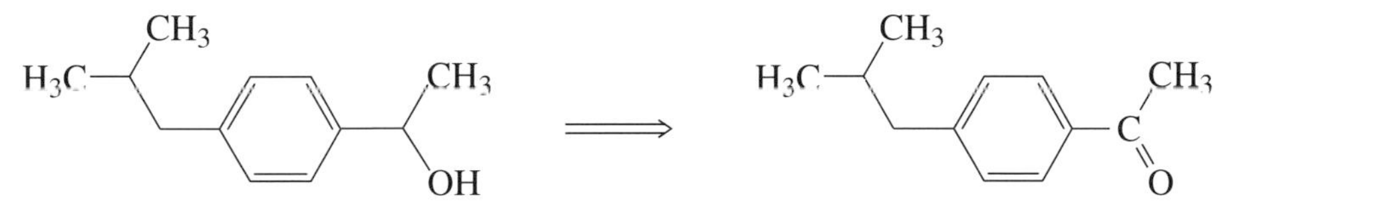

(E.12)

If I disconnect the C—C bond to the aromatic ring, I get synthons that correspond to (2-methylpropyl)benzene and ethanoyl chloride:

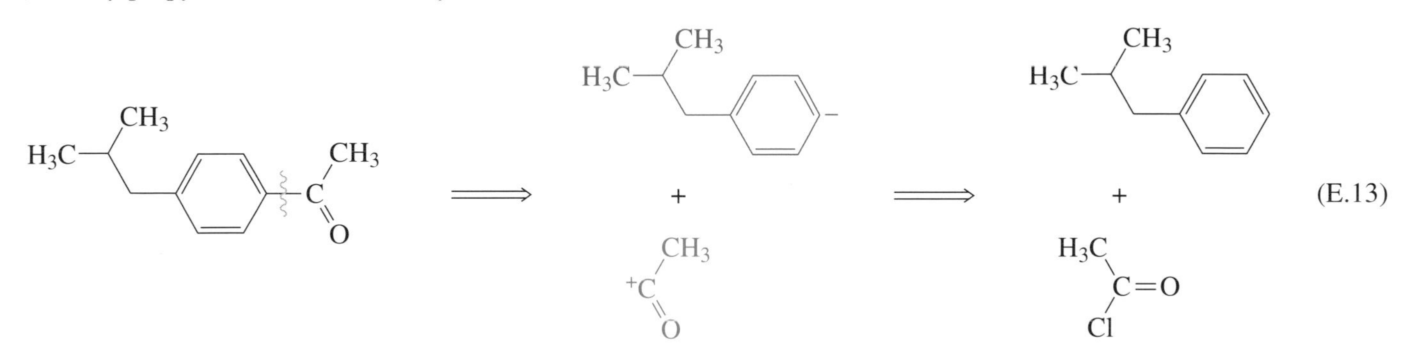

(E.13)

This Friedel–Crafts acylation needs a Lewis acid catalyst, and will give mainly the *para* isomer because the alkyl group is *ortho*/*para* directing. The alkyl group is also bulky, so the *para*-product is favoured:

(E.14)

This gives the overall synthesis shown in Scheme E.5.

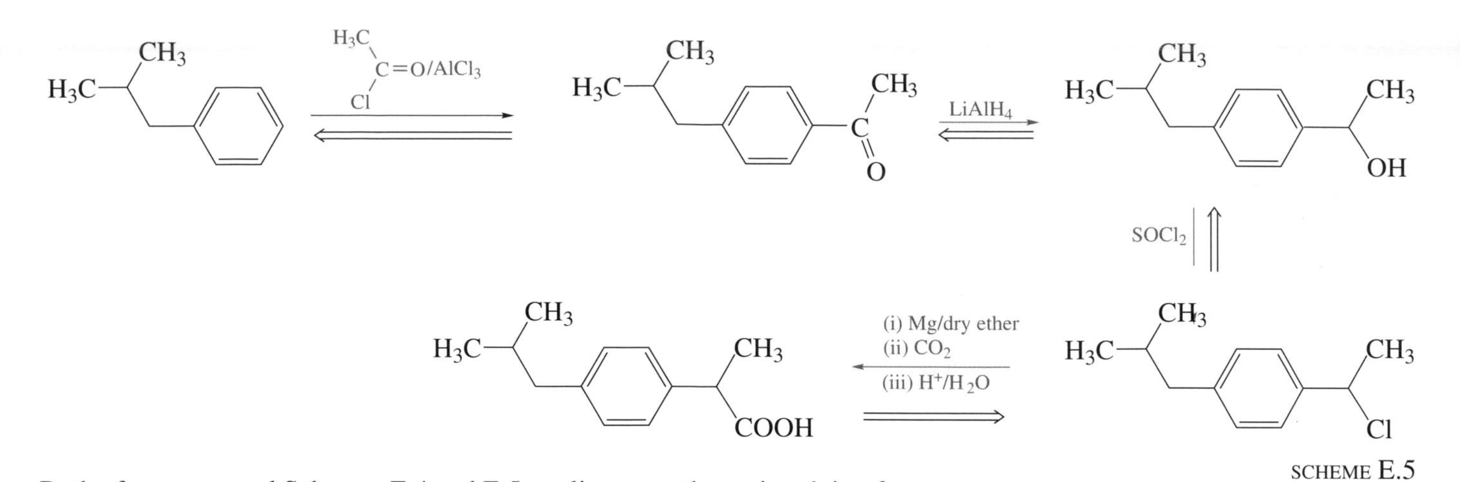


SCHEME E.5

Both of my proposed Schemes E.4 and E.5 are linear syntheses involving four steps. You should bear in mind that this Exercise has been devised to reinforce your understanding of the principles of retrosynthetic analysis. It does not mean that we have necessarily devised the best route to ibuprofen*.

**STUDY NOTE**

Open University students should now refer to *Video Notes for Book 10* and the associated video programme, which show how a retrosynthetic analysis is turned into reality in the laboratory, and then contrast this with a commercial synthesis.

* The commercial synthesis of ibuprofen is discussed in Part 3 of *Alkenes and Aromatics*[1].

# FURTHER READING

1 P. G. Taylor and J. M. F. Gagan (eds), *Alkenes and Aromatics*, The Open University and the Royal Society of Chemistry (2002).

2 M. Mortimer and P. G. Taylor (eds), *Chemical Kinetics and Mechanism*, The Open University and the Royal Society of Chemistry (2002).

3 L. E. Smart and J. M. F. Gagan (eds), *The Third Dimension*, The Open University and the Royal Society of Chemistry (2002).

4 L. E. Smart (ed.), *Separation, Purification and Identification*, The Open University and the Royal Society of Chemistry (2002).

# ACKNOWLEDGEMENTS

Grateful acknowledgement is made to the following sources for permission to reproduce material in this book:

*Figures 4.1, 5.2, 5.4 and 5.5*: © The Nobel Foundation; *Figure 5.1*: Pascal Nieto, Jerrican/Science Photo Library; *Figure 5.3*: Demi Brown/Oxford Scientific Films.

Every effort has been made to trace all the copyright owners, but if any has been inadvertently overlooked, the publishers will be pleased to make the necessary arrangements at the first opportunity.

*Part 5*

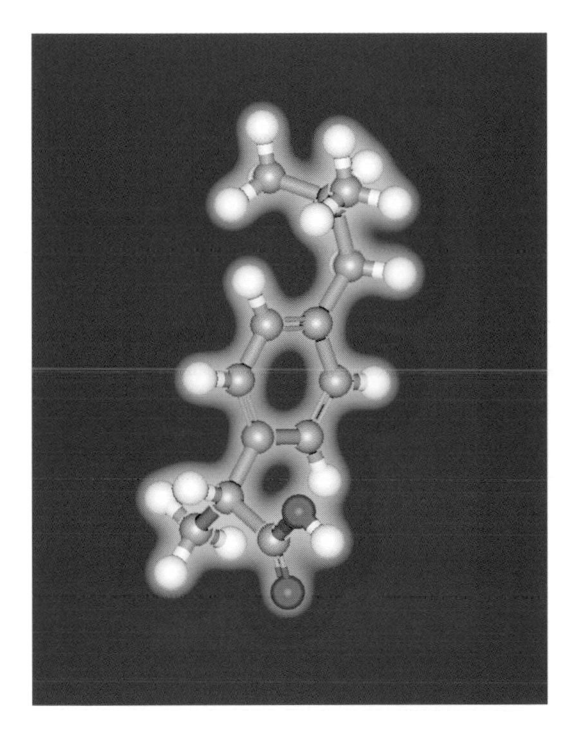

# Synthesis and biosynthesis: terpenes and steroids

edited by

Chris Falshaw and Peter Taylor

*based on*

Synthesis and biosynthesis: terpenes

*by Jim Iley and Richard Taylor*

1

# INTRODUCTION

## 1.1 Natural products

The isolation and identification of compounds from plant and animal sources is an activity common to both organic chemistry and biochemistry. Indeed, it was discoveries in these fields of research in the nineteenth century that led to the development of biochemistry as a separate subject. Today, it takes far less time to isolate and determine the structures of newly discovered compounds than it did in the nineteenth century, even though the structures are often more complex. One of the reasons for this is that there are now several very powerful chromatographic techniques available for the separation of mixtures and the purification of individual compounds*. Such techniques were not available in the nineteenth and early twentieth centuries, when chemists had to rely on extraction, distillation and recrystallization as techniques for purification. As for structure determination, there is now a wealth of spectroscopic methods available, but it was not so long ago that chemical analysis and functional group tests had to be combined with lengthy chemical degradations of the molecule under investigation, in order to obtain sufficient information to hazard a guess at the structure. Then the compound with the proposed structure had to be unambiguously synthesised and compared with the natural product in order to confirm the structural assignment. Not surprisingly, the structures elucidated in those days were fairly straightforward by present-day standards.

Organic chemists refer to all organic compounds isolated from living organisms as **natural products**, but this term is rarely, if ever, used in biochemistry. Biochemists label a naturally occurring compound according to the role it plays in the life cycle of the living organism. Many of the compounds isolated from natural sources are common to all living organisms, from bacteria and plants to humans. These compounds have been extensively studied by biochemists, who assumed, correctly, that their wide distribution is due to their involvement in the fundamental biochemical reactions common to all life. Biochemists refer to these fundamental reactions as **primary metabolic processes**, and so the compounds involved in these processes became known as **primary metabolites**.

Examples of primary metabolites are α-amino acids, nucleic acids and fatty acids. The molecular variety of these compounds can be gathered from the representative structures shown in Table 1.1.

* These techniques are discussed in *Separation, Purification and Identification*[1].

**Table 1.1** Representative primary metabolites

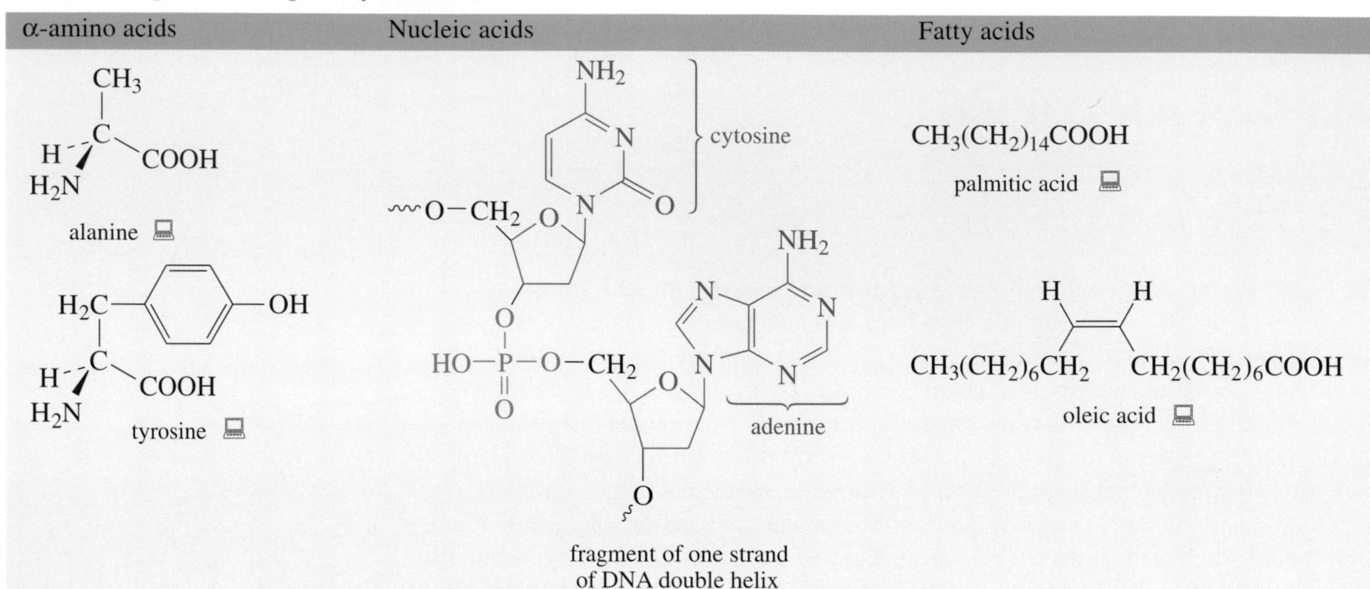


While biochemists concentrated on the role of primary metabolites in the life process, organic chemists continued to isolate new compounds from natural sources. Such compounds are usually peculiar to only one plant or species, and biochemists refer to them as **secondary metabolites**. Their biological function is not obvious, and they hold little interest for biochemists. However, these compounds do interest organic chemists, and research into the structure and properties of this type of natural product has been particularly fruitful. The range of natural products seems almost unlimited, and they come in all shapes and sizes. They can be classified in several ways: according to their origins (for example, mould metabolites), or their properties (for example, antibiotics and vitamins), or because of structural similarities (for example, terpenes and alkaloids). Several classes of natural products, along with some representative examples, are shown in Table 1.2.

Although the isolation of primary and secondary metabolites is still an important area of research, there is an increasing interest in their **biosynthesis** — that is, the way in which these compounds are synthesised by the chemical reactions that occur within the organism. However, you must remember that there are far greater physical constraints on reaction conditions for a transformation taking place in a living organism, such as the human body, than there are on a reaction performed in a laboratory.

- What might these greater physical constraints be?

- All reactions must be carried out in aqueous solution at 37 °C (in humans), at about atmospheric pressure, and at a pH of about 7.

Not only are biochemical reactions carried out under such mild conditions, but they also proceed very quickly and in high yield. In addition, many are highly selective: the required reaction is the only one to occur, even if there are several possible reaction sites, and often only one stereoisomer will be produced, even if more are possible. All together, this sounds like an organic chemist's dream.

**Table 1.2** Representative natural products (secondary metabolites)

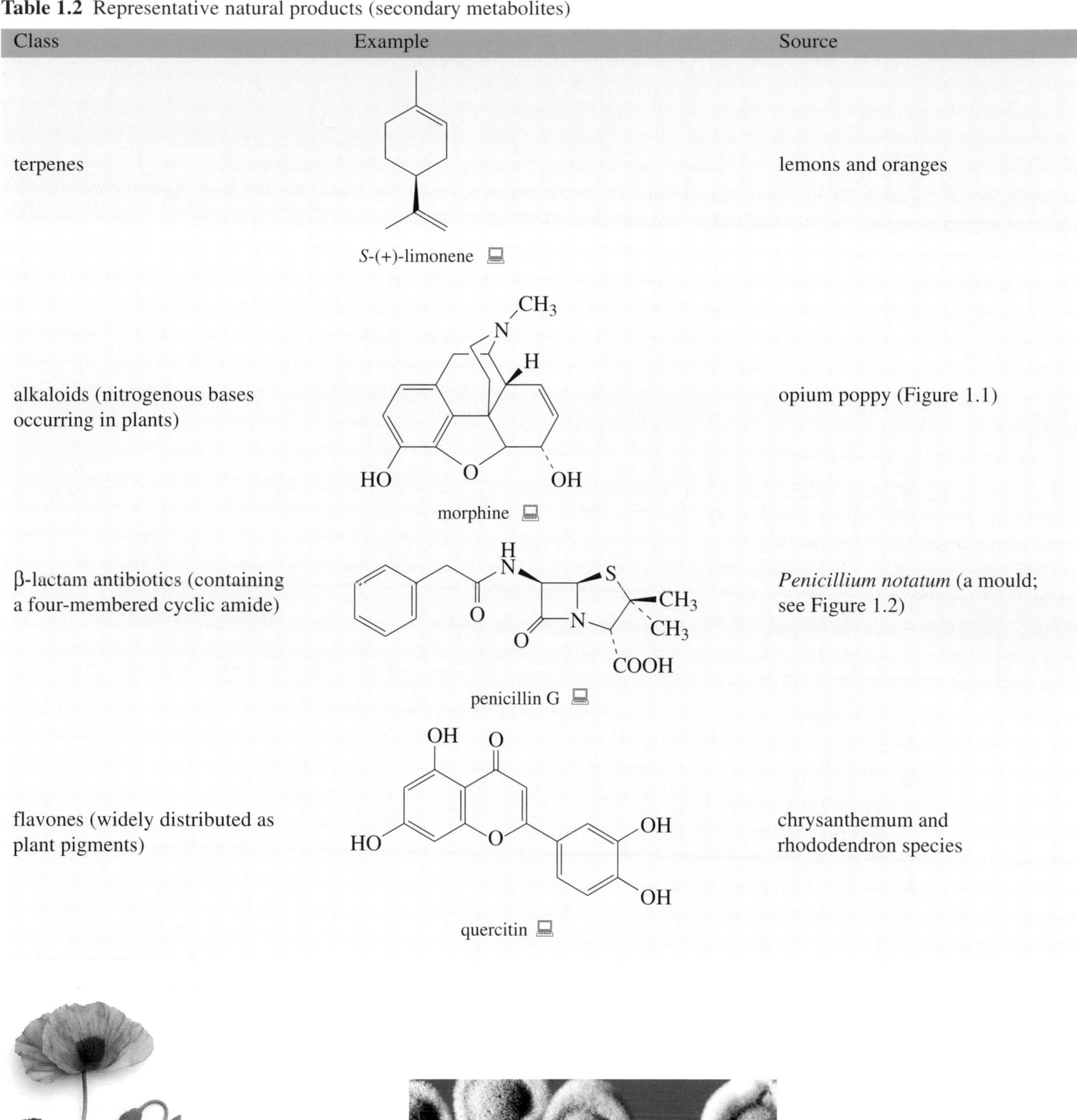

| Class | Example | Source |
|---|---|---|
| terpenes | *S*-(+)-limonene | lemons and oranges |
| alkaloids (nitrogenous bases occurring in plants) | morphine | opium poppy (Figure 1.1) |
| β-lactam antibiotics (containing a four-membered cyclic amide) | penicillin G | *Penicillium notatum* (a mould; see Figure 1.2) |
| flavones (widely distributed as plant pigments) | quercitin | chrysanthemum and rhododendron species |

**Figure 1.1**
The opium poppy.

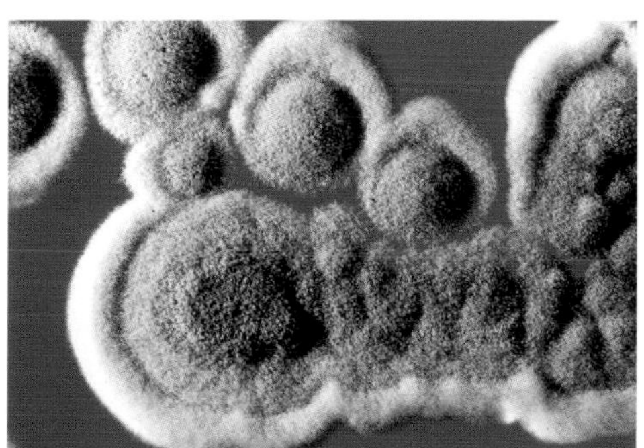

**Figure 1.2**
Penicillin mould.

The reason why biochemical reactions are so successful is that for each reaction there is a specific catalyst, called an *enzyme*, which is a large organic molecule, custom-built by nature to carry out the required chemical transformation. A catalyst is needed so that the reaction takes place at an appreciable rate within the organism. In addition to this, the enzyme also achieves selectivity by holding the reactants in the correct orientation so that the correct isomer of the product is obtained *. An enzyme is like any other catalyst in that it cannot make a thermodynamically unfavourable reaction go, but it can speed up a thermodynamically favourable reaction by providing an alternative pathway that has a lower activation energy.

In Part 5 of this Book, we have chosen to study the synthesis and biosynthesis of a group of natural products called terpenes. Terpenes have relatively simple structures, and they provide us with the possibility of comparing how we might synthesise such compounds in the laboratory with the ways in which they are made within living organisms. Terpenes are secondary metabolites, and their biosynthesis is fairly well understood. As it involves one of the main **biosynthetic pathways** by which many secondary metabolites are formed, terpene biosynthesis is worthy of further study. We shall show that the chemical reactions taking place in the human body or in a plant follow exactly the same principles as reactions carried out in the laboratory. The biosynthetic pathways in nature involve precisely the same kind of electron shifts that we have described in the mechanisms discussed in the earlier Parts of this Book, so the curly arrows that we use to depict them follow the usual rules.

## 1.2 Terpenes and the isoprene rule

The chemicals responsible for the fragrances of plants and flowers have always been of interest, not least to perfume manufacturers. Solvent extraction or distillation of fragrant plants with water (*steam distillation*) gives rise to oils that retain the characteristic scents of the plants. These oils were said to contain the 'essence' of the plants, and so they became known as 'essential oils'. Work in the late nineteenth and early twentieth centuries established the structures of many of the compounds present in these essential oils, and the majority of them appeared to be formed via similar biosynthetic pathways. Such compounds were named **terpenes**, the term being derived from turpentine (Figure 1.3), the oily liquid obtained by distillation of the resin exuded by certain coniferous trees. Table 1.3 gives the structure of some representative terpenes. You will gather that not all members of the terpene family are renowned for their scent, and, indeed, that not all terpenes originate from plant sources. From the examples given, you can see that terpenes range from complex alkenes, through cyclic compounds, to natural rubber, a polymer. In addition, some terpenes are hydrocarbons, whereas others contain hydroxyl or other functional groups. In fact, at first sight the compounds in Table 1.3 may seem to have little in common, and the claim that they all appear to result from similar biosynthetic pathways may seem far-fetched.

- Count the number of carbon atoms in each compound in Table 1.3. Is there a trend?
- All the compounds contain a multiple of five carbon atoms (10, 10, 10, 10, 15, 20, 30 and $5n$, respectively).

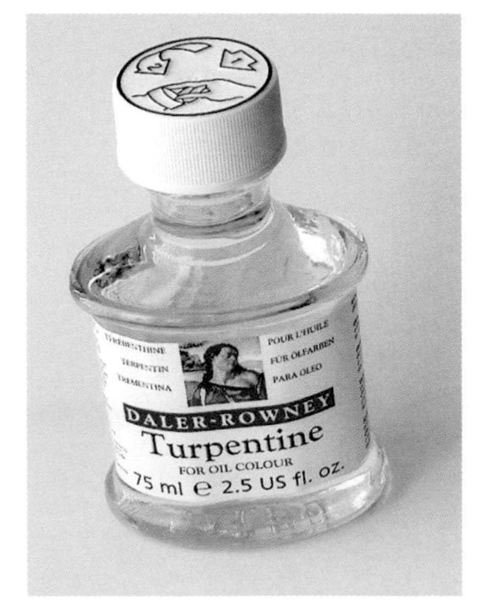


**Figure 1.3** Turpentine.

* How these interactions can be visualized is discussed in the Case Study of *Molecular Modelling and Bonding*[2].

**Table 1.3** Representative terpenes and their sources

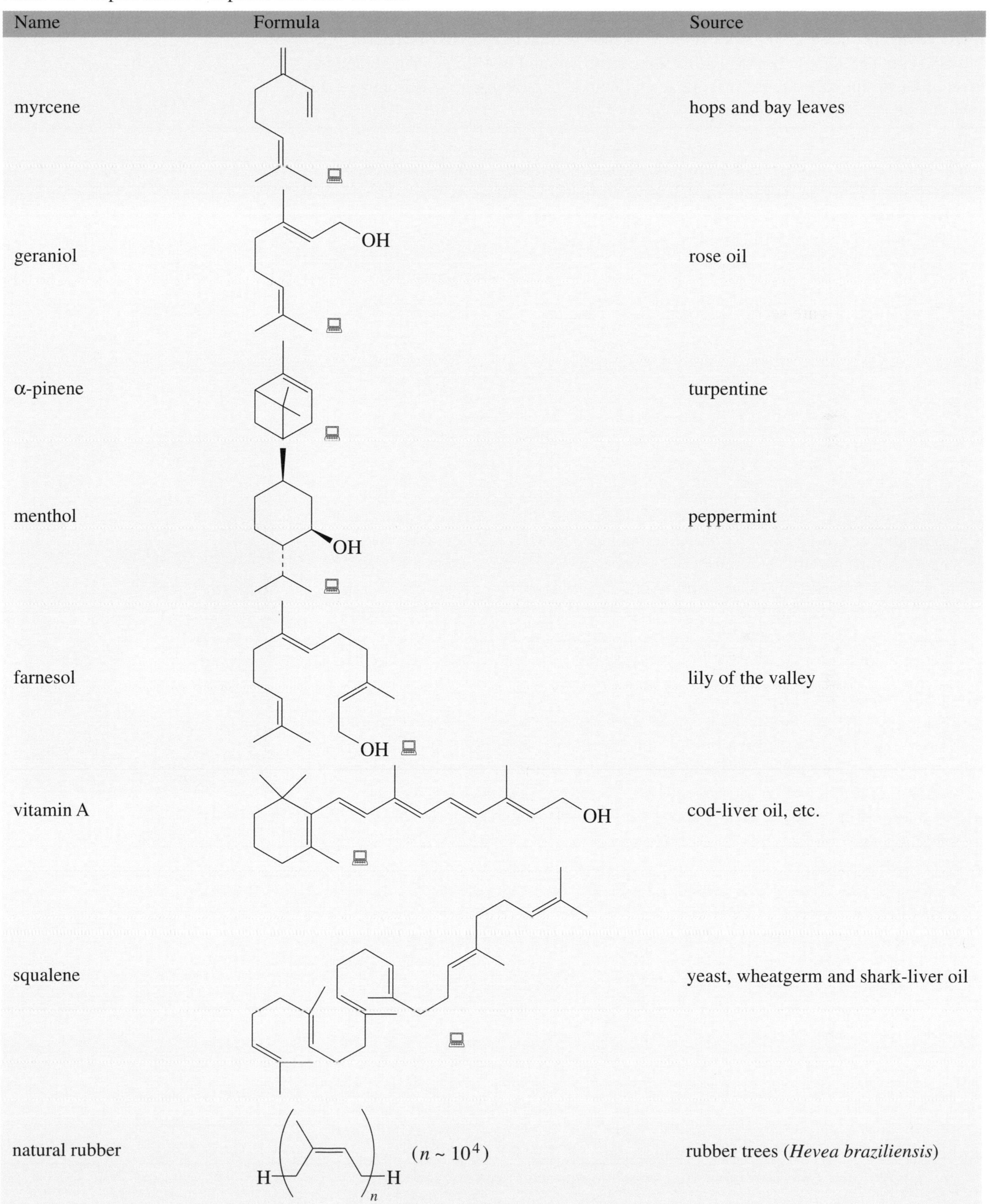

| Name | Formula | Source |
|---|---|---|
| myrcene | | hops and bay leaves |
| geraniol | | rose oil |
| α-pinene | | turpentine |
| menthol | | peppermint |
| farnesol | | lily of the valley |
| vitamin A | | cod-liver oil, etc. |
| squalene | | yeast, wheatgerm and shark-liver oil |
| natural rubber | $(n \sim 10^4)$ | rubber trees (*Hevea braziliensis*) |

Towards the end of the nineteenth century, the German chemist Otto Wallach (Box 1.1) analysed a great number of components of essential oils, and he found that many of them contained $5n$ carbon atoms, where $n$ was an integer, most frequently 2, but also sometimes 3, 4, 6 or 8. Wallach went further, and in 1887 he announced his famous **isoprene rule**. Isoprene is the trivial name for 2-methylbuta-1,3-diene, **1.1**, which contains five carbon atoms.

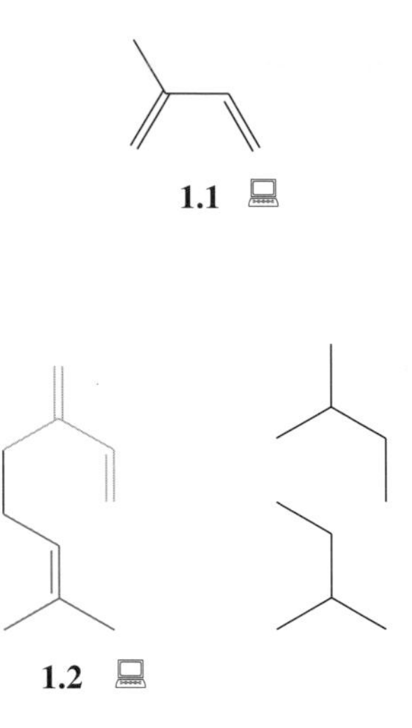


Wallach said that all terpenes could be considered as being built up from isoprene units. For example, myrcene, **1.2**, contains ten carbon atoms, and two five-carbon units (shown in blue and orange) with isoprene-like structures can easily be identified in its structure.

It is important to remember that the isoprene rule is merely a rule of thumb with which you can identify members of the terpene family by inspection of their structural formulae. The identification of two molecular fragments resembling isoprene is sufficient to categorize myrcene as a terpene.

## Box 1.1 Otto Wallach

Otto Wallach (Figure 1.4, born 27 March 1847 in Königsberg, East Prussia) was the son of a high-ranking civil servant. He went to school in Potsdam, and in 1867 began a course of chemistry with F. Wöhler at Göttingen, but soon after moved on to study under C. H. Wichelhaus at Berlin. He then returned to Göttingen, and worked so hard that he managed to get a PhD in under two years. Unfortunately, the laboratories were only open from 7am until 5pm, after which the gas was turned off, and he had to work by candlelight. While there, he examined the various *ortho*-, *meta*- and *para*-isomers of substituted toluenes. In 1869–70, he worked on the nitration of β-naphthol in Berlin, and then moved to Bonn to work with F. A. Kekulé. However, in late 1870 he had to leave for military service in the Franco-Prussian war.

**Figure 1.4**
Otto Wallach, 1847–1931.

After that war, he went to work for Aktien-Gesellschaft für Anilin-Fabrikation (which later became AGFA), but his delicate health could not cope with the factory fumes. He returned to Bonn, and in 1876 became Professor Extraordinary (in fact, despite its name, in mid-nineteenth century Germany this was the lowest grade of professorship). Kekulé suggested he should examine the contents of a cupboard containing essential oils, and thus started Wallach's interest in terpenes, on which he wrote 126 papers. Using only simple reagents such as HCl and HBr, he succeeded in characterizing many of these compounds. In 1889, he was appointed Director of the Chemical Institute in Göttingen. He retired in 1915, when six of his assistants were killed in action in the First World War. Wallach was awarded the Nobel Prize for Chemistry in 1910 for his work on terpenes. He remained a bachelor all his life, and died in 1931.

There is a 25 km-wide crater on the Moon named after him.

After the isoprene rule was proposed, many new compounds from various sources were soon added to the terpene family, and subdivisions were made according to the number of five-carbon units present. Monoterpenes, the largest class, contain ten carbon atoms, from two five-carbon units; myrcene, **1.2**, falls into this category. Many monoterpenes possess more functional groups than myrcene, and some are also cyclic, but it is still possible to identify the 'isoprene-like' carbon units.

Identify the two five-carbon units in another monoterpene in Table 1.3, menthol, **1.3**. (*Note* The methyl groups are usually the best place to start.)

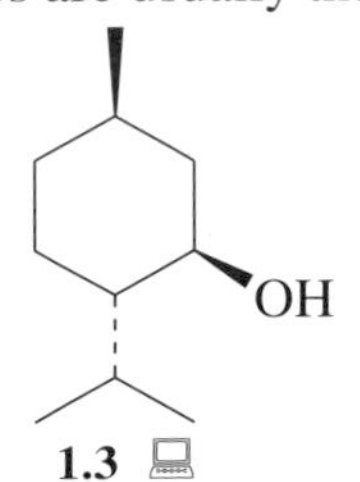


**1.3**

See the structures at the bottom of p. 290.

The presence of two identifiable five-carbon isoprene-like units in menthol confirms that it is a member of the terpene family.

Members of the terpene family containing more than ten carbon atoms are named as follows:

$C_{15}$ sesquiterpenes (*sesqui* is Latin for 'one and a half')

$C_{20}$ diterpenes

$C_{25}$ sesterterpenes

$C_{30}$ triterpenes

$C_{40}$ tetraterpenes

$> C_{40}$ polyterpenes (for example, rubber)

## 1.3 Summary of Section 1

1. Natural products may be categorized as primary or secondary metabolites. Primary metabolites are essential to life processes; secondary metabolites tend to be species-specific, and may have no obvious biological function.
2. Terpenes are a class of natural products responsible for many of the smells in the plant world. They have a wide variety of carbon skeletons and functional groups, but their molecular structures all contain a multiple of five carbon atoms. This observation led to the formulation of the isoprene rule, which suggested that terpenes could be considered to be formed by the coupling of isoprene-like molecules.

### QUESTION 1.1

Look at Table 1.3 and classify (i) geraniol, (ii) α-pinene, (iii) farnesol, (iv) vitamin A and (v) squalene as mono-, sesqui-, di-, tri-, tetra- or polyterpenes. Identify the isoprene-like units in each compound.

### QUESTION 1.2

Which of the molecules (i)–(iv) can be classified as terpenes?

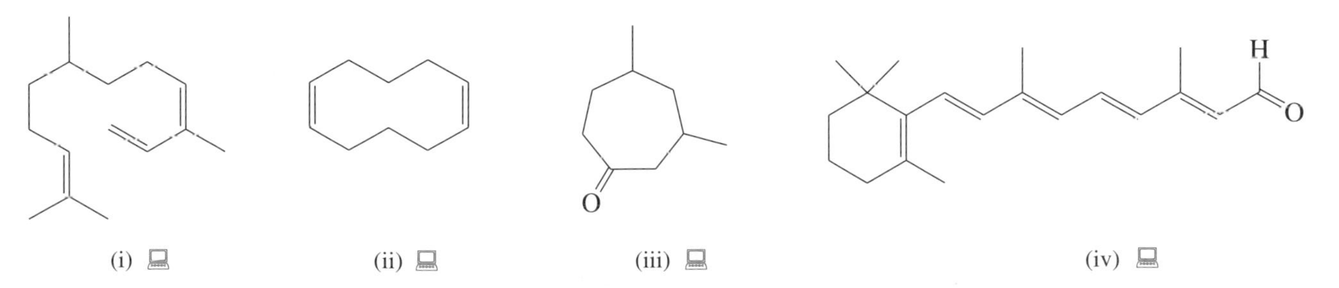

# THE LABORATORY SYNTHESIS OF MONOTERPENES

2

If monoterpenes appear to comprise two five-carbon units, it might seem logical that they can be made by somehow linking together two isoprene molecules. As you will discover in the next Section, something like this does occur during their synthesis in living systems. However, let's take a brief look at whether such a strategy is useful in the laboratory. To do this, we shall use retrosynthetic analysis. Imagine that we wanted to synthesise geraniol (**2.1**).

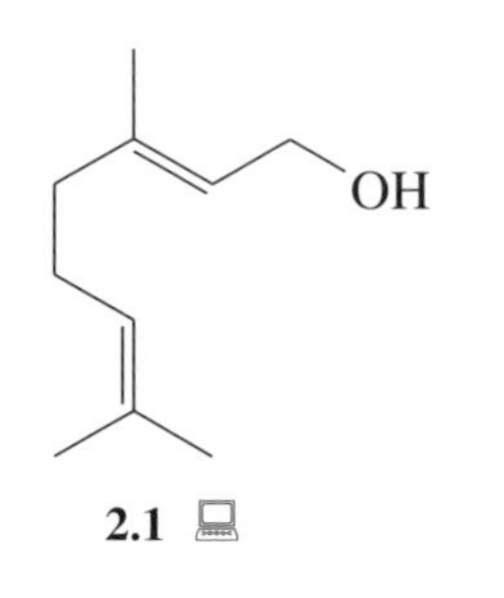


- If you were to attempt the synthesis of **2.1** directly from two five-carbon units, which bond would you disconnect to identify appropriate synthons, and therefore which bond would you need to form in the synthesis?

- Disconnection of **2.1** into two five-carbon fragments requires that the carbon–carbon single bond in the middle of the molecule must be broken.

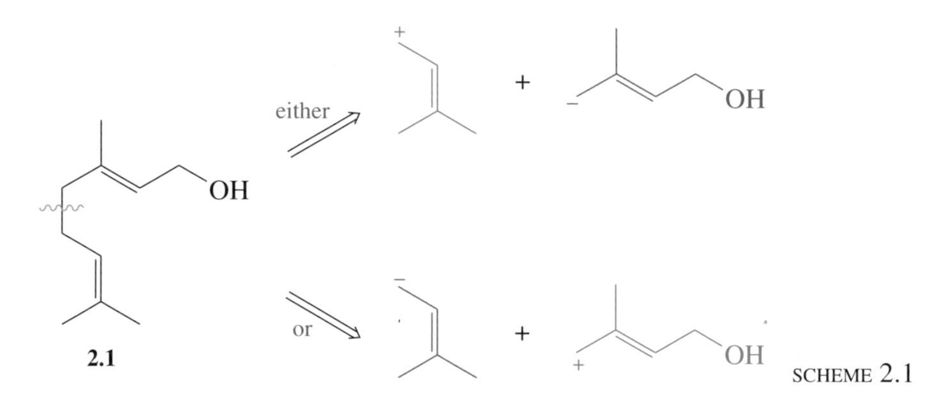


SCHEME 2.1

From what you've learned about retrosynthetic analysis in Part 4, this approach may appear rather strange, because it ignores the functional groups in the molecule, and instead focuses on the carbon atoms with least functionality. Despite this shortcoming, this type of strategy has some merit, because it is concise and requires only two five-carbon units that ultimately may come from the same starting material. In other words, it is highly convergent. For now, though, we shall return to our usual strategy, and concentrate on the functional groups. This will give us some laboratory syntheses of monoterpenes that follow a more linear approach.

We shall study the syntheses of two monoterpene alcohols, linalool and geraniol.

---

*Answer to question on p. 289*

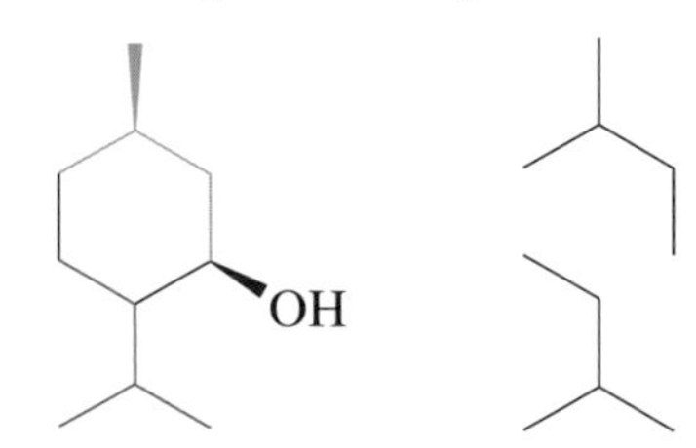

## 2.1 Synthesis of linalool

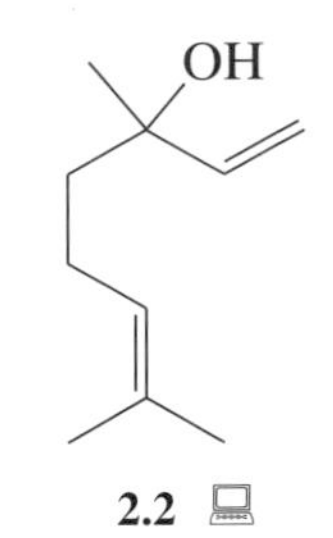


Linalool (**2.2**) contributes to the fragrance of lily of the valley. By applying the disconnection approach, let's attempt to identify potential routes for its synthesis.

- On which functional group in **2.2** would you focus your attention initially?
- The alcohol group seems most appropriate, since there are several potential carbon–carbon bond disconnections.

Indeed, we can disconnect each of the three C—C bonds to the carbon atom of the alcohol group, and this allows us to identify three pairs of nucleophilic and electrophilic carbon synthons.

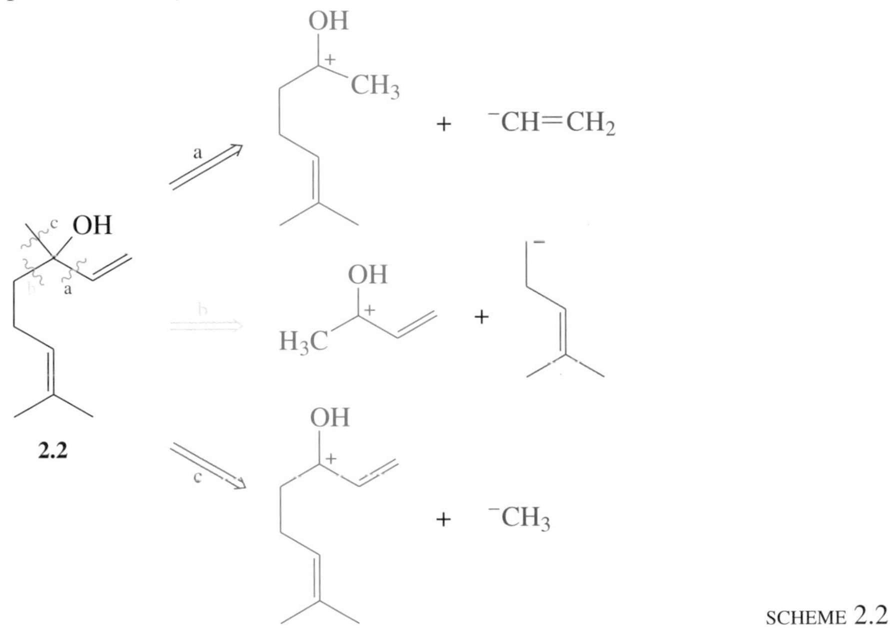


SCHEME 2.2

- What type of reagent could be used to provide the nucleophilic synthons identified by these carbon–carbon bond disconnections?
- Organometallic reagents (for example, Grignard reagents) are suitable sources of carbon nucleophiles.

So, our nucleophilic synthons can be provided by organomagnesium halides, which in turn are derived from the corresponding haloalkane or haloalkenes:

a $\quad {}^{-}CH{=}CH_2 \Longrightarrow XMgCH{=}CH_2 \Longrightarrow XCH{=}CH_2$ (2.1)

b MgX X (2.2)

c $\quad {}^{-}CH_3 \Longrightarrow CH_3MgX \Longrightarrow CH_3X$ (2.3)

Let's now turn our attention to the electrophilic synthons.

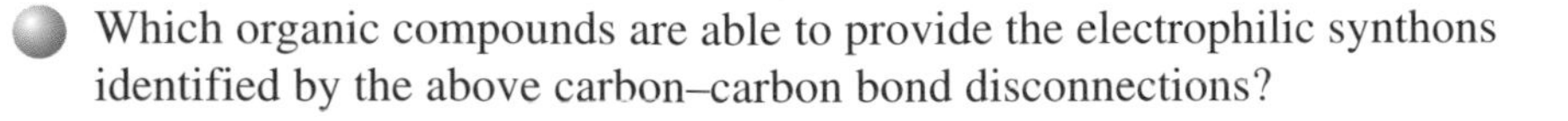

- Which organic compounds are able to provide the electrophilic synthons identified by the above carbon–carbon bond disconnections?

Each of the synthons may be related to the corresponding ketone:

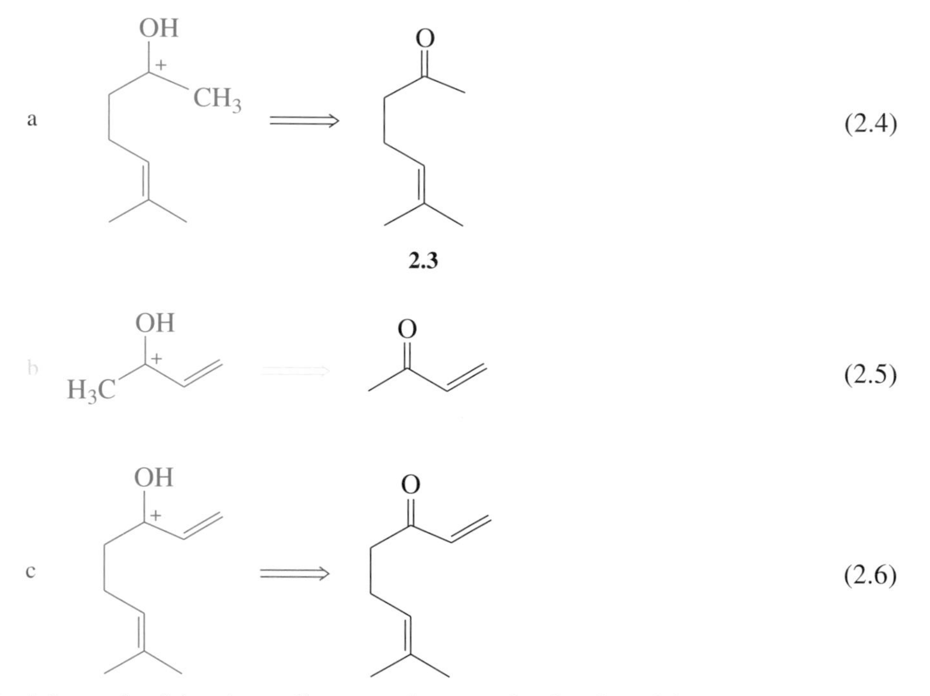


In principle, each of the above disconnections can be developed further to a full synthesis of linalool (**2.2**). However, we shall confine ourselves to developing the disconnection a:

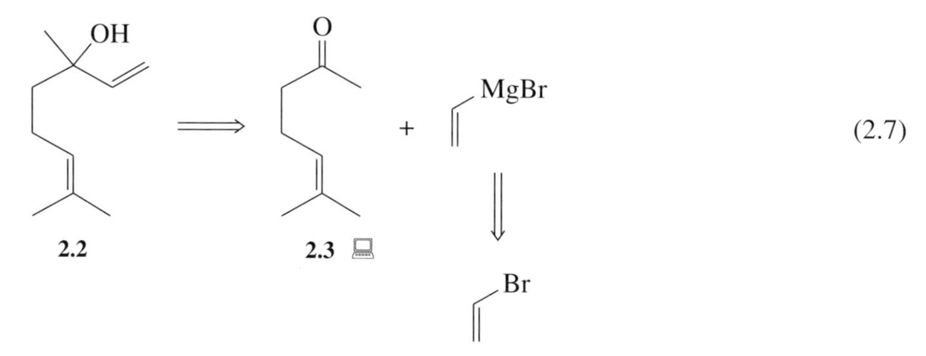


The compounds required are thus bromoethene, which is readily available, and the ketone **2.3**, which is not. In the synthetic direction we have:

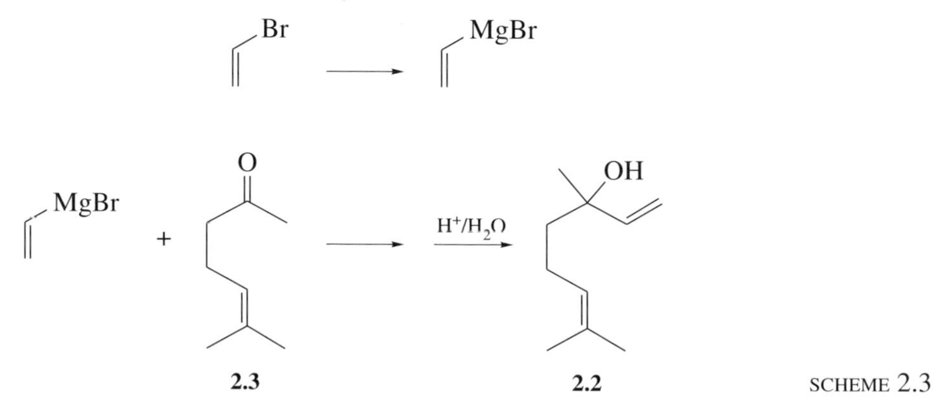


SCHEME 2.3

Since ketone **2.3** is unavailable commercially, it, in turn, must now be regarded as a target molecule, and two disconnections may be considered here:

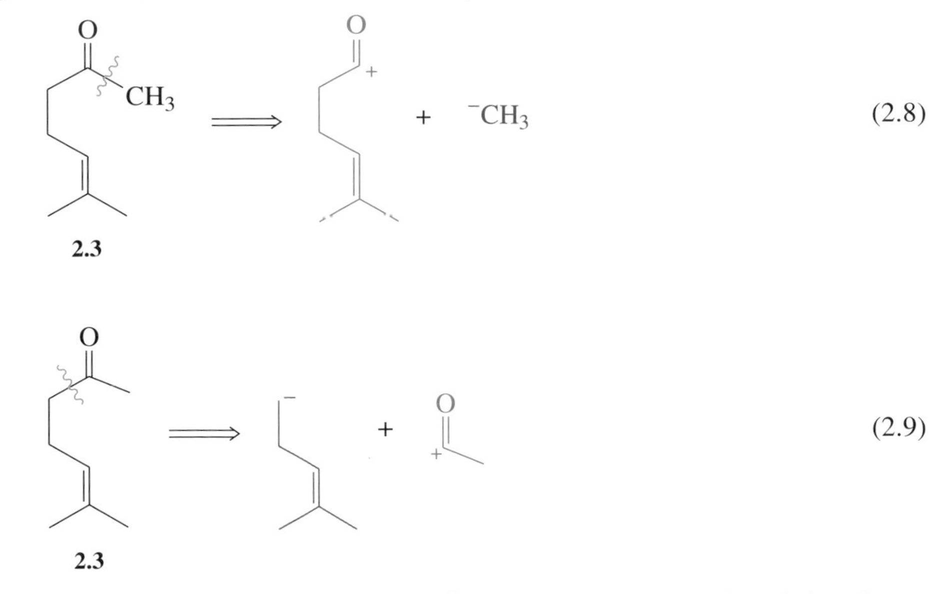


Of the two disconnections, the strategy in Reaction 2.9 is to be preferred since it breaks the molecule down into two more evenly sized chunks.

- To what reagents do the two synthons from this disconnection correspond?
- We are looking at the synthesis of a ketone, so a number of reagents are possible:

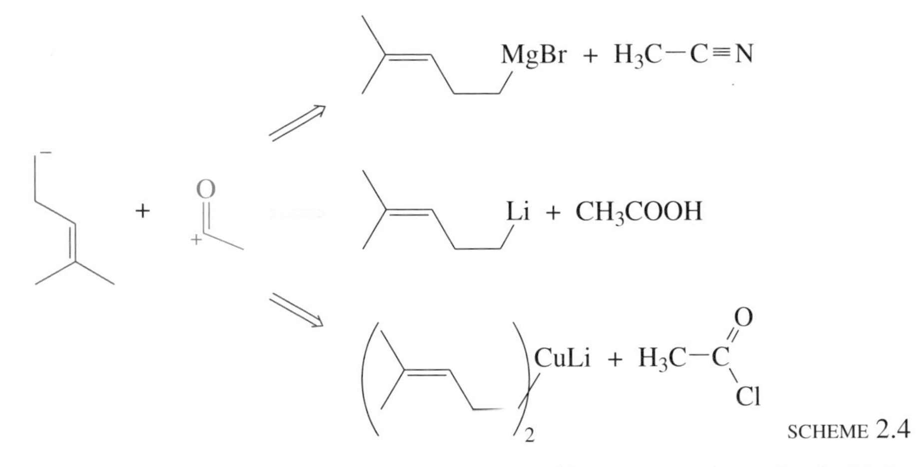


SCHEME 2.4

The nucleophilic synthon relates to three organometallic compounds, each of which can be synthesised from **2.4**. The organometallic reagents can be reacted with a range of carboxylic acid derivatives to give the ketone. Thus, the next step is to make the chloroalkene **2.4**. The chloro compound can be made from the corresponding alcohol, **2.5**, by a functional group interconversion:

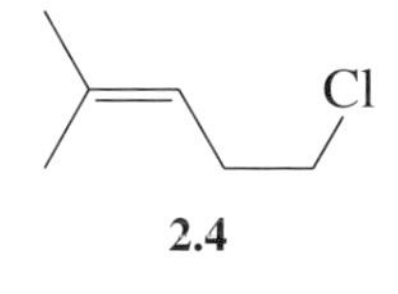


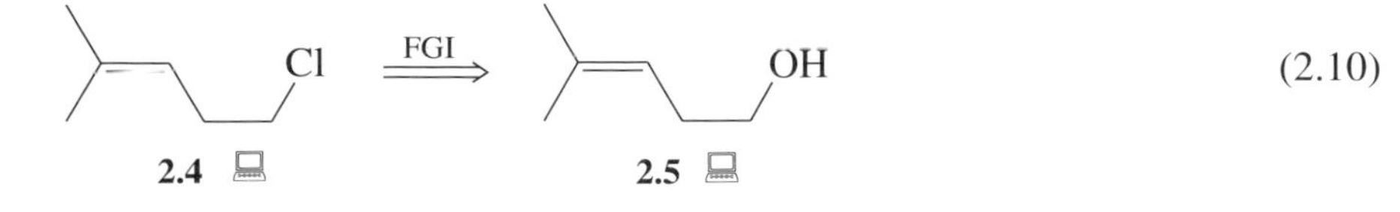

Disconnecting the carbon–carbon bond adjacent to the $-CH_2OH$ group in **2.5** gives:

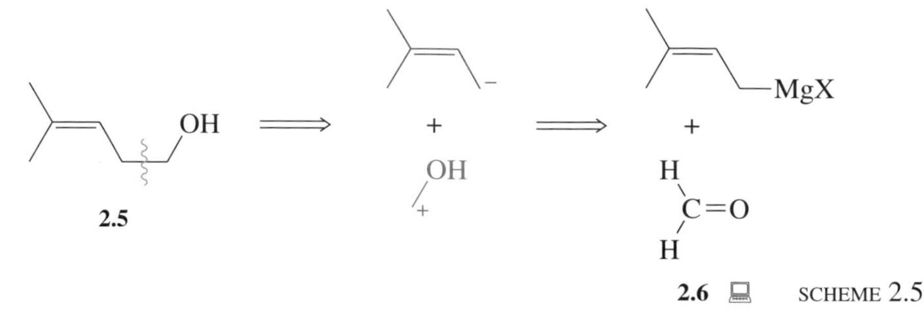


SCHEME 2.5

The electrophilic synthon corresponds to methanal (**2.6**), and the nucleophilic synthon to a Grignard reagent that can be made from the haloalkene **2.7**. Thus, one possible synthesis is:

X
2.7

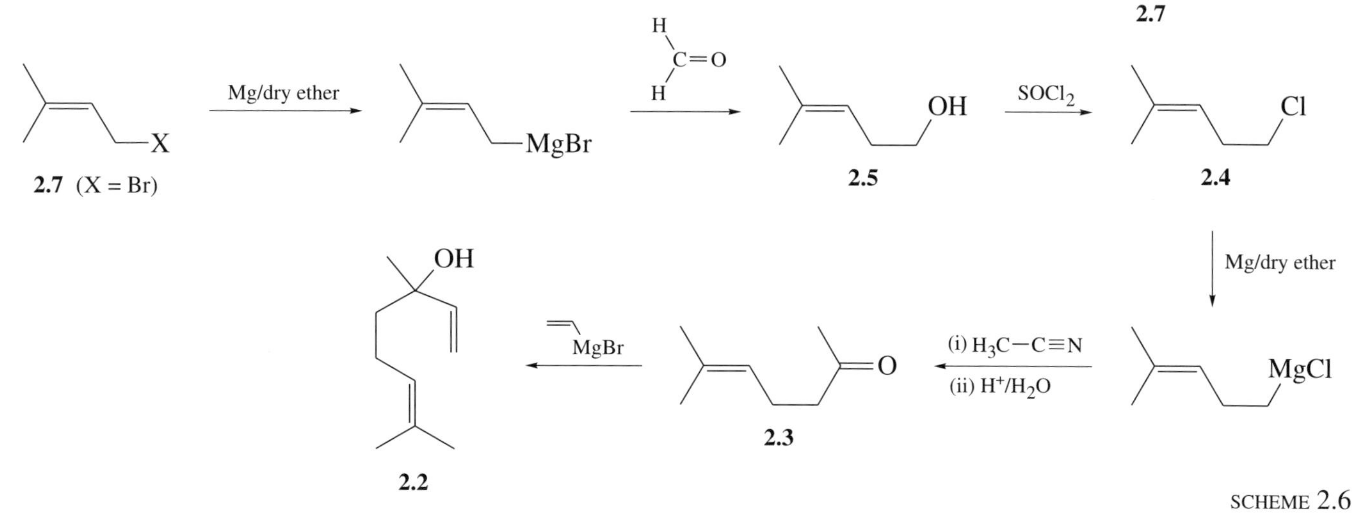


SCHEME 2.6

This six-step synthesis is a bit long winded. However, it can be simplified to a two-step reaction using a carbon nucleophile that we haven't discussed previously (Box 2.1)!

## BOX 2.1 Enolates — very useful carbanions!

In Part 2 of this Book, we saw how it was relatively easy to make carbanions from alkynes, because alkynes are a lot more acidic than alkanes:

$$R{-}C{\equiv}C{-}H \xrightarrow{\text{base}} R{-}C{\equiv}C^- \qquad (2.11)$$

Carbonyl compounds are also acidic, a process that again involves cleavage of a C—H bond. This was not discussed in Part 1 of this Book, but Table 2.1 gives you some idea of the acidity of such compounds.

Compared with alkanes, carbonyl compounds are relatively acidic (by a factor of more than $10^{25}$), because the carbanion formed ($X^-$ in Table 2.1 heading) is stabilized by resonance:

$$-\overset{\overset{\displaystyle O}{\|}}{C}-\overset{-}{C}H_2 \longleftrightarrow -\overset{\overset{\displaystyle O^-}{|}}{C}{=}CH_2 \qquad (2.12)$$

**Table 2.1** The acidity of some carbonyl compounds ($HX \underset{}{\overset{K_a}{\rightleftharpoons}} H^+ + X^-$)

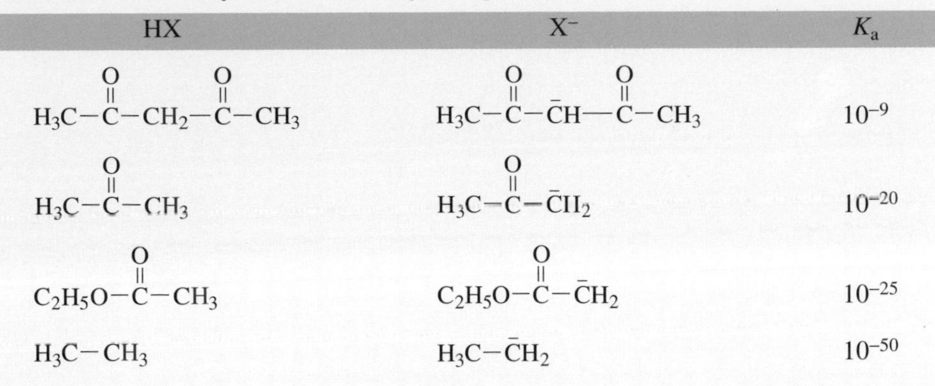

| HX | $X^-$ | $K_a$ |
|---|---|---|
| $H_3C{-}\overset{O}{\overset{\Vert}{C}}{-}CH_2{-}\overset{O}{\overset{\Vert}{C}}{-}CH_3$ | $H_3C{-}\overset{O}{\overset{\Vert}{C}}{-}\overset{-}{C}H{-}\overset{O}{\overset{\Vert}{C}}{-}CH_3$ | $10^{-9}$ |
| $H_3C{-}\overset{O}{\overset{\Vert}{C}}{-}CH_3$ | $H_3C{-}\overset{O}{\overset{\Vert}{C}}{-}\overset{-}{C}H_2$ | $10^{-20}$ |
| $C_2H_5O{-}\overset{O}{\overset{\Vert}{C}}{-}CH_3$ | $C_2H_5O{-}\overset{O}{\overset{\Vert}{C}}{-}\overset{-}{C}H_2$ | $10^{-25}$ |
| $H_3C{-}CH_3$ | $H_3C{-}\overset{-}{C}H_2$ | $10^{-50}$ |

Not only is the negative charge spread over the molecule, but the resonance form on the right of Reaction 2.12 has the charge residing on an electronegative oxygen atom. As a result, these carbonyl carbanions, called **enolates**, are reasonably stable, and can be formed fairly easily on treatment of the corresponding carbonyl compound with a base:

$$-\overset{O}{\overset{\Vert}{C}}{-}CH_3 \xrightarrow{\text{base}} -\overset{O}{\overset{\Vert}{C}}{-}\overset{-}{C}H_2 \qquad (2.13)$$

Of course, the base must be strong enough to remove a proton, but once formed, the resulting carbonyl carbanion can react with carbon electrophiles such as haloalkanes:

$$R^1{-}\overset{O}{\overset{\Vert}{C}}{-}\overset{-}{C}H_2 \quad R^2{-}CH_2{-}X \xrightarrow{-X^-} R^1{-}\overset{O}{\overset{\Vert}{C}}{-}CH_2{-}CH_2{-}R^2 \qquad (2.14)$$

$S_N2$ substitution of the haloalkane with the carbanion as the nucleophile, results in carbon–carbon bond formation — a very useful reaction known as **enolate alkylation**.

In terms of retrosynthetic analysis, this carbon–carbon bond-forming reaction in the synthetic direction corresponds to a carbon–carbon bond disconnection next-but-one to the carbonyl group in the retrosynthetic direction:

$$R^1{-}\overset{O}{\overset{\Vert}{C}}{-}CH_2 \wr CH_2{-}R^2 \Longrightarrow R^1{-}\overset{O}{\overset{\Vert}{C}}{-}\overset{-}{C}H_2 + \overset{+}{C}H_2{-}R^2 \qquad (2.15)$$

Returning to our synthesis of linalool (**2.2**), a more-effective disconnection of the ketone **2.3** is:

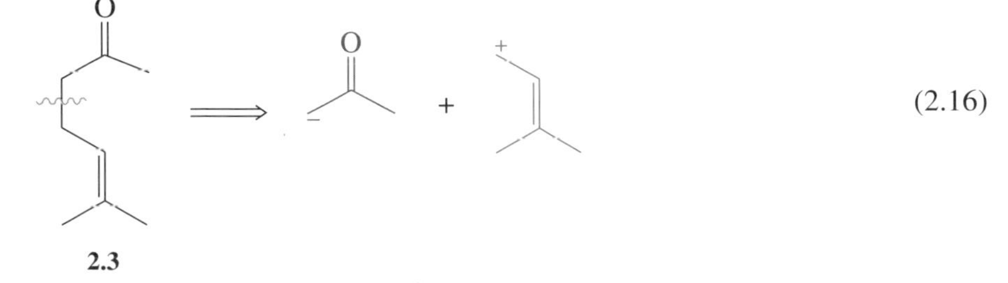


(2.16)

This generates a synthon in which a negative charge is adjacent to a carbonyl group — in other words, an enolate. So, for the synthesis, we can make use of the enolate alkylation reaction described in Box 2.1. This disconnection enables the target molecule **2.3** to be broken into two similar-sized fragments, which is likely to produce a more convergent synthesis.

The two precursors are therefore the ketone propanone (**2.8**) and the haloalkene **2.7**:

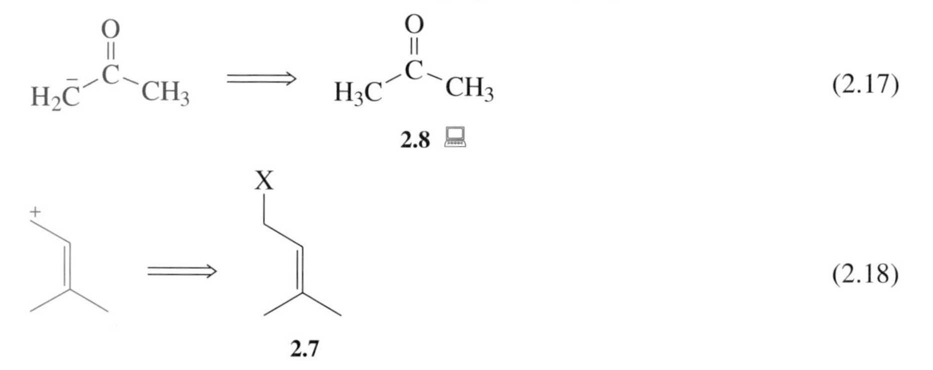


In fact, propanone and the bromoalkene **2.7** (X = Br) — which was also the starting material for our previous synthesis — are both readily available.

Our completed retrosynthetic strategy for linalool is illustrated in Scheme 2.7.

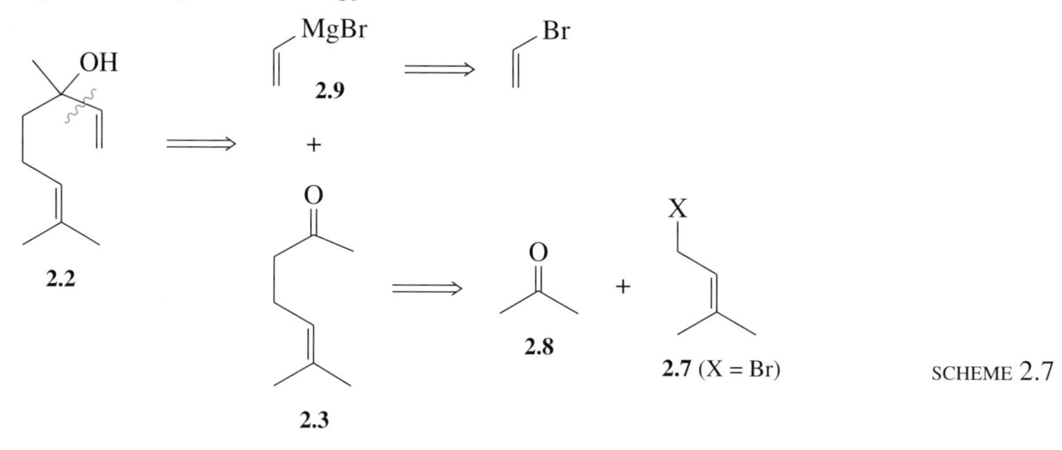


SCHEME 2.7

The synthesis of linalool thus resolves into just two steps: alkylation of propanone (**2.8**), using the bromoalkene **2.7** (X = Br), followed by treatment of the resulting ketone **2.3** with the Grignard reagent **2.9**:

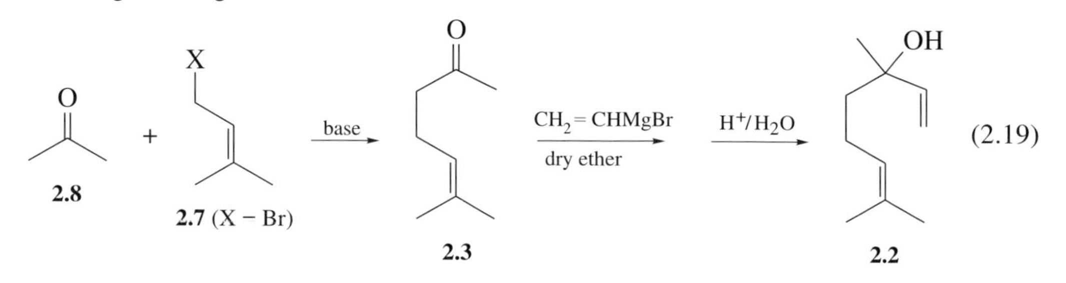

## 2.2 Synthesis of geraniol

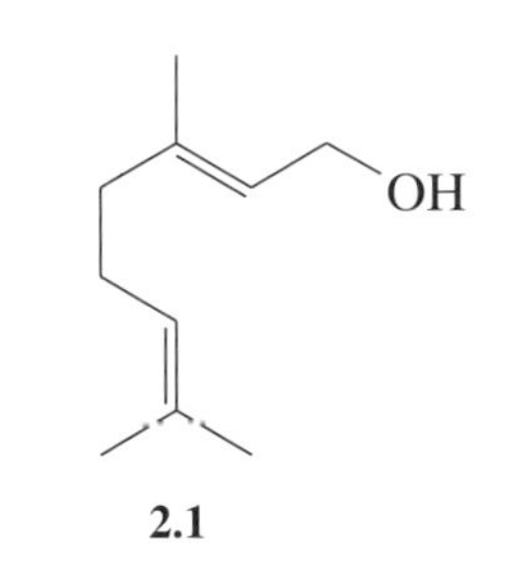


In Sections 3 and 4, we shall be looking at the way in which living systems synthesise geraniol (**2.1**), one of the compounds contributing to the scent of roses (Figure 2.1). So, let us attempt to develop a synthesis of **2.1** using the principles of retrosynthetic analysis.

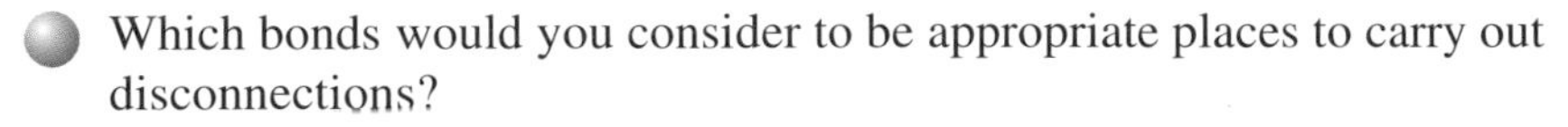

- Which bonds would you consider to be appropriate places to carry out disconnections?
- The two carbon–carbon double bonds.

**Figure 2.1**
Hybrid tea rose.

We could focus our attention on either of these bonds, but for now let's concentrate on the one nearer to the alcohol group in **2.1**. Although we could disconnect this double bond directly, it is more convenient to carry out an FGI on the alcohol group first, to form the carboxylic acid ester **2.10** (it will shortly become apparent why we carry out the FGI before disconnecting the carbon–carbon double bond):

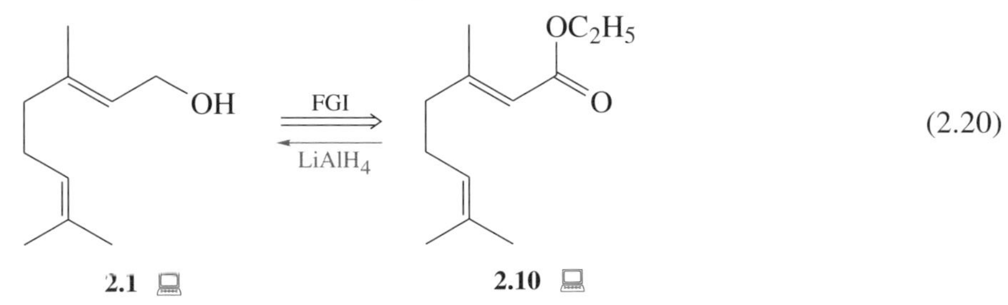


(2.20)

When we come to carry out the synthesis, the formation of the alcohol from the ester **2.10** will require a nucleophilic hydride reagent, such as lithium aluminium hydride to reduce the ester group to the primary alcohol function (Part 1).

Now let's focus on the double bond nearer the ester group in **2.10**. Since such double bonds are usually made by dehydration, we must first carry out a second FGI, in the retrosynthetic direction, to transform the C=C group into an alcohol. This gives us the choice of two alcohols, but the tertiary alcohol **2.11** is the more useful for our purposes.

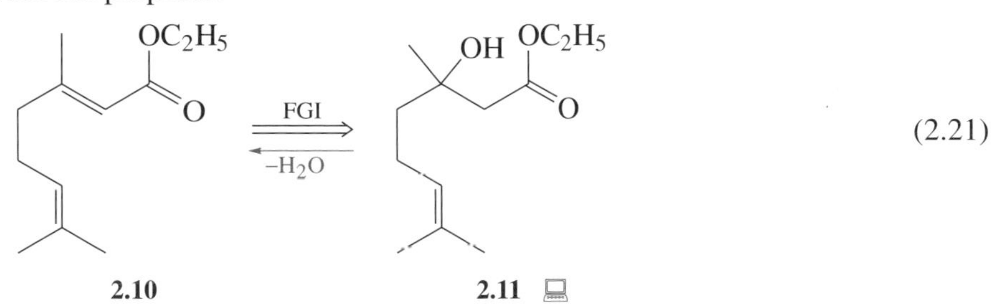


(2.21)

We can now disconnect the carbon–carbon single bond in the position where the carbon–carbon double bond was previously; this single bond is next-but-one to the carbonyl group!

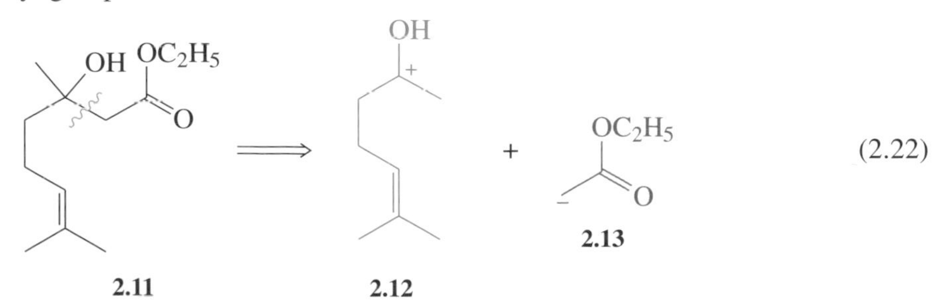


(2.22)

We can thus identify two synthons, **2.12** and **2.13**. The reagent that corresponds to synthon **2.12** is the ketone **2.3**, which is the same ketone that was used previously in the synthesis of linalool. Since we have already developed a strategy for its synthesis, we need go no further for synthon **2.12**.

Synthon **2.13**, on the other hand, contains a nucleophilic carbon centre adjacent to a carbonyl group. As we mentioned earlier, such carbanions can be formed from the corresponding carbonyl compound by treatment with a suitable base:

$$H_3C{-}C({=}O){-}R \xrightarrow{\text{base}} H_2\bar{C}{-}C({=}O){-}R \qquad (2.23)$$

Since R = $OC_2H_5$ in this case, the reagent corresponding to the synthon **2.13** is the readily available ester ethyl ethanoate, **2.14**:

$$\underset{\mathbf{2.13}}{H_2\bar{C}{-}C({=}O){-}OC_2H_5} \Longrightarrow \underset{\mathbf{2.14}}{H_3C{-}C({=}O){-}OC_2H_5} \qquad (2.24)$$

Now we can see why carrying out an FGI on the alcohol functional group of geraniol was so helpful in devising our strategy; it led us to identify a readily available reagent that can be used as a carbon nucleophile. In an actual synthesis of geraniol (**2.1**), the ester **2.10** was synthesised from the ketone **2.3**. The ester **2.10** is then converted into geraniol on reaction with lithium aluminium hydride:

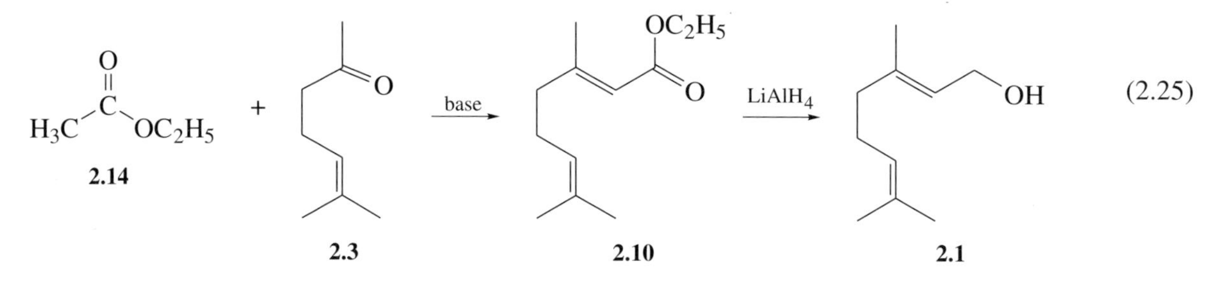


## 2.3 Summary of Section 2

Syntheses of monoterpenes may be devised using the principles of retrosynthetic analysis. If attention is concentrated on the functional groups, relatively simple syntheses can be devised.

### QUESTION 2.1

Concentrating solely on the alcohol functional group, identify the appropriate synthons and immediate precursors of each of the two monoterpenes **2.15** and **2.16**.

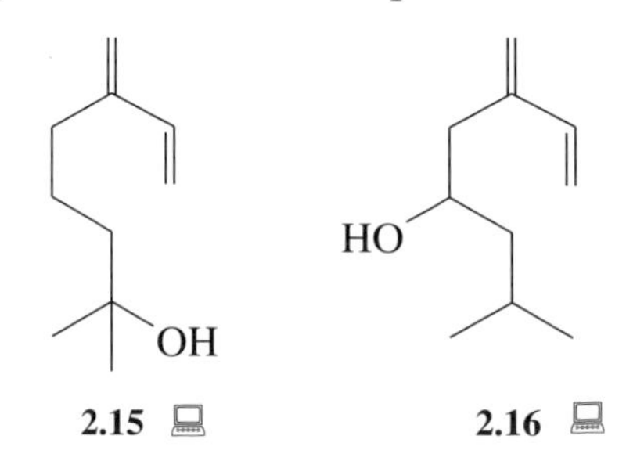

# A BIOCHEMICAL INTERLUDE

3

In this Section, we review some of the tools (reagents) that Nature uses in the synthesis of terpenes.

Most biochemical reactions occur through the mediation of an enzyme. Enzymes are proteins, and as such are constructed from chiral α-amino acids; hence, enzymes are chiral macromolecules. Enzymes contain an *active site*, which will selectively bind only one type of substrate; the analogy of a lock and key is a useful one here, in that only one type of substrate will fit the enzyme correctly (Figure 3.1). An enzyme can hold a substrate in just the right orientation such that when a reaction occurs on the substrate, just one stereoisomeric product results. Thus, the product may be an alkene with a particular geometry (*cis*- or *trans*-); alternatively, just one enantiomer of a chiral molecule will be produced. Often, in order to work satisfactorily, enzymes require a cofactor; usually a metal ion (for example, $Zn^{2+}$, $Mg^{2+}$), as well as a coenzyme, which carries out a reaction on the substrate. As a laboratory analogy, the enzyme can be thought of as the environment, like the reaction vessel, in which a reaction occurs, and the coenzyme plays the role of a reagent.

**STUDY NOTE**

**WARNING** Many large molecules are *good* for you, so please do not let them frighten you! Remember the functional group approach, and concentrate only on the sites in the molecule where reaction can occur. Most of a large molecule is simply a carbon framework, and so in many cases, abbreviations for the structures will be used.

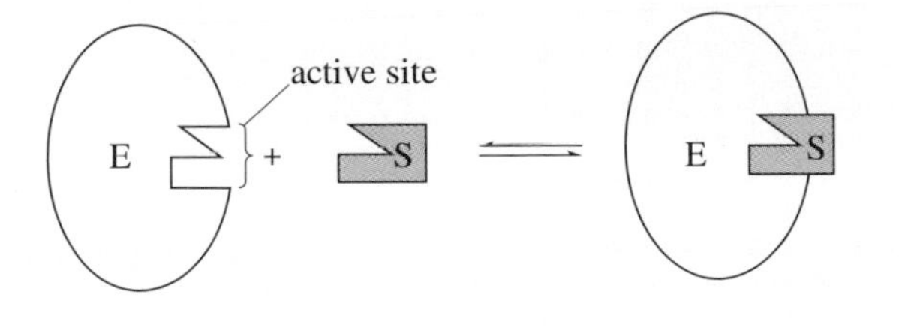


**Figure 3.1** Schematic relationship between an enzyme (E) and its substrates (S), showing the precise correlation in shape between them, which accounts for the observed specificity of a particular enzyme for an individual substrate.

Coenzyme A (**3.1**) is involved in the joining together of acetyl (ethanoyl, $C_2$ units); this is an extremely important biosynthetic reaction, which can lead to a multitude of natural products, ranging from aromatic systems to terpenes.

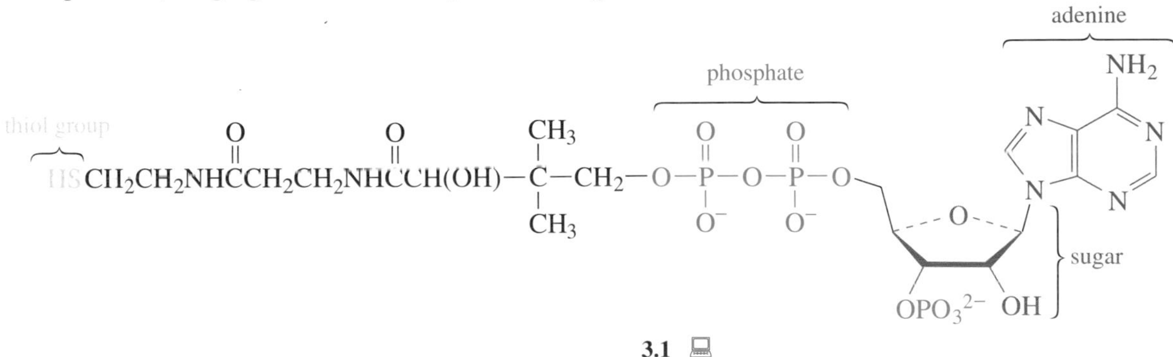


**3.1**

We shall use the abbreviation CoA—SH for **3.1**, which concentrates on the —SH group (thiol) and ignores the rest of the molecule. Thus, acetyl (ethanoyl) coenzyme A is the thioester of acetic acid (ethanoic acid), $CH_3CO—S—CoA$. Thioesters are much more reactive towards nucleophiles than esters, so addition–elimination reactions involving them proceed much more easily.

Adenosine triphosphate (ATP, **3.2**) is the source of phosphate groups (**3.3**) in biological systems. **3.4** is an abbreviated formula for ATP. Phosphate groups are extremely good leaving groups, and they are often found in nature when a good leaving group is required — for example, in nucleophilic substitution or elimination reactions.

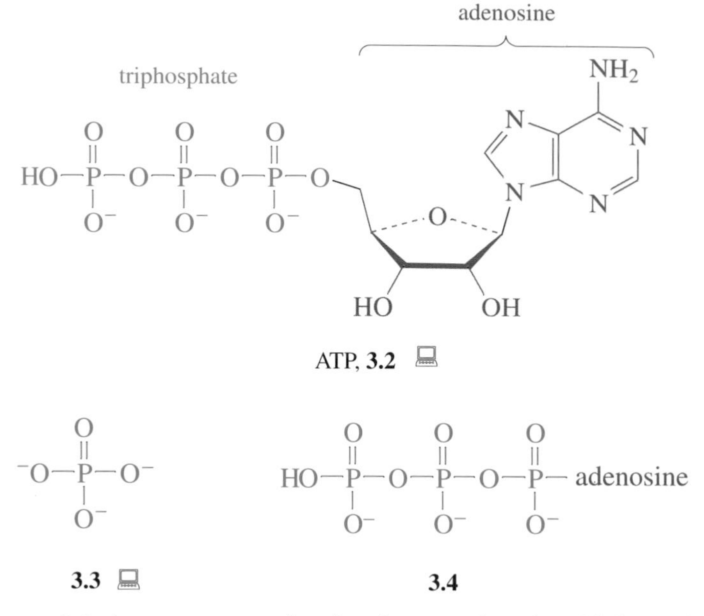


ATP, **3.2**

$^{-}O-P(=O)(O^{-})-O^{-}$

**3.3**

$HO-P(=O)(O^{-})-O-P(=O)(O^{-})-O-P(=O)(O^{-})-$adenosine

**3.4**

Scheme 3.1 shows an example of such a reaction, in which an alcohol reacts with ATP to give a phosphate ester and adenosine diphosphate (ADP, **3.5**). The phosphate ester may now react with a nucleophile to yield the required product. Phosphorylation of the alcohol hydroxyl group has converted the poor leaving group OH into the good leaving group $HPO_4^{2-}$.

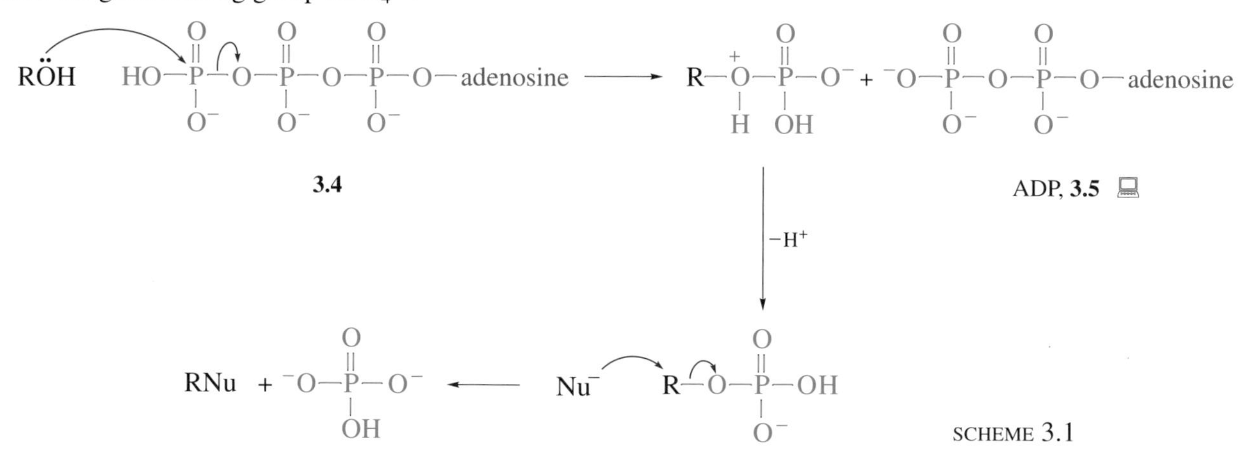


SCHEME 3.1

In some circumstances, a phosphate ester may react with another molecule of ATP to yield a pyrophosphate ester (for example, **3.6**, which is related to pyrophosphoric acid, **3.7**). Just like phosphate, pyrophosphate is also a good leaving group. These groups are rather cumbersome to deal with in complex diagrams, so they will be used in the shortened forms $ROP^{-}$ and $ROPOP^{2-}$ shown below:

$HO-P(=O)(OH)-O-P(=O)(OH)-OH$

**3.7**

$$R-O-P(=O)(O^{-})-OH + ATP \longrightarrow R-O-P(=O)(O^{-})-O-P(=O)(O^{-})-OH + ADP \quad (3.1)$$

phosphate ester, $ROP^{-}$

pyrophosphate ester, $ROPOP^{2-}$ (**3.6**)

Our third example of a coenzyme is nicotinamide adenine dinucleotide ($NAD^+$, **3.8**); this is the reagent that oxidizes alcohols to aldehydes in biosynthetic pathways:

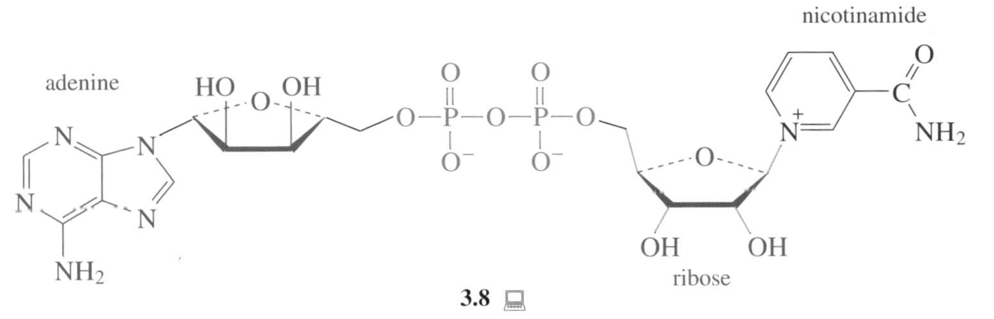


**3.8**

The formula may be simplified to **3.9** in order to show just the 'working end' (nicotinamide) of the molecule, where 'R' signifies the rest of $NAD^+$. Scheme 3.2 shows the involvement of $NAD^+$ in the oxidation of ethanol to ethanal:

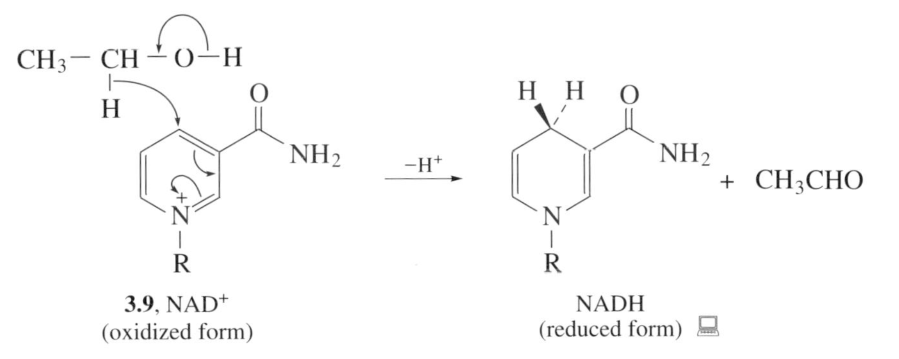

**3.9**, $NAD^+$ (oxidized form) NADH (reduced form)

SCHEME 3.2

In the human body, this conversion is the first step in the metabolism of ethanol in the liver by the enzyme liver alcohol dehydrogenase (LADH). Interestingly, the reverse reaction, reduction of ethanal to ethanol, is the ultimate step during the fermentation of sugar by yeast to yield ethanol, but for this reaction step, the reduced form of the coenzyme, NADH, is the reagent.

## 3.1 Summary of Section 3

1. Most biochemical reactions occur through the mediation of enzymes, which are made up of amino acids.
2. Many simple chemical reactions involve complex molecules when performed in living systems.
3. In biochemistry, complex molecules are often simplified using initials so that we focus only on the 'working end of the molecule'.
4. Phosphate groups, in particular diphosphates and triphosphates, are good leaving groups.

# THE SYNTHESIS OF TERPENES IN LIVING SYSTEMS

# 4

Laboratory syntheses of monoterpenes, such as linalool and geraniol, can be readily accomplished using the strategies identified in Section 2. However, they tend to suffer the drawback of following linear, rather than convergent, routes. Because all terpenes, without exception, are made up of isoprene-like units, a more convenient, and convergent, strategy would be one utilizing the isoprene rule. If such an approach were possible, we would disconnect the various molecules into common units containing five carbon atoms (corresponding to isoprene), which could be joined together in synthesis. For geraniol (**2.1**) this would entail disconnecting the bond involving the least functionalized carbon atoms in the molecule:

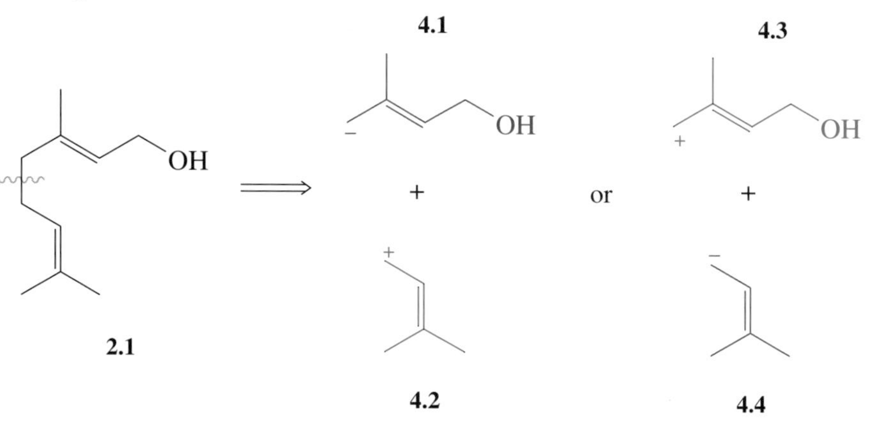


SCHEME 4.1

As revealed in this and the following Section, this is precisely the strategy adopted in nature. However, a major question immediately arises: 'what are the reagents that correspond to the nucleophilic and electrophilic synthons that are identified by such a bond disconnection?'. In view of Wallach's work, it is not surprising that the first candidate was isoprene itself. This hypothesis was strengthened by the observation that many members of the terpene family, including rubber, produce isoprene when heated strongly. However, experiments revealed that isoprene is *not* the natural precursor of monoterpenes.

The identification of the reagents corresponding to the nucleophilic and electrophilic synthons required by this strategy became a major research effort, and the search for the five-carbon biosynthetic precursor to terpenes, the so-called **active isoprene unit**, went on for many years. A host of compounds with an isoprene-like skeleton were suggested, and tried, but with no success.

It had been discovered that ethanoic (acetic) acid was incorporated into terpenes and steroids, and that this involved the acetyl (ethanoyl) coenzyme A complex, $CH_3CO{-}SCoA$, **4.5**. (Compare **4.5** with Structure **3.1**.) Thus, our active isoprene unit must be derived from acetyl CoA.

It was not until the 1950s that a breakthrough was achieved and, as is so often the case, it owed a great deal to a chance discovery in a very different field of research. In 1956, a group of chemists investigating the bacteria that cause turbidity in beer discovered a new natural product, 3,5-dihydroxy-3-methylpentanoic acid, which

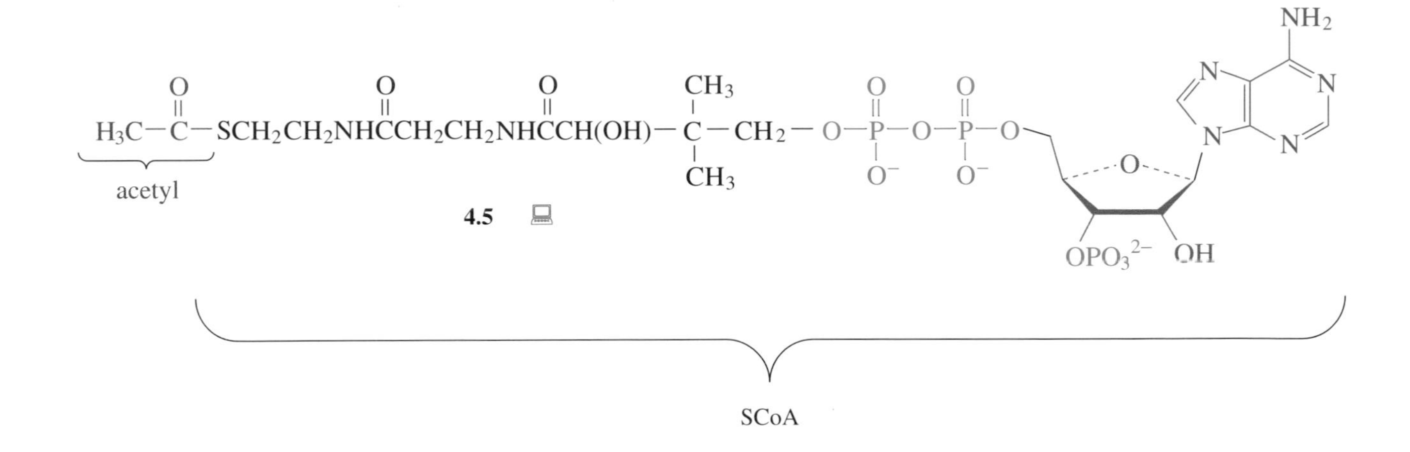


became known as mevalonic acid (**4.6**). Later, the biosynthetic steps between acetyl coenzyme A, and mevalonic acid were elucidated. Three molecules of acetyl coenzyme A (**4.5**) condensed in a series of steps to form hydroxymethylglutaryl coenzyme A (**4.7**), which was reduced in the presence of a specific enzyme (hydroxymethylglutaryl coenzyme A, HMG CoA, reductase) to give mevalonic acid (**4.6**). This coenzyme takes its name from glutaric acid (pentane-1,5-dioic acid, **4.8**).

HOOC COOH

**4.8**

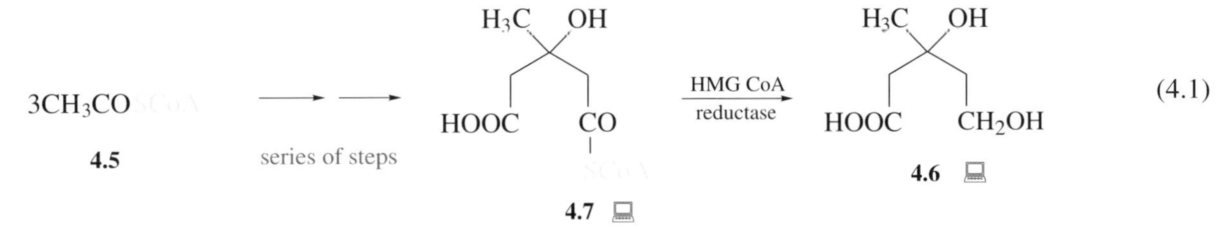


(4.1)

It so happened that chemists interested in terpene biosynthesis were working in the same building, and they realized that stripping the functional groups from mevalonic acid gave rise to a molecular framework similar to isoprene (**1.1**). They carried out various experiments, and established that mevalonic acid occupies an important position in the terpene biosynthetic pathway in plants.

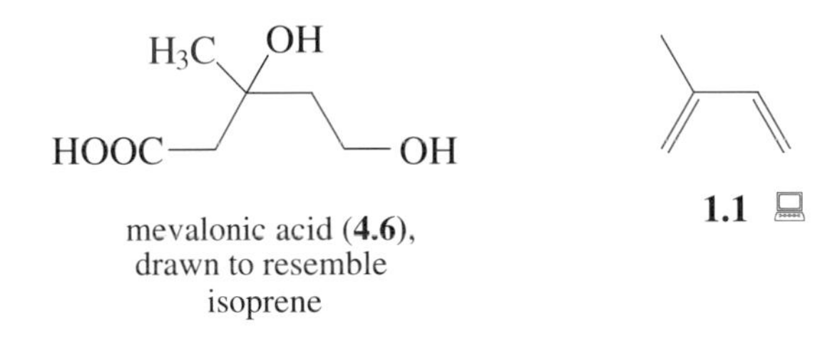


mevalonic acid (**4.6**), drawn to resemble isoprene

- Does mevalonic acid correspond to the synthons required, and to the sought-after 'active isoprene unit'?

- No. Mevalonic acid contains six carbon atoms, whereas the synthons and the active isoprene unit contain only five.

However, the discovery that mevalonic acid was an intermediate in the biosynthesis of terpenes was the lead that was needed. We are going to concentrate on the biosynthetic pathway linking mevalonic acid with the terpenes.

It is interesting to note that whereas nearly 70 years elapsed between the coining of the isoprene rule and the discovery of mevalonic acid, the remaining steps in the biosynthesis of terpenes were elucidated in only two or three years. First, let us examine the steps* that are involved in converting mevalonic acid (**4.6**) into reagents that correspond to the synthons **4.1** and **4.2** — isopentenyl pyrophosphate (3-methylbut-3-enyl pyrophosphate, **4.9**) and dimethylallyl pyrophosphate (3-methylbut-2-enyl pyrophosphate, **4.10**), respectively.

$CH_2{=}CH{-}CH_2{-}$
allyl group

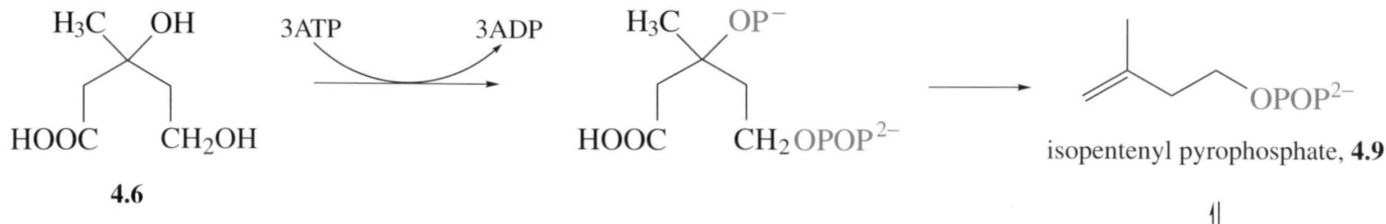


isopentenyl pyrophosphate, **4.9**

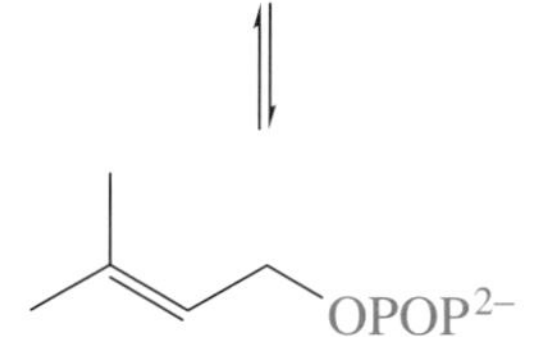


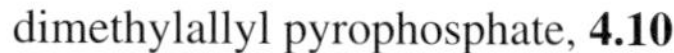

dimethylallyl pyrophosphate, **4.10**

SCHEME 4.2

At first sight, Scheme 4.2 looks daunting, but don't worry. It is included here to illustrate the chemistry involved. For now, all we need do is concentrate on the conversions that take place. In the next Section, we shall go on to look at the reaction mechanisms involved.

Scheme 4.2 shows that mevalonic acid (**4.6**) is transformed with the aid of adenosine triphosphate (Structure **3.2**, p. 300) via a series of phosphate esters into two isomeric alkenes, both containing five carbon atoms — isopentenyl pyrophosphate (**4.9**) and dimethylallyl pyrophosphate (**4.10**). In Section 3, we discussed why phosphate and pyrophosphate esters are involved: they make good leaving groups. It is these two compounds, **4.9** and **4.10**, that are the reagents that correspond to the synthons **4.1** and **4.2**, respectively. Isopentenyl pyrophosphate (**4.9**) corresponds to the nucleophilic synthon **4.1**, because **4.9** contains a terminal alkene group that can act as a nucleophile. Dimethylallyl pyrophosphate (**4.10**), on the other hand, corresponds to synthon **4.2**, because it contains a pyrophosphate leaving group on the terminal carbon atom.

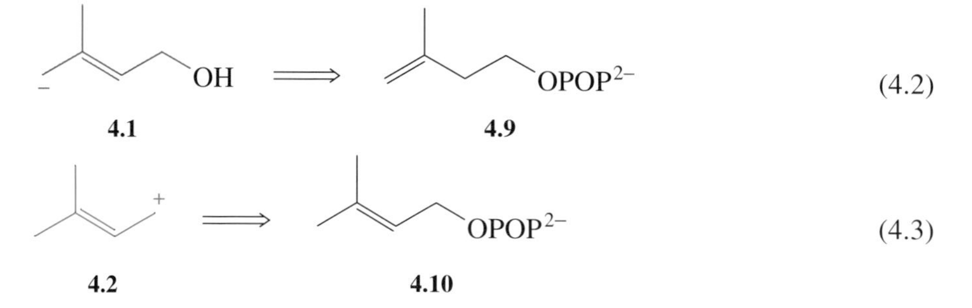


* In Scheme 4.2 the curved arrow attached to the conventional reaction arrow in the first step is a shorthand way of indicating what happens to the 'reagent' (ATP) here: it is converted to ADP.

Scheme 4.3 shows how isopentenyl and dimethylallyl pyrophosphates combine to give molecules with 10, 15, 20, 30 and 40 carbon atoms. (To enable you to follow the fate of the isopentenyl pyrophosphate and dimethylallyl pyrophosphate units through this and subsequent steps of the biosynthesis, we have shown the isopentenyl pyrophosphate unit in blue and the dimethylallyl pyrophosphate unit in red.)

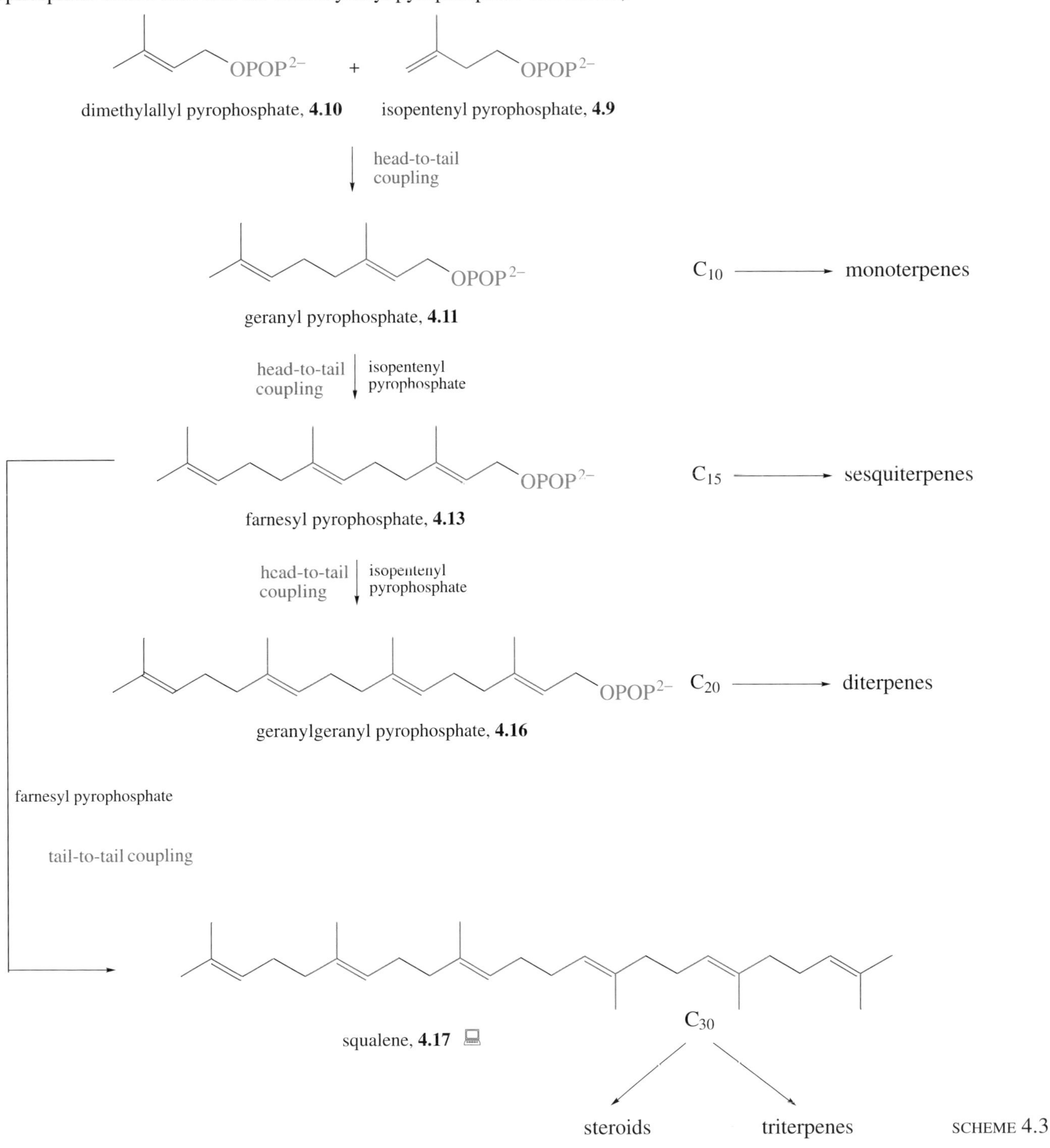


SCHEME 4.3

These pathways are common to all plants that produce terpenes. First, isopentenyl pyrophosphate and dimethylallyl pyrophosphate combine to form geranyl pyrophosphate (**4.11**). This reaction is often referred to as a **head-to-tail coupling** (see Reaction 4.4), the tail of each five-carbon molecule being the end bearing the pyrophosphate functional group:

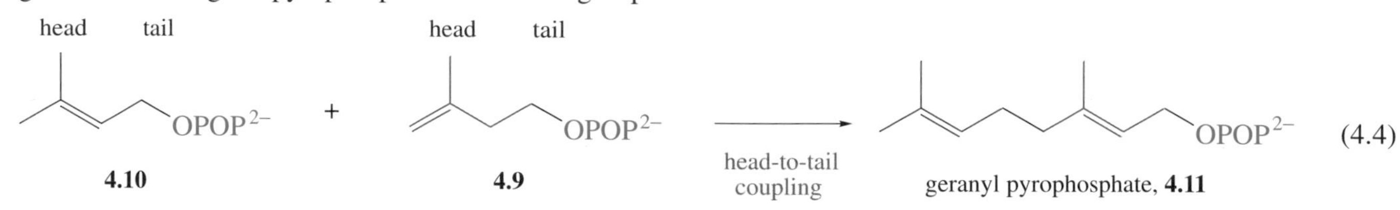


Geranyl pyrophosphate (**4.11**) is the intermediate involved in the production of all monoterpenes (see Scheme 4.4). The monoterpenes characteristic of each biological species are produced by specific enzyme-mediated modification of geranyl pyrophosphate. For example, geraniol (**2.1**), the monoterpene responsible for the fragrance of roses, is formed by straightforward enzyme-mediated hydrolysis of geranyl pyrophosphate. The monoterpene *S*-carvone (**4.12**), which is responsible for the smell of caraway seeds (Figure 4.1) is formed from geranyl pyrophosphate in a series of enzyme-mediated reactions.

**Figure 4.1**
Caraway seeds.

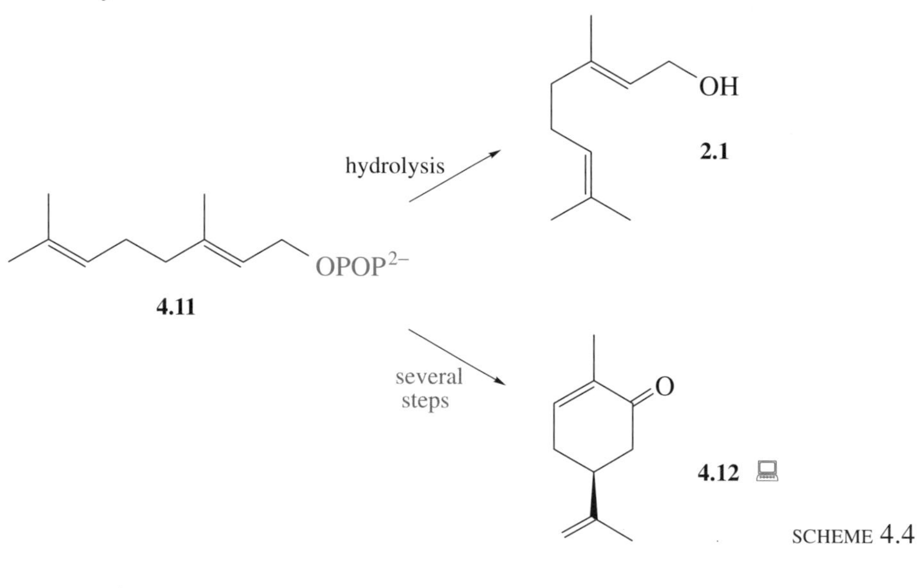


SCHEME 4.4

Plants that produce sesquiterpenes all do so by first forming farnesyl pyrophosphate (**4.13**) from the head-to-tail coupling of geranyl pyrophosphate (**4.11**) and isopentenyl pyrophosphate (**4.9**):

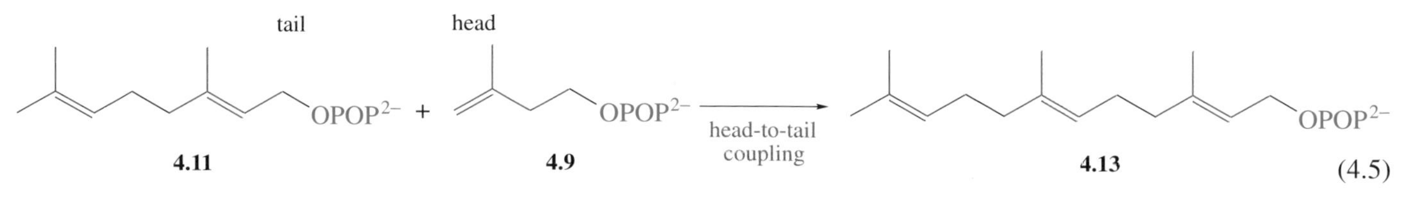

Subsequent enzyme-mediated reactions then produce the sesquiterpenes. The nature of the sesquiterpene produced depends on the plant in question: plants that produce different sesquiterpenes contain different enzymes. For example, hydrolysis of farnesyl pyrophosphate (**4.13**) produces the simplest sesquiterpene, farnesol (**4.14**), which is the main component in the scent of lily of the valley (Figure 4.2). More complicated enzyme-controlled modifications on the other hand, lead to complex sesquiterpenes, such as germacrene B (**4.15**).

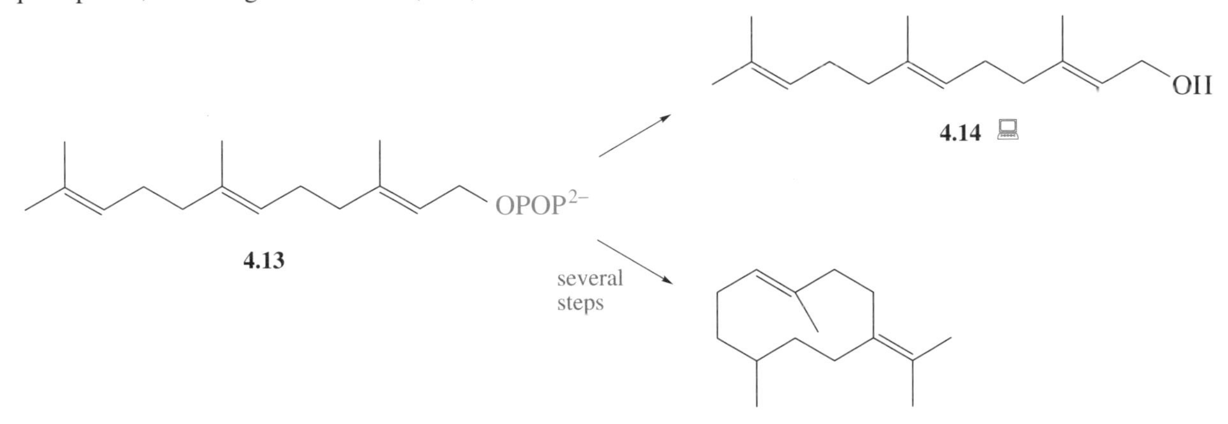

SCHEME 4.5

- Look at Scheme 4.3 again. Which compound is the main biosynthetic precursor of the diterpenes? How do you think it is formed?

- Geranylgeranyl pyrophosphate (**4.16**) is the 20-carbon precursor of the diterpenes. It is formed when farnesyl pyrophosphate (**4.13**) and a further molecule of isopentenyl pyrophosphate (**4.9**) undergo head-to-tail coupling.

All plants that produce diterpenes do so by first forming geranylgeranyl pyrophosphate. So, it is isopentenyl pyrophosphate, **4.9**, that deserves the label 'active isoprene unit', because it is involved in all the steps leading up to geranylgeranyl pyrophosphate (**4.16**). Notice the close similarity between **4.9** and isoprene (**1.1**). The 'paper chemistry' that predicted the 'active isoprene unit' was fairly accurate!

**Figure 4.2**
Lily of the valley.

Terpenes containing 25 carbon atoms (known as 'sesterterpenes') were not shown in Table 1.3. Although several compounds of this class are now known, biosynthetic routes leading to them are not fully understood even today.

An enzyme is present in certain plants and animals that catalyses the reaction between two molecules of farnesyl pyrophosphate (**4.13**) to give squalene (**4.17**). This reaction is often referred to as a **tail-to-tail coupling**, as the pyrophosphate functional group ends of the molecules are joined together during the reaction.

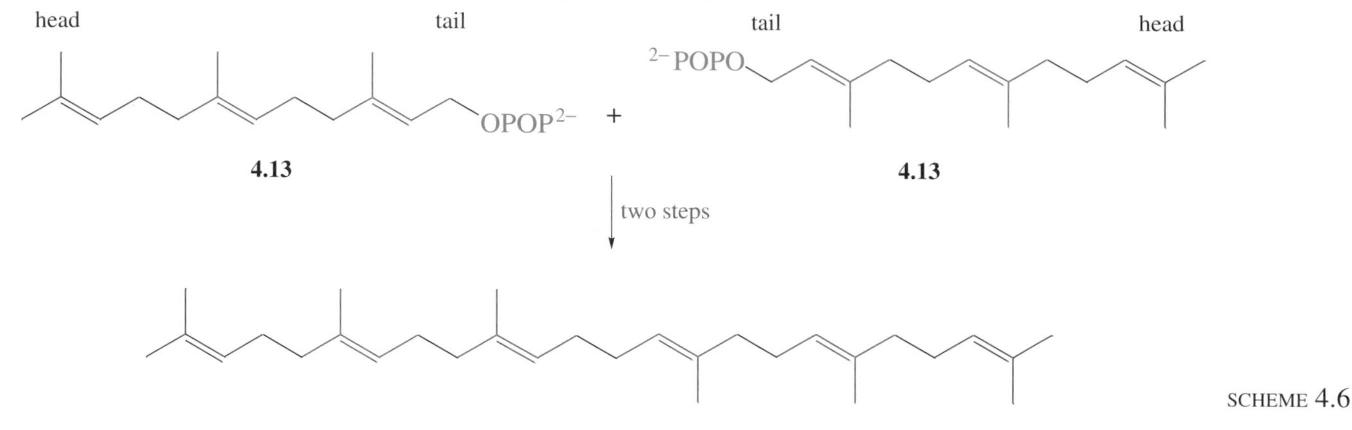


SCHEME 4.6

Squalene is the precursor of triterpenes and steroids. It was originally discovered in shark liver oil, and for a long time was considered to be of little interest. All this changed when it was discovered that squalene is the biological precursor of cholesterol and the steroid hormones. If the long squalene chain folds as shown in Reaction 4.6, you can see how the carbon skeleton of steroids like cholesterol (**4.18**) can be formed. Note that the steroids are not terpenes since some of the methyl groups have moved to different positions on the backbone or have been removed completely!

A small number of plants have enzymes that continue to catalyse the head-to-tail coupling reaction until 1 000–5 000 $C_5$ units are joined together. This naturally occurring polymeric terpene is known as gutta-percha. A similar biosynthetic process leads to natural rubber (Figure 4.3), which is also a polymeric member of the terpene family.

**Figure 4.3**
Tapping rubber trees for natural rubber.

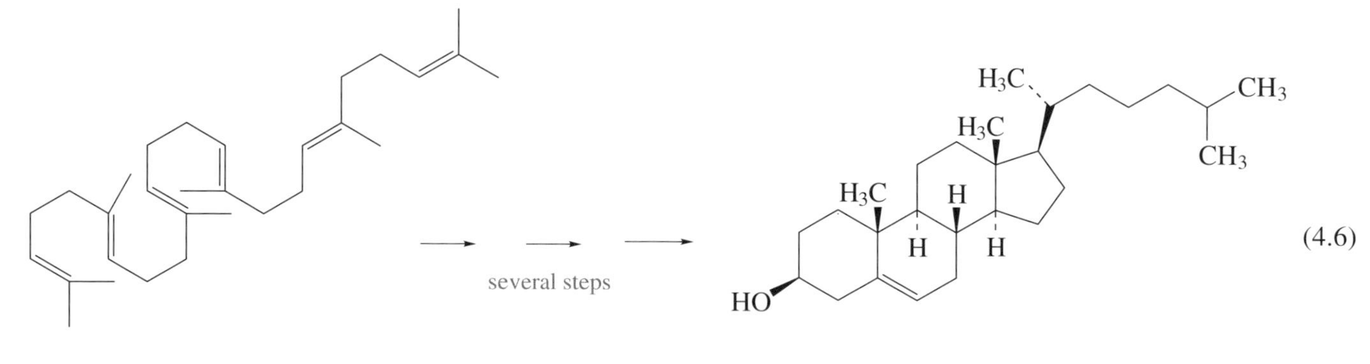


(4.6)

## 4.1 Summary of Section 4

1. Retrosynthetic analysis of the strategy for the biosynthesis of monoterpenes identifies two synthons, one nucleophilic (**4.1**) and one electrophilic (**4.2**), both of which contain five carbon atoms.
2. Isoprene itself is not a reagent involved in the biosynthesis of terpenes, although the key intermediates in the biosynthetic pathway do contain five carbon atoms.
3. In a series of enzyme-mediated reactions, mevalonic acid (which contains six carbon atoms) is converted into isopentenyl pyrophosphate and dimethylallyl pyrophosphate. These two compounds correspond to the nucleophilic and electrophilic synthons, respectively, identified by retrosynthetic analysis. Isopentenyl pyrophosphate is the 'active isoprene unit'.
4. Dimethylallyl pyrophosphate and isopentenyl pyrophosphate undergo a head-to-tail coupling reaction to produce geranyl pyrophosphate, the precursor of the monoterpenes.
5. Reaction of geranyl pyrophosphate with isopentenyl pyrophosphate leads to farnesyl pyrophosphate, the precursor of the sesquiterpenes.
6. In certain plants this chain continues, and geranylgeranyl pyrophosphate, the precursor of diterpenes, is produced.
7. Certain plants and animals possess an enzyme that catalyses the tail-to-tail coupling of two farnesyl pyrophosphate molecules to give squalene, the precursor of the triterpenes and the steroids.

### QUESTION 4.1

Are the following terpenes (i)–(iv) likely to be formed from geranyl pyrophosphate, farnesyl pyrophosphate, geranylgeranyl pyrophosphate or squalene?

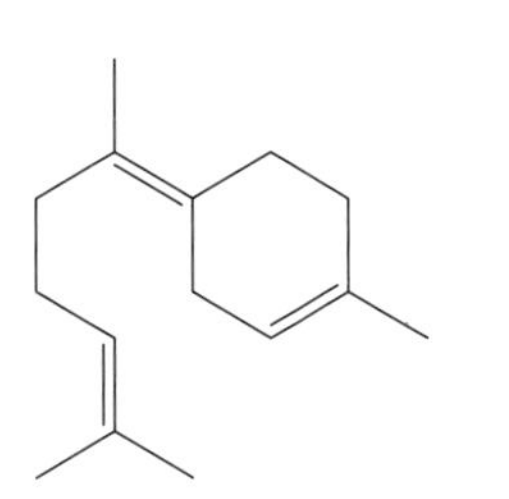

(i)

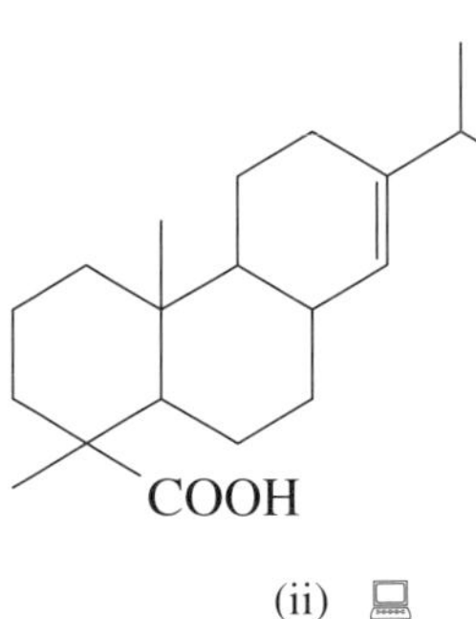


(ii)

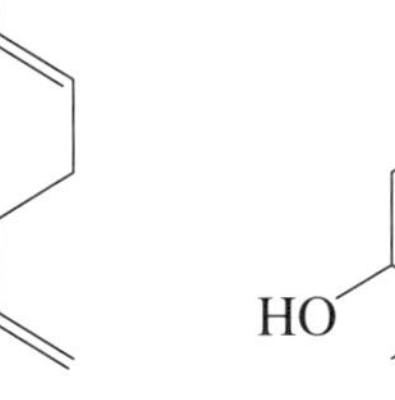

(iii)

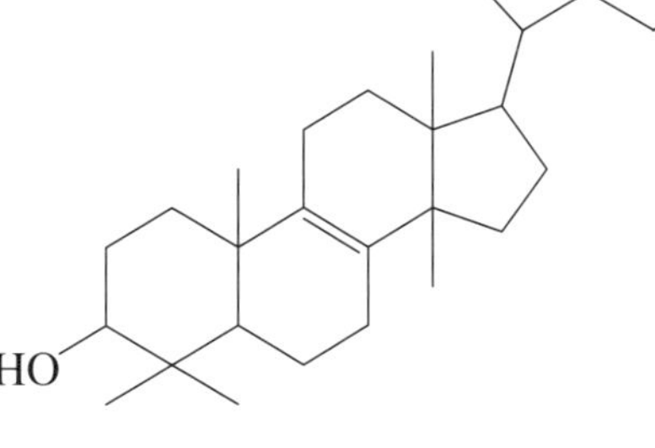


(iv)

# THE CHEMISTRY OF TERPENE BIOSYNTHESIS

# 5

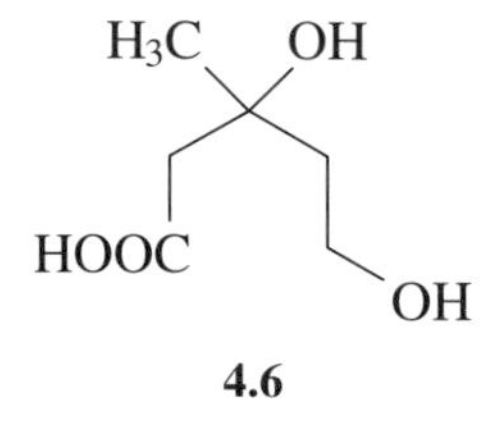


In this Section we shall look more closely at the reactions involved in the conversion of mevalonic acid (**4.6**) into geranyl pyrophosphate (**4.11**). We have reproduced the main points of the biosynthetic pathway in the Appendix. You will see that the Appendix contains some additional information relating to the first three steps of this conversion, but before discussing this, we shall concentrate on the structure of the starting material, mevalonic acid.

- Does mevalonic acid possess a chiral carbon atom?
- Yes. The carbon atom bonded to the methyl group is bonded to three other different groups, and so it is chiral.

There are therefore two enantiomers of mevalonic acid. In Section 3, we stated that as most enzymes are chiral, they can differentiate between stereoisomers, and will only catalyse the reaction of one of them. It has been shown that the *R*-isomer of mevalonic acid (**5.1**) is the one accepted by plants for incorporation in the biosynthesis of terpenes. As shown in Section 3, mevalonic acid reacts with adenosine triphosphate (ATP, **3.2**) to form successively mevalonic acid 5-phosphate (**5.2**), mevalonic acid 5-pyrophosphate (**5.3**), and finally mevalonic acid 3-phosphate-5-pyrophosphate (**5.4**). ATP is acting here as a **phosphorylating agent**.

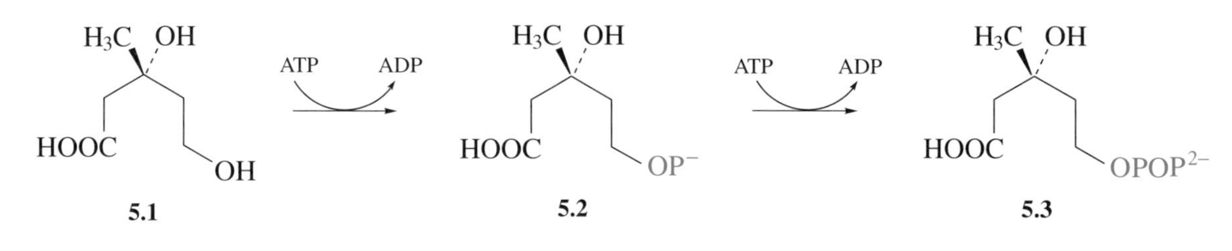


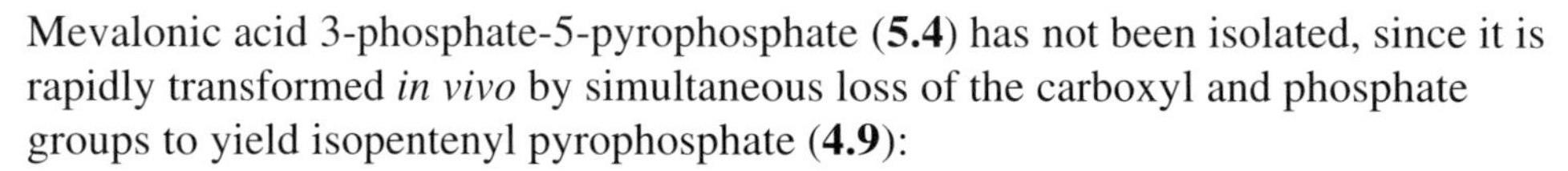

Mevalonic acid 3-phosphate-5-pyrophosphate (**5.4**) has not been isolated, since it is rapidly transformed *in vivo* by simultaneous loss of the carboxyl and phosphate groups to yield isopentenyl pyrophosphate (**4.9**):

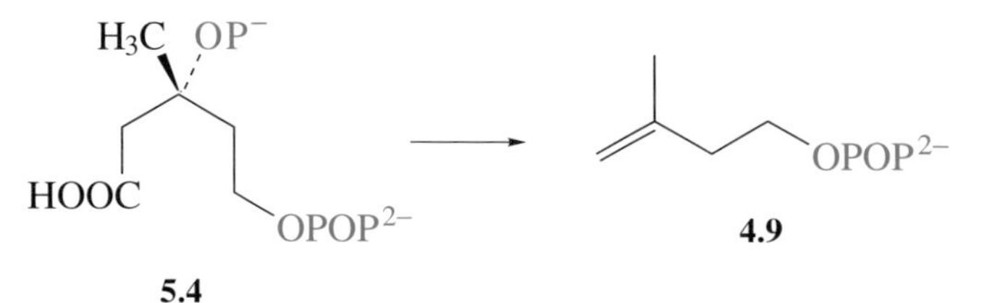


(5.1)

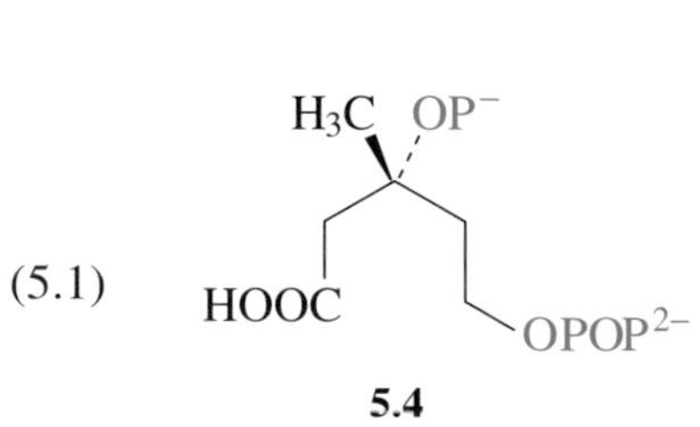


SCHEME 5.1

This type of reaction is not often encountered in enzyme chemistry, but it may seem familiar to you. It is similar to the decarboxylative elimination* of 3-bromocarboxylic acids; for example, 2,2-dimethyl-3-bromopropanoic acid (**5.5**) is decarboxylated by base as follows:

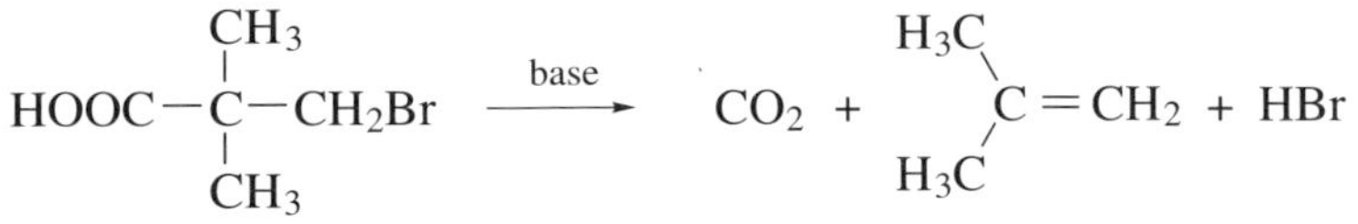

$$HOOC{-}C(CH_3)_2{-}CH_2Br \xrightarrow{\text{base}} CO_2 + (H_3C)_2C{=}CH_2 + HBr \qquad (5.2)$$

**5.5**

In the first step of this reaction, the carboxylate anion **5.6** is formed, which then loses carbon dioxide and a bromide anion, with concomitant formation of a double bond. Note the stereochemical requirement is the same as that for other E2 elimination reactions; that is, the departing groups must be antiperiplanar:

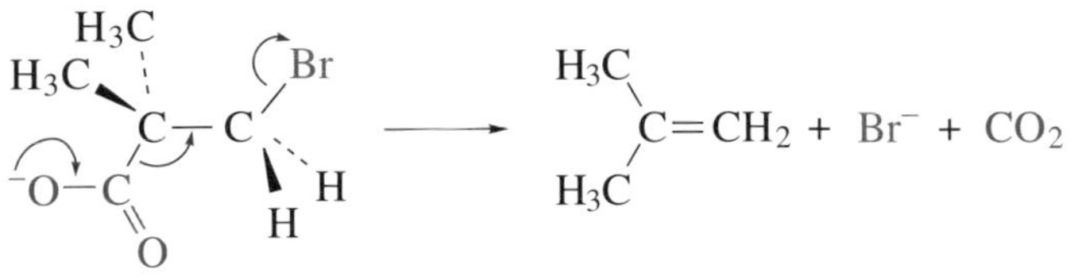

$$^{-}O{-}C(=O){-}C(CH_3)_2{-}CH_2Br \longrightarrow (H_3C)_2C{=}CH_2 + Br^- + CO_2 \qquad (5.3)$$

**5.6**

Is the conformation shown in **5.6** the preferred conformation?

Yes; in this antiperiplanar conformation, the two bulkiest groups on each carbon atom of the incipient double bond in **5.6** are furthest away from each other, as the Newman projection **5.7** shows:

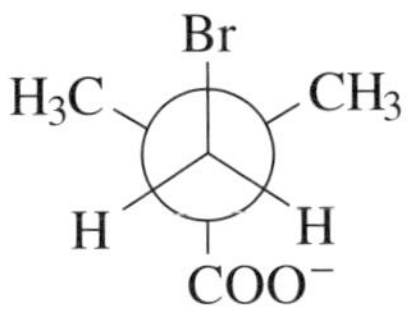

Br, $H_3C$, $CH_3$, H, H, $COO^-$

**5.7**

The point to note in Reaction 5.1 is that for a similar elimination to occur with the mevalonic acid derivative **5.2**, the 3-hydroxyl group has to be converted into a 3-phosphate group (via mevalonic acid 5-pyrophosphate), giving **5.4**.

Why do you think this is?

For this reaction to occur, there has to be a good leaving group on the carbon atom in a 3-position relative to the carboxylic acid group. In contrast to the hydroxyl group, phosphate is a good leaving group.

So, it appears that phosphorylation is Nature's way of converting hydroxyl groups into good leaving groups.

### QUESTION 5.1

Use curly arrows to illustrate the mechanism for the transformation of mevalonic acid 3-phosphate-5-pyrophosphate (**5.4**) into isopentenyl pyrophosphate (**4.9**; see Reaction 5.1).

* The mechanism of decarboxylative elimination is discussed in Part 3 of *Chemical Kinetics and Mechanism*[3].

The next process in the biosynthetic pathway involves the conversion of isopentenyl pyrophosphate (**4.9**) into dimethylallyl pyrophosphate (**4.10**); see Reaction 5.4. In this process, a disubstituted alkene (one with two alkyl substituents on one carbon, and two hydrogen substituents on the other carbon atom) is partially transformed into a trisubstituted alkene (three alkyl substituents and one hydrogen substituent), as an equilibrium mixture of the two compounds is established:

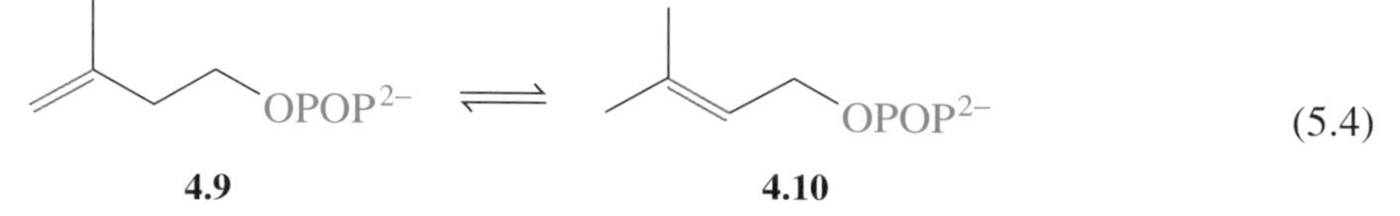


(5.4)

In the laboratory, this type of alkene isomerization involves addition of a proton, followed by elimination of a different proton. Thus, in Reaction 5.5, the less-stable disubstituted alkene **5.8** is isomerized into the thermodynamically more-stable trisubstituted compound **5.9**. In plants, an enzyme is present that catalyses the alkene isomerization shown in Reaction 5.4.

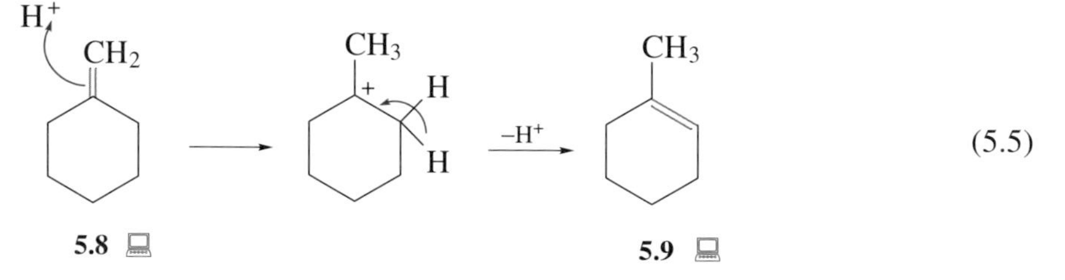


(5.5)

This enzyme-catalysed isomerization is particularly important in view of the next step in the biosynthetic pathway, which involves the formation of geranyl pyrophosphate (**4.11**) by the head-to-tail coupling of dimethylallyl pyrophosphate (**4.10**) and isopentenyl pyrophosphate (**4.9**). The new bond joining the reactants is shown in black in the product, **4.11**.

(5.6)

Look at this transformation with respect to dimethylallyl pyrophosphate, **4.10**:

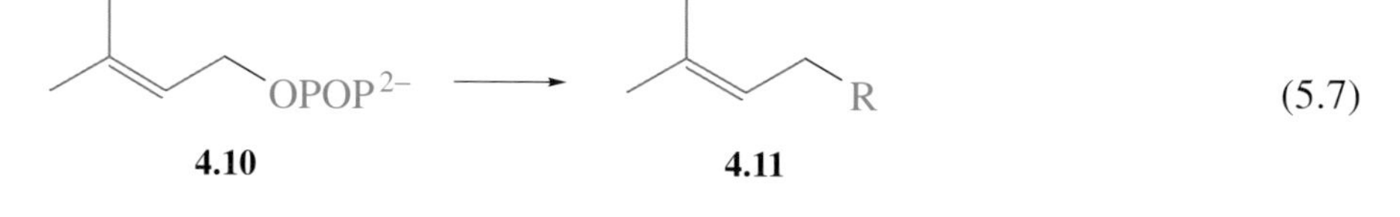


(5.7)

What type of reaction (substitution, addition or elimination) is this?

It is a substitution reaction.

However, when you try to classify Reaction 5.6 according to the bonding changes that take place as isopentenyl pyrophosphate (**4.9**) is converted into product, it is not so straightforward. At first sight, it appears that addition has occurred across the carbon–carbon double bond in the original alkene, but in the product a new double bond is present in a different position.

Let's concentrate on the idea that this is a nucleophilic substitution reaction with respect to dimethylallyl pyrophosphate (**4.10**):

$$\text{OPOP}^{2-} + \text{Nu}^- \longrightarrow \text{Nu} + \text{OPOP}^{3-} \qquad (5.8)$$

**4.10**

- How many distinct $S_N$ mechanisms are there?*
- There are two distinct mechanisms, $S_N1$ and $S_N2$, and, in principle, **4.10** could react by either route.

- By which mechanism would you expect **4.10** to react?
- You might expect it to be the one-step $S_N2$ mechanism, because the leaving group is attached to a primary carbon atom:

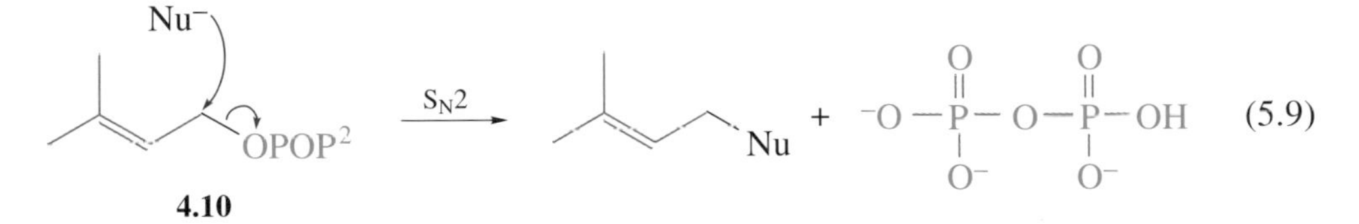


Such a mechanism *is* possible, but let's examine the alternative, $S_N1$, pathway.

- If the reaction proceeded by an $S_N1$ mechanism, would the carbocation intermediate from **4.10** be stabilized in any way? If so, how?
- The positive charge of the carbocation intermediate **5.10** derived from **4.10** would be resonance stabilized† by the adjacent double bond:

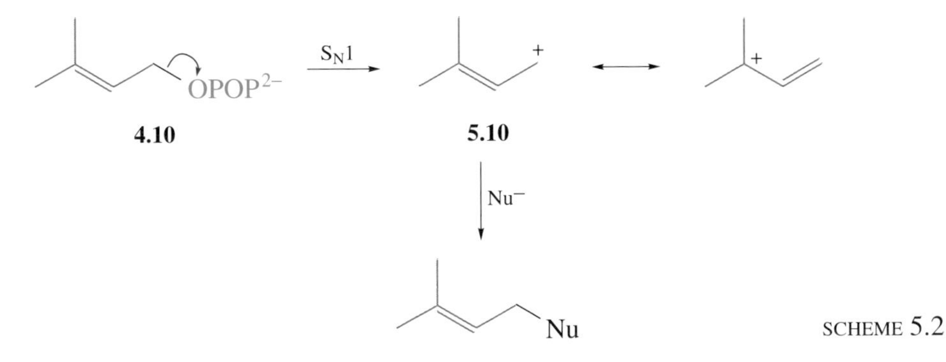


SCHEME 5.2

Dimethylallyl pyrophosphate (**4.10**) could therefore conceivably react by either an $S_N1$ or $S_N2$ type of mechanism, although it is now generally accepted that the $S_N1$ route is the more probable. Evidence for this has been obtained using the trifluoromethyl analogue **5.11** instead of **4.10** as the substrate for enzymes producing monoterpenes:

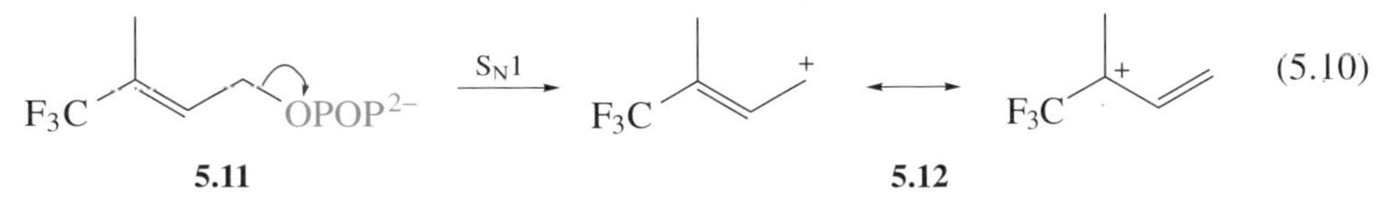


* The mechanisms of substitution reactions are discussed in Part 2 of *Chemical Kinetics and Mechanism*[3].

† Resonance stabilization is covered on the CD-ROM associated with *Chemical Kinetics and Mechanism*[3].

It was found that the rate of conversion of **5.11** was one million times slower than the natural substrate. This finding is understandable if the reaction proceeds via an $S_N1$ route, since it will involve the carbocation **5.12** as an intermediate. The trifluoromethyl group is a very strongly electron-withdrawing group. So a resonance form that places a positive charge on a carbon adjacent to the trifluoromethyl group, as in the right-hand resonance structure of **5.12**, would be substantially destabilized. This would significantly disfavour this resonance form, making the formation of the carbocation much more difficult.

In nature the double bond of isopentyl pyrophosphate **4.9** acts as the nucleophile to give a carbocation that eliminates $H^+$ as in the second step of an E1 mechanism.

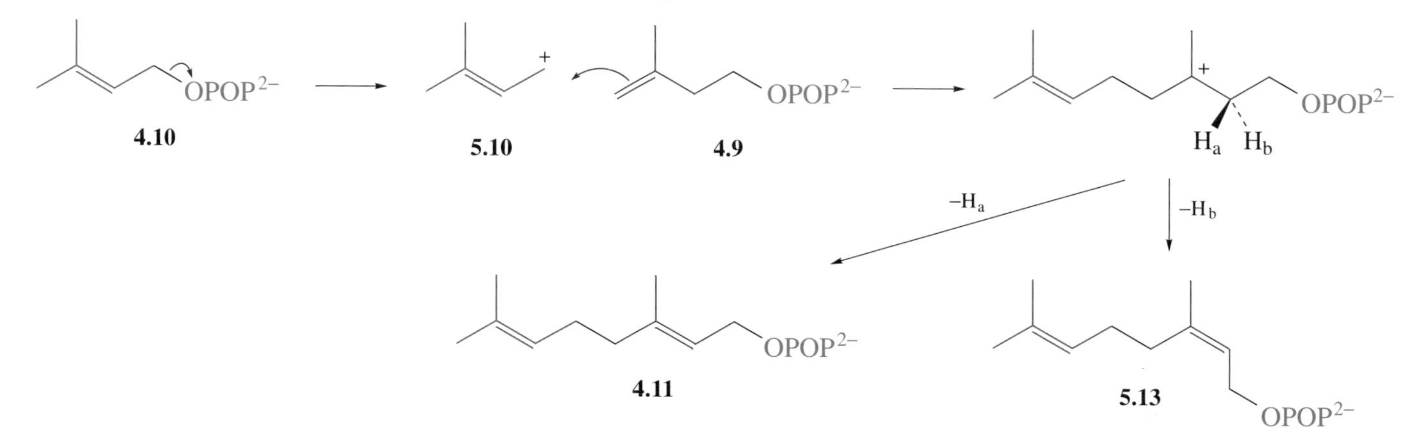


SCHEME 5.3

An enzyme determines which of the diastereoisomers, geranyl pyrophosphate (**4.11**) or neryl pyrophosphate (**5.13**) is produced; both types of elimination do occur, but in different species! Interestingly, the geometry of the resulting alkene is linked to which of the two protons labelled $H_a$ and $H_b$ in Scheme 5.3 is eliminated. In some species the enzyme ensures that the proton that is lost is $H_a$, leading to geranyl pyrophosphate (**4.11**), whereas other species have enzymes that remove proton $H_b$ to give its diastereoisomer, neryl pyrophosphate (**5.13**). This is an excellent demonstration of regio- and stereoselectivity in enzyme-catalysed reactions.

## 5.1 Summary of Section 5

1. The reactions linking mevalonic acid to geranyl pyrophosphate can be readily understood in terms of the chemical principles introduced in Sections 3 and 4.
2. By means of three successive reactions with adenosine triphosphate (ATP), mevalonic acid (**5.1**), is converted into mevalonic acid 3-phosphate-5-pyrophosphate (**5.4**).
3. Mevalonic acid 3-phosphate-5-pyrophosphate (**5.4**) spontaneously undergoes a decarboxylative elimination reaction, giving isopentenyl pyrophosphate (**4.9**). The phosphate and pyrophosphate groups are good leaving groups in biochemical reactions.
4. Isopentenyl pyrophosphate (**4.9**) is transformed into dimethylallyl pyrophosphate (**4.10**) by an enzyme-catalysed reaction.
5. The head-to-tail coupling reactions involving isopentenyl pyrophosphate could proceed by one of two substitution mechanisms. However, evidence points to the $S_N1$ route as the one that occurs in nature.

# FROM TRITERPENES TO STEROIDS

# 6

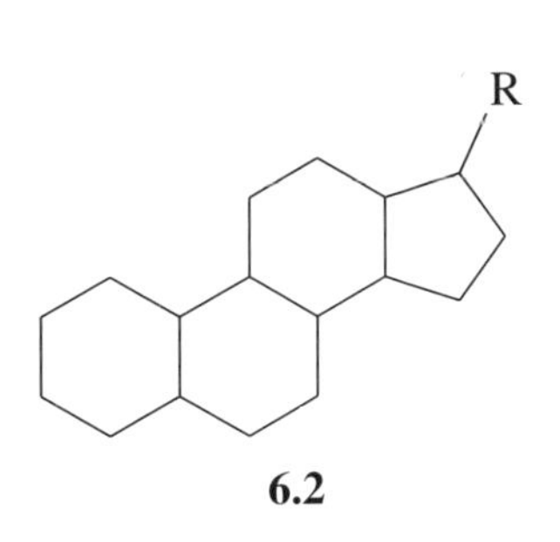


In Scheme 4.3, we indicated that squalene (**4.17**) was the biogenetic precursor to the $C_{30}$ triterpenes and steroids. In this Section, we shall explore this connection further. Extensive labelling studies (see Box 6.1) with the isotope labels, $^{2}H$, $^{13}C$ and $^{14}C$, in a large variety of both plants and animal cell and cell-free preparations, have led to an understanding of many of the steps involved in the complex bio-transformations that lead from squalene (**4.17**) to both triterpenes and steroids. We shall confine ourselves to studying the conversion of **4.17** into the $C_{30}$ triterpene lanosterol (**6.1**), a major constituent of sheep wool wax, and to the $C_{27}$ steroid cholesterol (**4.18**). All steroids are based on the same skeletal structure **6.2**.

4.17

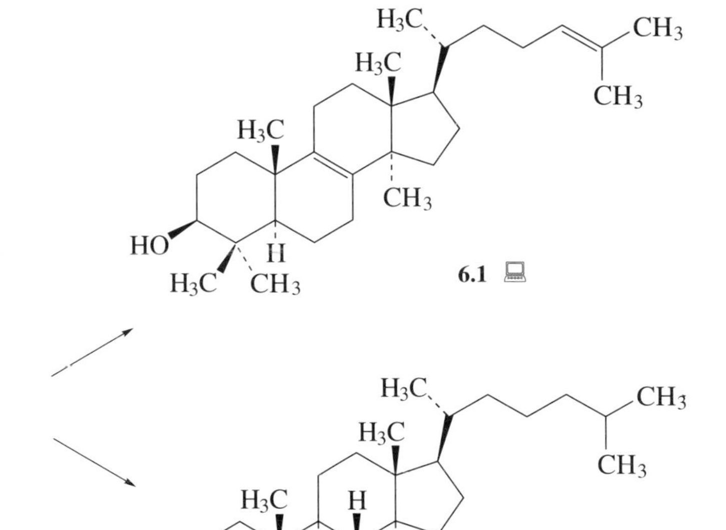


SCHEME 6.1

## BOX 6.1 Labelling studies

One important way in which biochemical pathways are elucidated is to use labelling studies. Effectively, a molecule that is known (or thought) to be involved in the biochemical pathway is prepared synthetically with one or more of its atoms replaced with an uncommon isotope of the atom. For hydrogen, deuterium ($^{2}H$) is used, and $^{12}C$ carbon atoms are replaced by $^{13}C$ or $^{14}C$ isotope labels. These labelled molecules are then fed to the organism (which doesn't notice any difference), and the products of the biochemical pathway are analysed to see where the isotope label ends up in the products. In this way, we can track which bits of the molecule end up where, and which parts are lost — usually as $CO_2$ and $H_2O$!

Let us begin by examining the conversion of squalene (**4.17**) into lanosterol (**6.1**). The initiation of this cyclization reaction puzzled chemists for quite a time, but it is now known that the first step involves an enzyme-mediated reaction of **4.17** with oxygen, to give the oxirane **6.3** (see Reaction 6.1). Oxiranes are strained and reactive ring systems, which are readily opened under a variety of conditions.

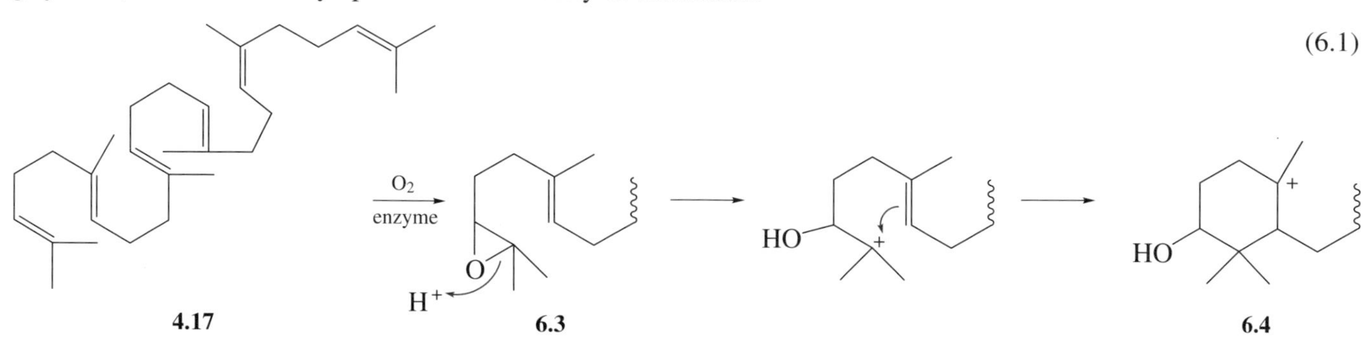


Oxiranes usually react by an $S_N2$ process:

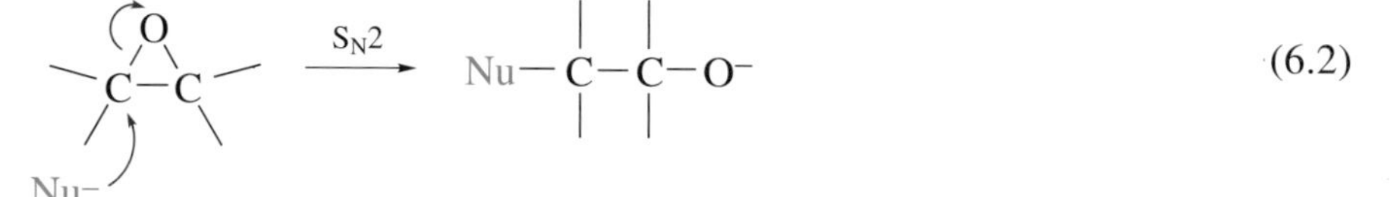


However, in the presence of an acid (either a Brønsted acid or a Lewis acid), they can react via an $S_N1$ process.

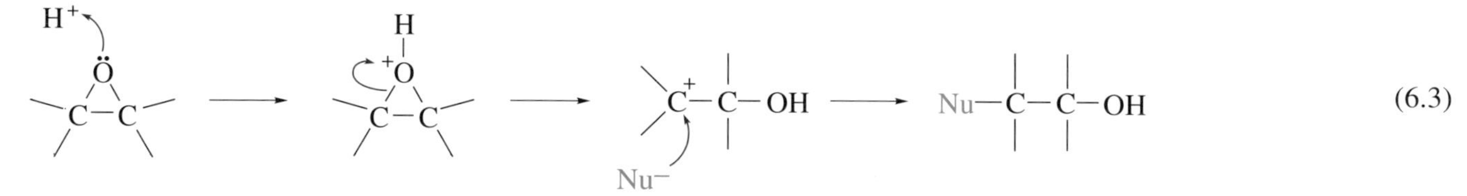


It is this oxirane ring-opening that triggers squalene cyclization. We shall confine ourselves firstly to the construction of just one cyclohexane ring, and use a partial structure for squalene (**4.17**) for clarity. Reaction 6.1 shows the consequence of ring-opening of the oxirane **6.3**, yielding a tertiary carbocation. The folding of the squalene molecule brings this carbocation in close proximity to a carbon–carbon double bond, allowing it to attack as a nucleophile. This results in formation of another tertiary carbocation, **6.4**.

The partial structure of the carbocation **6.4** may now be expanded to show the rest of the intermediate. The construction of the remaining rings then follows the same principles as those shown in Reaction 6.1. Successive carbocations are generated and then attacked by a 'nearby' double bond. The next steps in the cyclization of **6.4** are shown in full in Scheme 6.2.

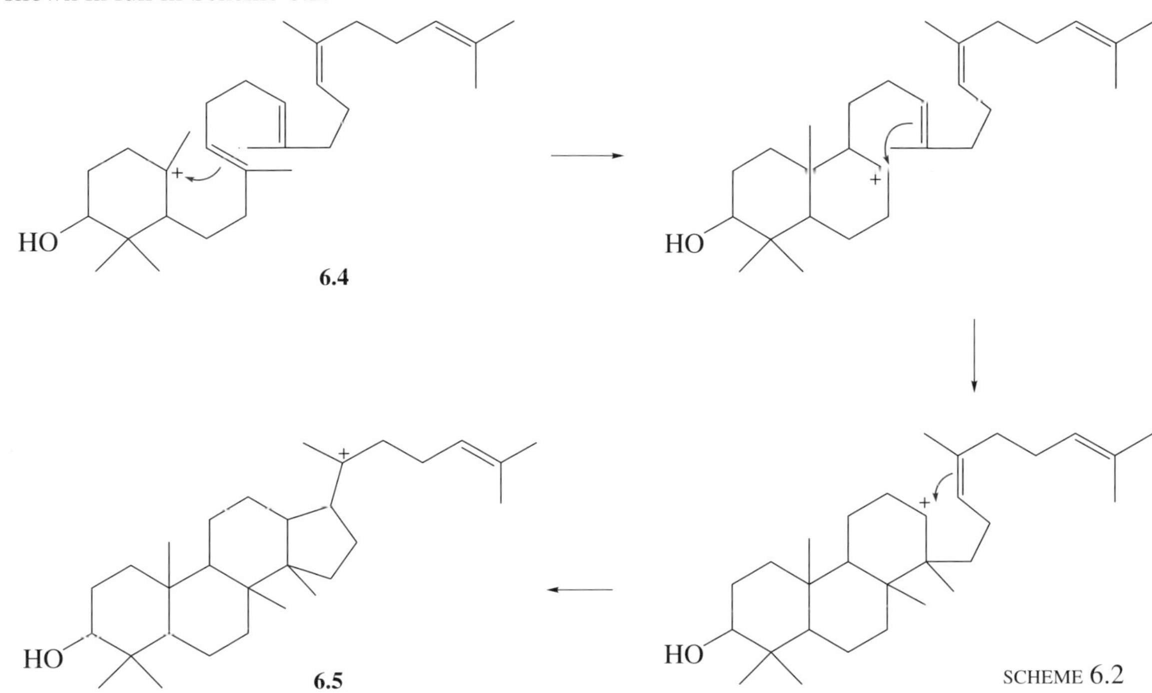


SCHEME 6.2

The final intermediate in Scheme 6.2 is **6.5**, which is still a carbocation, but we are not quite 'home and dry' yet! A series of two 1,2-hydride migrations (as in Reaction 6.4) occurs, which are followed by two 1,2-methyl shifts (as in Reaction 6.5); such rearrangements are discussed in more detail in Box 6.2. These shifts and migrations involve the movement or jump of an atom or group with a pair of bonding electrons. Such shifts and migrations are common in terpene chemistry, and they are also implicated in the biosynthesis of many steroids.

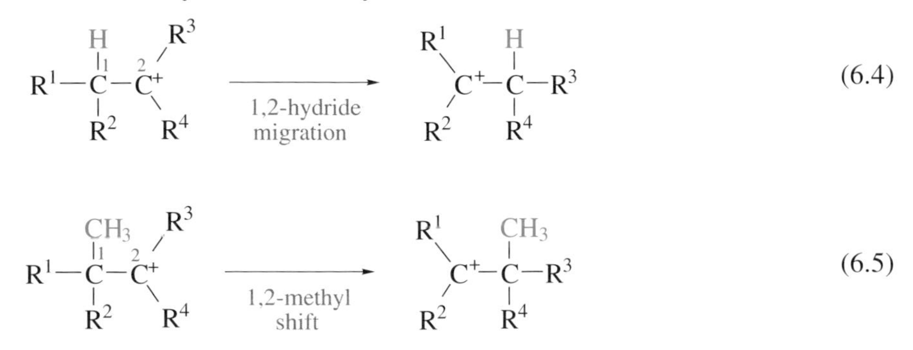

## BOX 6.2 1,2-hydride migration and 1,2-methyl shifts

You have probably met **1,2-hydride migrations** and 1,2-methyl shifts before. When $AlCl_3$ is used to catalyse a Friedel–Crafts alkylation, there is often rearrangement of the electrophile, as in the alkylation of benzene; for example:

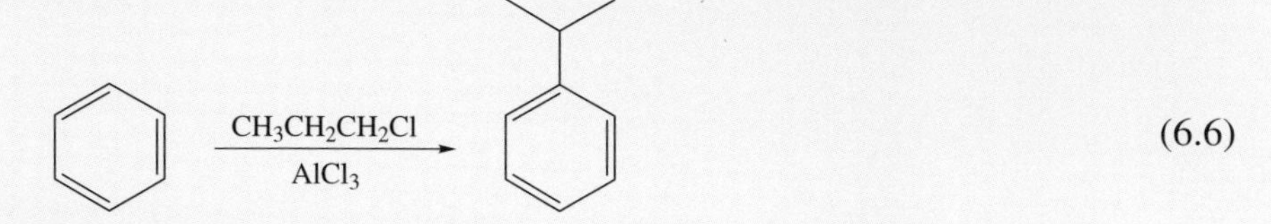


(6.6)

This is because reaction of $AlCl_3$ with the haloalkane produces a primary carbocation:

$$CH_3{-}CH_2{-}CH_2{-}Cl + AlCl_3 \longrightarrow CH_3{-}CH_2{-}\overset{+}{C}H_2 + {}^{-}AlCl_4 \qquad (6.7)$$

This primary carbocation rearranges to a more-stable secondary carbocation, via a 1,2-hydride migration:

$$H_3C{-}\overset{\overset{\displaystyle H}{|}}{C}H{-}\overset{+}{C}H_2 \longrightarrow H_3C{-}\overset{+}{C}H{-}\overset{\overset{\displaystyle H}{|}}{C}H_2 \qquad (6.8)$$

This is followed by attack on the benzene ring.

**1,2-methyl shifts** also occur in Friedel–Crafts reactions* to give a more-stable electrophile:

$$H_3C{-}\underset{\underset{\displaystyle CH_3}{|}}{\overset{\overset{\displaystyle CH_3}{|}}{C}}{-}\overset{+}{C}H_2 \longrightarrow H_3C{-}\underset{\underset{\displaystyle CH_3}{|}}{\overset{+}{C}}{-}\overset{\overset{\displaystyle CH_3}{|}}{C}H_2 \qquad (6.9)$$

So it is not surprising that such shifts are also observed to take place in biological systems.

In Scheme 6.3, the sequence of hydride and methyl migrations (shown in blue) which ensue from the carbocation **6.5** is indicated as a series of discrete steps. However, in nature, the sequence is probably concerted. Although the stereochemistry of these processes has not been shown, it is known; in Scheme 6.3, the absolute configuration of only the end product, lanosterol (**6.1**) is shown.

* The occurrence of 1,2-methyl shifts in Friedel–Crafts alkylation is discussed in Part 2 of *Alkenes and Aromatics*[4].

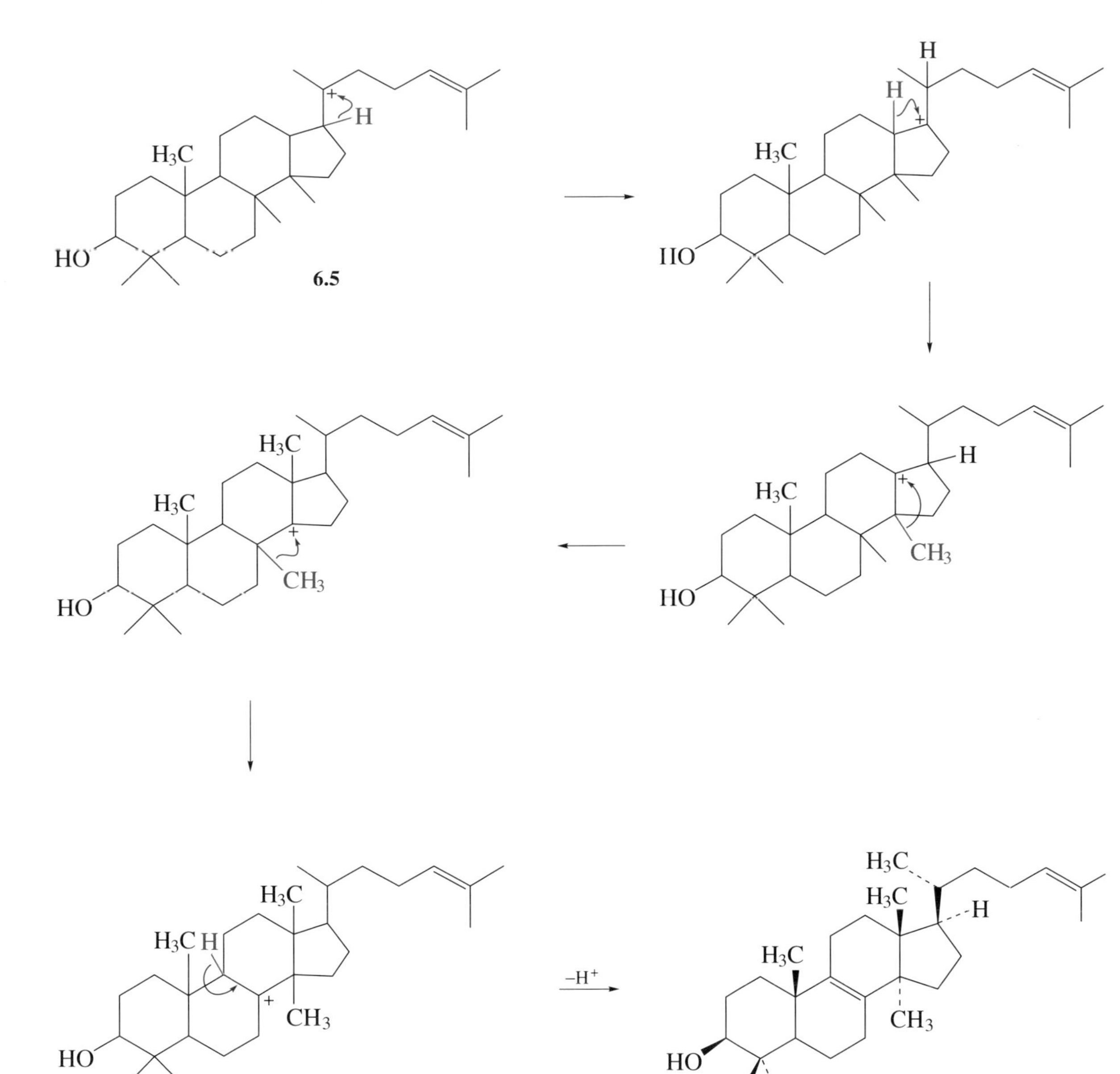


SCHEME 6.3

We can now turn briefly to the biosynthesis of cholesterol (**4.18**). A very extensive series of studies has shown that lanosterol, **6.1**, is the precursor. The conversion of **6.1** into **4.18** involves initial reduction of the side-chain double bond. A series of oxidations then ensues, resulting in the removal of three methyl groups (in red), and leading to cholesterol (**4.18**).

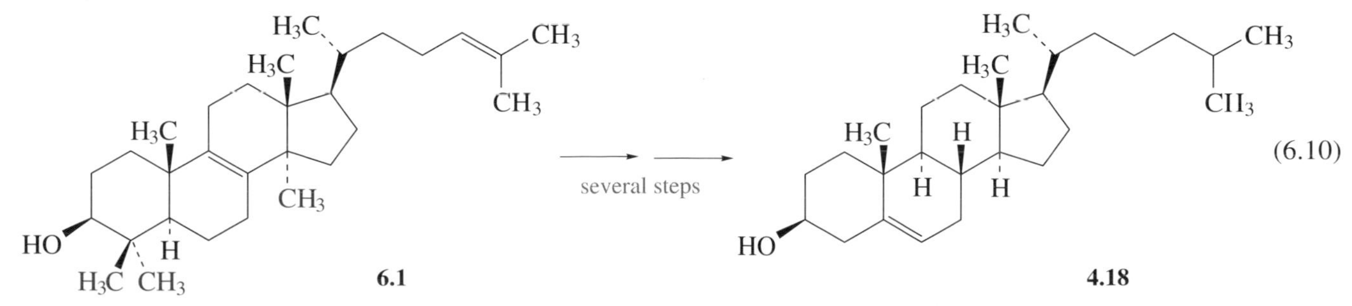


(6.10)

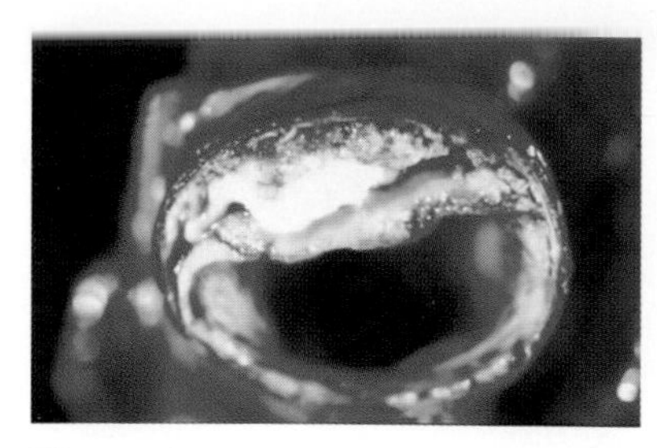

**Figure 6.1**
Build-up of plaque on artery walls.

Coronary heart disease is the major cause of death in the Western world. Atherosclerosis is a major factor, and it is caused by the build-up of plaque (fatty deposits) on artery walls (Figure 6.1).

Cholesterol (**4.18**) is a major component of this plaque, and, as outlined earlier, it is biosynthesised in the liver from acetyl coenzyme A. It was discovered that the rate-limiting step in cholesterol production lay very early in the biosynthetic pathway. It was, in fact, the step in which hydroxymethylglutaryl coenzyme A (**4.7**) was reduced in the presence of the enzyme hydroxymethylglutaryl coenzyme A reductase, to yield the mevalonic acid coenzyme A complex **6.6**. A search was made for substrates that would inhibit this stage in the cholesterol biosynthesis. A series of mould metabolites was eventually isolated; these were found to be inhibitors, and are now known loosely as 'statins'. Simvastatin (**6.7**, trade name Zocor$^{TM}$) is one of these.

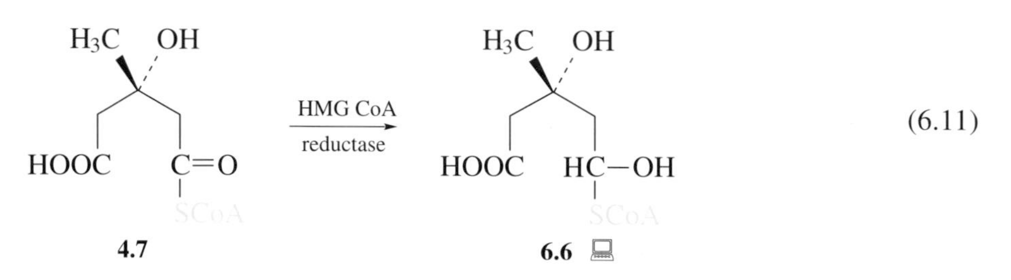


The relationship between simvastatin and hydroxymethylglutaryl coenzyme A complex (HMGCoA, **6.6**) may not be immediately obvious. However, if the hydrolysis product of simvastatin (**6.8**) is compared with that of HMG CoA (**6.6**), the resemblance becomes clearer. Simvastatin (**6.7**) is bound to the enzyme HMG CoA reductase, thus inhibiting access by the true substrate. Considerable lowering of blood cholesterol levels is observed with this drug in 'compromised patients'. Simvastatin (Figure 6.2) was third in the league of top-selling drugs in the Western world during 1998, with sales approaching $4 billion!

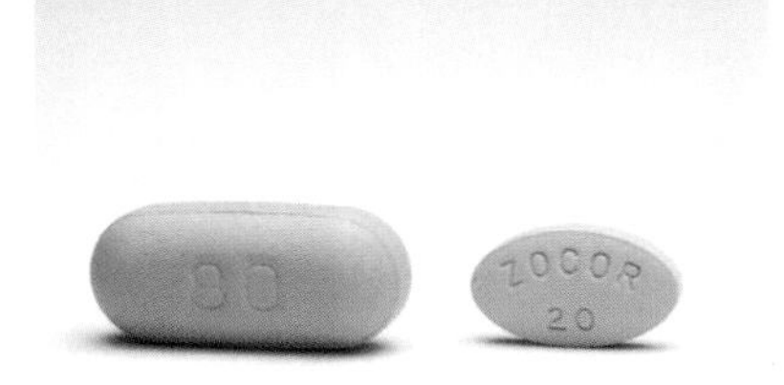


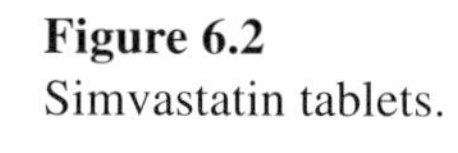

**Figure 6.2**
Simvastatin tablets.

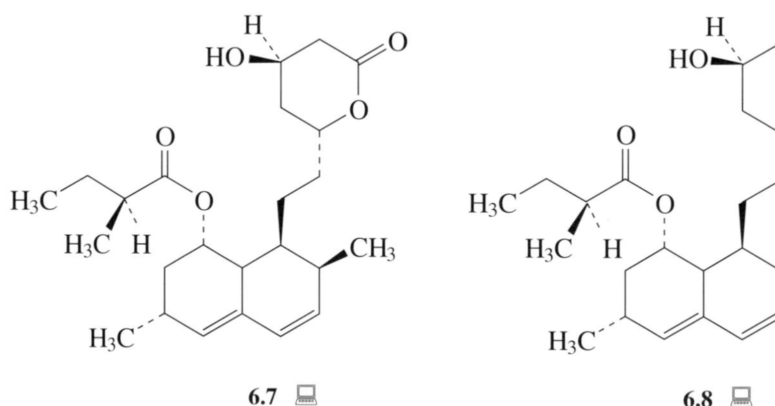

# APPENDIX TERPENE BIOSYNTHESIS FROM MEVALONIC ACID

## Biosynthesis of the active isoprene unit from mevalonic acid

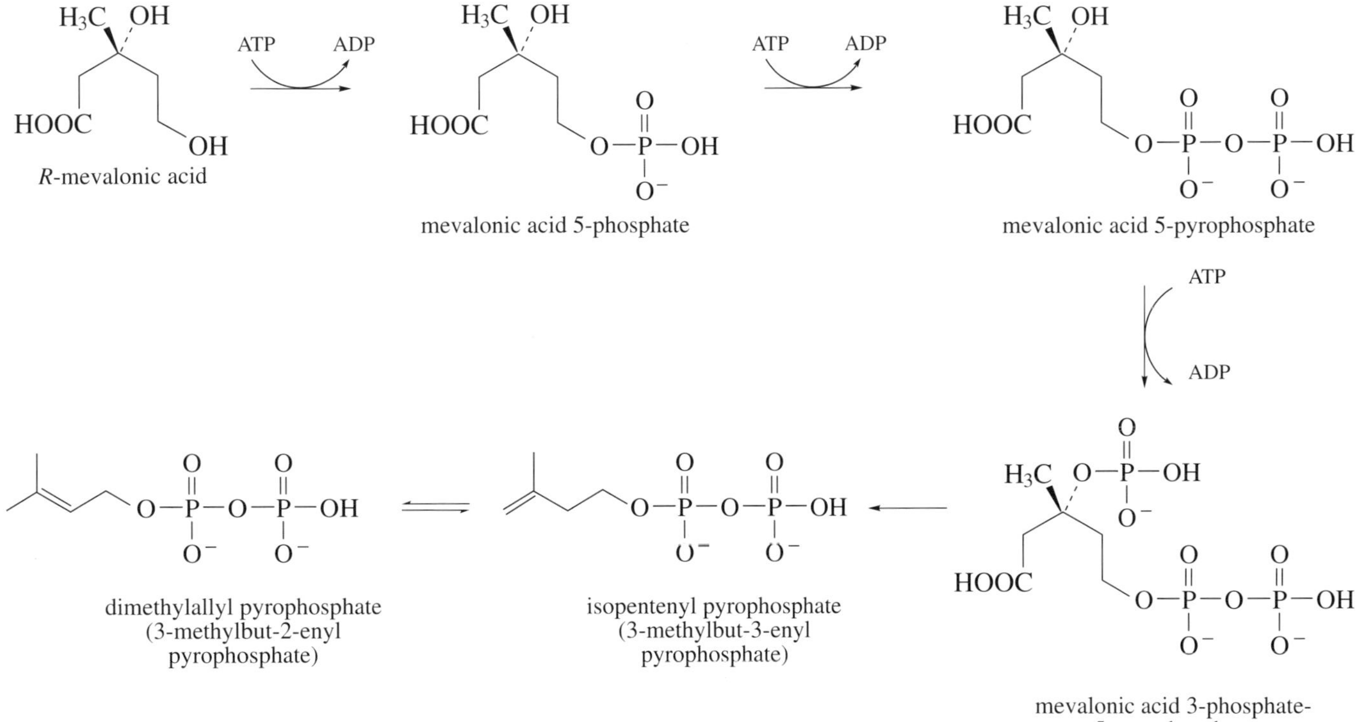


## Biosynthesis of monoterpenes

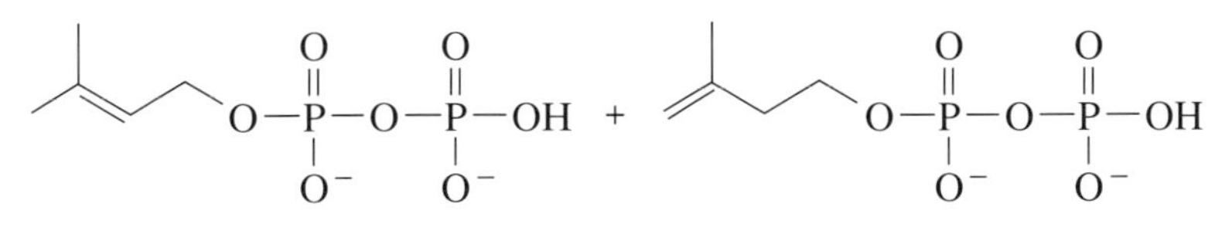


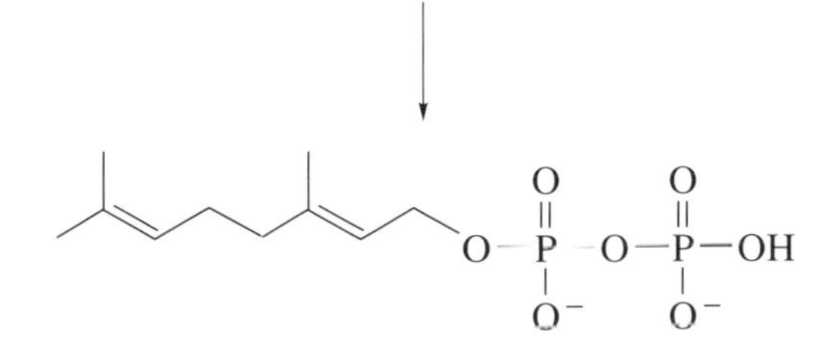

# LEARNING OUTCOMES

Now that you have completed *Mechanism and Synthesis: Part 5 — Synthesis and biosynthesis: terpenes and steroids*, you should be able to do the following things:

1 Recognize valid definitions of, and use in a correct context, the terms, concepts and principles listed in the following Table.

List of scientific terms, concepts and principles used in *Synthesis and biosynthesis: terpenes and steroids*

| Term | Page number | Term | Page number |
|---|---|---|---|
| active isoprene unit | 302 | 1,2-methyl shift | 318 |
| biosynthesis | 284 | natural product | 283 |
| biosynthetic pathway | 286 | phosphorylating agent | 310 |
| enolate | 295 | primary metabolic process | 283 |
| enolate alkylation | 295 | primary metabolite | 283 |
| head-to-tail coupling | 306 | secondary metabolite | 284 |
| 1,2-hydride migration | 318 | tail-to-tail coupling | 308 |
| isoprene rule | 288 | terpene | 286 |

2 Use the isoprene rule to identify given natural products as terpenes. (Question 1.2)

3 Classify given terpenes as mono-, sesqui-, di-, tri-, tetra-, or polyterpenes, and identify their five-carbon isoprene-like units. (Questions 1.1 and 4.1)

4 Apply the principles of retrosynthetic analysis to identify synthetic strategies for simple monoterpenes. (Question 2.1)

5 Understand the strategy of the synthesis of monoterpenes in living systems. (Question 4.1 and 5.1)

6 Give an account of the main biosynthetic pathway from mevalonic acid to geranylgeranyl pyrophosphate and squalene, and be able to specify which compound on the pathway is the precursor of a given terpene. (Questions 4.1 and 5.1)

7 Be familiar with the broad outline of terpene/steroid biosynthesis. (Questions 4.1 and 5.1)

# QUESTIONS: ANSWERS AND COMMENTS

## QUESTION 1.1 (*Learning Outcome 3*)

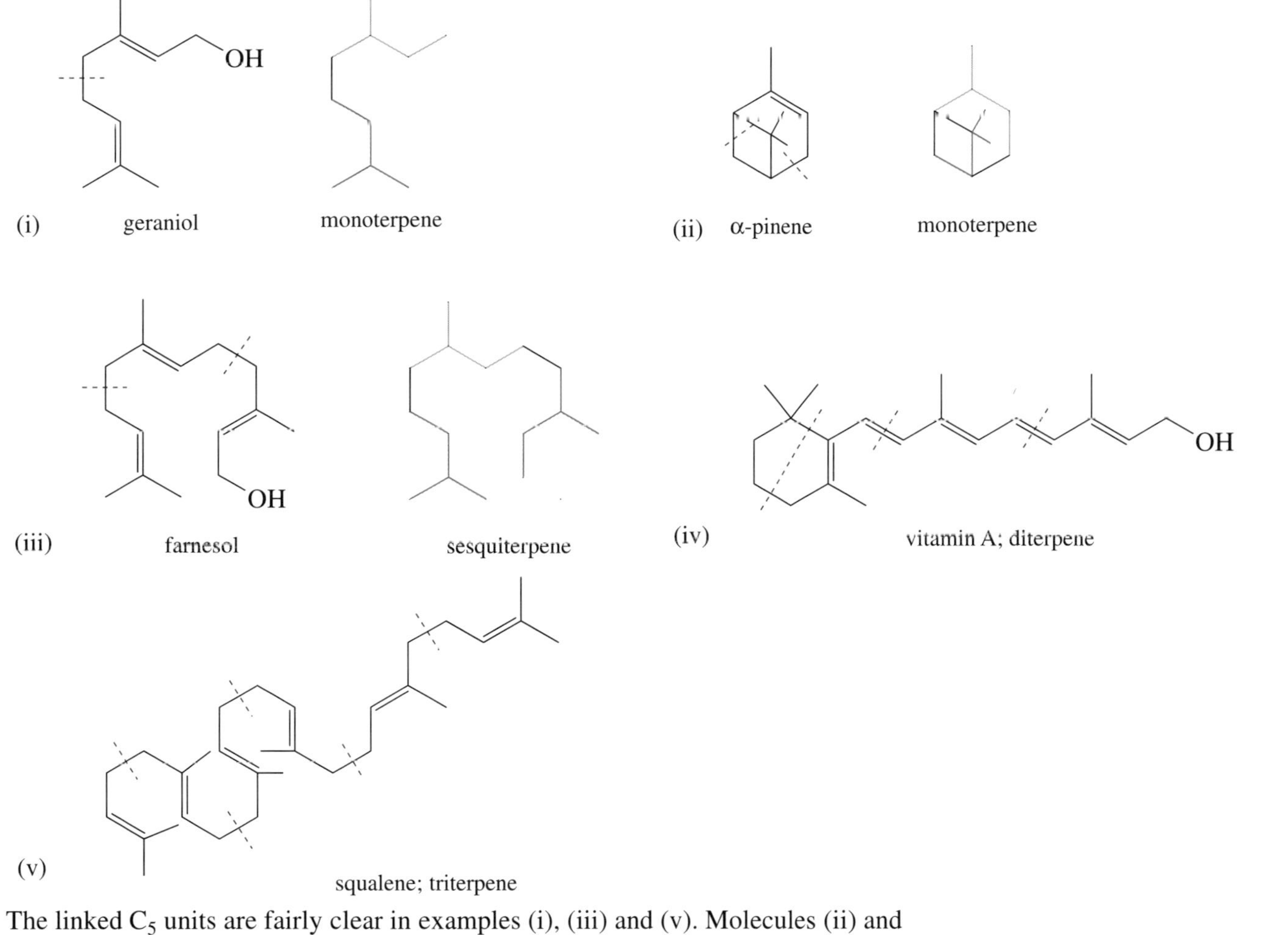


The linked $C_5$ units are fairly clear in examples (i), (iii) and (v). Molecules (ii) and (iv) contain ring systems, but with practice you should be able to identify the characteristic isoprene-like units as shown above.

## QUESTION 1.2 (*Learning Outcome 2*)

Molecules (i) and (iv) are terpenes; they both contain a multiple of five carbon atoms, and the $C_5$ isoprene-like units are readily observable, as shown right. Molecule (iv) is retinal, which is obtained by oxidation of vitamin A, and is important in the chemistry of vision.

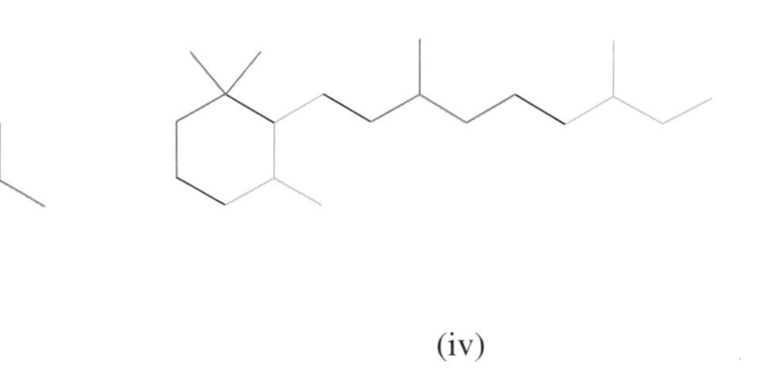


Molecule (ii), cyclodec-1,6-diene, contains ten atoms, but the isoprene-like skeleton is not present, so the compound is not a terpene (or even a natural product).

Molecule (iii), 3,5-dimethylcycloheptanone, contains only nine carbon atoms, so by our definition it is not a terpene.

## QUESTION 2.1 (Learning Outcome 4)

For **2.15**, either of two disconnections, a or b, may be made.

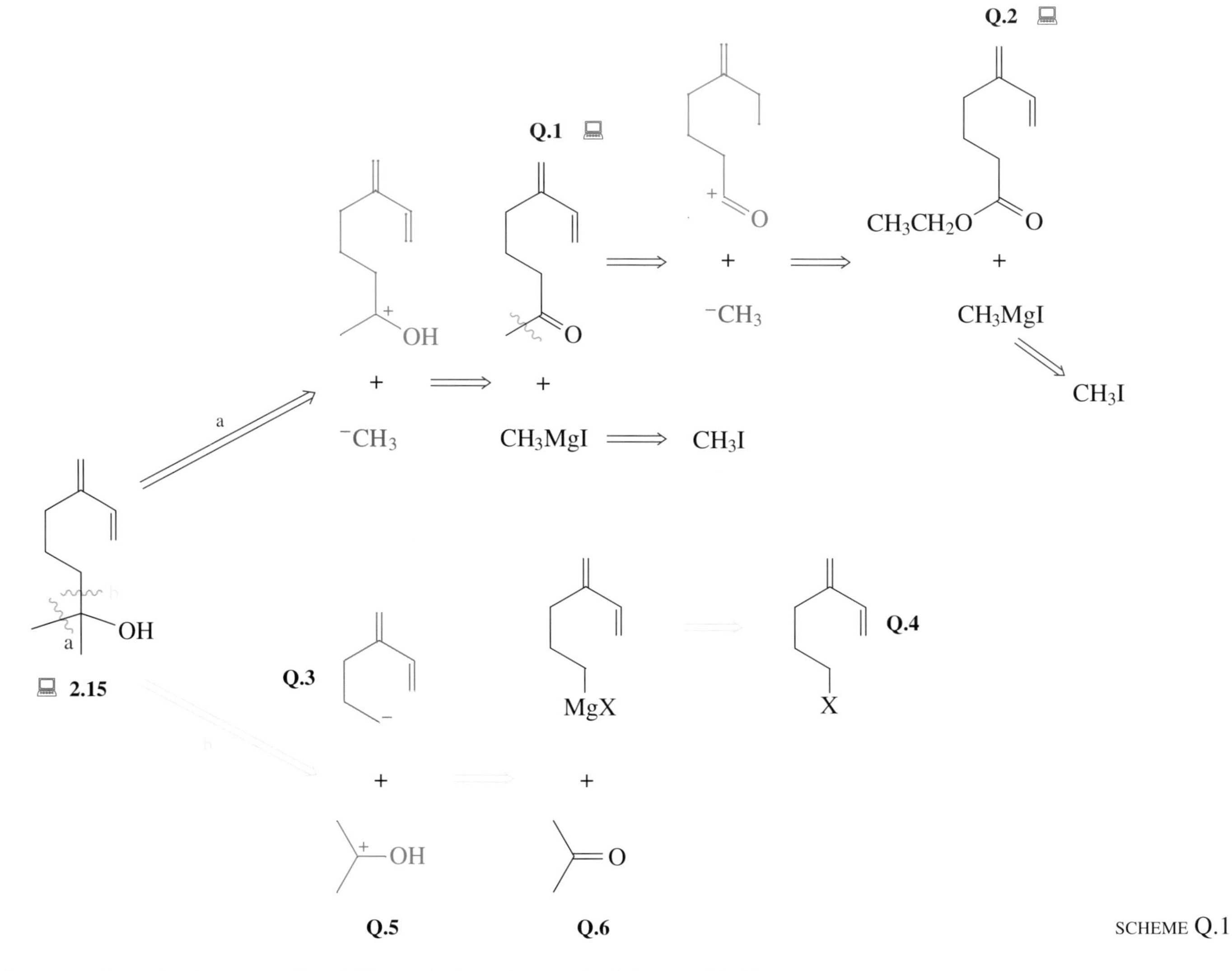


SCHEME Q.1

Disconnection a leads to a nucleophilic methyl synthon, which is provided by an appropriate organometallic compound, such as a Grignard reagent, and an electrophilic synthon that is provided by the ketone **Q.1**. Note that, because the target molecule **2.15** contains two methyl groups, **Q.1** can be further disconnected to produce an ester, **Q.2**, as the reagent.

Disconnection b leads to a nucleophilic carbon synthon, **Q.3**, which is provided by the haloalkene **Q.4** via an organometallic reagent, and an electrophilic synthon, **Q.5**, which is provided by propanone, **Q.6**.

For **2.16**, two disconnections, c and d, are possible.

Disconnection c gives the nucleophilic synthon **Q.7**, which leads to the haloalkene **Q.8** as a reagent, and an electrophilic synthon **Q.9**, which identifies the aldehyde **Q.10** as the other reagent. Disconnection d likewise generates two synthons **Q.11** and **Q.13**, which identify the aldehyde **Q.12** and the haloalkane **Q.14**, respectively, as the appropriate precursors.

Note that in each of the disconnections a–d, the disconnection is made such that the carbon atom bonded to the OH group is identified as the electrophilic centre.

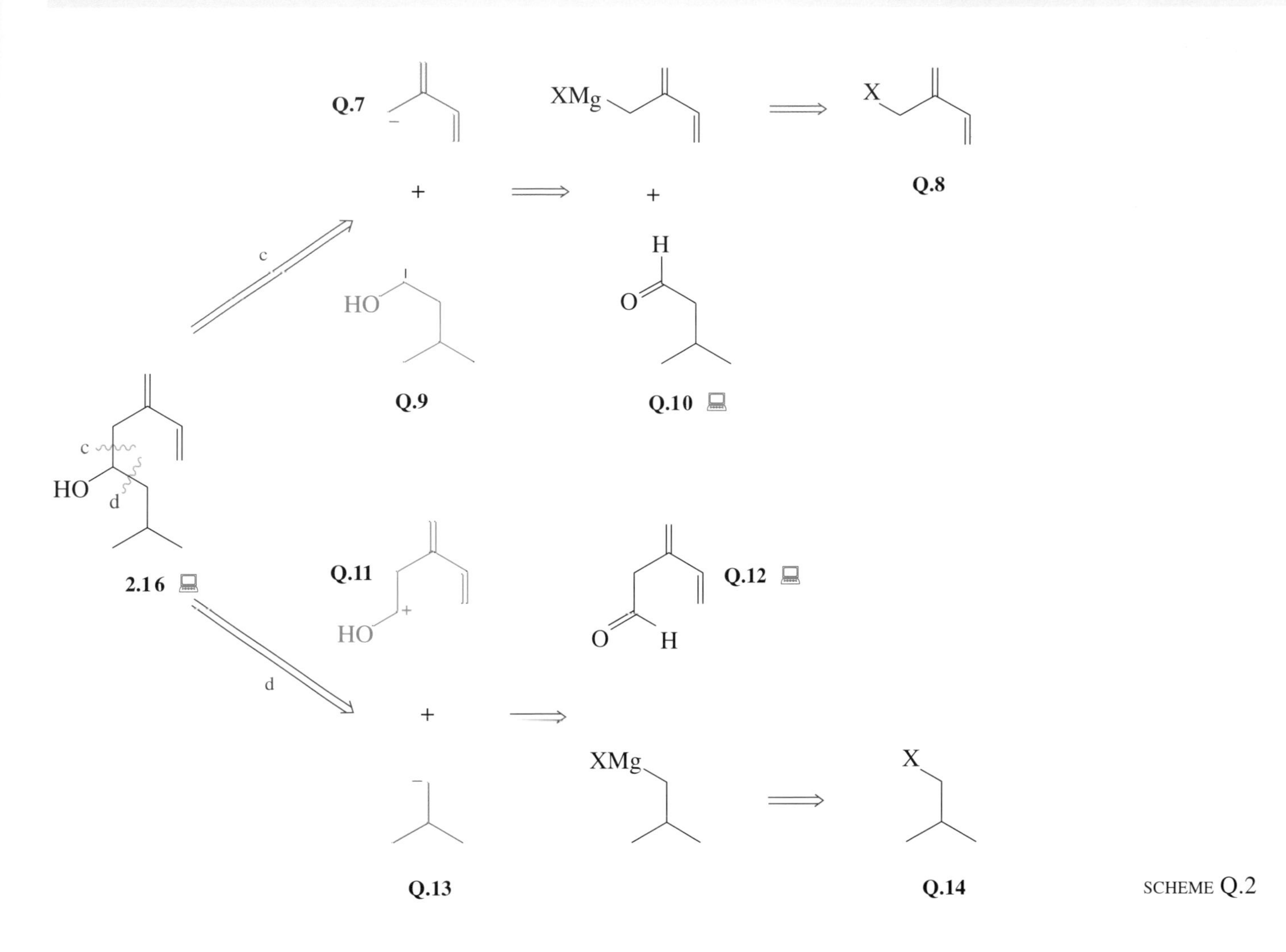


SCHEME Q.2

## QUESTION 4.1 *(Learning Outcomes 3 and 5)*

Compound (i) is γ-bisabolene, a sesquiterpene produced from farneysl pyrophosphate (**4.13**). Compound (ii) is abietic acid (a major constituent of rosin, which is the final product of the distillation of turpentine), a diterpene produced by further enzyme-mediated reactions of geranylgeranyl pyrophosphate, **4.16**. Compound (iii) is limonene, a monoterpene derived from geranyl pyrophosphate, **4.11**, which is present in citrus fruits. Compound (iv) is lanosterol, present in wool fat, and a key intermediate in cholesterol biosynthesis (see Section 6). It is a complex triterpene derived from squalene (**4.17**).

## QUESTION 5.1 *(Learning Outcome 6)*

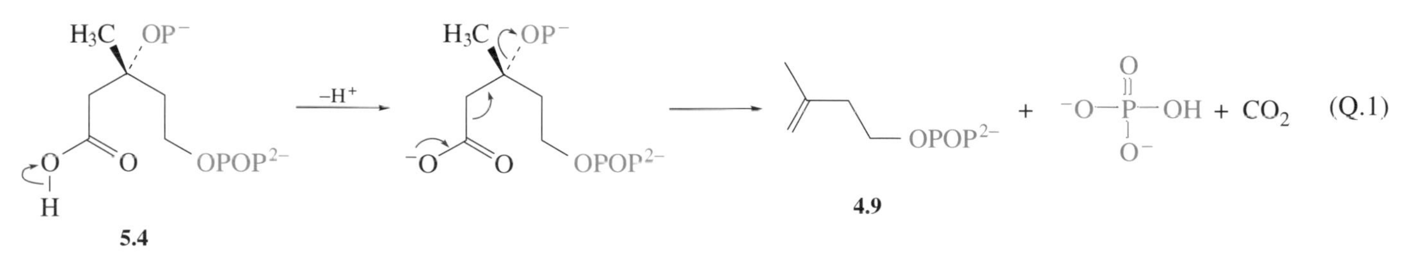

# FURTHER READING

1 L. E. Smart (ed.), *Separation, Purification and Identification*, The Open University and the Royal Society of Chemistry (2002).

2 E. A. Moore (ed.), *Molecular Modelling and Bonding*, The Open University and the Royal Society of Chemistry (2002).

3 M. Mortimer and P. G. Taylor (eds), *Chemical Kinetics and Mechanism*, The Open University and the Royal Society of Chemistry (2002).

4 P. G. Taylor and J. M. F. Gagan (eds), *Alkenes and Aromatics*, The Open University and the Royal Society of Chemistry (2002).

# ACKNOWLEDGEMENTS

Grateful acknowledgement is made to the following sources for permission to reproduce material in this Part of the book:

*Figures 1.1, 1.2, 2.1, 4.2 and 4.3*: Oxford Scientific Films; *Figure 1.4*: © The Nobel Foundation; *Figure 6.1*: Science Photo Library; *Figure 6.2*: courtesy of Merck Sharp and Dohme.

Every effort has been made to trace all the copyright owners, but if any has been inadvertently overlooked, the publishers will be pleased to make the necessary arrangements at the first opportunity.

*Case Study*

# Polymer chemistry

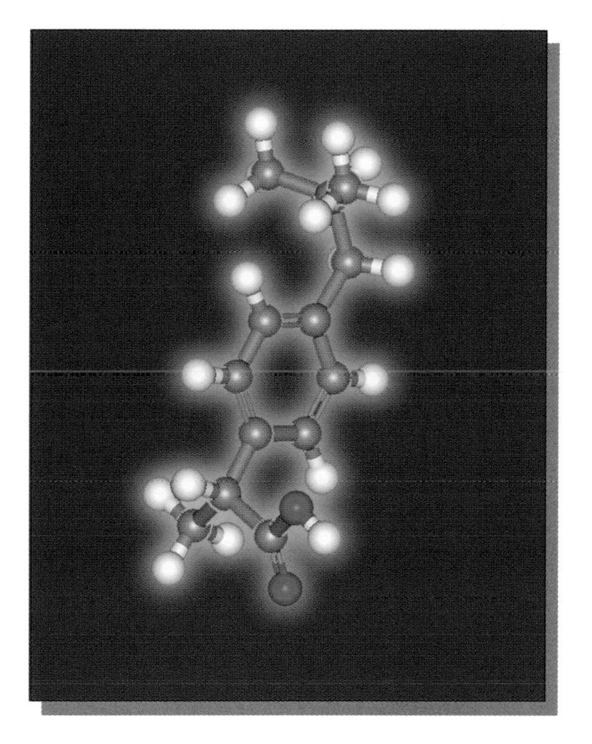

Bob Hill

# 1 HISTORICAL INTRODUCTION

To say that polymers are important in our modern world is a gross understatement. Our life depends, and indeed always has depended, on polymers of one sort or another. Our bodies contain many polymeric materials, such as proteins, polysaccharides and nucleic acids. Many of the natural materials that we use, for example, paper, wood, rubber, cotton, wool and silk, are polymeric. These materials were used for many centuries without any knowledge of their structure at the molecular level. However, having gained an understanding of their molecular structure, we have been able to develop more useful synthetic polymeric materials. The production of synthetic polymers has revolutionized our society. Much of our clothing is made from polymeric fibres such as nylon, polyesters and viscose. If you look around you and see how many articles are made from synthetic polymers, you can begin to appreciate how dependent we are on the large-scale production of polymers. In 1995, about one hundred million tonnes of polymers were manufactured world-wide.

Despite the modern world's reliance on polymers, the history of polymer chemistry has been rather chequered. The term 'polymer' was first used by the Swedish chemist J. J. Berzelius (Box 1.1) in 1833. It derives from the Greek *poly* meaning 'many' and *mer* meaning 'part' — hence, many parts. However, it is certain that Berzelius did *not* recognize polymers as long chains of repeating units. In fact, this was an idea that met with severe resistance from the majority of influential chemists in the nineteenth and early twentieth centuries. It was not until the 1920s that the idea of long chains of repeating units became accepted. Such was the opposition to the long-chain idea that one very eminent organic chemist was reported to have said that it was as if zoologists were to report the likelihood of there being somewhere in Africa an elephant which was fifteen hundred feet long and three hundred feet high.

## BOX 1.1 Jons Jacob Berzelius

Jons Jacob Berzelius (born on 20 August 1779 in Väversunda near Stockholm) was a chemist who followed the work of John Dalton. He developed the idea of ions and ionic compounds. He analysed many compounds and determined their formulae. In 1811, he introduced the classical system of chemical symbols. He published a table of atomic weights in 1826 which is still in good agreement with modern values.

His father, a teacher, died when he was young. His mother remarried but died shortly after. At 12 years old, he was sent to school in Linköping, where he supported himself by tutoring. He studied medicine at the age of seventeen, but stopped when his scholarship was withdrawn. He then studied as a pharmacist and thereafter resumed his medical studies. In 1807, he became professor at the Karolina Medico-Chirurgical Institute in Stockholm. When he was 56, he married a woman 32 years younger than himself and had 12 happy years of marriage.

**Figure 1.1**
Jons Jacob Berzelius, 1779–1848.

Let's look at the historical development in a little more detail.

The pungent, colourless, liquid styrene (phenylethene, **1.1**) was characterized by C. F. Gerhardt in 1839. On occasions, in quite unclear circumstances, it would convert to an odourless, clear, resinous solid. This was referred to as 'metastyrene'. Like styrene itself, the empirical formula of metastyrene was found to be CH (the molecular formula of styrene is $C_8H_8$). The long-chain idea had not even been put on the table at this time, so metastyrene and other mysterious polymerizations were classified as associations of smaller molecules held together by non-covalent forces. Natural rubber was another of these mysteries. The German chemist C. D. Harries spent much research activity on rubber, using mild degradative reactions in the hope of disassociating the molecular units. He was unsuccessful: if only he had been a little more forceful with his degradations!

$CH{=}CH_2$

**1.1**

Certainly, nineteenth-century chemists made the best use they could of naturally occurring polymers in complete ignorance of their structures. In 1820, the British chemist Thomas Hancock discovered that chewing rubber modified its properties. Twenty-three years later he patented a sulfur cross-linking process which greatly improved the elastic properties of natural rubber. In fact, *vulcanization*, as this was called, had also been developed by Charles Goodyear in the USA in 1839. This led to the development of the pneumatic tyre (Figure 1.2).

**Figure 1.2**
An early pneumatic tyre.

One of the first synthetic polymers was collodion, a material created by nitrating the natural polymer cellulose. Collodion was discovered in 1846 by Alexander Parkes, and articles made of collodion were shown at the Great International Exhibition in London in 1862.

Collodion's commercial possibilities remained largely unexplored until 1870, when, as a result of a shortage of elephants, a substitute for ivory had to be found for making billiard balls. J. W. Hyatt, in the USA, improved the Parkes formulation by mixing collodion with camphor: he named the new material 'celluloid', and it became a huge market success. Collodion was not the best of materials because it is related to gun cotton, and cannoning of the balls occasionally caused a mild explosion. This led to problems in the saloons of the American West, since it caused people to reach for their guns! Celluloid (Figure 1.3) held the market in polymers for about 30 years, but, because of its high flammability, a more-stable material was soon being sought. This led to the production of plastics based on cellulose acetate, and, in 1907, of Bakelite, probably the first truly synthetic polymer (Figure 1.4). Bakelite (named after its discoverer, the Belgian chemist Leo Baekeland) was developed using a phenol–formaldehyde (methanal) formulation. Bakelite remains in use to this day.

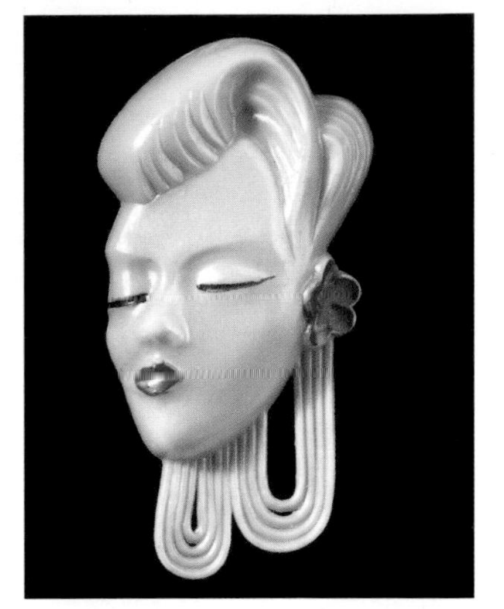

**Figure 1.3** A celluloid brooch.

Another synthetic polymer, polyethylene (polythene or poly(ethene), Figure 1.5) was discovered purely by chance in 1932 by Gibson and Swallow working for ICI. Then, in 1934 W. H. Carothers (Box 1.2), working for DuPont in the USA was the first to make the polyamide polymer nylon. Carothers went much further than this discovery by explaining the theoretical principles underlying the condensation polymerization reaction. In 1937, Rowland Hill and John Crawford, working for ICI, developed polymethyl methacrylate, PMMA (poly(methyl 2-methylpropenoate)), an addition polymer marketed as Perspex. Perspex soon replaced glass in many situations (its lightness favoured it) including the windows in Second World War fighter planes. This inadvertently led to PMMA being chosen as a material for contact lenses (Section 6), when it was discovered that fighter pilots were able to endure splinters of Perspex in their eyes when their cockpits had been damaged.

**Figure 1.4** A bakelite phone

**Figure 1.5**
Items made from polythene in the 1930s.

The opposition to the long-chain idea was only effectively crushed from about 1920, by the work of the outstanding German chemist Hermann Staudinger (Box 1.3). In a series of papers he explained the remarkable variances in the properties of different

## BOX 1.2 Wallace Hume Carothers

Wallace Hume Carothers was born in Burlington, Iowa, the oldest of four children. His sister became a radio star as part of a musical trio. He studied science at Tarkio college, Missouri and then obtained his MSc from the University of Illinois in 1924, followed by a PhD. He then taught at Harvard, but joined DuPont in 1928 to do basic research on polymers.

In 1936, he married Helen Sweetman who also worked at Dupont. Early in 1937 his sister died, and after a bout of depression he committed suicide by drinking lemon juice mixed with cyanide. Later that year his daughter was born.

**Figure 1.6**
Wallace Hume Carothers 1896–1937.

## BOX 1.3 Hermann Staudinger

Hermann Staudinger was born in Worms. After studying in Halle, Darmstadt and München, he became an assistant at Strasbourg in 1903. In 1907, he became Professor of Chemistry in Karlsruhe. He later took the Chair at Zurich in 1912 and at Freiburg in 1926. In 1912, he studied natural polymers, and began to show that these were long chain molecules. With his wife, Magda Staudinger-Woit, he studied the composition of proteins. He was awarded the Nobel Prize for Chemistry in 1953.

During the First World War, many German scientists were asked to sign a document supporting their country's military activities. Staudinger did not, and while in Switzerland wrote articles denouncing the use of poison gas as a battlefield weapon. On returning to Germany, his scientific contribution was never really recognized.

**Figure 1.7**
Hermann Staudinger 1881–1965.

polymers in terms of covalent links between repeat units with non-covalent bonds and/or cross-links, between these long chains. In 1953, he was awarded the Nobel Prize for Chemistry for this work. This was a fitting time for such an award, as it was the same year in which the structure of the most important natural polymer — DNA* — was elucidated by Rosalind Franklin, Maurice Wilkins, James Watson and Francis Crick.

In fact, the advances in polymer chemistry since 1900 have been very dependent on the two World Wars, showing the close relationship between science and technology, and international tension and economics. For example, Germany's decision in 1937 to become self-sufficient in rubber brought about the development of styrene–butadiene polymers. Then, in 1941, when the Japanese occupied the rubber plantations in the Far East, the United States had to develop the same synthetic rubber.

Since the Second World War, an acceleration in research and production has continued unabated, so that if our age were to be named after the materials that characterize it, as were the Stone, Bronze and Iron Ages, this might justifiably be called the Plastics Age.

* The discovery of the DNA structure is also discussed in Part 1 of *The Third Dimension*[1].

# WHAT IS A POLYMER? — SOME DEFINITIONS

# 2

A polymer is a long chain made up of many molecular units, each individual unit being formed from a **monomer**. The monomer units are linked through covalent chemical bonds. Many polymers are purely synthetic, but a wide range of natural polymers occurs in nature. The most famous examples of naturally occurring polymers are proteins, which consist of very long chains made from up to twenty types of monomers called amino acids.

Polymer structures are frequently very complex, not simply consisting of one linear chain of monomer units of fixed length. This is especially true of synthetic polymers, which are always a mix of many different chain lengths. However, this is rarely the case with natural polymers. For example, all molecules of a particular type of globular protein have the same chain length, with an exact and invariant order of amino acid monomer units. Many synthetic polymers also exhibit a feature known as *chain branching*. This is where a polymer chain will split, or branch, into two (or more) strands, which may (or may not) interact. Branching may occur many times along the length of a polymer. This will usually have been deliberately designed, but may also be a side-effect of the polymerization process.

In terms of overall polymer architecture, three broad categories are recognized — linear, branched and network.

**Linear polymer chains** should not be thought of as straight chains: they are more likely to be folded into intricate shapes, rather like lengths of string dropped randomly on top of each other.

**Branched polymers** may be similarly described. Figure 2.1 would represent just one possibility for one branched chain.

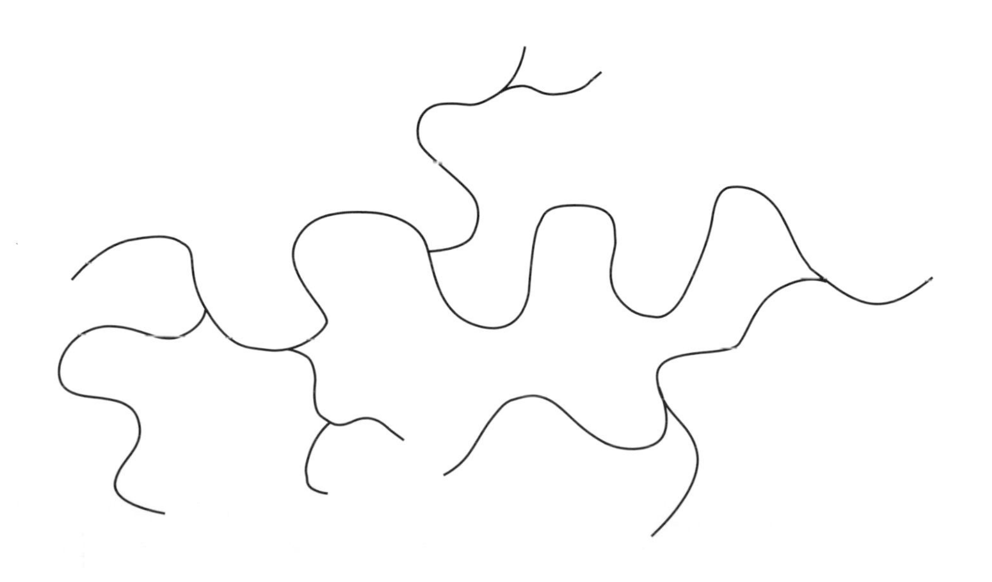

**Figure 2.1**
A branched chain polymer.

**Network polymers** present a still more complex picture of polymer chains interlinked in three dimensions forming a complex mesh. A highly organized network polymer may actually have a considerable organization of linked chains. Less-networked systems are likely to be disorganized. Figure 2.2 is a two-dimensional stylized representation of a small portion of one part of an organized three-dimensional network polymer.

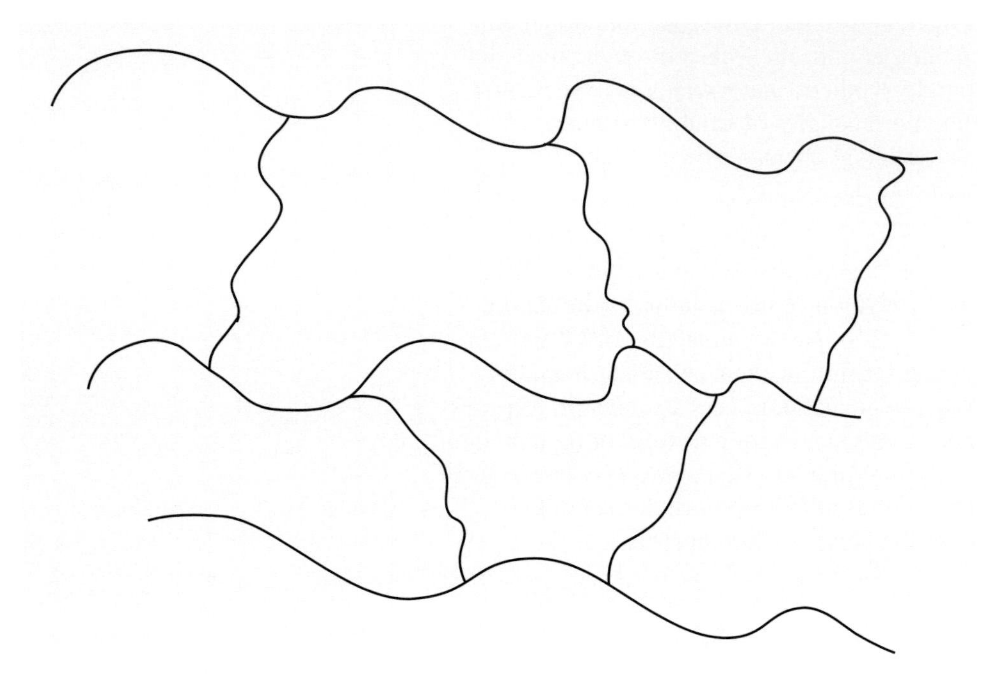

**Figure 2.2**
A network polymer.

The monomer units in a polymer chain may all be the same; in this case the polymer is called a **homopolymer**. Where two or more monomer units occur in a chain, the polymer is referred to as a **copolymer**. If we choose two monomer units, A and B, then we can recognize five broad categories of synthetic polymer, as shown in Figure 2.3.

```
~~~A—A—A—A—A~~~                      homopolymer

~~~A—B—A—B—A—B~~~                    alternating copolymer

~~~A—A—B—B—B—A—B—A—A~~~              random copolymer

~~~A—A—A—A—B—B—B—B~~~                block copolymer

~~~A—A—A—A—A—A—A~~~                  graft copolymer
         |
         B—B—B—B~~~
```

**Figure 2.3**
Different ways of organizing monomer units in polymers.

As we have already said, polymers are long chains of monomer units; but how long is long? If only two monomer units are linked, then the compound is often referred to as a dimer; three monomer units become a trimer and so on, until around eight units, which is called an octomer. Between approximately eight and twenty linked units is frequently termed an **oligomer**, and above this the chain becomes a **polymer**.

Table 2.1 (overleaf) shows the structures of the monomer units of a number of common polymers. Notice that compounds **2.1–2.3** are made by reactions leading to functional groups within the backbone. By contrast, compounds **2.4–2.11** have saturated carbon skeletons, even though they may have functional groups attached to the backbone; they are made from substituted ethenes.

Before we go any further, we must briefly address the issue of polymer nomenclature. The most systematic way of naming a polymer is to write the name of the monomer in brackets, combined with the prefix 'poly'. Hence we have names like poly(phenylethene), poly(hexanelactam), poly(ethene), etc., as shown in the right-hand column of Table 2.1. However, you may have already noticed that polymer chemists often resort to *non*-systematic (non-IUPAC) nomenclature. This is hardly surprising since many of the molecules they use will, when correctly named in accordance with IUPAC rules, take up a whole line of page or more! Therefore we will encounter many non-systematic names as we continue through the study.

**Table 2.1** Some common polymers

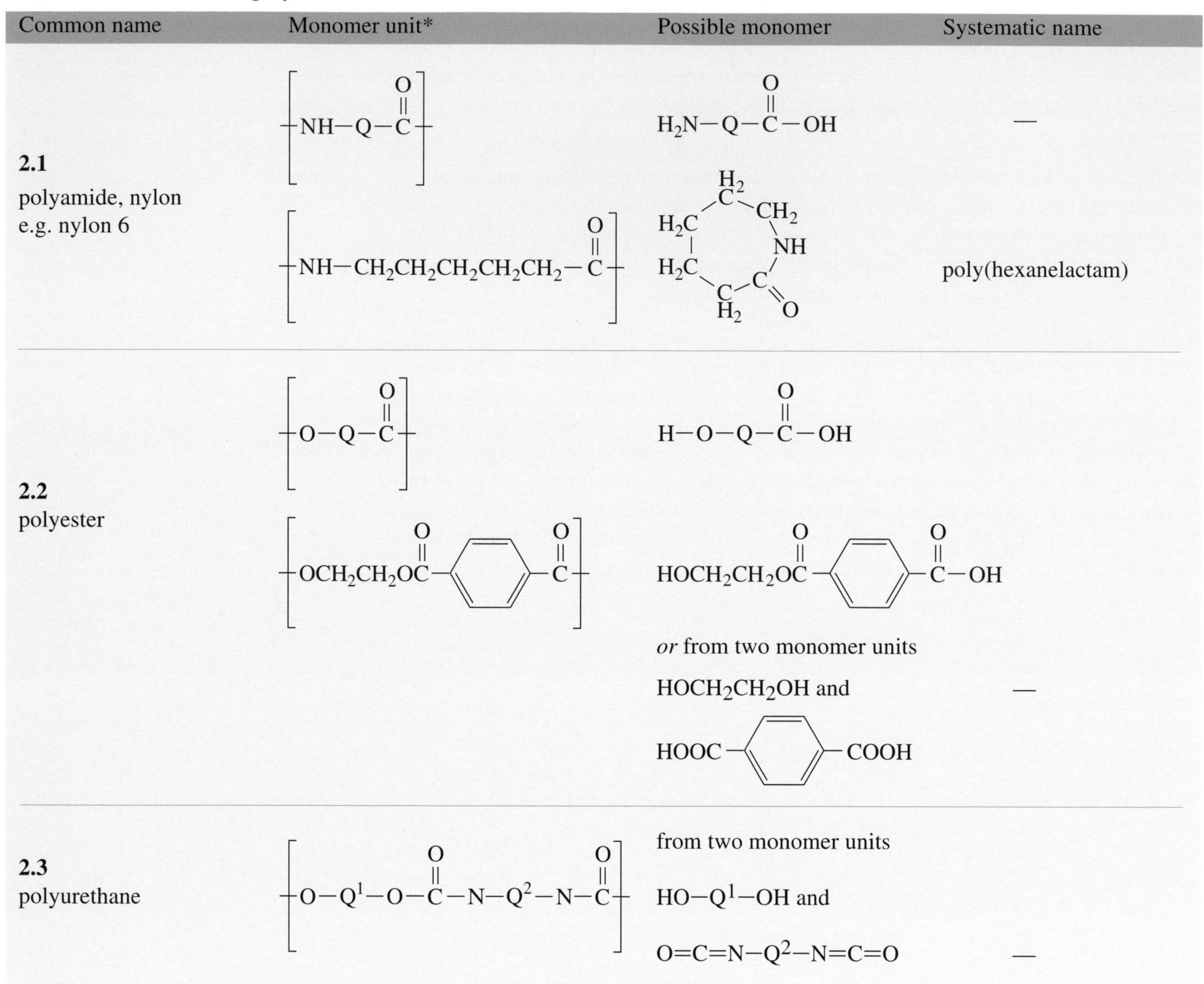

| Common name | Monomer unit* | Possible monomer | Systematic name |
|---|---|---|---|
| **2.1** polyamide, nylon | $[-NH-Q-C(=O)-]$ | $H_2N-Q-C(=O)-OH$ | — |
| e.g. nylon 6 | $[-NH-CH_2CH_2CH_2CH_2CH_2-C(=O)-]$ | cyclic lactam: $H_2C$, $C(H_2)$, $CH_2$, $NH$, $C(=O)$, $C(H_2)$, $H_2C$ ring | poly(hexanelactam) |
| **2.2** polyester | $[-O-Q-C(=O)-]$ | $H-O-Q-C(=O)-OH$ | |
| | $[-OCH_2CH_2OC(=O)-C_6H_4-C(=O)-]$ | $HOCH_2CH_2OC(=O)-C_6H_4-C(=O)-OH$ *or* from two monomer units $HOCH_2CH_2OH$ and $HOOC-C_6H_4-COOH$ | — |
| **2.3** polyurethane | $[-O-Q^1-O-C(=O)-N-Q^2-N-C(=O)-]$ | from two monomer units $HO-Q^1-OH$ and $O=C=N-Q^2-N=C=O$ | — |

* Q represents a structural unit connecting the two reactive functional groups of the monomer. This is often a short hydrocarbon chain or a benzene ring.

| Common name | Monomer unit | Possible monomer | Systematic name |
|---|---|---|---|
| **2.4** Polythene, Alkathene | $\text{-[}CH_2-CH_2\text{]-}$ | $CH_2{=}CH_2$ | poly(ethene) |
| **2.5** Teflon, Fluon | $\text{-[}CF_2-CF_2\text{]-}$ | $CF_2{=}CF_2$ | poly(tetrafluoroethene) |
| **2.6** polypropylene | $\text{-[}CH_2-CH(CH_3)\text{]-}$ | $CH_2{=}CH-CH_3$ | poly(propene) |
| **2.7** polyvinyl chloride (PVC) | $\text{-[}CH_2-CH(Cl)\text{]-}$ | $CH_2{=}CHCl$ | poly(chloroethene) |
| **2.8** polyvinyl acetate (PVA) | $\text{-[}CH_2-C(CH_3)(C(=O)OCH_3)\text{]-}$ | $CH_2{=}CH-O-C(=O)-CH_3$ | poly(ethenyl ethanoate) |
| **2.9** polyacrylonitrile | $\text{-[}CH_2-CH(CN)\text{]-}$ | $CH_2{=}CH-CN$ | poly(propenenitrile) |
| **2.10** polystyrene | $\text{-[}CH_2-CH(C_6H_5)\text{]-}$ | $CH_2{=}CH-C_6H_5$ | poly(phenylethene) |
| **2.11** Perspex, polymethyl methacrylate, Plexiglass | $\text{-[}CH_2-C(CH_3)(C(=O)OCH_3)\text{]-}$ | $CH_2{=}C(CH_3)-C(=O)-OCH_3$ | poly(methyl 2-methylpropenoate) |

# 3 POLYMER FORMATION

The process of linking monomer units together to form a polymer is called **polymerization**, and this is sub-divided into two categories — **step-growth polymerization** and **chain-growth polymerization**.

## 3.1 Step-growth polymerization

This type of polymerization involves a condensation reaction between monomers. A simple example of a condensation reaction is that between two molecules of 6-aminohexanoic acid, $H_2N(CH_2)_5COOH$. Here, the carboxylic acid group of one molecule condenses with the amino group of another, with the elimination of a molecule of water:

$$H_2N{-}(CH_2)_5{-}C({=}O)OH + H{-}NH{-}(CH_2)_5{-}COOH \longrightarrow H_2N{-}(CH_2)_5{-}C({=}O){-}NH{-}(CH_2)_5{-}COOH + H_2O \quad (3.1)$$

The key feature of this reaction is that the dimer produced has the same two groups at each end as the original monomer — that is, $-COOH$ at one end and $-NH_2$ at the other. Hence, the condensation reaction may occur again at either or both ends of the dimer, to build up longer units.

More generally, this can be drawn schematically as

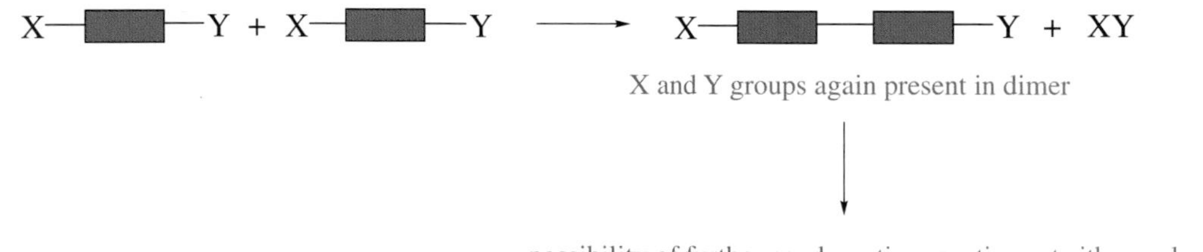


SCHEME 3.1

So further condensation reactions are possible for the dimer:

1 The dimer can react with another monomer molecule:

2 It could react with another dimer molecule:

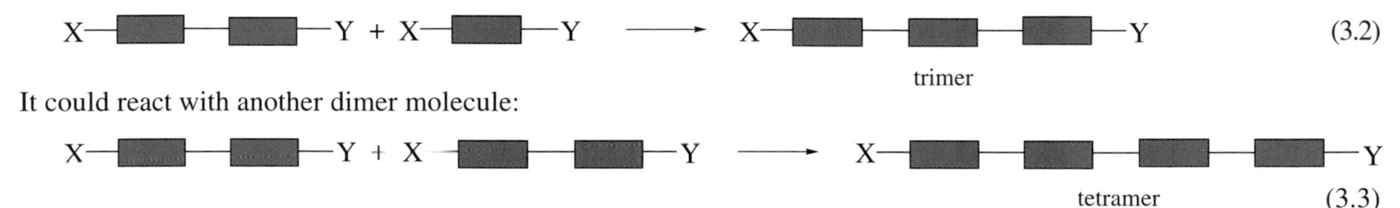


Combinations of either or both of these can then continue *ad infinitum*, in order to build the polymer chain.

In step-growth polymerization, reactions between monomers may occur without the need for any external agency such as heat or a catalyst, though most do require such agents.

## 3.2 Chain-growth polymerization

In contrast to step-growth polymerizations, all chain-growth polymerizations require at least one essential external factor called an *initiator*. Initiators are often chemical substances but can also be heat or radiation. Another contrasting feature is that the monomers taking part in chain-growth polymerization must possess a carbon–carbon double bond, since the reaction involved here is an alkene addition reaction. The majority of simple alkene addition reactions proceed by ionic mechanisms (such as the addition of bromine or HBr to an alkene*); in chain-growth polymerizations, both anionic and cationic mechanisms are well known. However, there is a third, more important, mode of polymerization which proceeds via radicals. You will recall from Part 3 that a radical is any atom, molecule or ion with an unpaired electron, and thus an odd number of electrons. As with most radical reactions, we must employ an initiator. For now, we shall refer to our initiator simply as '$\dot{\mathrm{In}}$' and not be concerned with its origin. Using the normal terminology, formation of the radical from the initiator is the *initiation stage*. This is followed by one of two possible *propagation steps* (for unsymmetrical alkenes):

$$\dot{\mathrm{In}} \quad \mathrm{H_2C{=}CHR} \longrightarrow \mathrm{In{-}CH_2\dot{C}HR} \tag{3.4}$$

*or*

$$\dot{\mathrm{In}} \quad \mathrm{RHC{=}CH_2} \longrightarrow \mathrm{In{-}CHR{-}\dot{C}H_2} \tag{3.5}$$

As R is usually alkyl or aryl, the favoured radical is that shown in Reaction 3.4, because primary alkyl radicals are less stable than secondary radicals. The resulting new radical can now react with another alkene molecule in a further propagation step:

$$\mathrm{In{-}CH_2\dot{C}HR} \quad \mathrm{H_2C{=}CHR} \longrightarrow \mathrm{In{-}CH_2CHR{-}CH_2{-}\dot{C}HR} \tag{3.6}$$

A series of many such propagation steps leads to long-chain formation (Structure **3.1**):

$$\mathrm{InCH_2{-}CHR}\left(\mathrm{CH_2CHR}\right)_n\mathrm{CH_2{-}\dot{C}HR}$$

**3.1**

So, what stops the chain growth? Well, it could be due to monomer running out. This would suggest formation of a polymer consisting of one immensely long chain. However, radical polymerization reactions, like all radical reactions, are subject to *termination reactions*. The most common is *radical dimerization*—simply the pairing of two radicals:

$$\mathrm{InCH_2CHR}\left(\mathrm{CH_2CHR}\right)_n\mathrm{CH_2{-}\dot{C}HR} \quad \mathrm{RH\dot{C}{-}CH_2}\left(\mathrm{CHRCH_2}\right)_m\mathrm{CHRCH_2In} \longrightarrow$$
$$\mathrm{InCH_2CHR}\left(\mathrm{CH_2CHR}\right)_n\mathrm{CH_2CHR{-}CHRCH_2}\left(\mathrm{CHRCH_2}\right)_m\mathrm{CHRCH_2In} \tag{3.7}$$

Since $n$ and $m$ could have any value, the final chain length will not be the same for all polymer molecules. So a mixture of polymer molecules with a range of chain lengths is obtained.

* The mechanism of alkene addition reactions is discussed in Part 1 of *Alkenes and Aromatics*[2].

## 3.3 A comparison of chain growth and step growth

Table 3.1 summarizes the main differences and similarities between chain-growth and step-growth polymerization. In general, chain-growth polymerizations involve unsaturated monomers (compounds **2.4–2.11** in Table 2.1), and step-growth polymerizations involve monomers with two separate functional groups that can form bonds to other monomers (compounds **2.1–2.3** in Table 2.1). You should remember that long average chain lengths are required to produce strong materials. It is for this reason that polymerizations are among the most demanding reactions of industrial chemistry, so far as purity of reagent and accurate control of conditions are concerned.

**Table 3.1** A comparison of chain-growth and step-growth polymerization

| Chain growth (as exemplified by radical polymerization) | Step growth |
| --- | --- |
| *Differences* | |
| 1 Only the reactive end can react with the monomer to give chain growth. | A monomer can react with any other monomer, oligomer or polymer. |
| 2 Distinct processes of initiation, propagation and termination can occur | All linking reactions are alike. |
| 3 The reaction mixture contains monomer, growing chains and dead chains. The monomer concentration decreases steadily during the reaction. | The monomer is almost entirely incorporated in a chain molecule near the beginning of the reaction, and the solution contains growing chains of various lengths. |
| 4 A long-chain polymer is formed very soon after the reaction starts, and, once formed, cannot be altered. The average polymer chain length varies little with time. | At the beginning, shorter chains are formed, and the average chain length increases steadily as the reaction proceeds. Long chains do not appear until late on in the reaction. |
| 5 Long reaction times increase the polymer yield but not the average chain length. | Long reaction times increase the average chain length. |
| *Similarities* | |
| 1 To produce long average chain lengths, the monomers must be very pure. | |
| 2 A strictly pure product is not obtained, because, although all the molecules are composed of one type of monomer unit, they are of various lengths. | |

## BOX 3.1 To stick or not to stick

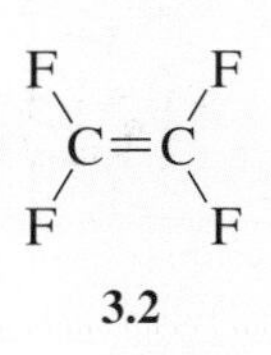

**3.2**

Teflon™ (Fluon™ or poly(tetrafluoroethene), PTFE) is made by a chain-growth polymerization of the monomer **3.2**, giving a polymer consisting of a carbon backbone with fluorine atoms attached. It was discovered by accident in 1938 by DuPont workers (Figure 3.1) who were looking for good refrigerants. It came to the fore during the Second World War when the Manhattan Project needed uranium hexafluoride to separate ${}^{235}_{92}U$ for the atomic bomb. Uranium hexafluoride is a toxic and corrosive substance, and it was proposed that Teflon™ could be used for reaction vessels because it wouldn't react. During the war, it was also found to be transparent to radar and made into nose cones for some types of bomb. After the war, DuPont developed cheaper routes to the polymer, and its use became widespread, from non-stick pans to 'frictionless' joints to Gore-tex™.

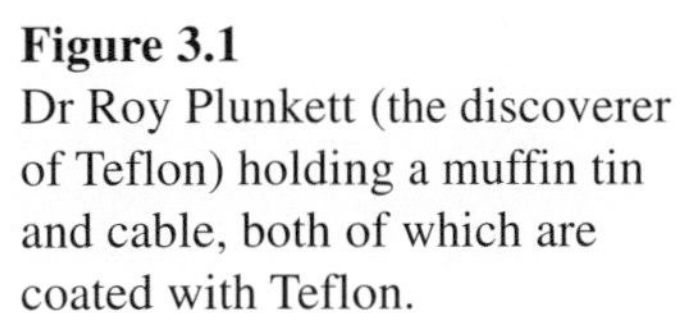
**Figure 3.1**
Dr Roy Plunkett (the discoverer of Teflon) holding a muffin tin and cable, both of which are coated with Teflon.

# POLYMER DESIGN AND DESIGNER POLYMERS

# 4

Consider the radical polymerization of the monomer *p*-ethenylbenzyl methacrylate (**4.1**):

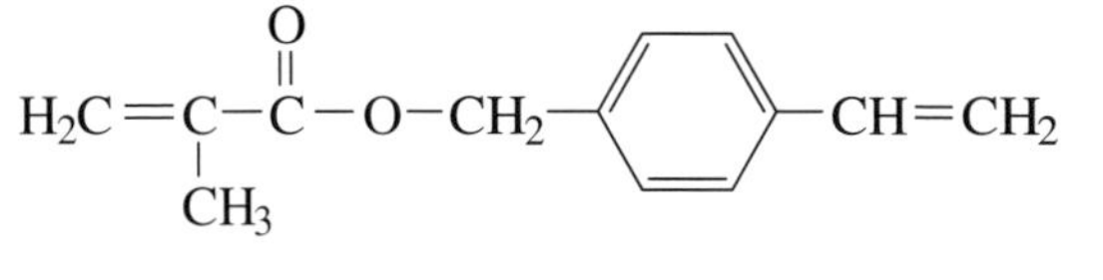

**4.1**

For simplicity, let us suppose that only one type of alkene group (the one on the left) polymerizes as follows (where Q represents the part of the molecule occurring between the alkene groups):

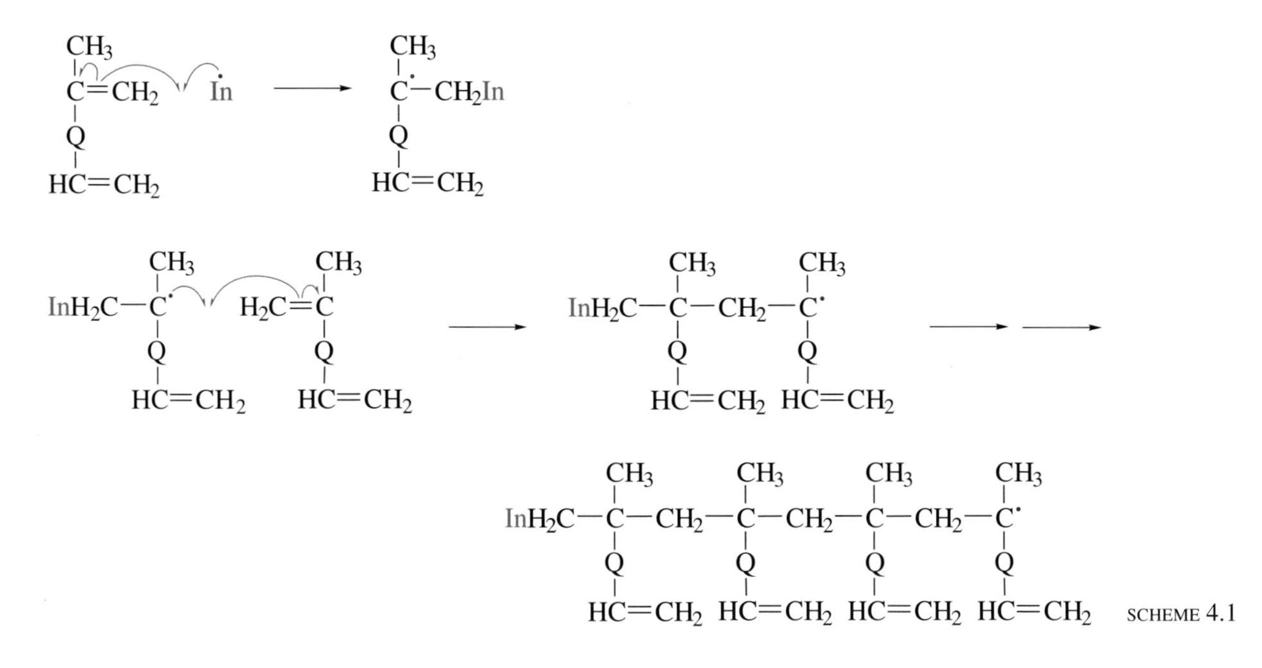

SCHEME 4.1

Now consider reaction of the second type of alkene group, this time those attached to the polymer backbone as shown in Scheme 4.2.

Clearly, many such cross-links could arise between developing chains, limited perhaps only by steric factors. The result would be a highly cross-linked three-dimensional mesh, often resulting in a very tough rigid polymer. The specific design of molecules with cross-linkages and a three-dimensional network is very important, and we shall be considering this for the remainder of the Case Study.

Designer polymers with electrical applications are discussed in Box 4.1.

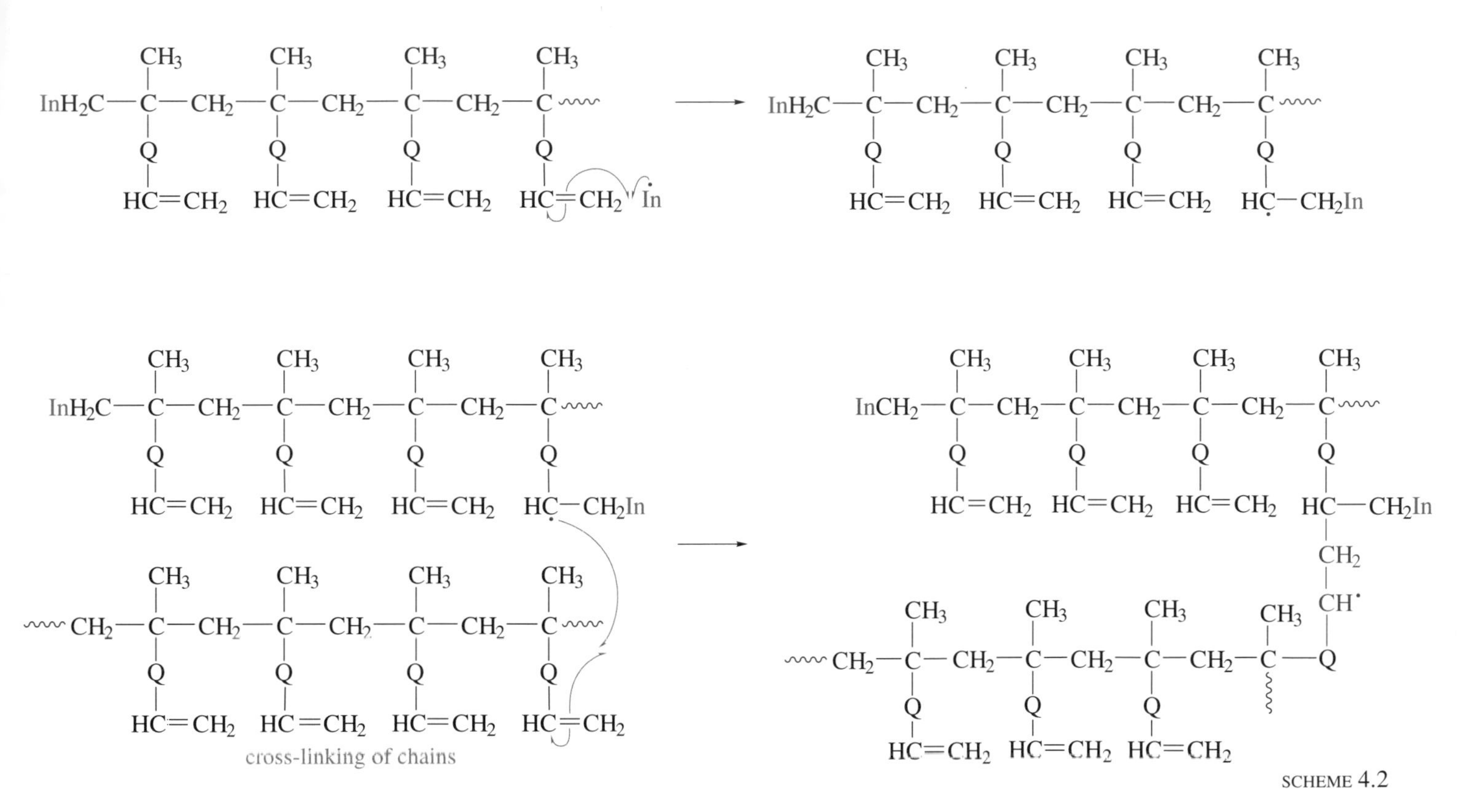


SCHEME 4.2

Our study of designer polymers is going to focus on those that form *gels* (specifically acrylic-type gels), and the two most common applications these have found: molecular separation in one case and contact lenses in the other. We should start by considering exactly what a gel is. A key piece of work in the study of gels is that of Toysich Tanaka and his group at MIT in the USA. Tanaka provides us with a clear definition of a gel:

> A gel is a form of matter intermediate between a solid and a liquid. It consists of polymers, or long chain molecules, cross-linked to create a tangled network and immersed in a liquid medium. The properties of the gel depend strongly on the interaction of these two components. The liquid prevents the polymer network from collapsing into a compact mass; the network prevents the liquid from flowing away. Depending on their chemical composition and other factors, gels vary in their consistency from viscous fluids to fairly rigid solids, but typically, they are soft and resilient or, in a word, jelly like.
>
> ***Toysich Tanaka,* Scientific American *(1981)***

Gels should be familiar to everyone, as there are many examples in the environment around us; for example, edible jellies (Figure 4.1) are derived from the animal protein gelatin. Only about 3% of the volume is gelatin: the rest is water, with sweetener and colouring agent. There are many gels in the human body — for example, the vitreous humour that fills the interior of the eye, the material of the cornea, and the synovial fluid that lubricates our joints.

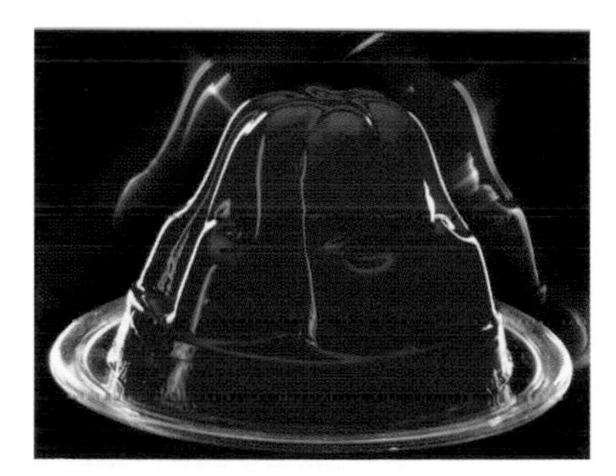

**Figure 4.1** A jelly!

## BOX 4.1 Conducting polymers

Plastics are commonly used as insulators; most metal cables have sheaths of plastic around them to prevent them from touching each other (and us!). However, it is possible to make plastics that conduct electricity. If ethyne is polymerized instead of ethene, a polymer is obtained that has alternating single and double bonds, namely poly(ethyne) (polyacetylene, Figure 4.2). Just as in benzene, this means the electrons are delocalized, allowing electrons to 'flow' from one end to the other, such that the material conducts electricity. Such polyacetylenes usually have to be made by special routes. Nowadays there is a range of different conducting polymers available, which are usually used in applications such as light-emitting diodes, solar cells and antistatic materials.

In 2000, the Nobel Prize for Chemistry was awarded to Alan J. Heeger, Alan G. MaDiarmid and Hideki Shirakawa for the discovery and development of conductive polymers. At the start of the 1970s, Shirikawa was working on the polymerization of ethyne, but always obtained black powders. A visiting researcher accidentally used one thousand times too much catalyst and obtained a beautiful silvery film. They thought that if it shone like a metal perhaps it conducted too. But it didn't. Heeger, MacDiarmid and Shirakawa then set about modifying the material to give a conducting plastic. Treating the film with iodine increased its conductivity by ten million-fold, and other modifications have developed a lightweight plastic conductor.

**Figure 4.2**
Professor James Feast holding a sample of polyacetylene (poly(ethyne)) synthesised in his laboratory at Durham University.

Tanaka's work was carried out on polyacrylamide gels. The homopolymer is made from the monomer acrylamide (propenamide, **4.2**) to give the polymer **4.3**:

$$H_2C{=}CH{-}C({=}O)NH_2$$

**4.2**

$$\mathrm{In}{-}\left(CH_2CH(CONH_2)\right)_n{-}CH_2CH(CONH_2){-}CH(CONH_2)CH_2{-}\left(CH(CONH_2)CH_2\right)_m{-}\mathrm{In}$$

**4.3**

Small-scale preparations are carried out in aqueous solution. We shall not give exact experimental details, but the procedure runs along the following lines. An approximately 12.5% aqueous solution of acrylamide is prepared. 500 $cm^3$ of this solution is placed in a three-necked one-litre round-bottomed flask equipped with a stirrer, gas inlet, thermometer and condenser. Carbon dioxide gas is passed over the solution (to exclude oxygen), which is then heated to around 70 °C. The reaction is exothermic, which increases the temperature to about 80 °C. This temperature is maintained for two hours. A clear viscous solution is obtained. Addition of methanol precipitates the polymer, which can be dried in a vacuum at 50 °C.

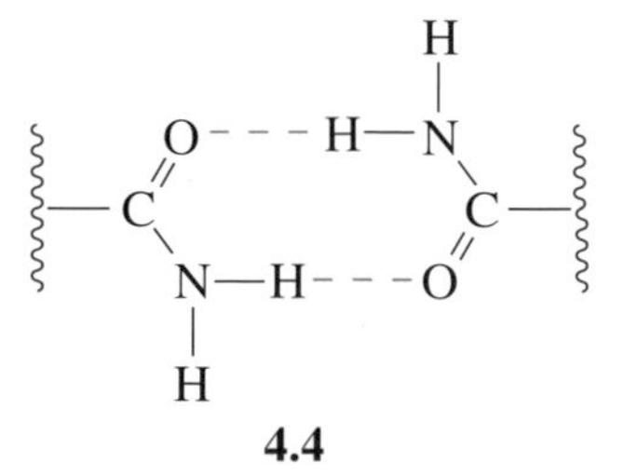


**4.4**

When moulded into shape, this polymer is very hard and brittle. This means that the chains cannot move across one another very freely, suggesting considerable interaction between them. This is most likely to be due to hydrogen-bonding between the amide groups of adjacent side-chains (**4.4**). However, when placed back into water, the polymer dissolves, restoring the viscous solution. This is again due to hydrogen-bonding. Hydrogen-bonding interactions of the amide group with water molecules are now favoured rather than those with another amide group in the chains (**4.5**).

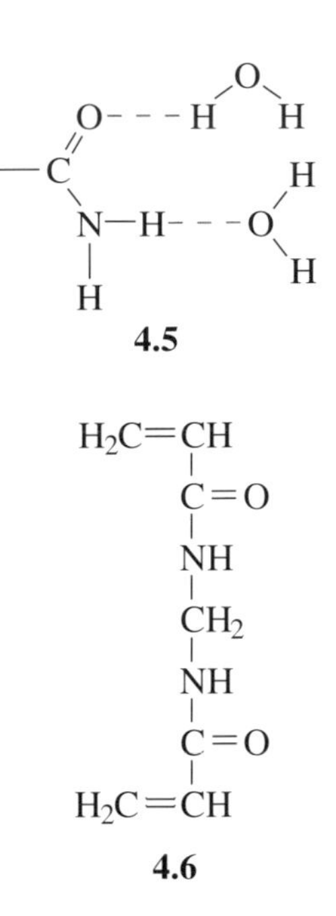


This homopolymer does not form a gel, however. As you have seen from Tanaka's definition of a gel, there has to be an especially intimate association between a solvent and the polymer chains; this is only achieved by creating a network polymer — that is, an extensively cross-linked polymer, as shown in Figure 2.2.

So, the problem to be solved was how to introduce cross-links into the polymer. One particular agent which has been found to give excellent results is *N,N′*-methylene-*bis*-acrylamide, normally abbreviated to *bis* (**4.6**).

The alkene groups at both ends of *bis* can take part in polymer chain formation, giving rise to links between two chains. The copolymerization of *bis* can be carefully controlled into the system to create a porous network polymer. Figure 4.3 helps to show what is happening

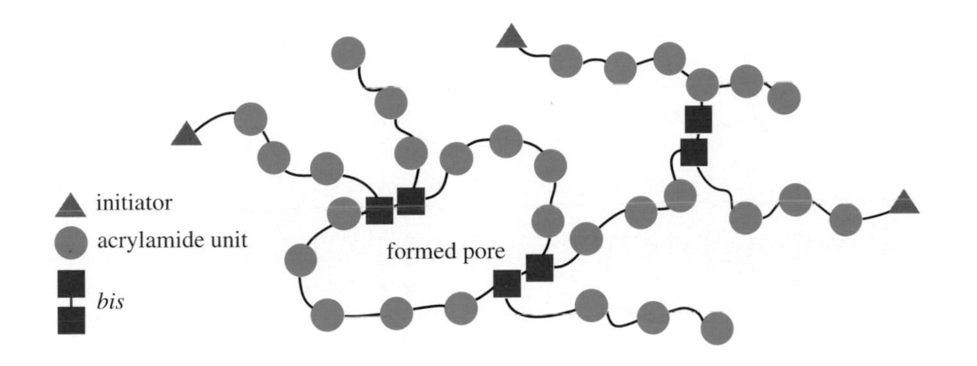

**Figure 4.3** Formation of a three-dimensional porous network using *bis*.

Depending on how much *bis* (red squares) is added, many pores with a range of dimensions will be formed in the network. Notice that the addition of *bis* would in itself help to increase attraction of water molecules because the amide groups in *bis* can participate in hydrogen-bonding. However, the amount of *bis* used rarely exceeds 15% and is usually far less.

Many of these gels can absorb as much as 70% by mass of water, and swell by up to three hundred times. The key to this phenomenal water entrapment, as Tanaka explained, is the intimate relationship between the solvent and the network. It is not simply the case that one amide hydrogen-bonding site in the network interacts with just one water molecule: it can order dozens of water molecules within a pore! Remember that water molecules also interact with each other through hydrogen-bonding. Thus, each amide group is hydrogen-bonded to a water molecule, which in turn is hydrogen-bonded to other water molecules, which in turn…, etc. This organization of water molecules around the amide groups of the network locks the water molecules together in the pores, giving a more rigid structure. Polyacrylamide gels or PAGs were discovered in the late 1950s, and were soon put to use in a technique called polyacrylamide gel electrophoresis (PAGE).

# POLYACRYLAMIDE GEL ELECTROPHORESIS (PAGE)

5

Electrophoresis is a major technique, used mainly by biochemists for separating mixtures of molecules derived from biological sources. These molecules can be as small as simple amino acids to massive molecules such as proteins and segments of DNA. As the name of the technique implies, the technique involves electricity, which, in turn, suggests that molecules capable of electrophoretic separation should interact electrically with a potential difference. Thus, only molecules capable of carrying some degree of positive or negative charge are amenable to the technique. Fortunately, this is a wide spectrum of biological molecules.

We shall explain the basics of the technique with reference to α-amino acids, which have the general form **5.1**. In an alkaline medium, the negatively charged carboxylate anion is formed:

$$R{-}COOH + OH^- \longrightarrow R{-}COO^- + H_2O \quad (5.1)$$

$$R{-}\underset{NH_2}{\overset{COOH}{C}}{-}H$$

**5.1**

Similarly, placing an amino group in a strong acid gives a positively charged ammonium ion:

$$R{-}NH_2 + H^+ \longrightarrow R\overset{+}{N}H_3 \quad (5.2)$$

Clearly, then, amino acids can be induced to possess positive or negative charges. But it is not necessary to use strong acid or alkali; milder buffer solutions (a mixture of a weak acid and its salt, which maintains a steady pH) are used. A commonly used buffer solution used in electrophoresis is one that gives a pH of 8.9 (mildly alkaline). It is still capable of converting a *proportion* of carboxylic groups to the carboxylate anion.

There are about twenty naturally occurring α-amino acids, each involving a different R group. The simplest is glycine, where R = H. In a buffer of pH 8.9, all 20 amino acids, HA, will dissociate into $H^+$ and $A^-$ ions to some extent, but each amino acid will dissociate to a slightly different extent from the others due to the differing electron-withdrawing/donating effect of the R groups. As dissociation is a dynamic process, any amino acid molecule will spend some time in the electrically neutral acid form and some in the anionic form. Averaged over time this means the amino acid will carry a partial negative charge. As a result of the electron-withdrawing/donating effect of the R group, each of the 20 amino acids will spend different amounts of time in the anionic form, and so on average have a different average partial negative charge. Hence, if a mixture of several α-amino acids is placed on a support-medium capable of electrical conduction, at pH 8.9, and a potential difference is applied, all the amino acids will migrate towards the positive electrode, but at slightly different rates depending on this average partial negative charge. The apparatus used is shown in Figure 5.1. On the basis of these different rates of migration, they can be separated.

The early support medium was paper, and this is still used; a very absorbent paper (like blotting paper) is needed, through which an aqueous solution can flow, and a potential difference can be applied.

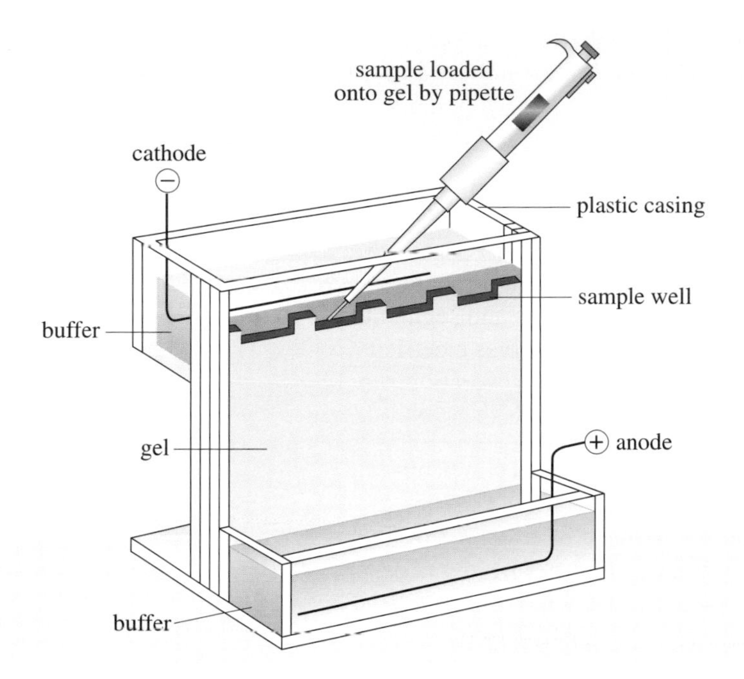


**Figure 5.1** Schematic diagram showing the set-up for gel electrophoresis, here shown capable of running four samples simultaneously. During its preparation the liquid gel has been allowed to harden with an appropriately shaped mould on top to make wells for the samples. During electrophoresis, an electric field is applied and negatively charged molecules move towards the positive electrode. After electrophoresis, the stained amino acids appear as bands.

Polyacrylamide gels or PAGs were introduced as an alternative support medium in the early 1960s. Other gel systems based on naturally occurring polysaccharides were already in use. However, PAGs showed one superior feature over the other gels, namely control over pore size. Thus PAGs are often considered as the first designer polymers.

Polyacrylamide electrophoresis (PAGE) is used to separate proteins and nucleic acids (RNA and DNA), since such species will carry a charge varying with the pH.

Note that the gels, like paper, can sustain a potential difference and permit migration of molecules through an essentially aqueous medium. What, then, was the significance of producing PAGs with fairly uniform pore sizes of differing dimensions? The answer is sieving. For a given pore size, some molecules could pass through readily, whereas larger ones would be held back. So two separatory factors were now available, the net charge and size/shape. Indeed, in many instances, sieving is the major separatory influence: the net charge is only required to drive migration through the porous gel.

Perhaps the most interesting use of PAGE is for *DNA fingerprinting**. The molecular variation that allows DNA to give an individual a particular genetic character depends on four bases: adenine (A), cytosine (C), guanine (G) and thymine (T). These are attached to a long polymer chain made up of deoxyribose (a sugar)–phosphate units; each base–deoxyribose–phosphate unit is known as a *nucleotide*. The order of the four bases on two intertwining long chains constitutes

* DNA fingerprinting is discussed in more detail in *Separation, Purification and Identification*[3].

the genetic code, and provides the blueprint from which proteins are constructed. Each base on one DNA chain hydrogen-bonds with another base on the second chain — T with A, and C with G. In other words, DNA consists of two strands, joined by base pairs, to give the famous double helix.

DNA polymers are extremely long: human DNA contains about $3 \times 10^9$ base pairs, 99% of which is identical in all individuals. However, by comparing the sequences that are different, it is possible to compare DNA samples found at crime scenes with those of individuals, and this has been used in many famous convictions. Usually, it is possible to quote a probability, between one in one million and one in three hundred million, that the samples originated from the same person.

The DNA samples are too large to compare directly, and so to analyse them, they are cut into smaller 'chunks' using enzymes obtained from bacteria and viruses. These enzymes only chop the DNA into fragments at specific sites corresponding to particular sequences of bases, chosen to highlight the differences between individuals. This means the pattern of DNA chunks obtained by fragmenting DNA will reflect the order of bases in the original DNA, and will differ from one person to the next. The chunks can be separated using PAGE, as shown in Figure 5.2a. This means that the patterns obtained by digesting the DNA and analysing the chunks by PAGE enable scientists to confirm whether or not the samples are the same. Features of the patterns of chunks are similar for families, and thus it is possible to identify offspring/parents, etc. This has been used in establishing maternity, paternity and proving family relationships. For example, it has been used to identify missing children in Argentina and soldiers killed in war.

PAGE has also been used in the Human Genome project. Firstly in separating out smaller chunks of DNA obtained by cutting the original DNA into manageable pieces and then in sequencing the order of the bases in these smaller pieces, as shown in Figure 5.2b. Since the human genome is so large, the key to analysis is automation and use of banks of PAGE instruments speeds up the process enormously. By 2001, the technology had advanced to the point where $2.5 \times 10^5$ base pairs a day could be analysed!

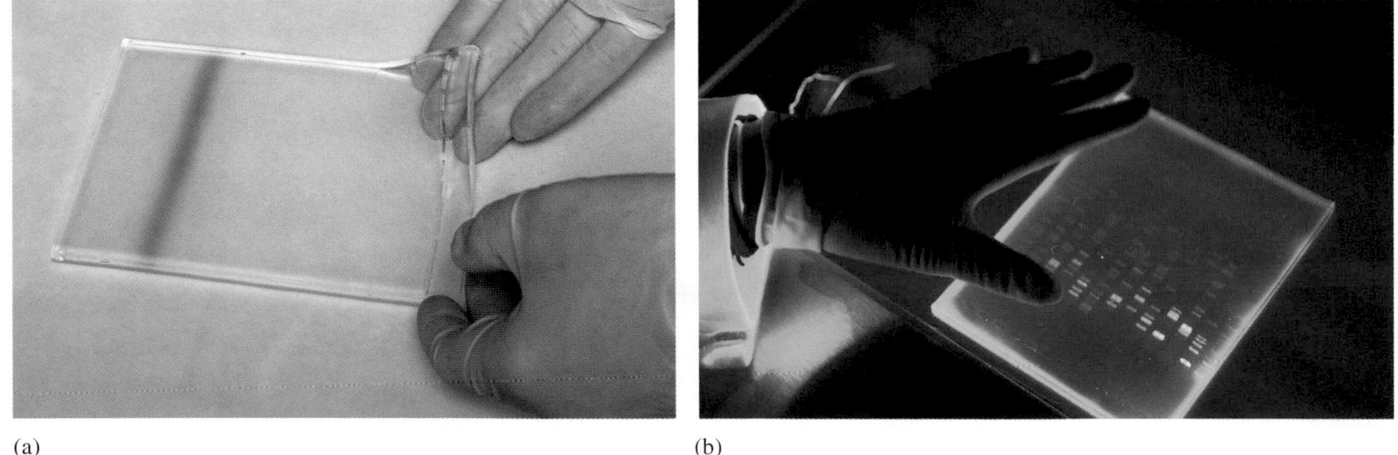

(a) (b)

**Figure 5.2** (a) Transfer of a polyacrylamide gel containing DNA into a tray of salt solution as part of a technique known as 'Southern blotting'. The purple dye shows how far electrophoresis has progressed from the wells on the right. (b) DNA sequencing by gel electrophoresis, as used in the Human Genome project. The banding pattern (fluorescent pink), showing fragments separated according to size and tagged with a radioactive label, is revealed under ultraviolet light. Each band consists of identical DNA fragments of a specific number of nucleotides.

# 6 CONTACT LENSES

The idea of contact lenses is believed to have begun with Leonardo da Vinci. It can be traced definitively to Sir John Herschel, who wrote an article in 1823 outlining the idea of grinding the inside curvature of a glass lens to fit as closely as possible to the cornea. Being an astronomer, he was an expert in optics and in lens construction. Astronomers at this time could not order telescopes, lenses, etc., from an optical apparatus manufacturer, so they had to build their own telescopes and ground their own lenses.

Herschel had studied the eye, in particular the role of the lens of the eye and cornea. Remember that the cornea is a gel. The cornea covers the central region of the eye and is the lens through which the reflected light from objects passes. Herschel understood that astigmatism (a defect of vision which prevents rays of light being brought to a common focus) was due to corneal irregularities. His suggestion was to construct a lens with equal but opposite irregularity to that on the surface of the cornea, thus cancelling out the irregularities. He even suggested that such a counteracting lens should be made from similar material to corneal material — for example, some sort of gel derived from an animal source (he did not, of course, use the word 'gel').

Little or no work was done about Herschel's ideas until 1888 when a Swiss physican, Adolf Fick, published some results of experiments on the fitting of contact lenses. These were made from glass and, as we will see, caused all manner of problems. Figure 6.1 shows one of the early examples of glass contact lenses.

**Figure 6.1**
Blown glass contact lens, c. 1930. made by Myller Sohne Wiesbaden of Germany, the pioneers of blown glass contact lenses.

The fundamental problem, which extended right through to the 1970s and still remains to some extent, was oxygen permeability. Corneal material is living tissue, and like all living tissue takes part in the process of respiration where oxygen is taken up and carbon dioxide expired. If respiring tissue is starved of oxygen, only anaerobic respiration can occur, leading to the production of lactic acid. Build up of lactic acid in the corneal material causes several metabolic disorders such as hypoxia (a build up of metabolites and, eventually, toxic shock). All these undesirable effects can combine to impede or even destroy corneal function. Symptoms include blurred vision, much discomfort and a condition called 'red eye'.

In 1889, August Müller reported extensively on experiments he conducted on himself with contact lenses. His lenses caused a gradual veiling of the visual field; this caused objects to appear surrounded by a mist. Eventually visual acuity decreased to a point where it could no longer be tolerated. Removal of the lenses did allow restoration of proper corneal function, but not immediately. Müller suggested that tears (produced by tear glands in the eye) were an essential nutrient for the cornea, and that tightly fitted glass prevented tears from reaching the cornea. It took until 1970 to verify this theory completely. This was the work of Patt and Hill, which proved that tears were the major agent of oxygen transport to the cornea.

The first plastic lenses, though still rigid like glass and still subject to the oxygen permeability problem, were developed in 1942. Improvements were made: the drilling of tiny holes, or cutting of thin groves improved wearability but only to a limited extent. These innovations were made in partial recognition that tear-mediated exchange might be important, but still in ignorance of the role of oxygen.

The first true corneal contact lenses, which are probably attributable to the American Kevin Tuohy, were developed in 1947. These were made from the polymer polymethyl methacrylate (commonly known as Perspex or PMMA; poly(methyl 2-methylpropenoate), Structure **2.11**, p. 337) derived from the monomer methyl methacrylate (**6.1**). PMMA was chosen because it was known to be biocompatible with the cornea — at least to a degree. The lenses were around 11 mm in diameter, covered nearly all of the cornea, and floated on a layer of tears.

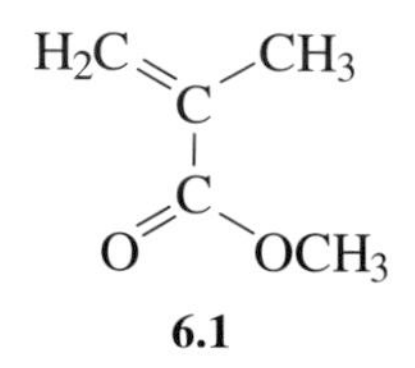


**6.1**

Even though the oxygen permeability of PMMA is near zero, contact lenses made from it were still prescribed in the 1970s.

With hindsight, it still seems astonishing that the oxygen permeability problem took so long to be realized. As far back as 1946, Edward Goodlow had suggested that contact lenses should allow oxygen in and carbon dioxide out, and in 1952 work by George Smelser and Victoria Ozanics showed that oxygenation of the cornea was crucial to its degree of hydration — that is, its gel status.

The revolution in contact lens technology can be traced to Czech chemists, O. Wichterle and D. Lim. In a 1960 paper in the journal *Nature*, they suggested that a gel based on poly(2-hydroxyethyl methacrylate) might be biocompatible with the cornea and serve as a material for *soft* contact lenses. More than 70% of the contact-lens using population now use these in preference to *hard* lenses. Even these new materials have the stubborn oxygen permeability problem, though much reduced. This means that soft lenses cannot be worn permanently. Ironically, new hard lens materials with 100% oxygen permeability have recently been developed, giving permanent wearability.

Our study is confined to soft lens materials — that is, **hydrogels**. A hydrogel is any macromolecular network that swells in, but does not dissolve in, water. Thus, the PAGs discussed in Section 5 are hydrogels.

Poly(2-hydroxyethyl methacrylate) — abbreviated to PHEMA, is the most studied of all hydrogels and still finds most application in contact lens manufacture, though recent research has led to even more effective hydrogels.

The monomer is 2-hydroxyethyl methacrylate (HEMA, **6.2**), which gives the polymer PHEMA, **6.3**.

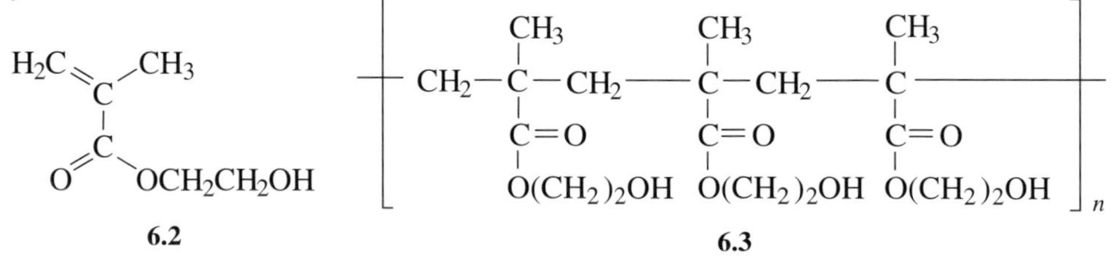


HEMA is made by the ester exchange reaction between ethane-1,2-diol (**6.4**), and methyl methacrylate (**6.1**):

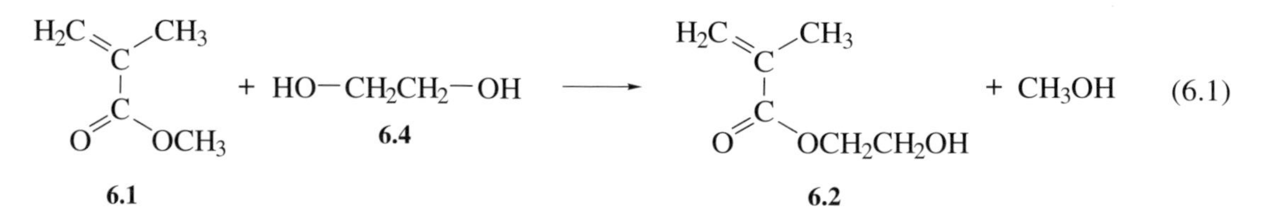

There is a second product from this reaction, which is also very useful. Notice that HEMA (**6.2**) contains a terminal primary alcohol group. This can also undergo an ester exchange reaction with a second molecule of methyl methacrylate:

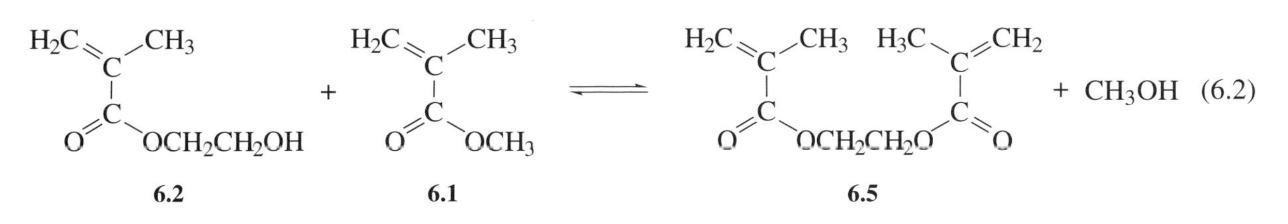


This product is ethylene glycol dimethacrylate (EGDMA, **6.5**). (You should note its similarity to *bis* in that there are alkene groups attached at each end of the molecule.)

Thus, the process to synthesise HEMA is carefully controlled to give HEMA as the major product, but also with a significant yield of EGDMA. Rather than going straight into the polymerization reaction, the HEMA and EGDMA must be separated. Although these compounds are liquids, fractional distillation cannot be used to separate them, since the high temperatures may initiate uncontrollable radical polymerization. Fortunately, HEMA and EGDMA turn out to have useful differences in solubility in diethyl ether and in hexane, HEMA being more soluble in the former and EGDMA in the latter. This property is used to separate the two products.

HEMA and EGDMA are copolymerized in carefully chosen ratios to yield the required network polymer with specified gel characteristics (Figure 6.2, overleaf).

Just as in Figure 4.3, a three-dimensional network is obtained in which water is organized in the pores to give a rigid gel. HEMA hydrogels contain many hydroxy groups (and some ester groups) that can hydrogen-bond with water, and thus the network can absorb about 40% by mass of water (Figure 6.3). However, for medical applications, equilibration is done in physiological saline (0.9% sodium chloride) such that hydration is lowered slightly from 40%. Hydration can be much increased by using TEGDMA (tetraethyleneglycol dimethacrylate, **6.6**) instead of EGDMA:

$$H_2C{=}C(CH_3){-}C({=}O){-}O{+}CH_2{-}CH_2{-}O{+}_4\,C({=}O){-}C(CH_3){=}CH_2$$

**6.6**

The additional three hydrophilic oxygen atoms in TEGDMA, compared with EGDMA, provide three extra hydrogen-bonding sites for water molecules, which, in turn, interact with yet more water molecules.

Why is the high hydration of these gels a desirable aim for contact lenses? As we have seen already, it is in connection with solving the oxygen permeability problem. Quite simply, the higher the hydration, the greater the oxygen permeability. Thus, the more we can get the lens to resemble a droplet of water, the better the compatibility. The gel network is there to maintain the shape and thus the optical properties.

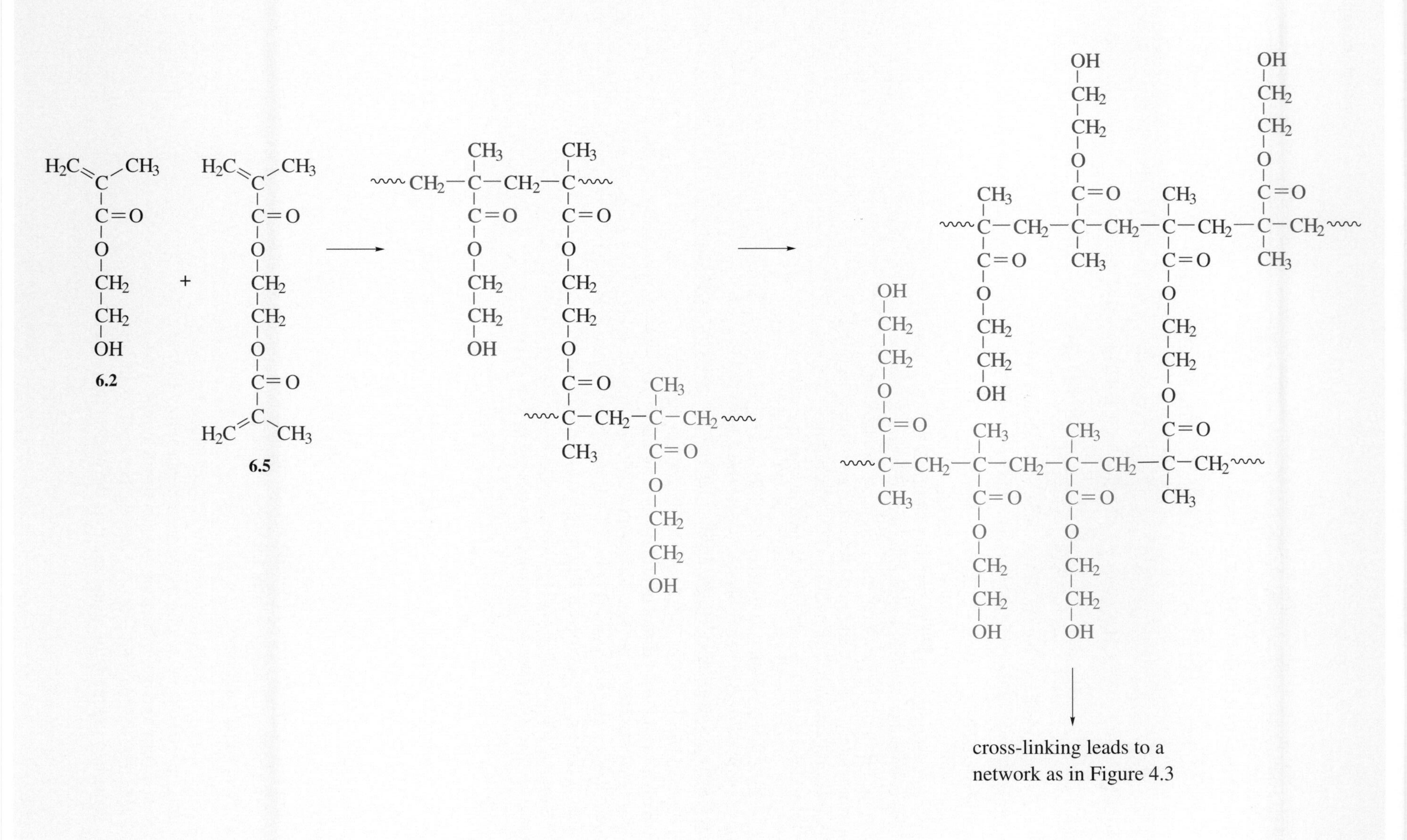


**Figure 6.2** Formation of a network copolymer from HEMA and EGDMA.

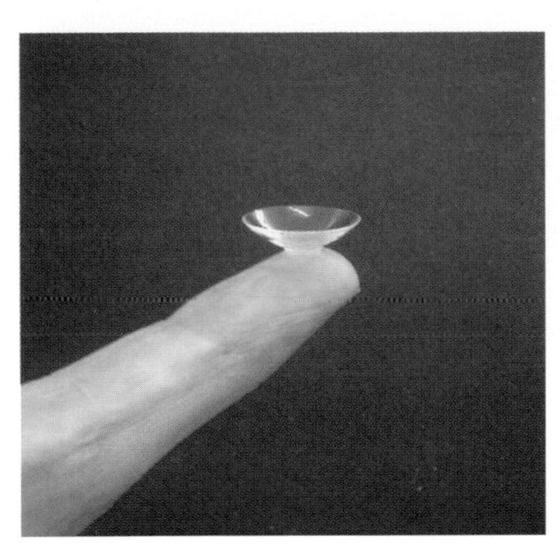

**Figure 6.3**
A soft contact lens after hydration.

**Figure 6.4**
Contact lenses now come in all manner of colours!

The HEMA-based soft contact lenses nevertheless do present some problems, in addition to mechanical strength and oxygen permeability. Ironically, high hydration, which is essential to good oxygen permeability, does itself present a problem. The cornea, being living tissue, produces proteins, and these are absorbed by the lenses — that is, by the gel water in them. Thus, over time, protein replaces water in the gel. This leads to cleaning and sterilization problems, which reduces the hydration and, in turn, lowers oxygen permeability. Research continues to find gel systems that can overcome all these problems. Research still focuses on polymerization of alkenes, but with other functional groups as well as hydroxy groups attached to the backbone.

Chemists have also turned their attention to more cosmetic problems. Polymerization of the HEMA in the presence of a water-insoluble dye, entraps the dye in the matrix, and since it is not soluble in water it is not leached out. This leads to coloured contact lenses (Figure 6.4).

The production and development of contact lenses is discussed further in the video sequence entitled *A glance at polymers* on the CD-ROM associated with this Book.

# FURTHER READING

1 L. E. Smart and J. M. F. Gagan (eds), *The Third Dimension*, The Open University and the Royal Society of Chemistry (2002).

2 P. G. Taylor and J. M. F. Gagan (eds), *Alkenes and Aromatics*, The Open University and the Royal Society of Chemistry (2002).

3 L. E. Smart (ed.), *Separation, Purification and Identification*, The Open University and the Royal Society of Chemistry (2002).

# ACKNOWLEDGEMENTS

Grateful acknowledgement is made to the following sources for permission to reproduce material in this Part of the book:

*Figure 1.1*: Courtesy of the Library and Information Centre, Royal Society of Chemistry; *Figure 1.2*: Courtesy of The Goodyear Tire and Rubber Company; *Figures 1.3 and 1.5*: Science Museum/Science and Society Photo Library; *Figure 1.4*: Science Photo Library; *Figure 1.6*: Courtesy of Hagley Museum and Library; *Figure 1.7*: The Nobel Foundation; *Figure 3.1*: Courtesy of Du Pont; *Figure 4.1*: Food Features Photographic; *Figure 4.2*: courtesy of Professor J. Feast, University of Durham; *Figure 5.2a*: James King-Holmes/Science Photo Library; *Figure 5.2b*: Philippe Plailly/Science Photo Library; *Figure 6.1*: Science Museum/ Science and Society Picture Library; *Figures 6.2 and 6.3*: courtesy of Cantor and Nissel.

Every effort has been made to trace all the copyright owners, but if any has been inadvertently overlooked, the publishers will be pleased to make the necessary arrangements at the first opportunity.

# INDEX

*Note* Principal references are given in bold type; picture and table references are shown in italics ; 'n' means footnote.

## A

## B

## C

## D



## E

## F



## G

## J



## K



## L



## M

## N



## O

# P

## W



## Y



## Z

# CD-ROM INFORMATION

## Computer specification

The CD-ROMs are designed for use on a PC running Windows 95, 98, ME or 2000. We recommend the following as the minimum hardware specification:

| | |
|---|---|
| processor | Pentium 400 MHz or compatible |
| memory (RAM) | 32 MB |
| hard disk free space | 100 MB |
| video resolution | 800 × 600 pixels at High Colour (16 bit) |
| CD-ROM speed | 8 × CD-ROM |
| sound card and speakers | Windows compatible |

Computers with higher specification components will provide a smoother presentation of the multimedia materials.

## Installing the CD-ROMs

Software must be installed onto your computer before you can access the applications. Please run INSTALL.EXE from either of the CD-ROMs.

This program may direct you to install other, third party, software applications. You will find the installation programs for these applications in the INSTALL folder on the CD-ROM. To access all the software on the CD-ROM, you must install, Isis/Draw, WebLab ViewerLite and Acrobat Reader.

## Running the applications on the CD-ROM

You can access *Mechanism and Synthesis* CD-ROM applications through a CD-ROM Guide (Figure C.1), which is created as part of the installation process. You may open this from the **Start** menu, by selecting **Programs** followed by **The Molecular World**. The CD-ROM Guide has the same title as this book.

The *Data Book* is accessed directly from the **Start | Programs | The Molecular World** menu (Figure C.2), and is supplied as an Adobe Acrobat™ document.

## Problem solving

The contents of this CD-ROM have been through many quality control checks at the Open University, and we do not anticipate that you will encounter difficulties in installing and running the software. However, a website will be maintained at

http://the-molecular-world.open.ac.uk

which records solutions to any faults that are reported to us.

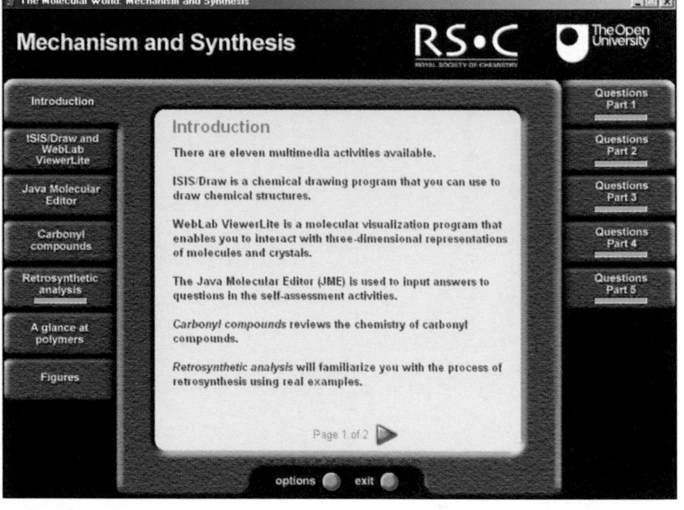


**Figure C.1** The CD-ROM Guide.

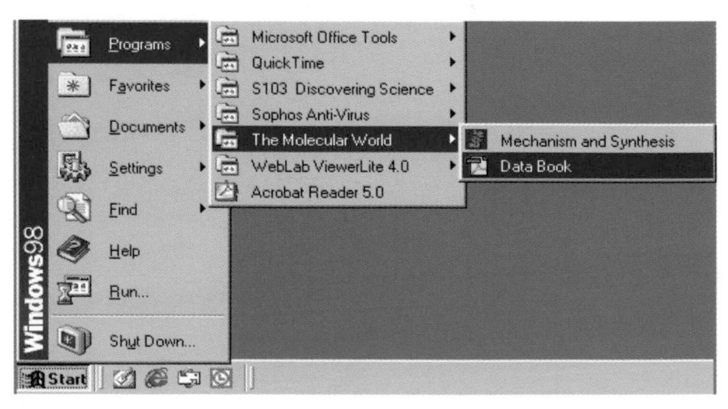


**Figure C.2** Accessing the *Data Book* and CD-ROM Guide.